Bloodstock Breeding

Sir Charles Leicester

BLOODSTOCK BREEDING

Revised by
Howard Wright

J. A. ALLEN & CO. LTD
LONDON

First published in 1957
© J. A. Allen & Co Ltd
Second Edition revised
published in 1983 by
J. A. Allen & Co Ltd

Reprinted 1989
Reprinted 1999

British Library Cataloguing in Publication Data

Leicester, Sir Charles
    Bloodstock Breeding – 2nd ed.
    1. Race horses
    2. Horse breeding
    I. Title
    636.1'2 SF291

ISBN 085131.349.3

Photoset in the Baskerville Series by
Rowland Phototypesetting Limited
Bury St Edmunds Suffolk
Printed in Malta by Interprint Limited

Foreword

THERE CAN BE few sports or pastimes which have been the subject of more progressive alterations since the Second World War than horse racing. So fast and fundamental have been the changes in organization that there has been a tendency to look upon racing as an industry. This is too sweeping a description, since racing is but one, relatively small branch of the entertainments industry, but its use is an indication of how the sport has been affected, most particularly by the twin forces of betting legislation and television.

Fortified by financial support from the Horserace Betting Levy Board, and projected to new audiences by the power of television, racing has gradually accepted that it must become a more sophisticated operation, founded on the keenest principles with the highest degree of security. These considerations are evident from the widescale introduction of such aids as starting stalls and photo-finish equipment, the use of such security measures as routine dope-testing and camera-patrol film evidence, and the implementation of various recommendations to give the punter more and better information daily.

While all this has gone on, the thoroughbred breeder has found that the business of providing the raw material for the sport has changed too, but in his, or her, case change has been effected on a much broader scale.

Once his horse has gone into training, the breeder's work is done, and whatever changes have been made in the day-to-day running of racing, they do not bother him in his breeding capacity. He is still concerned with evaluating what is best in thoroughbred terms and using that to his advantage, just as he has done throughout the century. Instead, the breeders' business has been most affected by changing patterns in international trade and economics.

Owner-breeders no longer operate in Britain on such a scale as they did before the War; taxation has largely seen to that. In their place have come the internationalists, men who from fortunes made in commodities such as minerals, oil and even football pools have been able to transfer considerable assets into bloodstock.

Bloodstock in the 1980s has become an international currency, which tends to rise in value when everything else goes down. This particularly applies to the best horses, or those considered to have the highest potential. Since these top-class, or potentially top-class horses are always in limited supply, even in world markets, a curiously accentuated financial imbalance has occurred. High auction prices at the top end of the scale have enabled a handful of breeders to make a handsome profit, though it is worth remembering that the recent appearance of worldwide syndicates in which the original vendor retains a share, provides an unnaturally inflated picture. But the depressed state at the bottom end of the market places a strain on breeders that was not evident in more equable times.

These are influences which can be traced in the main to the years since the late Sir Charles Leicester first saw "Bloodstock Breeding" published in 1957. They, and the inevitable appearance of new individuals, are the reason for bringing out this revised, up-dated version of the book.

So sound was the framework originally built by Sir Charles Leicester that it was not necessary to make adjustments to his approach, and the continued use of the first person—though the author died in May 1968—was chosen as evidence that the vast majority of his comments hold good, though they were first made over 25 years ago.

Since revision work takes time, and no-one has yet invented a way of making time stand still, even for revision work, it was decided that the main body of the new volume would use 1976 as its focal point. All statistics and all references to the produce of those horses discussed in any detail do not go beyond 1976. Chapter 30 was included as a postscript to cover the years since 1976.

In his original preface—an extract from which follows—Sir Charles Leicester wrote: "Whilst I have made every effort to be as accurate as possible, I am sensible to the fact that in a work of this nature I must have made some errors. Should any reader noticing a mistake care to drop me a line pointing it out, I would be most grateful". What applied in 1956 applies equally well now.

HOWARD WRIGHT

Sir Charles Leicester's Preface

TO THE ORIGINAL EDITION

THE WEALTH OF knowledge and experience required by a modern bloodstock breeder is so vast that no single volume can cover more than the barest fraction of the subject. Ignoring such vital matters as horsemastermanship, veterinary science, general farming, grass management, etc., etc., and confining attention purely to the breeding side of the undertaking, the necessary knowledge, in a rough and ready way, may be divided into two categories as follows:

FIRST: *The Theoretical and Practical.* Under this heading I include a grip of the basic principles of breeding and heredity, an understanding of all the various horse breeding theories, an appreciation of the results obtained by these, familiarity with the various lines of blood and particulars of influential parental stock, etc., and also a general knowledge of the management of stallions, broodmares and young stock.

SECOND: *The Historical.* A breeder must have a wide appreciation of specific horses and be able to weigh up the advantages or disadvantages of their names in a pedigree. There is no more unprofitable undertaking than to examine a family tree without the knowledge of what the various names recorded therein stand for.

I have therefore divided this book into two parts in conformity with these needs. I have expressed myself in non-technical terms—and sometimes homely similes— which will be easily understood by the normal horse breeder unfamiliar with biological expressions, by those interested in allied scientific or semi-scientific subjects unacquainted with veterinary and stud jargon, and by the ordinary reader with no specialist knowledge.

In Part II (Historical) I have taken the Derby winners of this century as my centre line and noted something of the histories of these and their relations. By this plan I have been enabled to cover some 2,500 well-known horses, including a very large number of those who have an important influence on modern blood lines in various parts of the world. I have, however, wandered far and wide from a strict recording of family events and have discussed a wide variety of matters which have cropped up as I have gone along.

In these days, probably the majority of bloodstock breeders in England and Ireland are what is generally termed "small breeders". They own half a dozen, or fewer, broodmares and very frequently cannot aspire to producing Classic winners or other horses of similar calibre. It will be found that in this work I have mostly mentioned in detail animals of the highest grade. These form the backbone of their breed, they are of more general interest than their less distinguished relatives as their names are constantly appearing in the pedigrees of all types of bloodstock. Furthermore their performances afford a definite clear cut yardstick on which conclusions may be based.

Their races are mostly run at level weights or weight for age, every breeder hopes—however secretly and however forlornly—to produce one of this class, and they are trained and ridden by men of outstanding ability in their professions. They are tended from earliest foalhood with the greatest care. They are thus all more or less on a level footing with the same objective in view. They offer a definite criterion of excellence so that their records can be taken as a reliable guide both by those interested in the production of horses of similar status and by those concerned with more humble class stock.

No sound deductions can usually be reached by the examination of the histories of the winners, etc., of unimportant races. Such are precluded by the diversity of these horses' upbringings, the variation of their opportunities, the operation of weight advantages, etc. Nevertheless there are a number of parental stock who started life in lowly circumstances and who have exerted very great influence on the Stud Book. To some of these I have referred.

Today, bloodstock breeding is an international undertaking and I have not forgotten to include notes on some famous horses who were bred in, or spent their stud careers in France, Italy, U.S.A., the Argentine, Australia, New Zealand, South Africa, etc.

Much of the information in this book has been gleaned over a number of years from the Sporting Press and particularly from *The British Race Horse, Horse and Hound, The Sporting Life* and *The Sporting Chronicle*. To these and others I must extend my thanks. I have also consulted the late Professor Keylock's works, *Flat Racing* (Lonsdale Library), *Horse Breeding in Theory and Practice* (Von Oettingen), *The Bloodstock Breeders' Review, Flat Racing Since 1900* (Ernest Bland) and other books. But I must express my primary gratitude to Messrs. Weatherby and Sons without whose wonderfully accurate records published in the *General Stud Book* and *Racing Calendars* no book on bloodstock could be written.

Introduction

EVERY ATTEMPT HAS been made to follow Sir Charles Leicester's original concept of a narrative within reach of readers who may not be familiar with the usual jargon and terminology of breeding and racing. Breeding manuals have been accused of blinding their readers with what they describe as science, usually with the hope of covering up deficiencies. The opposite was the intention here.

The following notations have been used throughout:

The date after a horse's name—as in St. Simon (1881)—means he was born in that year.

A monetary figure and/or the title of a race after the horse's name—as in Hugh Lupus (£15,232, Champion Stakes and Irish 2,000 Guineas)—indicates how much he won in first-prize money, and the important races in which he was successful. Stake money in Britain and Ireland generally refers to first-prize money on the Flat; elsewhere in the world, earnings for all placings and over jumps are included.

Half-brothers and half-sisters are horses by different sires out of the same dam. The expressions are never used to denote a combination of sire lines which may produce the same degree of relationship. Three-parts brothers and three-parts sisters are horses out of the same dam but by sires who are father and son.

The expression Premium Stallion refers to a thoroughbred sire who earns his premium under the Hunters Improvement Society scheme, and is led from place to place during the breeding season generally to serve "half-bred" mares. The term "half-bred" has been used in its normal stud jargon sense of "non-thoroughbred". It must not be taken to mean a horse who has one parent of one breed and the other of another breed.

Contents

xi

PART TWO:

Derby winners 1900–1976 and their relations

PART ONE

THEORY AND PRACTICE

CHAPTER 1

General Principles of Bloodstock Breeding

*Basic genetics: bloodstock breeding not an exact science:
advantages and disadvantages of the bloodstock breeder
compared to breeders of other pedigree stock*

I DO NOT intend to go deeply in the science of genetics, but to understand many of the problems which confront bloodstock breeders a knowledge of the basic principles of heredity is necessary.

The life of a species as opposed to that of the individual is continuous. It may be mentally compared to a river flowing on for ever unless some extraneous circumstances occur and the species dies out. Prehistoric animals are an example of this. Life is carried on from generation to generation by an actual living part of the parent becoming the first beginnings of the offspring. In the lowest forms of life this is effected by the parent splitting into two. The parent does not die in the normal sense of the word but loses its individuality and becomes two offspring. The offspring grow to maturity and repeat the process. Thus the continuity of the species is brought about.

In the case of the higher types of animal life instead of one parent splitting into two to form two offspring, two parents contribute a living part of themselves to form one new individual. The male subscribes the sperm and the female the egg. The new individual then grows up and in turn passes on part of him or herself to form the next generation, thereby the never-ending stream of life is maintained.

HEREDITARY FACTORS

At the time of mating each parent contributes an equal number of what are termed chromosomes to the embryo. Their number varies according to the species but is always constant for the same animals. In the case of the horse, the sire provides 30 and the dam 30. The chromosomes are living objects visible under the microscope and contain a whole pack of hereditary factors, or genes, as they are called. The genes are self-reproductive and are responsible for the build-up of the new individual. They are heterogeneous in function, some operating to produce

3

bone, others muscle, colour, blood, brain, organs, constitution, etc.—in fact, broadly speaking, all the basic factors of the new-born.

As development takes place, half the number of chromosomes given to the embryo collect in its germ plasm. The germ plasm is the well from which the new being, later in life, draws in his or her parental capacity. Thus the genes of parents are transmitted to their young. The young store some of these to pass on in due course to their descendants and so the hereditary factors are passed from generation to generation.

As the embryo receives genes of similar function from both parents, both cannot be commonly operative at the same time. For instance if a brown stallion and chestnut mare are mated, the embryo receives the genes which form both those colours. One becomes dominant and is visible in the new-born foal and the other is latent. The latent factor although not visible is nevertheless present in the foal and will be passed on by him when he goes to stud. Whether it will appear as dominant or latent in the next generation normally cannot be foretold. It may remain latent for generations and then suddenly reappear unheralded. A case in point familiar to many bloodstock breeders is the white ticks which sometimes occur in the coats of bays, browns, blacks and chesnuts. They are usually called Birdcatcher ticks. So far as I am aware every horse who carries this livery is a descendant of Birdcatcher, who was foaled so long ago as 1833. In a few cases they have been passed directly from a parent to a foal. The fairly well-known stallion Lesterlin (1892) sired a number of his progeny with these markings but frequently they are latent for many generations and only reappear very occasionally.

Another important point to be remembered when considering hereditary factors is that no two sperm cells of any sire and no two egg cells of any dam contain exactly similar genes. As a stallion discharges copious quantities of sperm when covering a mare and only one sperm cell fertilizes the egg, it would appear obvious that it is a matter of pure chance which cell is effective. Furthermore as full brothers and full sisters do not receive exactly the same genes they cannot be alike. They inherit factors from the same two germ plasm or wells, but which genes the lucky dip into the well will produce, no man can foretell. All that can be said is that in the wells there are certain ingredients, beyond that it is impossible to go.

To put the whole matter in simpler terms let us consider each parent as a dual personality. His or her first personality is his own make and shape, colour, constitution, soundness, racing ability, temperament, etc. These are visible to the naked eye and I will call them his visual attributes (technically termed somato-plasm). His second personality consists of corresponding qualities derived from his forebears. These are not visible in the parent himself, but are present in his body. These I will call his latent attributes (technically germ plasm). At the time of mating both parents transmit some of their latent attributes to the embryo. Which of these attributes will be visual and which will be latent in the new-born foal no man can accurately foretell. It is a secret which Nature reserves for herself.

I particularly draw attention to the fact that it is not the visual attributes which are passed on but the latent ones which the parent has received from his ancestors. This is important as it brings out the importance of breeding from well-bred stock. A horse may well—and commonly does—stamp his stock with certain characteristics which are amongst his visual attributes, but also these characteristics are latent attributes in his body and the latter are the ones he or she transmits.

REDUCED INFLUENCE OF RECEDING GENERATIONS

Results show that the visual attributes of the parents are more likely to appear as visual attributes in the foal than those of the grandparents. Those of the grandparents are more potent than those of the great-grandparents and so on. Each receding generation in the pedigree has a reduced chance of reviving its own peculiarities until the time comes when remote ancestors' influences have little or no practical portent. Yet it must be emphasized that although latent in the new-born these peculiarities are still there and may reappear as visual attributes in his or her descendants. This is sometimes referred to as a throw back, but in reality the particular characteristic which suddenly reappears has been latent in parents, grandparents, etc., possibly for generations. For example, from time to time, thoroughbreds appear whose make and shape, etc., show a remarkable resemblance to their Arabian ancestors of 200 or more years ago. The Derby winner and later great sire in U.S.A., Mahmoud (1933), is a case in point and readers are asked to compare photographs of him, his sire Blenheim and an Arab, which appear in the illustrations.

Some stallions and some mares have a preponderant capacity for transmitting certain qualities for good or evil. Consider two good, well-bred Derby winners. On retiring to stud both are mated to mares of approximately the same calibre, yet as likely as not one will pass on his desirable qualities and be a stud success whilst the other will transmit some latent fault of an ancestor and be a stud failure. No man can predict with accuracy which will be which. Again, it will remain one of Nature's secrets until the progeny are available to be sized up, and even then the reason will remain unknown.

TRANSMITTABLE FACTORS

We have seen that the genes or hereditary factors are transmitted from generation to generation and now the question arises as to what is hereditary and what is not. Broadly speaking, practically all the basic factors a foal is born with are hereditary, i.e, make and shape, constitution, bone muscle, organs, weaknesses, soundness, temperament, etc. Exception must be made in the case of certain diseases caused by micro-organisms and to certain deformities. Factors acquired during a parent's lifetime from outside causes are not hereditary. These include unsoundness or deformities due to accident or contracted disease, vices such as weaving and crib biting, bad temper due to mishandling, weaknesses due to improper feeding, etc.

From a bloodstock breeder's point of view the most important hereditary factors are those which go to produce racing merit. Strictly speaking, performance, whether on the Flat or in steeplechasing—or more obviously in hunters and polo ponies—is not hereditary. It is acquired as the result of training, muscle development, education, etc. But the individual must be born with the necessary attributes to allow him to excel in his allotted sphere of life as a result of this training. From this it must not be inferred that because a horse has the outward appearance of a racehorse he will necessarily be a good one. He may be, and frequently is, defective in his temperament, bodily organs, general ability to co-ordinate his movements, etc. It is common for horses who are sub-normal in size, and thus greatly handicapped by their limited length of stride, to be compensated by Nature with exceptional elasticity, ability to use themselves, soundness and hardiness. The good racer, whether stallion or mare, and to a large extent irrespective of personal appearance, etc., is more likely to be a success at stud than the bad one—other things being equal.

To the practical breeder the distinction between hereditary factors and the result of hereditary factors is of little moment. All he is concerned with is the fact that the necessary attributes to make a good racehorse are transmittable. Having made the point, I will refer to racing ability as hereditary in the remainder of this book and ask the indulgence of my more accurately-minded readers in the matter.

From the foregoing it will be seen that:

(*a*) The basic factors with which a foal is born are hereditary.

(*b*) No man can prophesy which of these factors will be visual and active in the foal and which will be latent.

(*c*) Neither can man predict with accuracy in what direction the foal will transmit his hereditary qualities when his (or her) time comes to go to stud. All man knows is that they are present and may or may not show themselves.

The practical results of this may be illustrated in a thoroughly unscientific but homely manner and in a rough and ready way, by taking a pack of playing cards. Select two cards to represent the parents of a proposed mating—an ace for a very good parent and a lower card for a not so good one. Then take four cards to represent the grandparents—kings for the best and lower ones for the others. Next take eight cards from queens downwards for the great-grandparents. The highest card in each receding generation is of a lower denomination to denote their reduced influence. Then place the 14 selected cards face downwards on the table and at random turn up one to represent the foal resulting from the mating. It will be obvious if the majority of cards face down are high ones, the turn-up is likely to be a high one and vice versa. On the other hand it may so happen that the turn-up from a good hand is the worst card therein.

Similarly in bloodstock breeding, if the names in a pedigree of a proposed mating stand for good winners and winner producers, a good foal is more likely to result than if the names indicate indifferent performers. But there is no certainty. A useless foal not infrequently comes from a good pedigree, and sometimes a good one from indifferent progenitors. There are probably proportionately more of the former than the latter, owing to a large extent to the incidence of numbers. The really well-bred parents are comparatively few whilst the indifferent ones are legion.

If Nature reserves for herself the secret of which hereditary factors are visual and active in the foal, and which are latent, it may be asked where the breeder's skill comes in and whether the whole business is not a pure lottery? The answer to this is that it is the breeder's job to place in Nature's hands the qualities he wishes reproduced and to arrange his matings so that undesirable traits are as remote as possible. Exactly as a farmer cannot expect a good crop from bad seed, so a breeder cannot expect useful winners from bad stock.

BLOODSTOCK BREEDING NOT AN EXACT SCIENCE

Bloodstock breeding is not an exact science, but a question of averages. We have seen that when a mating takes place, each parent dips into its well of hereditary factors and draws out those which are to become operative in the new-born. No man can foretell what will be the result. All he can estimate from a study of the ingredients of the well is that a particular stallion or mare will probably tend to produce stock with certain qualities. He can anticipate that the mating of a specific sire and a specific mare will on an average produce a result which would be unlikely from the union of two other named parents.

For example, a staying sire is likely to get staying stock and on an average does so. But there are numerous cases of sprinters sired by stayers and vice versa. The very brilliant Tudor Minstrel (1944) and Abernant (1946) with distance limitations of one mile were sired by Owen Tudor (1938), who was capable of staying double that journey. On the other hand Owen Tudor sired the Ascot Gold Cup (2½ miles) winner Elpenor (1950). Their pedigrees are worthy of comparison:

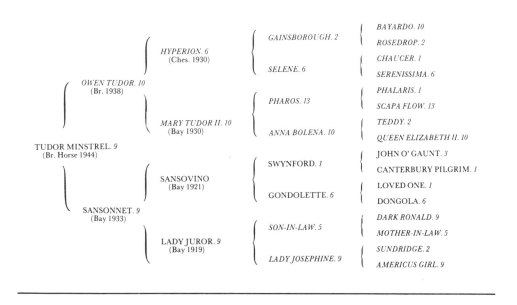

Tudor Minstrel won 8 races value £24,629 from 5 furlongs to one mile including the 2,000 Guineas over one mile.

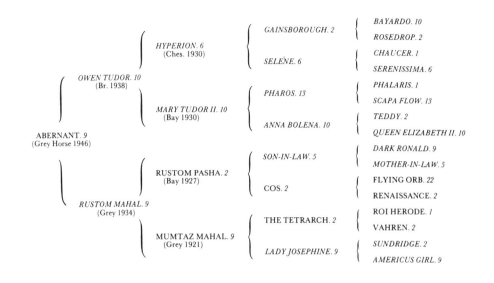

Abernant won 14 races value £26,394 from 5 furlongs to 7 furlongs. He was beaten a short head for the 2,000 Guineas.

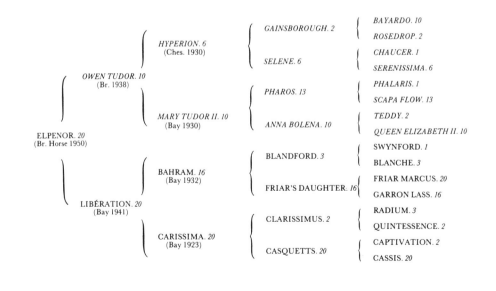

Elpenor won the Ascot Gold Cup, 2½ miles, and other races.

The italic names indicate mutual ancestors. It will be noticed that the general pattern of the sprinters Tudor Minstrel and Abernant's pedigrees are very similar. They are by the same sire and have a common tail female great-grandam, Lady Josephine (1912), reinforced with the staying Son-in-Law (1911) blood. The descendants of Lady Josephine are mostly notorious for their sprinting capacity. As will be seen from Tudor Minstrel's pedigree this influence is so potent that it has rendered negative the blood of the stayers Son-in-Law and Sansovino. Tudor Minstrel's dam, Sansonnet, was a 5 furlong racer. His grandam Lady Juror when mated to the St. Leger (1 mile 6½ furlongs) winner Fairway (1925) produced the brilliant but non-staying Fair Trial (1932) and so on.

On the other hand Elpenor's dam is an entirely different proposition. She has no common ancestry with the other two mares except for a somewhat remote strain of Swynford (1907), whose name also occurs in Sansonnet's tree. Libération's tail female line has shown itself receptive to the influence of the sires it is mated with. Carissima put to Pharos (1920) (full brother to the St. Leger winner Fairway and himself a sire and grandsire of St. Leger winners) produced the Grand Prix de Paris (1 mile 7 furlongs) winner Pharis II (1936). Carissima's daughter Caprifolia (1932) mated to the sprinter Fair Trial produced the non-staying The Solicitor (1941), whilst her granddaughter Castillane put to the middle-distance Goya II (1934) produced the middle-distance Nirgal (1943). It is therefore not surprising that mated to the Gold Cup winner Owen Tudor, Libération produced the staying Elpenor. Although at first sight it would appear that the results of these three matings are somewhat contradictory, in reality they are only what could be reasonably anticipated. In fact, broadly speaking, they are the outcome of averages. The average product of the Lady Josephine line are sprinters even when mated to staying sires, while the descendants of Carissima are flexible to their mates' influence. I do not imply from this that the female line is always paramount to the male line. Sometimes it is and sometimes it is not. Each individual case must be judged on average results.

Every year there are some 15,000 thoroughbred broodmares mated to various sires. The majority of these are of a class which renders them extremely unlikely to produce anything comparable to Tudor Minstrel or Abernant, who were nearly, if not quite, the two best sprinters since before the war. Yet the emergence of these two with very similar pedigrees in the face of great numerical competition repudiates the idea that bloodstock breeding is pure chance and amply justifies the prevailing system of parental selection. The intervening years have not changed this basic truth, and a study of the Stud Book and annual volumes of Races Past makes this fact abundantly clear.

RECORDS OF 200 YEARS

It is therefore necessary for a breeder to make a close study of these books of reference and so far as possible to form a judgement on individual horses by personal inspection. He must know what ingredients there are in the parental well of his stock. He must appreciate the strength and weaknesses of each animal in a pedigree and form an opinion regarding the average result that may be expected from a particular mating. The bloodstock breeder has a unique advantage over the breeders of other pedigree stock in this matter. The history of the development, matings and racing results of the thoroughbred have been recorded with accuracy for over 200 years—no other breeders have a corresponding mine of information on which to base their actions.

Incidentally, I always think that this mine is so vast that there must be innumerable lessons available from it which have never come to light. For example, we know that a certain once highly successful line of blood has died out, but we rarely know the reason. We form certain opinions, usually based on the deeds of a very few of its most famous scions and ignore its lesser known members. In a rough and ready way we say the line failed because it did not produce satisfactory results, but we do not know what its probable fate would have been if it had been mated entirely differently. We do not know the average overall incidence of inbreeding the thoroughbred will stand up to, or even the overall average amount of racing parents can undergo without harming their stud prospects—if indeed they are harmed at all by hard racing. This is a subject I will refer to later (see pages 107, 128). I cannot help feeling that if it was possible to hold a gigantic scientific investigation into the history of the breed, valuable concrete information would be produced, but the whole thing is so vast and involved that probably it will never be done, unless perhaps the advancement in the use of computers can be harnessed to provide a solution.

Another advantage the bloodstock breeder enjoys in common with breeders of racing pigeons, greyhounds, etc., is that his results are judged not only on facts but that both sexes compete *pari passu* and thus a comparison between male and female is available. In the case of some other pedigree stock the performance of one sex is purely theoretical—for example, milk yield in cattle, egg production in fowls, etc.

EFFECTS OF BREEDING FOR SALE

Until the second half of the present century the bloodstock breeder was not greatly influenced by the dictates of human fashion or opinion such as the breeder of the showyard animal is. He had one objective and one objective only—namely to breed winners. To an increasing extent the position has been modified in this respect, through the growing scale of breeding for sale. To obtain a remunerative price, a vendor of yearlings must offer for sale stock which are both good looking and fashionably bred to the whim of the moment. He therefore has not a free choice of the sires he uses. He may consider an unfashionable stallion most suitable for a particular mare but he dare not use him as the progeny will be unsaleable. The mare may therefore suffer as she is not mated to the best advantage whilst the unfashionable stallion, however good his latent stud prospects may be, gets little chance owing to lack of patronage.

If the breeder for sale cannot obtain shares in or nominations to the most fashionable stallions his next need is to produce good-looking yearlings which will appeal to the human eye. A trainer cannot reasonably be expected to advise his clients to buy the ugly ducklings, although horses of all shapes and sizes win races. I think there is a strata of danger in breeding with the specific object of producing good lookers above other considerations. I will refer again to this in the next chapter (see page 16).

In the more spacious days when there was a higher proportion of thoroughbred stock bred by the owner-breeder to race himself, he was entirely free from human opinion on the pedigree and appearance of his product. He could therefore devote his entire attention to trying to breed winners. Amongst the disadvantages a bloodstock breeder has to contend with are a heavy financial capital outlay, ever-increasing costs of production, the handling of a somewhat delicate and highly strung animal resulting in a heavy wastage incidence, and the relatively long time he has to wait for results.

Those with knowledge of the bloodstock market will be familiar with the heavy outlay required to form a stud, and to some extent with the heavy running expenses, which have multiplied in the 1970s through the effects of inflation. Whereas 20 years ago the cost of producing a yearling for sale was estimated at around £1,000, the figure is now approximately £3,500, which includes the stallion fee, the yearling's keep from birth, a share of the dam's keep, veterinary, shoeing, saddlery and travelling expenses, etc.

When we turn to the wastage account it must be remembered that the fertility level in Great Britain and Ireland is running at a little over 50 per cent. Returns for the General Stud Book show that in 1965 there were 10,797 mares reported at stud, and they produced 5,923 foals for 54.86 per cent fertility. In 1976 the figures were: 17,202 mares, 8,840 foals, 51.39 per cent fertility. It should be pointed out that these figures include mares who were not covered, or were covered by unregistered stallions, and so do not paint the whole picture about fertility. But even allowing for an intelligently-estimated figure of ten per cent for the "missing" mares, the lack of improvement is disturbing.

After a foal is born he is subject to continual risks of accident, and to a lesser degree disease. If he is offered for sale, it is to some extent a gamble whether he is bought to race at home or for export, or in Ireland for use later in life as a 'chaser. If he is not bought for Flat racing at home, he is generally a virtual write-off so far as enhancing his dam's reputation as a winner-producer goes. The market price of a yearling is greatly influenced by the racing record of its dam's previous progeny, and buyers do not always give the same status to winners overseas, except in instances where they involve a few of the best-known races in France or America.

The wastage incidence between birth and appearing on a racecourse in Britain and Ireland varies greatly and is much influenced by the class of yearling concerned. In the case of an unfashionable sire 50 per cent or more of his live foals may never see a Flat racecourse at home.

An interesting example of this was Montrose (1930), by the Derby and St. Leger winner Coronach out of Accalmie (won 7 races and was third in the Cesarewitch). He was a good, game racer, winning 13 races, value £10,886 including the City and Suburban, Atlantic Cup, etc. He went to stud at Newmarket for a couple of seasons and was moved to Ireland soon after the outbreak of war. When his early stock came up for sale in Dublin an astute agent bought one or two for export to South Africa, where they immediately made a name for themselves and their sire. In fact, he occupied the curious position of being the leading sire of winners in

South Africa when he himself was located in Ireland. So far as his stud record at home is concerned he must be regarded as a hopeless failure—simply due to the fact that he did not have sufficient runners to make any headway. He was undoubtedly a stallion above the average but his produce did little or nothing to enhance the value of their dams' subsequent foals.

Supposing there are no accidents, wastage, barrenness, etc.—in fact everything goes right for the breeder—the delay before obtaining results of a mating is illustrated below:

First year.   Mating arranged and nomination to stallion booked.

Nearly all stallions whose stock is in demand are fully booked up at least six months before the covering season.

Second year.   Mating takes place.

Third year.   Foal is born.

Fourth year.   Progeny is a yearling.

Fifth year.   Progeny probably runs as a 2-y-o.

Sixth year.   Progeny racing as a 3-y-o and probably gives a fair picture of its ability.

It will therefore be plain, that even with the early racing of two-year-olds, the breeder has a long, long wait before he can see the results of his matings. If the dam is barren, loses a foal or is unlucky in any way this period may easily be prolonged to seven, eight or more years. The breeder must steel himself against the disappointments caused by breeding a good horse that either falls into the hands of an ambitious owner who is always racing him in company just above his class, or, if of the highest class, happens to be foaled in a vintage year.

In the breeding and racing world a horse is very generally assessed either as a winner or as a non-winner. Comparatively little consideration is given to placed horses and I wonder how many people who can readily recall the Derby winners of the past 30 years can name their runners-up for even the past ten. Yet the placed horse is frequently only a fraction of a second slower than the winner and there can be little doubt that the placed horses in one Derby are superior to some winners of that race in other years. I venture to suggest that though Ballymoss, Gyr and Rheingold finished second to Crepello, Nijinsky and Roberto respectively at Epsom, they were superior to Larkspur, Blakeney, Morston, Snow Knight and Empery, who all won the Derby in the same period.

The highly competitive nature and chances of racing produce many vexations to which the breeders of some other pedigree stock are not subject. There are literally thousands of good, honest, well-made, sound thoroughbreds born every year, who are regarded as failures because they just lack the necessary speed to be winners, not infrequently by a very narrow margin. Yet broadly speaking they are good horses.

CHAPTER 2

The Development of the Thoroughbred

Parental selection: racehorses gradual improvement: importance of racing ability: inbreeding

AMONGST WILD ANIMALS nature operates a system of the survival of the fittest or natural selection to maintain the continuance of each species and to control the balance between them. The weak and subnormal either die or are killed and so do not reproduce themselves. Thus only those with the most suitable hereditable factors for the particular requirements of their species survive for parentage. Should any species be exempt from this thinning out it would increase and multiply to such an extent that it would become world paramount and oust all other species. If every rat which was born survived and produced to its limit, it would only be a question of time before they became so numerous that they would eat up all the food in the world and other animals would starve.

In the case of farm and domestic animals, man selects for parentage those specimens of the breed which show the qualities which he wishes reproduced. Going back some 270 years, the modern racehorse and the modern hackney trace to the same original stock, yet they have been developed by human selection into two distinct breeds. The thoroughbred breeder of old earmarked as his breeding stock those horses who showed the ability to gallop at speed, whilst the breeder of trotters concentrated his attention on those with knee action suitable for his requirements. In course of time by the exercise of an ever-increasing standard of selection, according to their different needs, the former developed the present-day thoroughbred and the latter the modern hackney.

The process of selection can be likened to a man sifting gravel through a series of sieves. The gravel represents the parent, the sieves the selection. The rougher specimens which fail to get through the first sieve are cast aside. Those which are passed through are then screened through a finer-meshed sieve and again the wastage put aside—and so on. Similarly in the development of the two breeds of horses mentioned, those specimens who failed to come up to the mounting requirements of their breeders were cast aside until the particular action and make and shape of the hackney has been eliminated from the thoroughbred and the galloping action and build of the thoroughbred have disappeared from the hackney. As time has gone on, the influence of the slower moving racehorse of bygone days has become more and more remote and the hereditary factors of his faster moving successor become more and more potent.

There comes a time, however, when improvement levels out. The initial surge of selection produces dramatic results, but in time the pursuit of one quality leads to a point where the law of diminishing returns comes into play. Such a stage seems to have been reached in thoroughbred development, where the aim of breeders has followed one path, namely the improvement of speed and thus racing performance.

I have little doubt that the average speed of the modern thoroughbred is superior to that of his forebears of 200 years ago; that Nijinsky, Tulyar, Blue Peter or Nearco are greatly faster than Eclipse (1764), Flying Childers (1715) or Highflyer (1774). But how the modern giants would compare with St. Simon (1881), Ormonde (1883) or Isinglass (1890), it is impossible to say with any certainty, since there is no acceptable standard of comparison to link individuals of widely different periods.

The time test is unreliable because of the pitfalls of its execution. Until recent years this was universally effected by a hand-operated stop-watch, whose use was influenced by an appreciable element of human indifference and error. One can have little faith in this system after being told by one experienced watch-holder that "most of the records were set with an egg-timer". Even the relatively recent introduction of automatic electrical timing devices has not allayed all the fears, since only 16 of the 35 British Flat courses are fitted with the system, thus putting over half the current picture at the risk of human error. The direction and strength of the wind, the state of the going, the presence of a pacemaker, whether the winning animal was all-out or could if pressed have improved his performance, are other elements which make timing an unreliable criterion. The harder the ground, the more likely it is that records will be set, as witnessed by the fact that of 467 records standing at the start of the 1977 Flat season, 69 were set on ground described as hard, 319 on firm, and 78 on good. The one remaining was notched on yielding going, a freak among freaks.

 Yet for all that I do not like the time factor as a barometer, its prominence in a paper read to the International Symposium on Genetics and Horse-breeding, in Dublin in 1975, by Professor Patrick Cunningham of Dublin University is important. Professor Cunningham examined times for the Derby, Oaks and St. Leger for more than 100 years, and after averaging the winning times for each ten-year period and graphing the results, showed that whereas there had been an improvement of two per cent per decade up to 1910, no improvement had been registered since then. Allowing for the factors of unreliability cited above, it would seem that the levelling-out stage has been reached.

The conformation of the modern racehorse is the result of breeding from parents who either themselves or their near relatives had racing merit above the normal. It is not the outcome of breeding to type. I cannot over-emphasize this vitally important point. The horse who is so made as to be able to use himself mechanically to the best advantage for galloping at speed has on an average during the long history of the Turf come out on top. The beautiful symmetry, balance and elegance of the high-class racehorse have come about as they afford the horse this advantage.

There are—and always have been—good winners whose conformation is faulty according to human judgement. These must have some latent factors such as gameness, ability to co-ordinate their movements, and general racing attributes above the normal to enable them to overcome their apparent handicaps. When these go to stud, provided they are not mated to partners with similar faults, there is every chance that their defects will not be reproduced and they may even transmit to their stock the little hidden extra which enabled them to be racecourse successes. Not only should the partner be free from the specific fault but also the partner's progenitors and relations; in fact it should not be in the blood.

Without going into such questions as a particular shape and formation of hock, shoulder, knee or cannon bone, let us take as an example the undersized horse. Other things being equal—and I stress this—he is manifestly at a disadvantage when racing against his full-sized opponents, since he has a shorter stride. Amongst such horses are Hyperion (1930), Chaucer (1900), Hampton (1872), Rabelais (1900), Colorado (1923), etc., who all measured about 15.2 or less. Yet after successful racing careers they all proved most influential stallions, who sired stock of normal size with perhaps some of the something extra which enabled them personally to overcome their handicap of size. They were mated to mares of good winning strains and they came from good winning strains themselves. The winning strains can be taken as synonymous mostly with average size. Their stock therefore had far more average-sized animals than small ones in their pedigrees, with the natural result that the lack of inches was not usually reproduced. The same principle holds good with most other apparent defects of conformation.

I think that the history of the thoroughbred shows the paramount importance of breeding only from stallions with racing merit. It may be asked if horses of apparent faulty conformation can and do sire good lookers, why those with faulty racing records should not, on the same principle, get good winners? The answer to this is that through necessity and the general lay-out of bloodstock breeding, many mares with inferior or no racing ability go to stud. If the useless racehorse was mated to the useless racemare, the project would be on the lines of the ugly sire being mated to a mare with the same defect of conformation. To try to breed a racehorse from both parents selected for their good looks rather than their racing records would be foolish; the outward appearance would be beautiful, but the performance far from it.

THE PLACE OF CONFORMATION

I remember around 1930 a number of German horses were imported to England and sold as hunters. They had been bred as remounts for the Germany Army. Their parents were selected, I was informed, for their good looks, weight-carrying capacity and endurance. Many of the importations were fine specimens of the medium weight-carrying hunter type, but after negotiating a few fields and fences of an English hunting country they just "packed up" and were unable to keep up with the hunt. Doubtlessly at the comparatively slow paces required for an army horse they were excellent at their job, but when asked to accelerate for any distance it was beyond them. This is but one example of the pitfalls of breeding to type rather than performance, and the same tendency is noticeable in various animals bred on these lines for the show ring. Some breeds of dogs bred for this purpose have totally lost their working ability. In the previous chapter I referred to the possible handicap to the whole breed of thoroughbreds from the necessity of sale yearlings being good lookers. I obviously do not imply that perfect conformation is in itself harmful, but rather that there is a risk of yearlings being bred on the lines of all looks and no performance.

Through the sheer force of economic conditions, breeding for sale has increased enormously in the last 30 years, with a corresponding decline in the number of influential owner-breeders. The large and powerful studs of the Lords Astor, Derby and Rosebery have been decimated, and no longer is it possible to say, as was the case 20 years ago, that on average among horses of the highest class the home-bred is superior to the sales yearling.

Whereas for the 20 years up to 1955 the five Classics had been won by some 70 per cent owner-bred horses and 30 per cent bought as yearlings, the figures for the 20 years after 1955 are 60 per cent owner-bred and 40 per cent bought yearlings. Of even greater significance is the fact that for the ten years 1966–75, the split was exactly 50–50. Unless the economic situation is suddenly to change, this trend will doubtless continue, and it will not be long before the figures show a bias in favour of bought yearlings.

Whether this trend will in the long run benefit the breed remains to be seen. I have sometimes noticed when looking round studs of high class that when the owner is breeding to race himself, some of the mares although bred in the purple are not hard to fault on confirmation. In other studs that are breeding for sale the reverse is the case and greater attention is apparently paid to the mares' personal make and shape than to the winning record of their ancestors.

A further point worth noting in this context is that the thoroughbred of today is not able to stand up to more racing than his predecessor. Bearing in mind the great advances veterinary science has made in recent years, it would not be unreasonable to expect this to be reflected in the horse's capacity for more work. I have examined the frequency of running of 1,000 horses of all ages taken at random for 1911, 1954 and 1975. I find that in 1911 and 1954 the figures were identical, and the average horse faced the starter 4.6 times. In 1975 the figure was 4.8, a miniscule improvement despite an increase of over 100 per cent in the number of races available to be contested compared with the early part of the century.

I think the lack of improvement may be due to the use of some parents who themselves could only stand light racing. But I also think breeding for sale might have something to do with it. One of the primary considerations for an owner-breeder is to produce an animal who can race and give him some sport. Although naturally all commercial breeders like to sell a yearling who will give every satisfaction, it is not always their worry if he breaks down.

All power to those engaged in the business of breeding for sale, for without them there would not be sufficient yearlings bred to keep racing going. But if we could do the impossible and go back to the piping times when more owner-breeders flourished, I think it would be in the best interests of the breed.

INBREEDING AMONGST WILD ANIMALS

The idea of close inbreeding amongst humans is repulsive to most of us, yet it must be remembered that it is almost universal amongst wild animals. Amongst herds of wild ponies running loose, provided man does not interfere, the stallion mates with his daughters, granddaughters, etc. When he becomes old he is either replaced by one of his own issue, or by a stranger from a neighbouring herd. If the new king of the harem is a descendant of the original sire more and yet more inbreeding takes place. This process of Nature is worthy of much thought as through it the survival of the species is maintained in the hard battle for existence. It may well be that the result in the long run produces an animal not in accordance with man's needs, but that is not a true criterion. The test of the efficiency of Nature's workings must be a natural one and not assessed by man's artificial standards.

Consider the wild rabbit, to whom generations of incestuous breeding have apparently had no ill effects. They certainly do not show any signs of loss of fertility—or lunacy—or other decadences often associated with inbreeding, which I attribute to the fact that their mentally sub-normal have failed to survive in the battle for life, and so this particular defect is absent in their hereditary make up. On the other hand mental illness is present in some breeds of domestic animals which do not have to fight a severe fight for existence and of which the parentage is the result of man's selection. Mad dogs, mad bulls, etc., are not uncommon. Many of these cases are probably due to extraneous circumstances, yet I cannot overlook the fact that the trouble is also in the blood. When a bull shows mental instability we probably slaughter him, but we do not slaughter or sterilize all his blood relations who carry the same hereditable factors.

The object of inbreeding is to increase the influence of the ancestor to whom we inbreed. It is the process of taking two dips into the same well of parental ingredients. The proportionate amount we pull out is dependent on the family relationship. For instance if we mate brother and sister it may be likened to taking two bucketfuls; if we mate cousins we are taking two half bucketfuls—and so on. When we take a dip into the well we withdraw both the good and the bad ingredients. We cannot escape this. It is therefore of the highest importance that we should only inbreed to progenitors who have shown an outstanding ability for transmitting the qualities we desire.

All racehorses are inbred. Every name in a pedigree traces in tail female to one of some 50 foundation mares and in tail male to one of three patriarchs, Eclipse (1764), Matchem (1748) or King Herod (1758). This is so well appreciated that we do not commonly consider a horse inbred unless he carries the same name in his pedigree at least twice in the first three or four removes: i.e. his parents are either (full or half) third or second cousins, cousins (or half-cousins) or closer. More remote inbreeding we usually refer to as line breeding. There is no difference in principle between inbreeding and line breeding—it is only a question of degree.

In times gone by, more close inbreeding was practised than today. A hundred years ago there were only 1,500 broodmares in the Stud Book compared to nearly 15,000 today. So the then position was something akin to the population of a village inter-marrying, whilst now it is like the population of a sizeable town doing the same.

From time to time I have heard breeders aver the advantages of inbreeding to sires and not to mares, whilst others have championed the reverse process. For my part I cannot see the difference in principle. If we inbreed to a particular stallion we must also from necessity inbreed to that stallion's dam and vice versa.

✗ INBREEDING OR OUTCROSSING?

In any discussion regarding inbreeding as opposed to outcrossing, it is possible to cite any number of exceptionally good horses (and bad ones) bred each way. I think the first requisite necessary to form an opinion is to balance opportunity against result and, with this in view, to ascertain the percentages of matings effected annually amongst thoroughbreds to produce various degrees of inbreeding compared to the relative successes achieved. If for instance it is usual for 50 per cent of the mares in the Stud Book to be mated to produce outcrosses and it is possible to show that only 25 per cent of good horses are outbred, it would be safe to say that inbreeding produces superior results. To obtain the necessary data for this would entail a prodigious research into all the facts recorded in the General Stud Book and annual volumes of Races Past since their inception. Even then it would not be very easy, arbitrarily, to decide which horses can be classified as good and which not up to standard. In the absence of this, the best thing seems to be to take a cross-section and from this to judge how matters stand.

Accordingly I have investigated the pedigrees up to and including the fourth generation of the following:

(1) 1,000 thoroughbreds taken at random, i.e., winners, non-winners, stallions, broodmares, etc.

(2) 77 Derby winners of this century (1900 to 1976). Further notes on this will be found on page 484 onwards.

(3) 25 horses who have exercised the greatest parental influence in the production of the above 77 Derby winners. The basis on which these horses have been pin-pointed and their particulars will be found on page 481 onwards.

(4) The ten leading sires of winners taken at ten-yearly intervals from 1910 to 1974, spread at regular intervals over a long period of years in order to obtain a broad basis.

(5) As in (4) there are cases of the same horse's name appearing amongst the ten leaders twice or more, I have eliminated duplication, leaving 94 different stallions for consideration.

Thus Category 1 represents opportunity and Categories (2) to (5) show successes.

I have classified as outbred, horses who have no name appearing twice in the first four generations of their pedigrees. The inbreds I have divided into four groups: (*a*) those inbred at the fourth generation; (*b*) those inbred with the same name occurring at the third and fourth generation; (*c*) those inbred at the third generation, which have seven instead of eight grandparents; and (*d*) those more closely inbred.

I then worked out the relative percentage in each case, which are given below.

	% of outbreds	% inbred at the 4th generation	% inbred with same name in 3rd + 4th generations	% inbred at the 3rd generation	% closer inbred
		A	B	C	D
(1) Cross-section of 1,000 horses taken at random	56.8	25.6	14.8	2.6	0.2
(2) 77 Derby winners (1900–1976)	53.2	26.0	15.6	5.2	.0
(3) 25 horses with the greatest parental influence in the production of (2)	24.0	36.0	32.0	8.0	.0
(4) 120 leading sires of winners at 10-yearly intervals (1910–1974)	47.5	26.7	21.7	3.3	0.8
(5) 94 separate leading sires of winners at 10-yearly intervals (1910–1974)	51.1	23.4	20.2	4.2	1.1

I have included Bayardo in categories (4) and (5) amongst those inbred at the third and fourth generation (Group B), whereas in fact he was inbred at the second and fourth remove to Galopin and at the fourth to Sterling.

I do not consider this investigation sufficiently extensive to form the basis for definite conclusions. The various categories (2) to (5) give both a wide foundation and standards which are almost free from weight advantages, or other outside influences, but the number of horses involved is too small to put the matter beyond dispute. Furthermore it is possible that if the pedigrees of 1,000 other miscellaneous horses (Category 1) were examined they might give a different "opportunity" ratio. To settle the matter satisfactorily it would be necessary to examine the pedigrees of thousands of horses in the achievement categories and to compare them with the inbreeding incidence of 15,000 or more thoroughbreds. But I think the accompanying percentages probably give some idea of the lie of the land. They show that there is some broad relationship between opportunity and achievements, although inbreds as a combined team (Groups A, B, C and D) have an appreciably better record in Category (3), the most influential horses in the production of Derby winners of this century. If readers care to turn to page 481 they will see that these are selected not on account of the immediate results that they have obtained, but because of their influence over a number of generations. Whether or not this supremacy would be maintained if an entirely different yard-stick, but on the same lines, was employed I confess that I do not know. But from my own observations, even in the case of outbreds I am always impressed by the frequency with which an inbred progenitor is present in the first few generations of the pedigrees of famous horses. Colonel Vuillier who invented the Dosage System (see page 34) investigated the pedigrees of 654 top grade racers between 1748 and 1901 and found that 54.3 per cent of these were inbred at the fourth generation or closer and that 45.7 per cent were outbreds. This seems to confirm that on balance the inbreds register the somewhat better results.

So far as modern blood lines are concerned I will now detail some important horses classified by the various degrees of inbreeding mentioned above.

OUTBRED

NAME AND YEAR OF BIRTH	REMARKS
1 Hampton (1872)	Champion sire and founder of the line which lived on through Gainsborough, Son-in-Law and Hyperion.
2 Chaucer (1900)	A great sire of broodmares.
3 Polymelus (1902)	Five times champion sire.
4 Swynford (1907)	St. Leger winner and champion sire.
5 Sir Gallahad III (1920)	French 2,000 Guineas winner, and champion sire in America.
6 Bull Dog (1927)	Brother to Sir Gallahad III, and also champion sire in America.
7 Brantome (1931)	Great French racehorse and sire.
8 Bahram (1932)	Unbeaten winner of the Triple Crown.
9 Fair Trial (1932)	A great sire of sprinters, and champion sire in 1950.
10 Mahmoud (1933)	Derby winner in record time, and a very successful sire in Europe and America.
11 Precipitation (1933)	Ascot Gold Cup winner and an important stallion.
12 Vatellor (1933)	A fine French stallion who sired two Epsom Derby winners.
13 Donatello II (1934)	An outstanding Italian-bred racehorse, and a significant sire.
14 Djebel (1937)	English and French 2,000 Guineas winner, and also won Arc de Triomphe. An exceptional stallion in France.
15 Owen Tudor (1938)	Winner of wartime Derby and substitute Ascot Gold Cup. Sire of Abernant and Tudor Minstrel.
16 Nasrullah (1940)	Champion sire in UK and America.
17 Chamossaire (1942)	Winner of the St. Leger, and champion sire in 1964.
18 Chanteur II (1942)	Winner of the Grand Prix de Paris, and sire of tough, long-distance horses.
19 Dante (1942)	Derby winner. Died in 1956 after a good start at stud.
20 Prince Chevalier (1943)	Winner of French Derby. Sire of Classic-standard.
21 Petition (1944)	Smart racehorse up to 1¼m. Sire of Petite Etoile and March Past.
22 Tudor Minstrel (1944)	Brilliant racehorse. Fairly successful sire after export to America.
23 Citation (1945)	Held the record for most stakes, 1,085,760 dollars up to 1956. Significant sire in America.

OUTBRED (*contd.*)

NAME AND YEAR OF BIRTH	REMARKS
24 Grey Sovereign (1948)	Top-class sprinter, and sire of speedy horses.
25 Aureole (1950)	Top-class racehorse, and twice champion sire.
26 Never Say Die (1951)	Winner of the Derby and St. Leger. Successful if not outstanding stallion.
27 Ribot (1952)	One of the outstanding racehorses and sires of the century.
28 Crepello (1954)	Winner of 2,000 Guineas and Derby before breaking down. Successful sire, though some of stock tended to be unsound.
29 Sea-Bird II (1962)	Outstanding Derby winner in the last 20 years. Died in 1973 after good start at stud.
30 Nijinsky (1967)	Triple Crown winner. Good start at stud.

INBRED AT THE FOURTH GENERATION

NAME AND YEAR OF BIRTH	REMARKS	INBRED TO
1 St. Simon (1881)	Unbeaten. Nine times champion sire.	Voltaire
2 Cyllene (1895)	Ascot Gold Cup winner, and one of the world's great sires.	Stockwell
3 Spearmint (1903)	Epsom Derby and Grand Prix de Paris winner.	Stockwell
4 The Tetrarch (1911)	Unbeaten. Champion sire in spite of a very short career.	(1) Speculum and (2) Rouge Rose
5 Son-in-Law (1911)	A great patriarch of stayers.	Blair Athol
6 Hurry On (1913)	Unbeaten. Champion sire.	Hermit
7 Tetratema (1917)	2,000 Guineas winner, and champion sire.	St. Simon
8 Blandford (1919)	Three times champion sire, whose get included four Derby winners.	Isonomy
9 Solario (1922)	St. Leger and Ascot Gold Cup winner, and champion sire.	Hampton
10 Asterus (1923)	French 2,000 Guineas winner, and an outstanding success at stud.	Hampton
11 Brown Jack (1924)	A gelding whose 64 races brought 18 wins value £21,646 on the Flat, including Ascot's Queen Alexandra Stakes in six successive years, and 6 over hurdles, including the Champion Hurdle.	St. Simon
12 Blenheim (1927)	Epsom Derby winner. Fine sire in America and at home.	Isinglass
13 Nearco (1935)	Unbeaten, winner of Grand Prix de Paris. Champion sire, and a great modern influence.	St. Simon
14 Pharis II (1936)	Unbeaten winner of the French Derby and Grand Prix de Paris. Champion sire in France despite being confiscated by the Germans during the war.	Cyllene
15 Court Martial (1942)	2,000 Guineas winner. Twice champion sire.	Polymelus
16 Tulyar (1949)	Set record for stakes won in Britain (£76,417), including the Derby.	Phalaris
17 Meld (1952)	Winner of the 1,000 Guineas, Oaks and St. Leger to set record stakes for a filly.	Blandford
18 Dahlia (1970)	European record-holder for stakes won. Twice winner of the King George VI and Queen Elizabeth Stakes, Ascot.	Hyperion
19 Brigadier Gerard (1972)	Outstanding racehorse, winner of 17 out of 18 races.	Fair Trial

INBRED WITH THE SAME NAME IN THE THIRD AND THE FOURTH GENERATION

NAME AND YEAR OF BIRTH	REMARKS	INBRED TO
20 Gallinule (1884)	Champion sire and also five times head of the list of maternal grandsires.	Stockwell
21 Carbine (1885)	Possibly the greatest horse bred in the Antipodes.	Brown Bess
22 Tredennis (1898)	Never won a race and started at stud in very humble circumstances, but a very successful sire.	(1) Stockwell (2) Also inbred to Touchstone at the 4th remove.
23 Prince Palatine (1908)	A great racehorse, winner of the St. Leger, Ascot Gold Cup (twice), Eclipse Stakes, etc.	Hampton
24 Phalaris (1913)	Twice champion sire, and a stallion of tremendous influence.	Springfield
25 Gainsborough (1915)	Triple Crown winner, and twice champion sire.	Galopin
26 Pharos (1920)	Champion sire both in England and France. Brother to Fairway.	St. Simon
27 Colorado (1923)	2,000 Guineas winner. An outstanding sire who died after only two seasons at stud.	St. Simon
28 Fairway (1925)	St. Leger winner, and four times champion sire.	St. Simon
29 Vatout (1926)	French 2,000 Guineas winner, achieved remarkable results during a comparatively short stud career.	Gallinule
30 Tourbillon (1928)	French Derby winner, and four times champion sire in France.	Omnium II
31 Hyperion (1930)	Derby and St. Leger winner, and six times champion sire.	(1) St. Simon (2) Also carries an additional line of Galopin not through St. Simon at the 4th remove.
32 Bois Roussel (1935)	Derby winner. Sire of Classic winners, and very successful sire of broodmares.	St. Simon
33 Big Game (1939)	2,000 Guineas winner. Outstanding broodmare sire.	White Eagle
34 Alycidon (1945)	Ascot Gold Cup winner, and champion sire in 1955.	Swynford
35 Mossborough (1947)	Useful racehorse just below Classic standard. Champion sire in 1958.	Chaucer
36 Sicambre (1947)	Winner of the French Derby and Grand Prix de Paris. Outstanding Classic sire.	Rabelais
37 Vaguely Noble (1965)	Arc de Triomphe winner. Outstanding early success at stud, and champion sire in 1973 and 1974 though standing in America.	(1) Hyperion (2) Also inbred to Bahram at the 4th remove.

INBRED AT THE THIRD GENERATION
(these horses have seven instead of eight great-grandparents)

NAME AND YEAR OF BIRTH	REMARKS	INBRED TO
38 Galopin (1872)	Derby winner. Twice champion sire whose get included St. Simon.	Voltaire
39 Petrarch (1873)	2,000 Guineas, Derby and Ascot Gold Cup winner, and sire of Classic winners.	Touchstone
40 Amphion (1886)	Sired the champion sire Sundridge and maintained the Speculum sire line.	Newminster
41 Sainfoin (1887)	Derby winner. Sire of the Triple Crown winner Rock Sand, and of the dams of Hurry On and Phalaris.	Stockwell
42 Symington (1893)	Sire of Junior, and the dam of Tetratema, etc.	Galopin
43 Marcovil (1903)	Sire of unbeaten Hurry On and the great chaser sire My Prince during a fairly short stud career.	Hermit
44 Pommern (1912)	Triple Crown winner.	Hampton
45 Gay Crusader (1914)	Triple Crown winner.	Galopin
46 Golden Myth (1918)	One of the few to win the Ascot Gold Cup (2½ miles) and Eclipse Stakes (1¼ miles) in the same year.	Bend Or
47 Filibert de Savoie (1920)	Winner of the Grand Prix de Paris and French Gold Cup.	Le Sancy
48 Sansovino (1921)	Derby winner.	Pilgrimage
49 Firdaussi (1929)	St. Leger winner.	Chaucer
50 Gold Bridge (1929)	Best sprinter of his day, and a fine sire of sprinters.	The Boss
51 Avenger (1943)	Grand Prix de Paris winner.	Swynford
52 Sunny Boy III (1944)	Champion sire in France in 1954, when his stock created a European stakes record, 91,267,885 francs.	Gainsborough

When we come to consider more closely-inbred horses who have a potent effect on the breed, we must keep in the forefront of our minds the fact that matings arranged on these lines are rare.

The most significant example of recent years was M. Marcel Boussac's brilliant but nervous filly Coronation V (1946), who hated travelling. Despite her temperamental failings, Coronation V dead-heated with her stablemate Galgala for the French 1,000 Guineas, and later made hacks of some of the best horses in Europe—including the winners of the French Derby, 2,000 Guineas, Oaks and Grand Prix de Paris, and the Epsom Derby runner-up Amour Drake—to win the Prix de l'Arc de Triomphe with the greatest ease.

Her short pedigree is detailed overleaf.

It will be seen that not only were her parents by Tourbillon, but of her six (instead of the normal eight) great-grandparents two were by Teddy. She therefore only had 11 great-great-grandparents. I draw special attention to her dual great-grandsire Ksar, as he also was an incestuously-bred horse, his dam being by Omnium II (1892) and his sire Brûleur's dam being by Omnium II also. Ksar won the French Derby, St. Leger, Arc de Triomphe (twice) and other good

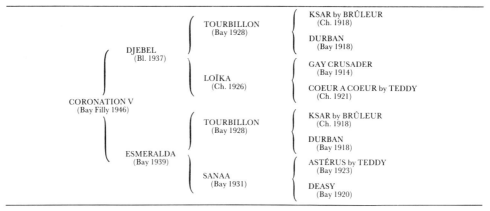

races. Not only was he an outstanding racehorse, but he was also an outstanding sire in France.

The notable racecourse success of Coronation V made it all the more sad that at stud she was a complete failure. She was covered for 14 consecutive years, by Marsyas, Pharis (four times), Owen Tudor, Auriban (seven times) and Iron Liege, and was barren on each occasion. Critics of inbreeding would be unwise to cite Coronation V as an example of the bad effects it can have, since her failure at stud was the result of a veterinary disability, though not for the want of trying.

On the other hand, an example of inbreeding which produced an unraced mare who starred at stud can be found in Flagette (1951), whose sire Escamillo and dam Fidgette were both by Firdaussi. In turn Firdaussi was inbred at the third generation to Chaucer, who is accompanied in the fifth generation of Flagette's pedigree by three appearances from Polymelus, and two from Flying Fox. Flagette apparently grew too big to be trained and went straight to stud, where she produced Herbager (1956), winner of the French Derby and beaten only once as a three-year-old.

It has been reported that Flagette's breeding was influenced by the example of Coronation V. How interesting therefore that the two results should differ so widely—Coronation V being a racecourse success who failed at stud, and Flagette a mare who could not be raced but took high honour in the paddocks.

The successful American stallion High Time was also bred on very close lines. His sire Ultimus was by a son of Domino out of a daughter of Domino, whilst High Time's dam was by Domino out of a daughter of Domino. His pedigree therefore was as follows:

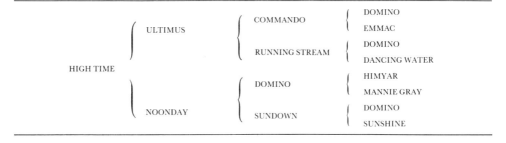

Domino was a great racehorse and was most successful during his short stud career of two seasons. His daughter Cap and Bells II (1898) was the first American-bred filly to win the Epsom Oaks. Commando was also an outstanding racer, unlucky ever to know defeat. He died young, having had only four years at stud, during which he sired unbeaten Colin, Peter Pan, Celt and others of high merit in American racing. Ultimus never ran and High Time ran only once, but both were excellent sires. High Time is especially notable as the sire of the mare Dinner Time, whose son Eight Thirty (1936) won 16 races worth 155,475 dollars and became a prominent sire himself.

We have already seen that Galopin was a closely inbred horse—his grandsire and grandam both being by Voltaire (1826). The Triple Crown winner and highly successful sire in France, Flying Fox (1896), was incestuously inbred to him, Flying Fox's dam being by Galopin whilst his sire Orme (1889) was out of a Galopin mare. The tremendous world-wide influence of Flying Fox will be familiar to all as he established a sire line which has produced such as Ajax, Teddy, Astérus, Sir Gallahad III, Bull Dog, Citation, Coaltown, Ortello (six times champion sire in Italy and then sent to U.S.A.), Borealis, etc.

Barcaldine (1878) was another great racehorse and important stallion who was incestuously bred, his sire Solon being out of Darling's Dam whilst his own grandam Bon Accord (1867) was out of the same mare. Barcaldine never knew defeat on the racecourse and on retiring to stud got several Classic winners. Moreover, he succeeded in carrying on the Matchem line to its representation by Precipitation (1933), Chamossaire (1942), Airborne (1943), Niccolo Dell 'Arca (1938), etc. But for Barcaldine this sire line would have died out in Europe, although in America there is another branch of it handed down by exported Australian (1858) through Man o' War (1917) to his various sons and grandsons.

Wisdom (1873), whose name is familiar as his son Love Wisely (1893) sired Fairway's grandam Anchora (1905), was a closely inbred horse. His paternal grandsire Rataplan (1850) and his maternal grandsire Stockwell (1849) were full brothers—so he had only six great-grandparents.

One can search the list of winners of big races for Wisdom's name in vain, for he never won a race of any sort and finished seventh of the ten starters in the Derby of his year. In fact he was thought so little of that he was once sold for 50 guineas. Nevertheless he proved a highly successful stallion and ere he died he got the Derby winner Sir Hugo, the 2,000 Guineas winner Surefoot, the Oaks winner La Sagasse, the Ascot Gold Cup winner Love Wisely, and other good racers, whilst his broodmares were particularly valuable.

There have been 15 winners of the Triple Crown. Honours amongst inbred and outbreds are divided six to nine. They are:

Nijinsky (1967), outbred.
Bahram (1932), outbred.
Gainsborough (1915), inbred to Galopin at the third and fourth removes.
Gay Crusader (1914), inbred to Galopin at the third remove.
Pommern (1912), inbred to Hampton at the third remove.
Rock Sand (1900), inbred to Stockwell at the fourth remove.

Diamond Jubilee (1897), outbred.
Flying Fox (1896), inbred to Galopin at the second and third removes.
Galtee More (1894), inbred to Stockwell at the fourth remove.
Isinglass (1890), outbred.
Common (1888), outbred.
Ormonde (1883), outbred.
Lord Lyon (1863), outbred.
Gladiateur (1862), outbred.
West Australian (1850), outbred.

It will be noted that three—Gay Crusader, Pommern and Flying Fox—were inbred at the third or closer removes, which is remarkable considering the very few horses so bred.

To the extent which inbreeding amongst thoroughbreds is practised I have not been able to detect any derogatory results. Some of the finest physical specimens with outstanding staying ability have been so bred.

Barcaldine and Gay Crusader are examples. The redoubtable Brown Jack (1924) although not a particularly good-looking horse was an exceptional battler and he was inbred to St. Simon at his fourth remove. Neither have I noticed any reduction in fertility. The notorious cases of Ormonde (1883), Caligula (1917), Call Boy (1924) and Guinea Gap (1931), who were practically sterile, were all outbred, although Blue Train (1944) and The Tetrarch (1911), who both became sterile after a very few seasons at stud, were both inbred at the fourth generation.

A MATTER OF TEMPERAMENT

It is sometimes said that inbreeding produces an excitable temperament. I think this may be the case when the inbreeding is to an excitable ancestor. In my view the opinion is largely based on the results of inbreeding to St. Simon, who was so temperamental that when his owner paid him a visit at Welbeck the stud-groom used to put his billy-cock hat on the end of a stick to subdue him for inspection. It was thought that a cat as a companion might have a soothing influence on St. Simon's nerves but the great horse seized poor pussy in his mouth, threw him against the stable ceiling and killed him. I do not imply that inbreeding to St. Simon or any other temperamental horse necessarily produces an excitable foal. The peculiarity is transmitted like any other hereditable factor. It may or may not show itself in the produce but the fact of inbreeding to it increases the probability of its appearance.

I am certain of one thing, which is that when a temperamental animal who is inbred appears on the racecourse, the world and his wife will blame the inbreeding for the trouble, but when the more frequent cases of excitable outbreds appear, no remarks are made.

I have already mentioned the incestuously bred Coronation V (her pedigree is on page 24). She was decidedly temperamental due no doubt to the fact that she was inbred to Brûleur, who sired a number of his stock with the same failing. Yet on innumerable occasions I have heard Coronation V held up as an example of the dangers of close inbreeding, whilst such as Sun Chariot (1939), Nasrullah (1940),

Mât de Cocagne (1946), Aquino II (1948), Zucchero (1948), etc., who were far more excitable and all outbred, were passed over without comment.

Anyone familiar with pedigrees of this century must have noticed how common inbreeding to St. Simon is. Apart from his excellence as a racehorse he was champion sire for nine seasons and his sons held a similar position for six more seasons. Furthermore, he had other sons such as Chaucer (1900), St. Serf (1887), William the Third (1898), Florizel II (1891), etc., who, although never at the head of the list of sires of winners, sired Classic winners. In fact for a period of 30 years he carried all before him. Breeders in those days were to some extent faced with the alternative of inbreeding to St. Simon or of using blood which had proved markedly inferior to his. Is it to be wondered that many went for the best and chose inbreeding?

It is a natural human tendency to breed from the best, and when one line of blood shows an overwhelming superiority over all rivals, it can only be expected that breeders will utilize it to the full.

CHAPTER 3

Breeding Systems and Theories

Galton's law of ancestral contribution: stamina indices:
average stamina of sire's progeny:
Vuillier Dosage System: Varola's Typology: Bruce Lowe's Figure System

SIR FRANCIS GALTON expounded a rule of ancestral contribution which laid down that on an average both parents between them contribute one-half to the offspring's make-up, the four grandparents subscribe one-quarter between them, the great-grandparents one-eighth and so on indefinitely. Thus each parent is responsible for one-quarter, each grandparent one-sixteenth, each great-grandparent one sixty-fourth and so on. It must be emphasized that this is not a hard and fast rule, but a guide based on the average. As we have already seen in Chapter 1 the hereditary factors of remote ancestors sometimes remain latent for several generations only to reappear with a potency which has no relation to their position in the ancestral tree.

STAMINA INDEX

Galton's Law is commonly used amongst bloodstock breeders to determine the probable staying capacity of an unraced horse or to estimate the probable distance ability of a foal which will result from a certain mating. It is known as the animal's stamina index.

Although exceptional cases occur, the stamina index is a pretty accurate estimate of a horse's actual staying powers and so is invaluable.

As an example of the method of calculation I set out opposite the pedigree of the mare Rainstorm (1924), dam of the smart sprinter and sire of sprinters Golden Cloud (1941).

When a name in the pedigree stands for a horse who never ran, or only as a two-year-old, it is necessary to work out his stamina index and substitute it for his racing distance. The average staying capacity of a parent's progeny should not be substituted for his or her own racing distance as this is subject to its partner's influence.

If we estimate without the help of Galton's Law the probable staying capacity of a foal by the 5 furlongs Gold Bridge out of the 9½ furlongs Rainstorm, it would be reasonable to put it at slightly over 7 furlongs—the mean between the two parents' distances. We should be wrong. The foal Golden Cloud (1941) could only stay 6 furlongs. Gold Bridge's pedigree abounds in sprinting blood which must be taken into account, and if this is done on the lines indicated it will be found that Golden Cloud's stamina index approximates to his actual distance capacity.

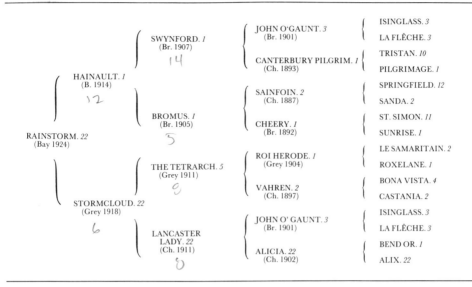

The calculation of the average stamina of a parent's progeny is usually confined to the progeny of stallions only and not mares. The reason for this is that a stallion's runners may add up to hundreds, whereas mares have very few and therefore can readily be treated individually. The basis of calculation is simply the average distance of all races won by the get of a particular stallion excluding two-year-olds. Two-year-olds are not included, as first they are not fully developed, and second the distances they race over are limited by the Rules of Racing and not by their own capacity.

The average stamina of a sire's progeny gives a rough and ready but useful guide to the distance that his stock may be expected to stay. It is influenced by the staying capacity of the mares he is mated to, so should not be taken as a basis for

NAME OF PROGENITOR PARENTS	Racing Distance in Furlongs	NAME OF PROGENITOR GRANDPARENTS	Racing Distance in Furlongs
Hainault	12	Swynford	14
Stormcloud	6	Bromus	5
		The Tetrarch	8
		Lancaster Lady	8
	2)18		4)35
Average distance of first generation	9	Average distance of second generation	8.75

NAME OF PROGENITOR GREAT GRANDPARENTS	Racing Distance in Furlongs	NAME OF PROGENITOR GREAT-GREAT GRANDPARENTS	Racing Distance in Furlongs
John o' Gaunt	12	Isinglass	20
Canterbury Pilgrim	18	La Flêche	20
Sainfoin	12	Tristan	20
Cheery	5	Pilgrimage	12
Roi Herode	15	Springfield	10
Vahren	11	Sanda	5
John o' Gaunt	12	St. Simon	20
Alicia	5	Sunrise	8
		Le Samaritain	15
		Roxelane	10
		Bona Vista	8
		Castania	5
		Isinglass	20
		La Flêche	20
		Bend Or	12
		Alix	5
	8)90		16)210

Average distance of third generation	11.25	Average distance of fourth generation	13.123

	Average distance in furlongs	Influence		
First generation	9.0	½	= 4.5	furlongs
Second ,,	8.75	¼	= 2.187	,,
Third ,,	11.25	⅛	= 1.406	,,
Fourth ,,	13.123	1/16	= 0.820	,,
			8.913	,, "A"
Add 1/16 of "A" to complete			0.557	,,
		1	9.470	,,

The stamina index of Rainstorm therefore is 9.47 furlongs.

calculation in arranging a particular union. For this purpose the stamina index of the foal-to-be should be worked out.

Below will be found the average stamina of the progeny of some of the most important stallions of recent years.

NAME OF SIRE	AVERAGE IN FURLONGS	NAME OF SIRE	AVERAGE IN FURLONGS
Sing Sing	5.89	High Treason	6.71
Indigenous	6.04	Welsh Abbot	6.74
Como	6.24	Constable	6.91
Vilmorin	6.32	Vilmoray	6.94
Bleep-Bleep	6.36	Goldhill	7.03
Sound Track	6.43	Vigo	7.09
Whistler	6.57	Matador	7.13
Democratic	6.67	Right Boy	7.14

NAME OF SIRE	AVERAGE IN FURLONGS	NAME OF SIRE	AVERAGE IN FURLONGS
Princely Gift	7.23	Neron	9.60
Gratitude	7.24	Hill Gail	9.62
Skymaster	7.25	Honeyway	9.67
Native Dancer	7.28	Sica Boy	9.76
Grey Sovereign	7.30	Darius	9.83
Polly's Jet	7.37	Tamerlane	9.83
Abernant	7.52	Big Game	9.93
Floribunda	7.56	Infatuation	9.96
Hook Money	7.66	Ossian II	9.97
Tudor Melody	7.66	Mustavon	9.98
Epaulette	7.67	King of the Tudors	10.02
Crocket	7.76	Henry the Seventh	10.05
Red God	7.79	Gilles de Retz	10.06
Derring-Do	7.80	St. Paddy	10.10
March Past	7.84	Ragusa	10.22
Fairey Fulmar	7.86	Supreme Court	10.22
Ratification	7.86	Hugh Lupus	10.26
Dumbarnie	7.89	Damremont	10.32
Will Somers	7.89	Rockefella	10.34
Roan Rocket	7.91	Primera	10.36
Hard Tack	7.93	Pardal	10.49
King's Troop	7.97	Hethersett	10.50
Monet	8.00	Le Levenstell	10.50
Palestine	8.03	Crepello	10.61
Donore	8.12	Chamier	10.64
King's Bench	8.22	Pinza	10.66
Kelly	8.24	Arctic Storm	10.67
Queen's Hussar	8.29	Faberge II	10.67
Royal Palm	8.34	Saint Crespin III	10.68
Sovereign Path	8.44	Hard Ridden	10.82
Buisson Ardent	8.49	Santa Claus	10.86
Fighting Don	8.51	Pirate King	10.87
Firestreak	8.65	Immortality	10.88
Pall Mall	8.66	Parthia	10.93
Sir Gaylord	8.67	Vienna	10.95
Relic	8.72	Ribot	10.99
Blast	8.78	Preciptic	11.04
Petition	8.78	Relko	11.07
Eudaemon	8.97	Psidium	11.10
Rustam	8.97	Pampered King	11.12
Klairon	9.10	Premonition	11.14
Counsel	9.12	Anwar	11.34
Fortino II	9.13	Narrator	11.34
Hard Sauce	9.16	Acropolis	11.35
Major Portion	9.23	Sayajirao	11.38
Northern Dancer	9.25	Match III	11.40
Quorum	9.25	Hornbeam	11.44
Milesian	9.27	Mossborough	11.50
Romulus	9.33	Colonist II	11.54
Proud Chieftain	9.49	Aggressor	11.57
Cash and Courage	9.53	Sicambre	11.58
Sea-Bird II	9.55	Above Suspicion	11.63

NAME OF SIRE	AVERAGE IN FURLONGS	NAME OF SIRE	AVERAGE IN FURLONGS
Ballymoss	11.68	Right Royal V	12.21
Fidalgo	11.68	Worden II	12.31
Panaslipper	11.68	Javelot	12.38
French Beige	11.69	Vimy	12.55
Aureole	11.81	Pandofell	12.82
Straight Deal	11.85	Tanerko	13.03
Alcide	11.86	Black Tarquin	13.53
High Hat	11.92	Arctic Slave	13.82
Charlottesville	12.06	Bowsprit	14.69
Exbury	12.09	Vulgan	14.73
Never Say Die	12.10	Fortina	15.73
Zarathustra	12.10	Even Money	16.10
Sheshoon	12.19	Straight Lad	16.55

The following stallions, listed alphabetically, had sired the winners of over 150 races (not including two-years-olds) but were ineligible for the above list because they had not sired a winner in the previous two years.

NAME OF SIRE	AVERAGE IN FURLONGS	NAME OF SIRE	AVERAGE IN FURLONGS
Abbots Trace	8.43	Donatello II	11.47
Achtoi	10.97	Epigram	10.51
Admiral's Walk	8.58	Fair Trial	6.96
Alycidon	12.51	Fairway	8.96
Apron	9.62	Felstead	9.90
Arctic Star	10.49	Friar Marcus	7.56
Argosy	6.60	Fun Fair	7.78
Bachelor's Double	8.44	Gainsborough	10.29
Ballyogan	6.99	Galloper Light	9.87
Beau Sabreur	10.75	Gay Crusader	10.48
Beresford	8.39	Gold Bridge	5.44
Blandford	9.70	Golden Cloud	6.70
Blue Peter	10.36	Golden Sun	6.51
Bobsleigh	10.16	Grandmaster	9.03
Bois Roussel	11.11	Grand Parade	7.89
Bold Archer	6.40	High Highness	8.68
Borealis	10.89	Hurry On	9.80
Bridge of Earn	9.84	Hycinthus	8.90
Buchan	10.14	Hyperion	10.07
Chamossaire	11.89	Jackdaw	11.29
Chanteur II	11.70	Junior	9.23
Charles O'Malley	10.61	Knight of the Garter	8.86
Chaucer	9.36	Lemberg	9.95
Colombo	8.56	Lighthouse II	11.32
Combat	9.25	Lomond	8.93
Coup de Lyon	9.56	Manna	8.71
Court Martial	7.59	Mieuxce	11.36
Dante	9.67	My Babu	8.68
Denturius	5.52	Nasrullah	8.84
Devonian	10.59	Nearco	9.78

NAME OF SIRE	AVERAGE IN FURLONGS	NAME OF SIRE	AVERAGE IN FURLONGS
Nimbus	10.07	Solario	10.97
Obliterate	9.77	Soldennis	9.21
Orthodox	9.57	Solonaway	8.60
Owen Tudor	10.22	Son-in-Law	11.00
Panorama	5.90	Spearmint	10.74
Pappageno II	11.47	Spion Kop	10.32
Papyrus	9.07	Stardust	8.59
Persian Gulf	11.25	Steadfast	9.71
Phalaris	8.01	St. Frusquin	10.64
Pharos	9.00	Stratford	7.32
Phideas	10.83	Sunstar	8.52
Pink Flower	7.90	Swynford	9.45
Polymelus	8.68	Tehran	10.99
Pommern	9.30	Tetrameter	7.35
Precipitation	11.67	Tetratema	6.43
Prince Chevalier	10.62	The Phoenix	8.50
Prince Galahad	9.35	The Tetrarch	7.66
Rochester	9.21	Tracery	9.64
Roi Herode	9.54	Tredennis	8.49
Sansovino	9.87	Tudor Minstrel	8.09
Santoi	11.35	Turkhan	10.66
Scottish Union	11.03	Umidwar	9.83
Signal Light	9.10	Watling Street	8.59
Simon Square	8.39	Winalot	10.26
Sir Cosmo	6.57		

The above figures are taken from the *Bloodstock Breeders' Review* 1971, and I must pay tribute to the publishers of that invaluable annual.

Practically all the stallions who were themselves stayers show an average distance won by their stock well below their own distance. This is to be expected, since the number of long-distance races in the Calendar is much smaller than shorter distances, thus stallions who themselves were stayers have a limited number of runners concentrating on long-distance races.

Out-and-out sprinters show an average above their own distance limitations. This is due to the simple fact that there are no races below 5 furlongs, so their winner must either be five-furlong or longer runners.

A case for reflection, if not conclusive explanation, concerns certain steeple-chasing sires. Cottage, who had very few winners on the Flat so his average was not handicapped by a considerable number who had to race over short distances through lack of long races, shows an average for his Flat-racers of 1¼ miles; yet he got three winners of the Grand National, over 4½ miles. And Quorum, who himself did not stay beyond a mile and shows an average of 9.25 furlongs, is responsible for the great Aintree hero Red Rum.

I remember the owner of an earlier great sire of chasers, Drinmore (1908), telling me that when his horse was first put to stud, he was tried as a Flat-race sire but failed. It seems his stock frequently would not stay the minimum distance of five furlongs!

Finally I will draw attention to the difference between the full brothers

Tetratema, (6.43 furlongs) and The Satrap (10.15). The discrepancy is probably due to the great difference in their stud histories and the type of mares to which they were mated. As individuals they were very much alike—both greys—showing a distinct resemblance to their Arabian ancestors. On the racecourse they were both the best two-year-olds in England of their respective generations. Tetratema proved his ability to stay a mile by winning the 2,000 Guineas. The Satrap was exported to U.S.A. at the end of his two-year-old days, returning as a 12-year-old when he received a somewhat cool reception. His chances at stud in England were further jeopardized by the outbreak of war after his first two seasons, yet he sired a number of useful winners, including the Doncaster Cup, Ascot Gold Vase, Goodwood Stakes, etc., winner Auralia (1943). There was nothing in The Satrap's style of racing and general make-up which would warrant the assumption that he himself was a better stayer than his elder brother.

THE VUILLIER DOSAGE SYSTEM

Colonel Jean-Joseph Vuillier, the inventor of the Dosage System which bears his name, was a French Cavalry officer who made an extensive study of thoroughbred pedigrees. He propounded his system in the 1920s in *Les Croisements Rationnels*. It hinges on Galton's Law of Ancestral Contribution. Vuillier traced the pedigrees of some hundreds of good horses to the twelfth generation and then analysed their blood mixtures. He found that there was a marked similarity between them all in certain respects.

At the twelfth generation a horse has 4,096 ancestors. Vuillier based his calculations as follows:

An ancestor in the twelfth generation counted as 1 i.e. $\dfrac{1}{4096}$

                eleventh   ,,   ,,   ,,   2 i.e. $\dfrac{2}{4096}$

                tenth   ,,   ,,   ,,   3 i.e. $\dfrac{3}{4096}$

and so on.

The whole system is sometimes adversely criticized from the fact that the above method of calculation is mathematically erroneous. If each of the 4,096 ancestors of the twelfth generation count 1/4096, the total influence of this generation comes to 1/4096 × 4096 or 1, i.e. the complete build up of the horse, and so on all through the pedigree. Thus each generation is self sufficient, leaving no room for the influence of any other generation. Galton's Law laid down that *both* parents *between* them subscribed half, *all four* grandparents quarter, and so on, to the progeny. Although Vuillier's plan of reckoning is undoubtedly at fault, I do not consider that this has any bearing on the principles of his system. The figure of influence, generation by generation, is but a symbol and as he used the same set of symbols throughout his calculations, the mathematical inexactitude is immaterial.

Working on this basis, Vuillier computed that 15 stallions and one mare appeared in the various pedigrees with approximately the same frequency. He then ascertained the average influence of each of these 16 names in the numerous pedigrees which he dissected. He thus fixed a standard dosage desirable in the make-up of a first-class racehorse. These are as follows:

Pantaloon (1824)	140	Voltaire (1826)	186
Touchstone (1831)	351	Gladiator (1833)	95
Birdcatcher (1833)	288	Bay Middleton (1833)	127
*Pocahontas (1837)	313	Melbourne (1834)	184
Newminster (1848)	295	Stockwell (1849)	340
Hermit (1864)	235	Hampton (1872)	260
Galopin (1872)	405	Isonomy (1975)	280
Bend Or (1877)	210	St. Simon (1881)	420

* Pocahontas is the only mare in the list.

The general outline of the system is somewhat akin to a doctor's prescription. It tabulates the ratio of ingredients required to produce on the average a desirable result. It makes no claim that parents of poor calibre can be turned into Classic winner producers simply by following the formula, but merely that if the principles are followed over a wide spread the results are likely to be superior to those bred with disregard to them.

The system provides that if a parent has an excess dosage of any of the 16 named progenitors he or she should be mated to a partner who has a deficiency of the same blood and so the foal would receive approximately the standard dosage. Similarly, if a parent has a deficiency, his or her mate should have a compensating excess.

So far as I am aware no horse has exactly the correct dosages and the best that can be achieved is an approximation. Should it be possible to find two parents whose combined strains would produce exactly the correct figures it is long odds that there would be other factors, such as lack of racing merit, wide difference in staying capacity, conformation, etc., which would make their union undesirable.

In order to explain the method of calculating a horse's dosage readers will find in the folder at the back of the book Hyperion's pedigree. So as not to make the example too cumbersome let us consider his dosage of St. Simon, Galopin, Bend Or, Hampton and Isonomy only. This can be done in a six-generation pedigree.

Each time one of the named horses appears in the pedigree it counts as follows:

In the	twelfth	generation	1 strain
,,	eleventh	,,	2 strains
,,	tenth	,,	4 ,,
,,	ninth	,,	8 ,,
,,	eighth	,,	16 ,,
,,	seventh	,,	32 ,,
,,	sixth	,,	64 ,,
,,	fifth	,,	128 ,,
,,	fourth	,,	256 ,,
,,	third	,,	512 ,,
,,	second	,,	1,024 ,,
,,	first	,,	2,048 ,,

Thus if St. Simon's name appears in the third generation he scores 512; if it also appears in the fifth he scores a further 128 bringing his total score up to 640. He scores nothing further for his various sons, daughters, grandsons, granddaughters, etc. which appear in the pedigree. His influence through them is taken into consideration in his own personal computation.

It will be seen in Hyperion's case St. Simon's name appears

Once in third generation so, as above, this counts	512 strains
Once in fourth ,, ,, ,, ,, ,, ,,	256 ,,
Add together	768 ,,

and this is Hyperion's St. Simon dosage.

Galopin's name appears

Twice in the fourth generation each counting 256 as above	=	512
Once ,, ,, fifth ,, ,, ,, 128 ,, ,,	=	128
Twice ,, ,, sixth ,, ,, ,, 64 ,, ,,	=	128
Add together to obtain Hyperion's Galopin dosage		768

Bend Or's name appears

Once in the fifth generation counting	128
,, ,, ,, sixth ,, ,,	64
Hyperion's Bend Or dosage	192

Hampton's name appears

Once in the fourth generation counting	256
Hyperion's Hampton dosage	256

Isonomy's name appears

Once in the fifth generation counting	128
,, ,, ,, sixth ,, ,,	64
Hyperion's Isonomy dosage	192

Let us compare Hyperion's dosages of these five horses with the standard dosages which I have already given and we find.

	St. Simon	Galopin	Bend Or	Hampton	Isonomy
Hyperion	768	768	192	256	192
Standard	420	405	210	260	280

According to the Vuillier system therefore Hyperion has a heavy excess of both St. Simon and Galopin blood—the Bend Or, Hampton and Isonomy dosages are approximately correct.

Further to illustrate the system I have set out Hyperion's son Aureole's pedigree which will also be found in the folder at the back of the book.

Aureole's dosage compilation will be found thereon and reads:

St. Simon	Galopin	Bend Or	Hampton	Isonomy
592	624	224	240	280

It will be noted Hyperion's excess of St. Simon—348 strains—has been reduced to an excess of only 172 strains above the standard dosage in the case of Aureole, whilst Hyperion's Galopin surplus 363 has come down to 219 in Aureole's pedigree. Aureole is near the best—if not actually the best—son of Hyperion and his case serves as a useful illustration of the general workings of the Vuillier dosage system.

Amongst European bloodstock it would be hard to find many mares of class so free from St. Simon blood as to offset completely Hyperion's excess. The whole system affords breeders a most valuable means of assessing, on a rational basis, the amount of blood of a particular ancestor, a horse or mare possesses. For example, in Aureole's normal six-generation pedigree Galopin's name only appears three times in somewhat remote incidence and on the face of it he does not appear to have an excess of this blood. But when we add four strains at the seventh remove, nine at the eighth and four at the ninth it puts a different complexion on matters.

On the other hand whether or not there is any particular significance in adhering even approximately to the standard dosage figures is extremely debatable. Taking into consideration both their racing merit and their stud careers, among the most important horses of recent years have been Hyperion, Fairway, Pharos and Nearco—all of whom carry St. Simon blood greatly in excess to the laid down quota. Other outstanding horses who show a wide discrepancy are Blandford (over 100 per cent too much Isonomy), Gainsborough (over 100 per cent too much Galopin), Hurry On (no St. Simon at all), Tourbillon (no Hampton) and so on.

I have just remarked that the system is somewhat on the lines of a doctor's prescription but it must be borne in mind that a doctor is dealing with drugs whose reactions to one another are constant, whilst a horse's parental contribution to his issue is varied. As we have already seen (page 4) a parent never transmits exactly the same genes to any two of his stock. At one mating a union may produce a Classic winner, but repeat matings often result in totally useless progeny. So when weighing up the pros and cons of the whole scheme this vitally important aspect must not be overlooked. Furthermore, consideration must be given to the percentage of successes and failures compared to matings carried out in conformity with the system.

It is all very well to say that a number of good horses are bred in accordance with its provisions, but also many utterly useless specimens are similarly bred. The first thing the practical breeder will want to know is whether the Vuillier bred horses, in relation to the numbers, have better records than those bred on some other plan. In other words if 1,000 matings are carried out, half on Vuillier's principles and half with disregard of them, which will have the highest percentage of clear-cut winners (not due to a weight advantage) and which will have the highest incidence of bad selling platers and non-winners? I confess that I cannot answer this but I have always observed that the advocates of most theories show an inclination to base their claims on the achievements of a limited number of well-known horses and to ignore the usually more plentiful quite unknown and thoroughly bad specimens.

I had better explain what is known as a horse's (or mare's) ecarts and their significance. Vuillier divided the 16 basic progenitors, given on page 35, into three categories, in accordance with their year of birth, as follows:

(1) The Birdcatcher Series. These were foaled between 1824 and 1837.

Progenitor	Standard Dosage	Progenitor	Standard Dosage
Pantaloon	140	Gladiator	95
Voltaire	186	Bay Middleton	127
Touchstone	351	Melbourne	184
Birdcatcher	288	Pocahontas	313
Totals	965		719
			965

Total standard dosage for the whole Birdcatcher series   1,684

(2) The Stockwell Series. These were foaled in 1848 and 1849.

Progenitor	Standard Dosage
Newminster	295
Stockwell	340
Total standard dosage for the whole of the Stockwell series	635

(3) The St. Simon Series. These were foaled between 1864 and 1881.

Progenitor	Standard Dosage	Progenitor	Standard Dosage
Hermit	235	Isonomy	280
Hampton	260	Bend Or	210
Galopin	405	St. Simon	420
Totals	900		910
			900

Total standard dosage for the whole of the St. Simon series   1,810

The expression ecarts used in the system can be taken to mean the degree of deviation from the standard number of dosage strains given on page 35. It is important to appreciate that this applies equally to a surplus and to a deficiency. The ecart figure is obtained in each of the above series (Birdcatcher, Stockwell and St. Simon) by adding up both a horse's plus and his minus strains of each individual progenitor and by bringing them into line with the standard dosages. In order to illustrate my meaning I will take at random the example of the well-known Vatellor and set out his position in respect to the Birdcatcher Series.

Progenitor	Vatellor's Dosage	Standard Dosage	Vatellor's surplus or deficiency
Pantaloon	94	140	− 46
Voltaire	179	186	− 7
Touchstone	376	351	+ 25
Birdcatcher	208	288	− 80
Gladiator	98	95	+ 3
Bay Middleton	149	127	+ 22
Melbourne	165	184	− 19
Pocahontas	360	313	+ 47
		Total	249

Vatellor's deviation from the standard dosages for the Birdcatcher Series is therefore 249 and this is his ecart for the group. Working on exactly the same lines it will be found that his ecart for the Stockwell Series is 49 and for the St. Simon Group 568. In order to render the position clearer the late Professor H. E. Keylock suggested instead of using a bare figure to indicate the variance from the standard that this should be done on a percentage basis. In Vatellor's case this would read:

	Birdcatcher Series	Stockwell Series	St. Simon Series
1. Standard dosage	1,684	635	1,810
2. Vatellor's ecarts	249	49	568
Vatellor's ecarts shown on a percentage basis, i.e. % of *1* above shown in *2*	14.8%	7.7%	31.3%

Keylock took things a step further. It will be noticed that the standard dosage for the whole Birdcatcher series is 1,684, for the Stockwell's 635 and for the St. Simon's 1,810, making a grand total of 4,129 strains. Vatellor's ecarts are 249, 49 and 568 respectively which add up to 866. 866 is 21 per cent of 4,219 and this he computed represents Vatellor's overall deviation from the standard. Colonel Vuillier perhaps paid greater attention to the ecarts than to a difference from one specific ancestor's standard dosage. A variation in one strain may be compensated by the superior incidence of ancestry in another. In fact what you lose on the swings you may pick up on the roundabouts.

VAROLA'S TYPOLOGY

Having spent considerable space explaining the Vuillier Dosage System, it must be pointed out that there is much criticism of the theory, over and above the mathematical considerations already cited, and the fact that full brothers or sisters can demonstrate enormous differences in ability.

If there is, in fact, any portent in the standard figure, I suggest that they need continual revision to keep them up to date. Touchstone, Pocahontas, Voltaire and the other lights of over 100 years ago can have but insignificant influence on present-day matings.

With this in mind, the Italian Dr Franco Varola undertook a revision and extension of Vuillier's work, his eventual findings being published in *Typology of the Racehorse* (J. A. Allen, 1974). Whereas Vuillier divided his 16 progenitors into three series, Varola has brought the subject into the 20th Century by employing 120 sires, covering the international scene and which he has split into four series of "chefs-de-race", the last 20 representatives of which embrace the second half of the 20th Century.

Having selected those 20th Century sires which have had most impact on the evolution of the modern thoroughbred, Varola digressed from Vuillier's framework by classifying his chefs-de-race into five aptitudinal groups—namely Brilliant (providing brilliance, which is manifested through speed); Intermediate (whose members are notable as "mixers", or in Varola's words, whose "function in the breed is to provide connective tissue"); Classic (representing ability by an elite group to run in the Classic races); Stout (who, again by Varola's own definition, "are particularly charged with preserving the so-called moral qualities of the racehorse: courage, endurance, will to win"); and Professional (the residue of outstanding names where stamina stands out). Once Varola had introduced the 20 most recently significant influences to his list of chefs-de-race, he found it prudent further to refine his system by splitting the Brilliant group into Brilliant and Transbrilliant, and the Stout group into Solid and Rough, thus giving a final analysis based on seven aptitudinal groups.

Varola's exhaustive work thus differs from the purely mathematical interpretation of Vuillier's Dosage System, since it introduces the idea that certain qualities essential to the racehorse are propagated by various groups. However, just as Vuillier's mathematical inadequacy is easily spotted, so Varola's injection of personal opinion to his classification of individuals stands out. Unless this important factor in either work is acknowledged and accepted, it would be pointless delving deeper into their value. And in Varola's case that would mean overlooking the importance of his work in its introduction of international stallions to the spectrum of thoroughbred breeding, and its practical value in helping towards the achievement of a balanced mating.

THE BRUCE LOWE FIGURE SYSTEM

The Bruce Lowe Figure System has nothing to do with mathematical calculations in the same way that Vuillier's System has.

Bruce Lowe was an Australian who during the latter part of the 19th Century traced every mare in the current General Stud Book of his time back, in tail female, to her root or foundation mare in the original volume of that compendium. During the same period the German Hermann Goos did exactly the same. They worked entirely independently but naturally reached the same result—namely that every single mare traced to one of some 50 matrons in the original volume.

The original Stud Book was published between 1791 and 1814 and listed about 100 mares. Approximately 50 of these had no survivors in tail female by the time Lowe and Goos made their investigations. Owing to the rules of entry in the Stud Book these moribund lines can never be revived and have in fact died out in the process of competitive breeding. Breeders found either that they were not producing stock up to the required standard or that they had a history of hereditary infertility and so did not persevere with them.

Goos published the result of his work in his *Family Tables of English Thoroughbred Stock* and thereby laid down the general outline of various publications of the same kind which have since appeared from time to time. Goos specially drew attention to the fact that the descendants of a few of the foundation mares had a greatly superior record for producing winners than others.

Bruce Lowe having completed his research into the various mares' tail female origin then proceeded to give each of the foundation mares a, shall I say, surname. This surname took the form of a figure—1, 2, 3, 4, etc. The Bruce Lowe figures therefore are nothing more or less than a mare's family name. As opposed to the human race which perpetuate their surnames in tail male, thoroughbreds carry theirs on in tail female. The reason for this is that a stallion may have several hundred progeny, so classification in the male line would only produce the greatest confusion in a very few generations.

Bruce Lowe allotted the name No. 1 to the foundation mare who, up to his time, was the ancestress in tail female of the largest number of winners of the Derby, St. Leger and Oaks. No. 2 Family were the descendants of the foundation mare who had the second best winning record in these races and so on. The families were thus named in order of merit 1 to 43. When he came to some of the higher numbered families he found they had produced no Classic winners and he numbered them according to his estimate of their value. Since his time a few families have appeared in the Stud Book who do not trace to any of his 43 named mares. These are mostly due to importations from abroad and have no bearing on his system. Bruce Lowe ignored the 2,000 Guineas and 1,000 Guineas results as he considered that they were preliminaries to the more important Classics.

Whatever system of assessment is used—the three major Classics, all the Classics, the greatest number of winners of all races, the largest amount in stakes won, etc.—the result is always approximately the same, subject to slight seasonal fluctuations. The low numbered families or horses inbred to them always come out on top. In some racing seasons No. 2 Family register better results than No. 1 and so on, but taking it as a whole Bruce Lowe's seniority classification is correct. The reason is not that these families are endowed with special racing ability denied to the other families but that they are more prolific. In addition to producing far more winners than the high figured families they also produce far more "also rans". Both high and low numbered families all trace back to one specific foundation mare and the fact that the low numbers are in such great numerical superiority shows that they at least have a long history of superior fertility. And a broodmare's first duty is to produce foals with regularity. If she is continually barren she can have but few, if any, runners and if she has no runners she cannot be a winner producer. A mare of No. 1 Family is not necessarily easier to get in foal than a mare of No. 43 Family but the overall history of all members of No. 1 Family proves that they are far more productive than their counterparts of No. 43 Family. This is not alone due to a greater hereditary fecundity but is also influenced by man's selection. Those mares which were not producing stock up to the required standard were not persevered with and so their lines have died out.

Having completed his Family Classification, Bruce Lowe then designated Families 1 to 5 inclusive "Running Families", and Families 3, 8, 11, 12 and 14, "Sire Families". The Running Families were selected as they had produced the greatest number of winners of the three major Classics. The Sire Families were chosen as in Bruce Lowe's estimation horses from them, or inbred to them, had a superior record as stallions. He then evolved a number of theories based on his researches, several of which are quite untenable in the light of modern scientific knowledge of genetics.

From the system of classification it will be clear that the low numbered families are broadly speaking synonymous with success. Bruce Lowe was thus advocating in different terms a system racehorse breeders of all time have followed—namely the principle of breeding from and inbreeding to winner-producing stock. It would be totally incorrect to say that a parent of a low numbered family is of necessity likely to be a stud success. There are innumerable quite useless mares and stallions who emanate from these sources. But it would be accurate to say that on an average members of the high numbered families (or poor winning strains) are unlikely to do well at stud unless either they themselves are inbred to low numbered families (or good winning strains) or they are mated almost exclusively to partners rich in the more successful blood.

Bruce Lowe remarked that he had been unable to discover a single top-class racehorse or sire in the whole of the 19th Century who did not carry a strain of either one of his Running Families (1–5) or of one of his Sire Families (3, 8, 11, 12 and 14) in its first three removes. So far as I am aware there has not been a case since. As there has been considerable misunderstanding on the point I will repeat that this result is not brought about by members of the low numbered families having exceptional hereditary racing merit. This may be so in certain specific cases but the overall result depends largely on their superiority in numbers. Let us suppose for sake of argument that a top-class racer suddenly appeared in No. 42 Family whose nearest progenitors came from, let us say, Families 41, 40, 39 and 38. When he or she went to stud we would be forced to mate him to partners carrying some of the low numbered blood simply because it would be practically impossible to find many free of it. The produce therefore would be in accordance with Bruce Lowe's claim and carry at least some strain from either the Running or the Sire Families.

Lowe's theory that Families 3, 8, 11, 12 and 14 carried some element which is vital in the make-up of a good sire cannot be substantiated. One needs to look no further than Nasrullah (1940) and Ribot (1952), two of the greatest modern influences, to find stallions who have no element of this blood in the first three removes.

Nowadays the Bruce Lowe system, while retained in certain publications and private stud books for identification purposes, has largely been discarded as a practical means of bringing about success in the world of thoroughbred breeding. However, the harm which originally resulted from his book *Breeding Race Horses by the Figure System* was in some degree due to a misinterpretation of its contents. Some breeders cast aside the normal principles of breeding, such as racecourse performance, family record, etc., and concentrated almost exclusively on producing foals who had the maximum amount of Running and Sire Blood, regardless of anything else.

In America the effects seem to have been particularly unfortunate, as publication of the book more or less coincided with the introduction of a wave of anti-gambling laws. These very seriously interfered with American racing and breeders drastically reduced their studs. Unhappily, when doing so, some of them pinned their faith in mares from Running Families—many with indifferent records—and weeded out mares from good American families which were not mentioned by Bruce Lowe.

Despite these criticisms and misunderstandings, the book did a certain amount of good. It placed at breeders' disposal a practical and easy method of identifying the various strains in a pedigree, and Lowe's research unearthed an enormous amount of detail which helped to show the way in which the thoroughbred had developed from the 43 original mares he designated outstanding.

Today, however, Lowe's work has little practical significance, since the science of genetics has exploded his views that qualities such as prepotency in sires can be guaranteed by way of blood-lines or "families". His numbered families have now been superseded by named families emanating from the 20th Century as a means of cataloguing achievement, and these will be referred to in Chapter Nine.

CHAPTER 4

Further Theories and Some Fallacies

Superiority of the male; importance of grandparents: telegony: mental impression

WE HAVE SEEN that at the time of mating each parent subscribes 30 chromosomes to the embryo and that half of these form its germ plasm. We have also seen that when in due course the newborn goes to stud he or she will pass on hereditable factors drawn from this germ plasm. This focusses attention on the question as to whether when arranging a mating at least equal consideration should be given to the four grandparents as to the proposed sire and dam who, after all, are only custodians of the hereditable ancestral make-up. That well-known authority the late Professor J. B. Robertson on page 7 of his work *The Principles of Hereditary applied to the Racehorse* (1906) uses these words ". . . it may safely be asserted that no horse or mare ever reproduces exactly the same combination of characters as are present in the somatic cells, hence the practical necessity of paying more regard to the four prospective grandparents than to the two individuals you propose to mate". Various other very experienced breeders or students of genetics have expressed similar opinions including the famous Director of the former German National Stud, the late Burchard von Oettingen.

I think few will dispute the logic of the point but its practical application is a slightly different matter. If we take it at face value a stallion or broodmare's racing record is of secondary importance to his or her parents' ability. The history of the breed effectively repudiates such a premise.

SUPERIORITY OF THOROUGHBRED MALES

It is easier to illustrate the position in the case of stallions than of broodmares as the breeding results of a stallion are usually much better known than those of a mare and moreover, as he has far more offspring, a greatly wider spread is available on which to base conclusions. There can be no possible doubt that on an average the male thoroughbred is a better racehorse than the female. A filly who is capable of winning the 2,000 Guineas, Derby, St. Leger or Ascot Gold Cup against the colts of her generation, in spite of the advantage of a weight allowance, is extremely rare.

If the grandparents of a prospective foal are more important than the parents, it would be reasonable to anticipate that a sire who is by a Classic winner out of a Classic winner should be a stud success, more or less regardless of his own racing merit, provided, of course, that he is mated to mares of approximately the same calibre as the other stallions with which he is in competition for an honourable position in the list of sires of winners. The poor performers so bred would be in a position to subscribe to the foal the genes of two parents who were amongst the best racers of their time, whilst his rivals, who themselves may have won a Classic, would normally only be able to subscribe the genes of one parent of corresponding racing merit. It is rare, but by no means unknown, for a Classic winning colt to be able to claim a Classic winning filly as his dam. If we investigate the stud records of the sires who were indifferent racers themselves but were the product of two Classic winners we find almost without exception that they were stud failures.

I suppose it would be hard to find a better bred horse than Grosvenor (1913) who only managed to win one race. His sire Cicero, who was a Derby winner, was by the Ascot Gold Cup winner and great stallion Cyllene out of Gas (sister in blood to the Derby winner Ladas and half-sister to the 1,000 Guineas winner Chelándry). Grosvenor's dam Sceptre won all the Classics except the Derby and was by Persimmon—a great racehorse and sire—out of Ornament (full sister to the unbeaten Triple Crown winner Ormonde). Here indeed was a horse of flawless pedigree who could supply a foal-to-be with the blood of two grandparents probably unmatched by any other sire of his time. Grosvenor retired to stud with a considerable flourish of trumpets and his success in that métier was chiefly noteworthy for the large number of selling plate class stock he sired. In one racing season alone (1923) 21 of them won collectively 39 races of a total value of £3,559, or an average of £91 a race. This large number of small winners is an indication that he did not suffer from a shortage of mates, neither do they appear to have been sub-normal in quality. Grosvenor sired nothing commensurate with his parentage and was eventually exported to Australia, where also he was a failure.

CLASSIC WINNING FILLIES

But one swallow does not make a summer so I have set out below all the Classic winning fillies of this century who have had a son at stud in the British Isles together with a note against each son regarding his success or otherwise.

DERBY WINNERS

1 WINNING FILLY	*2* SON AT STUD	*3* RACES WON BY SON	*4* STUD RESULTS BY SON
(1) Signorinetta (1905) Also won the Oaks	The Winter King (1918) by Son-in-Law	4 value £3,376	Sired the Grand Prix de Paris winner Barneveldt (1928). Otherwise a failure.

DERBY WINNERS (*contd.*)

1 WINNING FILLY	*2* SON AT STUD	*3* RACES WON BY SON	*4* STUD RESULTS BY SON
(2) Tagalie (1909) Also won the 1,000 Guineas	(A) Tagrag (1914) by Chaucer	7 value £1,799	Failure.
	(B) Allenby (1917) by Bayardo	3 value £5,815 Second in the 2,000 Guineas	Failure.
(3) Fifinella (1913) Also won the Oaks	Press Gang (1927) by Hurry On	5 value £10,115	Failure. Sent to Russia.

ST. LEGER WINNERS

1 WINNING FILLY	*2* SON AT STUD	*3* RACES WON BY SON	*4* STUD RESULTS BY SON
(4) Sceptre (1899) Also won the 1,000 Guineas, 2,000 Guineas and Oaks	Grosvenor (1913) by Cicero	1 value £850	Failure.
(5) Pretty Polly (1901) Also won the 1,000 Guineas and Oaks	Passchendaele (1916) by Polymelus	Nil	Had only a few mates and then was sent to Belgium. Failure.
(6) Tranquil (1920) Also won the 1,000 Guineas	Headway (1935) by Fairway	Nil	Failure.
(7) Book Law (1924)	(A) Canon Law (1930) by Colorado	3 value £5,630	Failure.
	(B) Rhodes Scholar (1933) by Pharos	3 value £14,326	Sent to U.S.A. after two seasons in U.K. and did moderately well.
	(C) Archive (1941) by Nearco	Nil	Sired smart sprinter Arcandy (1953) but made name under NH Rules, notably as sire of great chaser Arkle (1957).

ST. LEGER WINNERS (*contd.*)

1 WINNING FILLY	2 SON AT STUD	3 RACES WON BY SON	4 STUD RESULTS BY SON
(8) Sun Chariot (1939) Also won the Oaks and 1,000 Guineas	(A) Blue Train (1944) by Blue Peter	3 value £3,735 Unbeaten.	Showed promise of success but became sterile.
	(B) Gigantic (1946) by Big Game	2 value £2,473	Failure in U.K. but gained success when sent to Australia.
	(C) Pindari (1956) by Pinza	4 value £18,458	Failure, and exported to India.
	(D) Javelin (1957) by Tulyar	1 value £133	Failure.
(9) Meld (1952) Also won the 1,000 Guineas and Oaks.	Charlottown (1963) by Charlottesville	7 value £101,211 including the Derby	Moderate success from half a dozen crops in England, and exported in 1976 to Australia, where his unraced half-brother Mellay (by Never Say Die) had done well.

OAKS WINNERS

1 WINNING FILLY	2 SON AT STUD	3 RACES WON BY SON	4 STUD RESULTS BY SON
(10) La Roche (1897)	(A) Cannobie (1913) by Polymelus	3 value £7,929	After three seasons in England sent to France and did well.
	(B) Sir Berkeley (1915) by Sunstar	5½ value £2,832	Failure.
Sceptre (1899)	See (4) above.		
Pretty Polly (1901)	See (5) above.		
Signorinetta (1905)	See (1) above.		
(11) Rosedrop (1907)	(A) Gainsborough (1915) by Bayardo	5 value £14,080 including the Triple Crown.	One of the best sires of this century.
	(B) Baydrop (1918) by Bayardo	Nil	Failure.

OAKS WINNERS (*contd.*)

1 WINNING FILLY	*2* SON AT STUD	*3* RACES WON BY SON	*4* STUD RESULTS BY SON
(12) Cherimoya (1908)	The Cheerful Abbot (1925) by Abbots Trace	9½ value £4,431	Only had a few mares. Failure.
(13) Jest (1910) Also won the 1,000 Guineas	Chief Ruler (1920) by The Tetrarch	Nil	Only had a few mares in U.K. with no good result. Sent to New Zealand where he was twice champion sire.
(14) Princess Dorrie (1911). Also won the 1000 Guineas Fifinella (1913).	Magnus (1922) by The Tetrarch See (3) above.	1 value £243	Failure.
(15) My Dear (1915)	Hartford (1925) by Swynford	2 value £1,035	Failure.
(16) Brownhylda (1920)	(A) Far and Sure (1928) by Phalaris	Nil	Failure.
	(B) Firdaussi (1929) by Pharos	8 value £21,500 including the St. Leger.	At stud in England, France, Rumania and behind the Iron Curtain. Did well in France.
(17) Straitlace (1921)	(A) Lovelace (1927) by La Farina	7 value 715,964 frs.	A moderate sire in France.
	(B) Interlace (1930) by Hurry On	2 value £2,625	Failure.
(18) Saucy Sue (1922)	Truculent (1928) by Teddy	1 value £3,375	Failure.
(19) Toboggan (1925)	Bobsleigh (1932) by Gainsborough	3 value £3,593	Fairly successful.
(20) Pennycomequick (1926)	High Profit (1941) by Hyperion	1 value £266	Failure. Sent to U.S.A.

OAKS WINNERS (*contd.*)

1 WINNING FILLY	*2* SON AT STUD	*3* RACES WON BY SON	*4* STUD RESULTS BY SON
(21) Rose of England (1927)	(A) Chulmleigh (1934) by Singapore	2 value £10,565 including the St. Leger.	Fair, but disappointing. Sent to Argentina.
	(B) Rangoon (1939) by Singapore	Nil	Failure.
	(C) Coastal Traffic (1941) by Hyperion	1 value £522	At stud in Ireland, France, U.S.A., and France again. Did well in France first time.
(22) Brulette (1928)	(A) Airway (1937) by Fairway	1 value £127	Failure.
	(B) Thoroughfare (1938) by Fairway	1 value £206	Failure.
(23) Udaipur (1929)	Umiddad (1940) by Dastur	7 value £3,942, including a wartime Gold Cup.	At stud in France with disappointing results.
(24) Quashed (1932) Also won the Ascot Gold Cup. Family not admitted to the General Stud Book until 1969.	(A) Perion (1939) by Hyperion	Nil	Failure.
	(B) Prince Titon (1943) by Hyperion	Nil	Failure.
(25) Exhibitionnist (1934). Also won the 1,000 Guineas	Monty (1948) by Bellacose	6 value £2,340	Failure.
(26) Rockfel (1935) Also won the 1,000 Guineas	Rockefella (1941) by Hyperion	3 value £1,212, but a better racer than winnings indicate.	Did fairly well, siring Rockavon (2,000 Guineas) and Classic-placed Gay Time, Bounteous and West Side Story.
(27) Commotion (1938)	(A) Combat (1944) by Big Game	9 value £7,725. Unbeaten.	Only fairly successful in his own right, but notable as sire of broodmares.

OAKS WINNERS (*contd.*)

1 WINNING FILLY	*2* SON AT STUD	*3* RACES WON BY SON	*4* STUD RESULTS BY SON
	(B) Rage Royal (1956) by Flush Royal	Nil	Failure.
	(C) Fury Royal (1958) by Flush Royal	5 value £2,393.	Moderately successful.
Sun Chariot (1939) (28)	See (8) above.		
Sun Cap (1951)	(A) Polaroid (1964) by Crepello	4 value £1,967	To stud in 1972; too early to assess.
	(B) Imperial Crown (1966) by Aureole	5 value £5,106.	To stud in 1976.

1,000 GUINEAS WINNERS

1 WINNING FILLY	*2* SON AT STUD	*3* RACES WON BY SON	*4* STUD RESULTS BY SON
(29) Aida (1898)	(A) Alcanzor (1903) by Sainfoin	2 value £1,550	Failure.
	(B) Winstanley (1909) by Gallinule	2 value £1,271	Failure.
Sceptre (1899)	See (4) above.		
(30) Quintessence (1900)	(A) Clarissimus (1913) by Radium	3 value £6,907 including the 2,000 Guineas.	Did very well in France.
	(B) Mountaineer (1916) by Polymelus	2 value £689	Failure.
Pretty Polly (1901)	See (5) above.		
(31) Witch Elm (1904)	Sir Benedict (1921) by Great Sport	Nil	Failure.
(32) Electra (1906)	Orpheus (1917) by Orby	11 value £11,971	Got a fair num- ber of winners but not a good sire.
(33) Winkipop (1907)	Blink (1915) by Sunstar	2 value £1,647, second in the Derby, third in the 2,000 Guineas	Sired a useful colt called Twink (U.S.A.). Died young. Failure.

1,000 GUINEAS WINNERS (*contd.*)

1 WINNING FILLY	2 SON AT STUD	3 RACES WON BY SON	4 STUD RESULTS BY SON
Tagalie (1909)	See (2) above.		
Jest (1910)	See (13) above.		
Princess Dorrie (1911)	See (14) above.		
(34) Vaucluse (1912)	Forerunner (1918) by Chaucer	6 value £2,206	Failure.
(35) Canyon (1913)	(A) Colorado (1923) by Phalaris	11 value £30,358 including the 2,000 Guineas	A great success but died after two seasons at stud.
	(B) Caerleon (1927) by Phalaris	3 value £11,210	Moderately successful. Died young.
	(C) Overthrow (1934) by Bosworth	Nil	Failure.
(36) Roseway (1916)	Heatherway (1924) by Craig an Eran	1 value £256	A hunter sire.
(37) Cinna (1917)	(A) Buckler (1923) by Buchan	1 value £2,385	Failure.
	(B) Beau Père (1927) by Son-in-Law	3 value £974	Covered only a few mares at home with no good result. A most successful sire in the Antipodes and U.S.A.
	(C) Hot Haste (1928) by Hurry On	Nil	Failure.
	(D) Mr. Standfast (1931) by Buchan	1 value £100	Failure at home. A great success in New Zealand and Australia.
(38) Silver Urn (1919)	Golden Chalice (1924) by Abbots Trace	6 value £3,105	Failure.
Tranquil (1920)	See (6) above.		
Saucy Sue (1922)	See (18) above.		

1,000 GUINEAS WINNERS (*contd.*)

1 WINNING FILLY	*2* SON AT STUD	*3* RACES WON BY SON	*4* STUD RESULTS BY SON
(39) Taj Mah (1926)	Rajah II (1955) by Blandford	1 small race in France	Failure.
(40) Mesa (1932)	Historic (1939) by Solario	6 value £2,391	Failure.
Exhibitionnist (1934)	See (25) above.		
Rockfel (1935)	See (26) above.		
(41) Dancing Time (1938)	Arctic Time (1952) by Arctic Star	3 value £1,613	Moderately successful, siring Arctic Vale (Irish St. Leger) before export to South Africa.
Sun Chariot (1939)	See (8) above.		
(42) Herringbone (1940)	Entente Cordiale (1951) by Djebe	12 value £9,566	Failure in short career.
(43) Queenpot (1945)	Ace of Clubs (late Hitryon) (1959) by Djebe	1 value £343, and 4 value £882 over hurdles	Failure.
(44) Belle of All (1948)	Pendragon (1955) by Alycidon	Nil (left at start in only race)	Failure on Flat, but moderately successful under NH Rules.
(45) Honeylight (1953)	Hessonite (1965) by Princely Gift	Nil	To stud in Scot- land in 1970 and stood first season at £48. Too early to assess, but portents not good.
(46) Pourparler (1961)	Avocat (1969) by Relko	5 value £2,881	To stud in 1974 at £100.
(47) Glad Rags (1963)	Pumps (1968) by Native Dancer	Nil	To stud in 1971 at £48.

With two or three unimportant exceptions (The Cheerful Abbot, Monty, Golden Chalice and Sir Benedict) all the stallions in the above list have two things in common. First all were sired by horses of, or near Classic class and second they were all out of Classic winning fillies. So broadly speaking it may be said that they all had equal chances of having a germ plasm of approximately the same calibre. If we are to rely for results on grandparents rather than parents it could therefore reasonably be expected that not only should they as a class be stud successes—as they were in a position to subscribe the influence of two very good grandparents to their foals—but also that their records as stallions should show some resemblance to one another. This is far from the case.

It will be noticed that usually the better racing record the colts had, the better sires they proved to be. The Classic winners Gainsborough, Colorado, Clarissimus and Charlottown all did comparatively well, as did unbeaten Combat. The St. Leger winners Firdaussi and Chulmleigh were sent abroad comparatively early in their stud careers and although they made no great show as sires of winners at home their fillies have proved good broodmares.

On the other hand the colts in the list who only won races of small average value, and therefore showed their limitations as racehorses, on the whole were stud failures. In a rough and ready way I think we may infer from this that usually—but by no means always—the somatic or bodily cells, which give rise to racing merit or the reverse, and the all important germ plasm cells of a parent are somewhat akin. We obviously cannot take out and examine under a microscope a parent's germ plasm, so we test the somatic cells on the racecourse and hope that the germ plasm will contain a somewhat similar combination of genes.

Usually the bad racehorse has inherited the poor quality hereditary elements from his ancestors, both in his bodily make-up and also in his germ plasm, whilst the good racer has received the reverse. But this is not always so. Press Gang, Orpheus and Allenby were all high-class performers but poor sires, whilst Beau Père and Mr. Standfast were bad racehorses but stud successes. The failure of these two, whilst at home, I attribute entirely to the shortage and poor quality of their mates. In all these cases the hereditable elements which found their way into the germ plasm were entirely different from those in the horse's own make-up. I therefore think that whilst the grandparents of a proposed mating are important, the performances of the parents, both on the racecourse and at stud, are more so, as they will give a guide as to what factors the parent holds, or is likely to hold, in his or her germ plasm.

If breeders had given more weight to stallions' parents than to the horse himself they would have cold shouldered such as Carbine, unbeaten The Tetrarch and unbeaten Hurry On, who were all sired by second-class racehorses out of worse mares. But they were all without doubt great sires who have exerted beneficial influences on the modern thoroughbred. Yet every foal they sired had at least two and probably three (i.e. the maternal grandam) grandparents who were nowhere near top grade.

The ultimate and only true test of a parent's worth is what he or she produces and this no man can foretell accurately until the progeny show their merits on the racecourse. But as a general guide to probabilities a parent's record on the Turf is of great importance.

Before leaving the subject of stallions produced from Classic-winning mares, it is worth recording the value of some of these sires to overseas breeding interests. Australasian breeders have shown particular keenness in this direction, and have benefited from the purchase of such as Atilla (1961, by Alcide out of Festoon, a leading sire in Australia), Battle-Waggon (1962, by Never Say Die out of Carrozza, a leading sire in New Zealand) and Without Fear (1967, by Baldric II out of Never Too Late, whose first crop of racing age in Australia in 1975–76 was a record-breaking success). And the export to Argentina of Aristophanes (1948, by Hyperion out of Commotion) resulted in his siring Forli as well as a number of highly successful broodmares.

Again these sires tend to bear out the theory regarding the importance of racecourse merit, for with the exception of Battle-Waggon, who won one race value £528, they were above-average performers.

PRINCIPLES OF TELEGONY

A theory which was quite commonly held by breeders of old was that of telegony—or saturation as it is sometimes called. It is known also as infection. The idea was that when a mare is carrying a foal she sometimes absorbs some elements which the embryo has derived from its sire. She then stores these in her body and may transmit them to a subsequent foal. In other words if a mare is mated to say Hyperion, he may have some influence on the foals which that mare produces subsequently. Belief in the theory was not confined to bloodstock breeders but was held by many of the world's greatest experts on eugenics and genetics.

Bruce Lowe used it to account for the difference in merit of full brothers and sisters. He maintained that if a mating produced good results it should not be repeated, as the subsequent foals to the good one would have too much of the sire's blood and thus, as it were, tip over the scales. For example the suitability of the blood mixture of Northern Dancer and Flaming Page has proved itself by being responsible for the production of Nijinsky. In Bruce Lowe's view, the subsequent foal from a re-mating of Flaming Page to Northern Dancer would not receive the same dosage as Nijinsky, since Flaming Page would be carrying an additional element of Northern Dancer's blood from the first mating.

It is true that just such a re-mating, the result of which was a colt called Minsky, did not produce a horse of equal merit to Nijinsky, but Minsky was nevertheless a very useful performer, and as I have previously indicated (see page 4), there are other reasons to account for the difference in full brothers and sisters.

Nowadays the theory has been discarded by breeders of racehorses and good class farm stock but I understand that it is still steadfastly championed by the desert breeders of Arabs—and they and their predecessors for hundreds of years have bred horses of outstanding worth in their own sphere. I have heard breeders of hunters in England and Ireland aver that if they mated their broodmares to a Shire or other heavy horse in the first instance, the subsequent foals by thoroughbred sires would be up to more weight than if they had not followed this course. They have pointed out to me a youngster, by a thoroughbred sire, with the hairy feathered legs indicative of the heavy horse's influence in proof of this contention. In most cases they knew little or nothing of their broodmare's history and I have little doubt that the hairy heels and big frame were of tail female descent. Their broodmares, considered of sufficient substance to breed hunters, traced back— and only a few generations back—to cart mares. The idea that the foals would have been smaller if a heavy horse had not been used in the original mating I attribute to a rumour passed from one breeder to another, from father to son, etc.

It was so ingrained amongst farmer breeders of hunters in some districts that a thoroughbred sire as a first mate was rarely tried.

The most famous case of alleged telegony was the following. In about 1815 Lord Morton mated a quagga stallion (a quagga is of the zebra species) to a mare of seven eighths Arabian blood with the result that she foaled a hybrid quagga/horse. He then sold the mare to Sir Gore Ougeley who put her to an Arab stallion. From this mating she produced a bay foal with a black dorsal line and a few dark (quagga like) stripes on its forehand and legs. To a subsequent mating to the same horse she produced a similarly marked foal. On discovering this Lord Morton sent details to the Royal Society, whereupon many learned members of this institution accepted the facts as indisputable proof of the existence of telegony.

Pictures of the mare's produce by the quagga and by the Arab sire were painted by Agasse and found their way into the Royal College of Surgeons' Museum. They show the animals by the latter to be of undoubted Arab type with no detectable sign of quagga conformation but the peculiar quagga-like stripes are plainly discernible. These, although well marked, are very few, and later I myself saw photographs of two different pony foals with similar but more abundant stripes. Neither of the dams of these two pony foals had ever been near a quagga or zebra. The fact is that in some breeds of horses—and I believe especially amongst Norwegian ponies—these zebra stripes are sometimes seen. It was pure coincidence that the peculiarity showed itself in the two horses bred from Lord Morton's mare.

Since those far-off days there have been innumerable attempts to produce a similar result both by crossing zebras, etc., with mares and by other hybrid matings but all have failed. The Lord Morton case produced discussion for 100 years or more and amongst others Darwin appears to have accepted it as evidence of telegony. There are two relative points regarding it which are sometimes misunderstood. First the mare the quagga was put to was not a pure-bred Arab or anywhere near it. Of her tail female ancestors no one knows anything except that they were not pure bred. It is probable that the markings came from one or more of these. We have seen that the modern thoroughbred sometimes shows remarkable physical resemblance to his Arabian ancestors of 200 or more years ago (see page 5), thus if this factor suddenly reappears after many generations, so also can unusual liveries. The second point is that the mare's subsequent foals by the Arab sire showed no indication of the quagga's influence in their general make and shape.

PROOF AGAINST TELEGONY

The most tell-tale proof against telegony is the simplest. If a mare is mated to a donkey a mule results, but if she is then covered by a horse the foal will show no sign of donkey inheritance. It may be that occasionally the produce has big ears, donkey feet, etc., but then, the mares generally used for this purpose are from under-bred ancestors to whom these unsightly features can be attributed.

Some 70 years ago the late Professor Ewart made a number of practical experiments regarding the theory. He used a zebra stallion and a miscellaneous stud of mares and in no case did results justify any belief in its correctness. His book *The Pencuik Experiments* published in 1899 should be read by those interested in the subject. More recently a technique was developed of removing by surgical operation the fertilized ova from one female and transporting it to another to complete development. Large Border Leicester ewes were covered by rams of the same breed. The fœtuses were then surgically removed from the dams and transferred to small Welsh ewes. The resultant lambs were Border Leicesters with no sign of Welsh influence, in spite of the fact that in some cases the Welsh "acting" mothers had previously had lambs by Welsh rams. Further the "acting" mothers were then sometimes mated to their own breed again and the lambs were entirely free of Border Leicester taints. I think it is plain that so far as the practical bloodstock breeder is concerned he will be wise to dismiss from his mind any thoughts about telegony and saturation.

MENTAL IMPRESSION

Another myth which I have sometimes come across amongst a few small farmer/breeders is that of Mental Impression. This is based on the Biblical story of Jacob's cattle-dealing transactions which are recorded in Chapter 30 of the Book of Genesis and may be summarized as follows. Jacob was employed as a cattleman to look after Laban's livestock. When the time came to settle up his wages he asked for, and was given, all the spotted and speckled cattle and goats amongst the stock in lieu of a direct payment. Laban, who appears to have known a thing or two, slipped out on the very day the bargain was made and removed all those so marked to a far place. So when Jacob came around to take his entitlements there were none. But Jacob was a man of resource. He continued to look after the depleted stock but set up before them striped trees and they produced speckled and spotted stock. When he had had satisfactory results from this technique he reserved its use for the stronger cattle, leaving the weaker ones to carry on without its aid. Thus the produce of the strong cattle were speckled and were Jacob's and the produce of the weaklings were whole coloured and therefore Laban's. The story serves to show that the higher ethics of cattle dealing are no invention of the modern world and I am even suspicious that Jacob had up his sleeve his Plan for the Production of Striped Cattle before he asked for them in lieu of wages.

The usual tale of present-day believers in the cult is that a specific whole-coloured mare suddenly produced a foal with white markings almost identical with a similarly marked mare she was running with, in spite of the fact no other mares of the family had produced foals of this colour. There are various variations to the story but that is the general outline. A particularly favourite one is the birth of a foal with a wall eye when there was a wall-eyed animal in the field with the dam. The explanation to these occurrences is that the peculiarity has been handed down latently from a progenitor on one side or the other of the pedigree, possibly for generations, and it is pure chance that it suddenly comes to the surface. No man living can tell the exact markings, etc., of every animal in his mare's pedigree for five, ten or more generations. If the theory of mental impression was sound, we could reasonably expect that if an in-foal thoroughbred mare is given a donkey as a companion, her produce should be something akin to a mule!

The practice, still sometimes followed by superstitious stud workers, of walking a stallion around a mare before service originated in this theory, the idea being that if she gets a good look at her handsome mate she will produce a foal like him.

CHAPTER 5

Blood Affinities and Other Matters

"Nicks": equality of parental ages: Lord Wavertree's astrology

CERTAIN LINES OF blood appear to have an affinity for, and to produce exceptional results when mated to other specific lines. The phenomenon is usually termed a nick. Bend Or mated to mares by Macaroni got the unbeaten Triple Crown winner Ormonde and his sister Ornament (dam of Sceptre); the brothers Laveno and Orvieto; the Goodwood Cup winner Martagon; Kendal and his sister Rydal (grandam of the St. Leger winner Troutbeck); Bona Vista (2,000 Guineas and sire of Cyllene); Ortegal (dam of the successful American sire Octagon); Medora (dam of the Ascot Gold Cup winner Zinfandel); Golden Iris (grandam of Sans Souci II); Doremi (grandam of Teddy) and so on. Again Isonomy had great success with Hermit mares and from them produced the brothers and sister Seabreeze (ten wins value £24,266, including the Oaks and St. Leger), Le Var (£10,000), Riviera (ten wins value £12,237) and Antibes; the Doncaster Cup winner Prisoner; Gallinule (a great sire); Ravensbury (Hardwicke and Ascot Stakes); Fortunio; Son o' Mine and his sister Alimony (dam of Winkfield's Pride); Themis and others.

The apparent success of these nicks presents us with a problem of vital interest to all breeders. It is easy to understand that good results are likely to be produced by eliminating inferior ancestors from breeding stock and using only the best, or that duplication of the blood of a particularly desirable progenitor by inbreeding to him or her, can reasonably be expected to be rewarded, but I have heard no satisfactory explanation as to why the mixture of two specific lines should be successful in the case of pure breds. The ancestral elements which go into the melting pot when we carry out a mating cannot be considered a constant similar to the ingredients of a doctor's prescription, so we cannot attribute the success of nicks to a correct compounding of the blood mixture.

The various dams, grandams and female-line sires of the great horses produced from them are widely different. Furthermore we find that although a specific sire seems to nick with mares by another sire, a reversal of the process generally results in disappointment. I have always heard that the late Duke of Portland was influenced in his purchase of Carbine from Australia by the fact that he was by a son of Toxophilite. St. Simon sired La Flêche and Memoir, who between them won two St. Legers, two Oaks, one 1,000 Guineas and one Ascot Gold Cup, out of a Toxophilite mare, so it was not unreasonable to expect that Carbine would do well with St. Simon mares. In point of fact the reversal of the nick did not work and the only Classic winner the importation got was Spearmint, whose dam was entirely free of both St. Simon and his sire Galopin's blood. However, the rather moderate Ascot Gold Cup winner Bomba (1906), who was by Carbine out of St. Neophyte by St. Simon, shows there may have been something in His Grace's idea. Yet considering the wealth of St. Simon mares mated to Carbine the cross must be regarded as a failure.

This brings up the question as to whether the success of nicks is largely the result of opportunity—or as it has been termed more recently, propinquity—by the continual mating of one sire to mares by another specific sire.

In more recent times the value of a particular nick has been seen as the result of careful selection, whereby a certain type of sire has been chosen because he suits a certain type of mare. The nick is therefore of important practical assistance in attempting to bear fruit from the hope that the two will complement each other. This is a logical expression of the term nick, and its practical application will be shown later in this chapter in relation to the Nasrullah-Princequillo cross.

However, no such logical considerations applied to an earlier nick which had a most potent effect on modern bloodstock, that of Phalaris on Chaucer mares, to which I propose to refer at some length because of its significance.

I shall commence by setting out below the winners of £2,000 or over sired by Phalaris out of Chaucer mares (Table 1) and comparing these to similar winners by Phalaris out of other mares (Table 2).

TABLE 1

Winners of £2,000 or over in stakes by Phalaris out of Chaucer mares.
(N.B.—Chaucer was by St. Simon)

Fairway	£42,722	Meadow Rhu	£4,175
Colorado	30,358	Sickle	3,915
Pharos	15,694	Burnt Sienna	3,705
Fair Isle	13,219	Pharamond	3,695
Caerleon	11,210	Sargasso	2,160
Warden of the Marches	8,422	Pharalope	2,116
	£121,625		£19,766
	£19,766		
Total amount won by good winners	£141,391		

TABLE 2

Winners of £2,000 or over in Stakes by Phalaris out of other mares.

CATEGORY "A" DAMS BY OTHER SONS OF St. Simon		CATEGORY "B" GRANDAMS BY St. Simon	
Christopher Robin	£5,440	Manna	£23,534
Devonie	3,525	Moabite	8,155
Pondicherry	2,096	Parwiz	3,987
Phanarite	2,040	Legionnaire	3,725
		Bold Archer	2,096
Total Category "A"	£13,101	Total Category "B"	£41,497

CATEGORY "C" DAMS WITH St. Simon IN OTHER REMOVES		CATEGORY "D" DAMS WITH NO St. Simon BLOOD	
Le Phare	£6,151	Plantago	£9,217
Phalaros	5,081	Chatelaine	8,332
Avalanche	4,479	Priscilla	5,602
Herbalist	2,820	Canfield	4,989
Sarum	2,752	Dian	4,207
Trinidad	2,545	Silver Grass	2,892
Torlonia	2,412	Anthurium	2,024
Holyrood	2,237		
Shambles	2,076		
Gynerium (grandam by Chaucer)	2,039		
Total Category "C"	£32,592	Total Category "D"	£37,263

Total Category "A"	£13,101
Total Category "B"	£41,497
	£54,598
Total Category "C"	£32,529
Total Category "D"	£37,263
Grand Total of stakes won by good winners by Phalaris out of mares by sires other than Chaucer	£124,453
Grand Total of stakes won by good winners by Phalaris out of Chaucer mares (Table 1)	£141,391
Balance in favour of Chaucer mares	£16,938

I will explain later my purport for dividing the winners not out of Chaucer mares into the various categories shown above and for the moment will draw attention to two facts. First it is most remarkable that the value of stakes won by important winners by Phalaris out of Chaucer mares is materially greater than the corresponding figures for all the various other mares he was mated to added together. Second it is plain that without the aid of Chaucer, Phalaris could only be regarded as a moderate stud success. His only Classic winners free from Chaucer blood were the 2,000 Guineas and Derby winner Manna and the Oaks winner Chatelaine—beyond that we come down to handicap class stock, many of them not very high-class performers in that field.

Now let us turn to the Chaucer mares. Phalaris had his first runners in 1922 and died in 1931. So when considering the results obtained it would not give a true picture for our purpose if winners from these mares before 1922 or after 1936 were considered. I therefore set out below in Table 3 winners of £2,000 in stakes or over by sires other than Phalaris out of all the Chaucer mares in the Stud Book during the period under review.

TABLE 3

Winners of £2,000 or over in stakes by other sires out of Chaucer mares.

WINNER	SIRE	PATERNAL GRANDSIRE	STAKES WON
Hyperion	Gainsborough	Bayardo	£29,509
Colombo	Manna	Phalaris	26,228
Crème Brulée	Brûleur	Chouberski	9,435
Betty	Teddy	Ajax	7,041
Highlander	Coronach	Hurry On	6,123
Sierra Leone	Great Sport	Gallinule	5,924
Spithead	John o' Gaunt	Isinglass	5,641
Hunter's Moon	Hurry On	Marcovil	4,999
Owenstown	Apron	Son-in-Law	4,234
Teacup	Tetratema	The Tetrarch	3,720
Inflation	Bulger	Bridge of Earn	3,555
The Font	Son and Heir	Son-in-Law	3,513
St. Oswald	Son-in-Law	Dark Ronald	3,513
Winandermere	Beresford	Friar Marcus	3,370
Vermail II	Blenheim	Blandford	3,187
Dick Turpin	Diligence	Hurry On	3,182
Guiscard	Gay Crusader	Bayardo	3,132
The Yank	Cylgad	Cyllene	3,069
Boethius	Dark Legend	Dark Ronald	3,007
Priok	Astérus	Teddy	2,975
Cho-Sen	Prince Galahad	Prince Palatine	2,802
Helter Skelter	Sky-rocket	Swynford	2,535
D'Oraine	Haine	Hainault	2,231
Jamnagar	Black Jester	Polymelus	2,254

Grand Total of stakes won by good winners by all other sires except Phalaris out of Chaucer mares — £145,179

Grand Total of stakes won by good winners by Phalaris only out of Chaucer mares (see Table 1) — 141,391

Balance in favour of the other sires — £3,788

Chaucer rightly enjoys a very high reputation as a sire of broodmares so it is most noteworthy that, during this period, approximately 50 per cent of their best winners were got by one horse alone—Phalaris. It is plain that not only was the latter largely dependent on Chaucer mares for his success as a sire of winners, but also Chaucer to a great extent was dependent on Phalaris for his reputation as a sire of broodmares. If we eliminate the Phalaris element, except in the case of Hyperion, Chaucer has had comparatively little influence on modern pedigrees.

Some of the horses figuring in the above lists were brothers and sisters (or half such) so the number of mares under consideration is fewer than the number of winners. The particulars are as follows:

	NUMBER OF WINNERS OF £2,000 IN STAKES OR OVER	NUMBER OF DAMS OF WINNERS OF £2,000 IN STAKES OR OVER
By Phalaris out of Chaucer mares (Table 1)	12	8
By Phalaris out of other mares (Table 2)	26	24
By other sires out of Chaucer mares (Table 3)	24	18

Some Chaucer mares produced winners both by Phalaris and by other sires.

I do not think that there is anything to be garnered from this but I mention it because of the fact that amongst the best produced by the Phalaris/Chaucer cross were the full brothers and sister Fairway, Pharos and Fair Isle—and also Caerleon and Colorado. Readers may therefore be inclined to attribute the success of the nick to the suitability of the blood of two or three mares to Phalaris's rather than to the general effect of the cross. In my opinion the above figures do not confirm this view. Some of the best winners by other sires out of Chaucer mares equally were brothers and sisters (or half such) such as Hyperion, Hunter's Moon and Guiscard; Crème Brulée and Betty; Highlander and Spithead; etc.

A very relative point is whether or not the apparent nick was due to an exceptionally high proportion of Chaucer mares being mated to Phalaris. I have no record of the various matings of every Chaucer mare in the Stud Book throughout the whole of Phalaris's 12½ years at stud but I think a glance down the various sires in List 3 shows this was not the case. They were mated to all sorts of high-class sires and I regard the list as particularly informative as it brings out that except for Phalaris they showed no affinity to any other particular sire or sire line.

It is not possible to examine how Chaucer as a sire fared with Phalaris mares as he was dead before the latter had retired to the paddocks. Chaucer left no sons who were stallions of sufficient merit in England to draw conclusions. The only one to sire a Classic winner was Stedfast who got the Oaks winner Brownhylda. Incidentally she was mated to Pharos (by Phalaris) and produced the St. Leger winner Firdaussi. Other useful winners by sons of Phalaris out of Stedfast mares were Fairplay (£6,026) and Phakos (£5,696) whilst the reverse mating (Stedfast on Phalaris mares) produced those two hardy brothers Garnock and Gareloch. I do not consider these isolated results supply any justification for believing that the nick was carried on between the descendants of the two sires under consideration.

The final aspect of the matter to which I will refer is whether or not these nicks are due to inbreeding at a certain incidence to a particularly desirable ancestor. On the next three pages will be found the pedigrees produced by mating (A) Isonomy to Hermit mares (B) Bend Or to Macaroni mares and (C) Phalaris to Chaucer mares.

(A)
ISONOMY TO HERMIT MARES

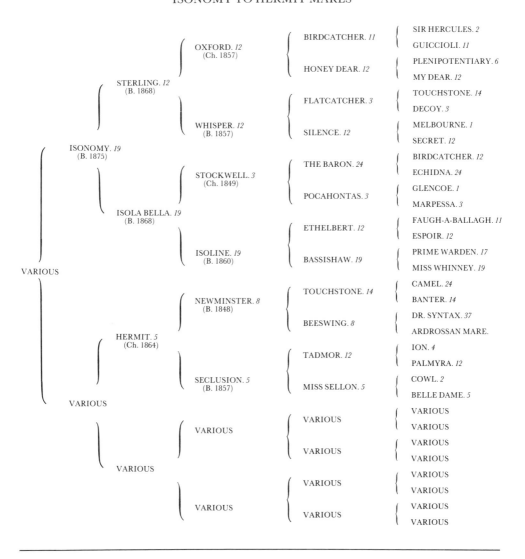

ISONOMY. *19* (B. 1875)	STERLING. *12* (B. 1868)	OXFORD. *12* (Ch. 1857)	BIRDCATCHER. *11* { SIR HERCULES. *2* / GUICCIOLI. *11*
			HONEY DEAR. *12* { PLENIPOTENTIARY. *6* / MY DEAR. *12*
		WHISPER. *12* (B. 1857)	FLATCATCHER. *3* { TOUCHSTONE. *14* / DECOY. *3*
			SILENCE. *12* { MELBOURNE. *1* / SECRET. *12*
	ISOLA BELLA. *19* (B. 1868)	STOCKWELL. *3* (Ch. 1849)	THE BARON. *24* { BIRDCATCHER. *12* / ECHIDNA. *24*
			POCAHONTAS. *3* { GLENCOE. *1* / MARPESSA. *3*
		ISOLINE. *19* (B. 1860)	ETHELBERT. *12* { FAUGH-A-BALLAGH. *11* / ESPOIR. *12*
			BASSISHAW. *19* { PRIME WARDEN. *17* / MISS WHINNEY. *19*
HERMIT. *5* (Ch. 1864)	NEWMINSTER. *8* (B. 1848)	TOUCHSTONE. *14*	{ CAMEL. *24* / BANTER. *14*
		BEESWING. *8*	{ DR. SYNTAX. *37* / ARDROSSAN MARE.
	SECLUSION. *5* (B. 1857)	TADMOR. *12*	{ ION. *4* / PALMYRA. *12*
		MISS SELLON. *5*	{ COWL. *2* / BELLE DAME. *5*

VARIOUS

VARIOUS

VARIOUS — VARIOUS — VARIOUS { VARIOUS / VARIOUS

VARIOUS { VARIOUS / VARIOUS

VARIOUS — VARIOUS { VARIOUS / VARIOUS

VARIOUS { VARIOUS / VARIOUS

(B)
BEND OR TO MACARONI MARES

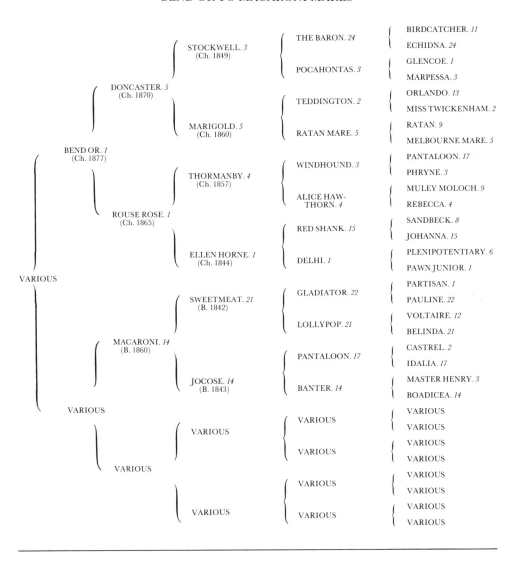

		THE BARON. *24*	BIRDCATCHER. *11*
	STOCKWELL. *3* (Ch. 1849)		ECHIDNA. *24*
		POCAHONTAS. *3*	GLENCOE. *1*
DONCASTER. *5* (Ch. 1870)			MARPESSA. *3*
		TEDDINGTON. *2*	ORLANDO. *13*
	MARIGOLD. *5* (Ch. 1860)		MISS TWICKENHAM. *2*
		RATAN MARE. *5*	RATAN. *9*
BEND OR. *1* (Ch. 1877)			MELBOURNE MARE. *5*
		WINDHOUND. *3*	PANTALOON. *17*
	THORMANBY. *4* (Ch. 1857)		PHRYNE. *3*
		ALICE HAW-THORN. *4*	MULEY MOLOCH. *9*
ROUSE ROSE. *1* (Ch. 1865)			REBECCA. *4*
		RED SHANK. *15*	SANDBECK. *8*
	ELLEN HORNE. *1* (Ch. 1844)		JOHANNA. *15*
		DELHI. *1*	PLENIPOTENTIARY. *6*
VARIOUS			PAWN JUNIOR. *1*
		GLADIATOR. *22*	PARTISAN. *1*
	SWEETMEAT. *21* (B. 1842)		PAULINE. *22*
		LOLLYPOP. *21*	VOLTAIRE. *12*
MACARONI. *14* (B. 1860)			BELINDA. *21*
		PANTALOON. *17*	CASTREL. *2*
	JOCOSE. *14* (B. 1843)		IDALIA. *17*
		BANTER. *14*	MASTER HENRY. *3*
VARIOUS			BOADICEA. *14*
		VARIOUS	VARIOUS
	VARIOUS		VARIOUS
		VARIOUS	VARIOUS
VARIOUS			VARIOUS
		VARIOUS	VARIOUS
	VARIOUS		VARIOUS
		VARIOUS	VARIOUS
			VARIOUS

(C)
PHALARIS TO CHAUCER MARES

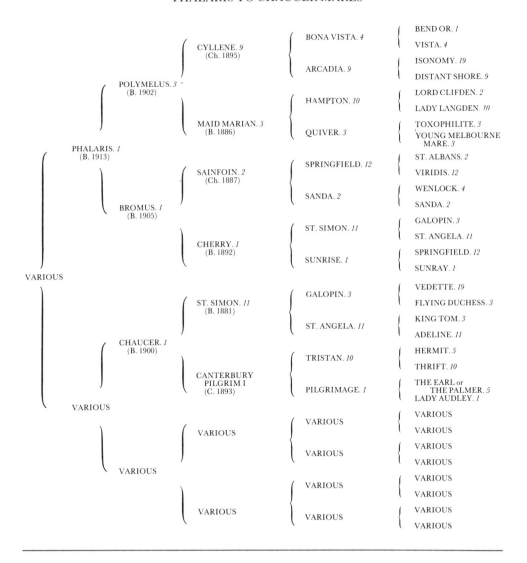

It will be seen that both the Isonomy/Hermit and the Bend Or/Macaroni crosses produced no close inbreeding to any individuals. In the former case the names of Birdcatcher and Touchstone occur once in the fourth and once in the fifth removes, whilst in the latter only Pantaloon's name is duplicated in similar positions. Literally tens of thousands of other matings have been effected to produce a corresponding line breeding to various desirable horses without any uniformly satisfactory results. We must therefore reject the idea that inbreeding has anything to do with the success of these two nicks.

In the case of Phalaris/Chaucer matings there is an inbreeding to St. Simon at the third and fourth removes, plus a subsidiary duplication in Phalaris's own ancestry of Springfield at his third and fourth removes. We can dismiss the latter from our considerations as it was common to all his progeny whether out of Chaucer mares or not. I have already noted that inbreeding to a desirable common ancestor at the third and fourth or fourth removes has produced many outstanding horses (see pages 19–22). I will now refer to this specific case of inbreeding to St. Simon.

Readers are asked to turn back to Table 2 on page 60 and they will see that I have separated the good Phalaris winners into those from mares by sons of St. Simon other than Chaucer (Category "A") and those with grandams by St. Simon (Category "B"). In both cases the inbreeding to St. Simon will be of exactly similar incidence to the Phalaris/Chaucer cross yet they produced the winners of only £13,101 and £41,497 in stakes respectively, compared to the £141,391 emanating from the latter.

When Phalaris went to stud there were 66 Chaucer mares in the Stud Book but several hundred by other sons of St. Simon including 76 by William the Third, 76 by St. Frusquin, and 73 by Desmond—all of whom had a far better record as sires of winners than Chaucer. If the success of the nick was due to the correct dosage of the St. Simon blood it ought to have shown itself through some of these numerous mares. In fact, Phalaris sired nothing of special note from them except perhaps Christopher Robin. In addition there were many mares then at stud whose grandams were by St. Simon and again Phalaris produced only one good winner (Manna) from these. Quite apart from Phalaris, at that period matings which effected an inbreeding to the Duke of Portland's great horse at the third and fourth remove were common but produced no comparable results. Furthermore Phalaris did not do well with mares by Chaucer's half-brother Swynford in spite of the fact that the latter's sire was out of a St. Simon mare.

Isonomy won the Ascot Gold Cup twice and Hermit was a Derby winner whilst both Bend Or and Macaroni were Derby winners, so in the case of their nicks two very good racers were involved. But neither Phalaris or Chaucer was above good handicap class, the former winning 15 races value £4,578 (First World War values) and the latter eight value £5,633. They both raced hard for four seasons and were both noteworthy for their soundness, constitution and gameness. Most stallions, even very good ones, have a tendency to sire stock whose racing merit is inferior to their own. Great racehorses sire one or two who are at least their equals if not their superiors, such as Gainsborough's sons Hyperion and Solario, Bend Or's Ormonde, Galopin's St. Simon, Isonomy's Isinglass, etc., yet the Phalaris/

Chaucer cross produced Fairway, Pharos, Colorado and Fair Isle, who were all materially better performers than their sire or maternal grandsire. To sum up I think the following points are clear:

(A) The Phalaris/Chaucer cross produced exceptional results.

(B) There is no proof that this was due to unusually frequent matings between Phalaris and Chaucer mares.

(C) There is no proof that inbreeding to St. Simon—or any other horse—was a relative matter.

(D) Both Phalaris as a sire of winners and Chaucer as a sire of broodmares are largely dependent for their reputations on one another.

(E) There is no proof that the blood affinity was carried on by their descendants, and last but not least

(F) So far as I am aware there is no satisfactory explanation to the phenomenon in the field of genetics—and here I need hardly point out that we are not considering cross breds but parents of pure blood for some 200 years.

I have dealt with the subject at considerable length as I feel it is one of the utmost importance to all practical breeders. If a breeder considers that in some unexplained way these nicks occur then he can arrange his matings accordingly. Not only will he be on the look-out for crosses on similar lines to those I have cited, but also he must rationally be receptive to the potential advantages of breeding to produce brothers and sisters in blood to proved good racers, and to the whole general principle of blood affinities. If, on the other hand, he forms the opinion that the idea of the genes of two lines of blood co-operating for no known reason to produce superior results is mere fantasy, and that their apparent success is brought about only by chance or opportunity, his rational action will be to confine his attention to arranging his matings so as to eliminate as far as possible inferior ancestors and to inbreeding to desirable ones—or to some other plan of campaign in which he has faith. If he works on the principle of eliminating the weak and inbreeding to the strong he will at least be on sound commonsense ground but whether or not by so doing he will miss opportunities is another matter.

Naturally the occurrence of apparent nicks cannot be foretold. They are only discernible by the careful observation of the results produced by various crosses and weighing these against the number of similar matings carried out. They are most usually apparent in the pedigrees of brothers and sisters in blood but it must not be overlooked that alliances of this nature are common—so the question of unusual opportunity needs careful investigation. Apparent nicks between sire and maternal grandsire are extremely rare and I deprecate the claims that the phenomenon has made itself manifest every time a stallion sires two or three winners from a specific cross. In such cases, not infrequently, the result is due either to exceptional opportunities or to a weight advantage in favour of the winners themselves which enabled them to score. To my mind unless outstanding results can be observed all thoughts of a nick are best discountenanced.

NASRULLAH AND PRINCEQUILLO

While there is a certain element of mystery about the success of the Phalaris/ Chaucer cross, there seems a more logical explanation for the modern triumph of the Nasrullah/Princequillo nick, for which propinquity and selection can be put forward as two major reasons.

Both Nasrullah, who was exported to America in 1951 and died ten years later, and Princequillo, who died in 1964, were born in 1940 and spent most of their stud careers at Claiborne Farm in Kentucky, where Princequillo's famous son Round Table also stood. This proximity, both in time and distance, meant that the opportunity for Princequillo mares to be mated with Nasrullah, or later his sons, was greater than if they had been separated by either years or miles. However, the concept of careful selection cannot be overlooked.

Claiborne's owner, the late "Bull" Hancock, believed in the practicality of a nick, and explained: "You like to drill where oil has been found. Nick may be a bad word, but it gives you an outcross in which some things in the stallion compensate for their absence in the mare. The Round Table–Nasrullah cross might be considered a nick. Nasrullah was a very fiery horse; Round Table was rather phlegmatic. Round Table was not too big; Nasrullah mares are big and rangy."

Hancock's reasoning was therefore based on the practicality of selection, an extension of which manifested itself in the Nasrullah–Princequillo cross. To the fire and vitality of Nasrullah were added the toughness, stamina and soberness of Princequillo, and in several notable cases the two complemented each other to perfection; to such extent that their individual qualities persist through several generations, and have established the most important international nick in modern times.

Some outstanding examples of this success are:

Bold Bidder (1962). Best handicap horse of 1966 in U.S.A. and won 478,021 dollars. By Nasrullah's son Bold Ruler out of High Bid, out of Stepping Stone, by Princequillo.

Bold Lad (U.S.A.) (1962). Won 14 races and champion two-year-old in 1964. By Nasrullah's son Bold Ruler out of Misty Morn, by Princequillo.

Successor (1964). Champion U.S.A. two-year-old in 1966. Brother to Bold Lad.

Mill Reef (1968). Winner of Derby and Arc de Triomphe. By Nasrullah's son Never Bend out of Milan Mill, by Princequillo.

Riverman (1969). Winner of French 2,000 Guineas in 1972. By Nasrullah's son Never Bend out of River Lady, by Prince John, by Princequillo.

San-San (1969). Winner of Arc de Triomphe in 1972. By Nasrullah's son Bald Eagle out of Sail Navy, by Princequillo.

Key to the Kingdom (1970). Sold for a record public auction price of 730,000 dollars at the end of his racing career in 1975. By Nasrullah's son Bold Ruler out of Key Bridge, by Princequillo.

Secretariat (1970). American Triple Crown winner of 1973. By Nasrullah's son Bold Ruler out of Somethingroyal, by Princequillo.

Revidere (1973). Leading U.S.A. three-year-old filly of 1976 and winner of Coaching Club American Oaks. By Nasrullah's grandson Reviewer out of Quillesian, by Princequillo.

Seattle Slew (1974). Leading U.S.A. two-year-old of 1976. By Nasrullah's great-grandson Bold Reasoning out of My Charmer, by Princequillo's grandson Poker. Glamour, the dam of Poker, was by Nasrullah.

THEORY OF PARENTAGE AGE EQUALITY

To turn to another matter. It is sometimes claimed that if parents of equal age or near equal age are mated more satisfactory results can be expected than if there is a considerable disparity. The idea is that in the former case the parents have approximately the same vitality to transmit to the foal. Needless to say it is not suggested that if two low-grade parents are allied they are likely to produce a Derby winner purely on account of their age equality, but rather that, if all other considerations are level, a mare is more likely to produce satisfactorily to a stallion of her own generation.

There is no doubt that an impressive list of important horses so bred can be cited in support of the theory. In the history of the British Turf there have been 15 winners of the Triple Crown (2,000 Guineas, Derby and St. Leger) including those who scored in the First World War substitute races. Three of these were bred from parents of equal age—Common, Flying Fox and Diamond Jubilee; two—Galtee More and Gainsborough—from parents of one year's difference; three—Gladiateur, Pommern and Bahram from parents of two years' difference and four—Lord Lyon, Isinglass, Gay Crusader and Nijinsky—from parents of three years' difference. Thus of these 15 outstanding racers 80 per cent had this feature in common in their breeding.

If we investigate similarly the parental age of Derby winners we find nearly 50 per cent were the produce of sires and dams who were foaled within three years of each other. When we consider parents of equal age it is one thing but when we expand to those of approximately equal age it is another. A discrepancy of three years can be regarded in the latter category but it must be borne in mind that this embraces a parent who is 1, 2, or 3 years younger, of equal age, or 1, 2, or 3 years older than her partner. The span therefore is a great deal wider than appears at first sight. If we consider a stallion in his prime—say a ten-year-old—a three year discrepancy will cover all his 7, 8, 9, 10, 11, 12 and 13-year-old mates—in fact the majority. It is not surprising therefore with the door so wide open that many good horses are bred on these lines. On the other hand as there are more non-winning racehorses bred than winners the plan will produce more of the former than the latter.

To form any reasonable opinion as to the soundness or otherwise of the theory it is, as usual, necessary to balance results against opportunity. We must compare the number of good winners so bred with the number of mares so mated. It would be an almost impossible task to carry out such an investigation covering all the mares and all the good winners in Turf annals but as a guide I set out below the differences in ages of the parents of every Derby winner from 1780 to 1976 inclusive and compare these with similar information relative to 2,000 other thoroughbreds. I have compiled the latter at random from sales' catalogues and it gives a fair picture of the prevailing frequency of mating mares to sires younger, older and of equal age to themselves. These I have labelled "Opportunity" in the table on page 70.

NS

BOLD
REASONING
MY
COMMON
— SEATTLE SLEW

PQ

DIFFERENCE IN AGE BETWEEN SIRE & DAM	NUMBER OF DERBY WINNERS	NUMBER GIVEN OPPOR- TUNITY	PERCENTAGE OF DERBY WINNERS	PERCENTAGE OF OPPOR- TUNITY
Equal	21	144	10.6	7.2
1	33	262	16.7	13.1
2	24	298	12.1	14.9
3	22	258	11.1	12.9
4	17	214	8.6	10.7
5	21	214	10.6	10.7
6	12	162	6.1	8.1
7	7	112	3.5	5.6
8	8	98	4.0	4.9
9	9	62	4.6	3.1
10	4	58	2.0	2.9
11	9	40	4.6	2.0
12	3	34	1.5	1.7
13	1	18	0.5	0.9
14	4	12	2.0	0.6
15	3	6	1.5	0.3
16	–	6	–	0.3
17	–	2	–	0.1
	198	2,000	100%	100%

Although there are some minor variations, taking it fine and large the percentage of Derby winners in the various parental age groups corresponds with the percentage of mares mated within these groups. The discrepancies would probably be eliminated or reversed if another cross section of thoroughbreds was taken. I think we may conclude that the apparent good results of the plan are an illusion and merely the natural result of the frequency of matings.

LORD WAVERTREE'S ASTROLOGY

The late Lord Wavertree was influenced in his breeding plans by astrology. This is a subject I know nothing whatsoever about so it would be an impertinence for me to discountenance it. The idea may seem fantastic to the vast majority of practical breeders but then they, like myself, are totally in the dark as to its possible advantages or disadvantages. But it must not be overlooked that His Lordship was a man with a first-class brain and a highly successful bloodstock breeder. I cannot, however, refrain from recounting a relative story which always amuses me. A very good mare at Tully had a difficult foaling. When the accouchement was completed a hot and bothered stud hand emerged from the foaling box and was asked what the foal was like. He replied "The foal is or'right—I 'ope the Colonel's bloody 'oroscope is". Perhaps in that sentiment there may be some clue to the superior results Lord Wavertree thought he obtained when a foal was born under a lucky star. If they were not, were they regarded as failures from the start and not given a fair chance? Anyhow that was the opinion of many of the men working on the stud. Apart from astrology I often wonder if some of us subconsciously give preferential treatment to a foal bred according to our pet ideas or out of a particularly favourite mare. It is but human nature to do so.

CHAPTER 6

Sires of Winners Lists

Interpretation of the winning sires list: American average earning index figure system: the Jersey Act: the 1949 and 1969 General Stud Book Rules: difficulty of aligning biological books with practical breeding

A STALLION'S REPUTATION is commonly assessed on the number and class of winners he sires, combined with the position he holds in the list of sires of winners. This list is published annually, with interim results, and in its simplest form merely consists of a plain addition of the amount of stakes won by all the progeny of each sire during the Flat-racing season. More detailed data is given in various sporting publications and embraces the names of the winners, their age and sex, the number and total value of the races won by each winner, the total value of placings, etc. The example below is taken from the *British Bloodstock Breeders Review* 1971.

Season 1971

BLAST bay 1957 by Djebe

WINNERS	SIRE OF DAM	RACES WON	VALUE £	TIMES PLACED	TOTAL £	DISTANCE IN FURLONGS IN RACES WON
Black Sky (5) br h . . .						
	Court Martial	2	1,135	2	1,544	12, 11½
Blastavon (2) b c . . .						
	Rockavon	2	1,087	2	1,281	5, 5
Blastie (2) b g . . .						
	Dumbarnie	1	285	–	285	5
Broken Secret (3) b f . . .						
	Bewildered	1	356	1	412	5
Dam N' Blast (3) ch f . . .						
	Talgo	1	490	–	490	8
Gold Ribbon (2) ro f . . .						
	Tenterhooks	2	1,208	–	1,208	7, 8
Oceanic (2) gr c . . .						
	Maharaj Kumar	1	597	1	820	5
Regal Fanfare (6) br m . . .						
	Fairey Fulmar	1	133	1	156	17½
Remraf (6) gr h . . .						
	Prince Chevalier	4	4,661	4	5,149	8, 8, 8, 8

Season 1971 (*contd.*)

BLAST bay 1957 by Djebe

WINNERS	SIRE OF DAM	RACES WON	VALUE £	TIMES PLACED	TOTAL £	DISTANCE IN FURLONGS IN RACES WON
Shiver My Timbers (4) b g						
	Prince Canarina	1	452	2	638	16
Waggy (6) gr h . . .						
	Pappageno II	1	473	3	729	7
Whirlwind (4) ch c . . .						
	Dumbarnie	1	498	2	738	7
Others placings . . .		–	–	5	480	
		18	11,375	23	13,930	

27 runners – 8.78 fur
Stud 1963; 81 races, £59,447

For brevity's sake I have cited an established stallion with an average number of winners during the season. From the details given I think we are entitled to draw the following probable, but not certain, conclusions:

1   He has been mated to mares of average merit. The maternal grandsires mostly were good, but not great stallions.
2   He shows no specially good results from the mares of any particular sire line.
3   His stock show a tendency to stand up to work and train on. He has winners up to six years old.
4   He is capable of getting two-year-old winners.
5   The colts and fillies bred by him are probably equally good. The fact that in this year he had twice as many male winners as female winners (eight to four) is natural, since throughout the breed more of the former score every racing season than of the latter. Moreover, some of the fillies will be sent to stud early in life.
6   The average value of each race won by his stock in 1971 was £630, and £740 for the whole of his stud career. Therefore, although he has sired the winners of an average number of races (81 from six crops), his stock are lacking in class. The average value of each race won by the get of the top 20 stallions in 1971 was £1,863.
7   Only one of his 12 winners in 1971 accumulated more than £1,300 in winning stakes, which appears to confirm relative mediocrity.
8   In 1971, 44 per cent of his 27 runners were winners. This compares favourably with the year's top 20 sires, who averaged 42 per cent of successes from their runners.
9   As he had a reasonable number of runners, he was probably mated to a normal number of mares and his stock do not appear to suffer from any unsoundness or defect which prevents them from racing.
10   The average winning distance of his stock of 8.78 furlongs is itself an average figure compared with the progeny of other sires (see pages 30–33), but he is capable of siring horses which stay much further, since two of his winners in 1971 won over two miles or more.

Accurately to sum up a stallion's merit, it is necessary to investigate his whole stud career, but from the brief details of one season it appears possible to say of Blast that he has a creditable record for siring winners not in the top class, and that his stock are hardy and satisfactory within this limitation, and are capable of staying well. Statistics can be dull things, but a careful examination of them often discloses much of interest.

There are, however, always points to bear in mind when studying statistics, and the assessment of a sire's merit purely on the position he holds in the list of sires of winners has various drawbacks.

One of the imperfections has been eliminated in recent years by the addition of place money to the amount of win money earned by a stallion's progeny, thus providing a more comprehensive table of merit. This scheme was pioneered by the Thoroughbred Breeders' Association in its annual *Statistical Abstract*, and has enabled a clearer picture of each season to emerge, since it embraces substantial earnings gained by those horses of sufficiently high merit to be placed in the major weight-for-age or level-weight races. Under this new system Ribot, not Ballymoss, was champion sire in 1967, and Crepello, not Hethersett, champion in 1969. The wider implications can be observed by reference to the following tables for 1971. The top 20 sires are listed on the left in order of earnings, including place money; on the right is the order based on win money alone:

	SIRE	EARNINGS £ (INC PLACE MONEY)		SIRE	EARNINGS £ (WIN ONLY)
1	Never Bend	133,160	1	Never Bend	125,022
2	Queen's Hussar	94,709	2	Queen's Hussar	89,715
3	Saint Crespin III	89,153	3	Saint Crespin III	83,233
4	Baldric II	87,343	4	Baldric II	76,489
5	Tudor Melody	65,129	5	Celtic Ash	53,471
6	Ragusa	63,845	6	Derring-Do	43,027
7	St. Paddy	59,566	7	Pirate King	40,906
8	Celtic Ash	55,990	8	St. Paddy	40,077
9	Fortino II	55,513	9	Tudor Melody	38,982
10	Le Levenstell	55,157	10	Fortino II	38,753
11	Pirate King	51,437	11	Bold Lad	37,629
12	Derring-Do	50,231	12	Le Levenstell	35,513
13	Crepello	47,212	13	Relko	34,843
14	Relko	44,141	14	Klairon	34,552
15	Bold Lad	42,663	15	Primera	33,995
16	Roan Rocket	41,829	16	Pampered King	31,723
17	Klairon	39,688	17	Ragusa	30,176
17	Pampered King	39,688	18	Atan	29,560
19	Skymaster	39,254	19	Red God	29,346
20	Primera	38,622	20	Skymaster	29,020

The major observation in making a comparison between the tables is that Tudor Melody, Ragusa, Crepello and to a lesser extent Roan Rocket benefited from the inclusion of place money to the win-money earnings of their progeny . . . and rightly so if one is to get a better-balanced picture of their merit. Both Ragusa (who drops from sixth place in the major list to 17th in the win-only table) and

Crepello had the distinction of gaining more from places than from wins in 1971. Ragusa, who earned £30,176 in wins, collected £33,669 from places, of which Homeric's second in the St. Leger and Lombard's second in the Irish Sweeps Derby were the most significant. Crepello, who does not figure in the win-only top 20, earned £20,580 from wins and £26,632 from places, of which Linden Tree's second in the Epsom Derby was the most notable. Tudor Melody gained his high place in the major list by virtue of placings from Philip of Spain (who himself earned more from three seconds than from his single victory in the New Stakes at Royal Ascot) and Welsh Pageant (third in both the Eclipse Stakes and Champion Stakes). All three stallions were represented in 1971 by horses of outstanding merit, and the new system of compiling the list of leading sires acknowledges that fact.

However, even the revised compilation has its drawbacks, such as:

a) no account can be taken of the number or quality of his mates. A horse who serves a full complement of good mares patently should show better results than one who has to put up with a limited number of underbred ones.
b) no account is taken of the number of runners. A stallion represented by 50 horses in training ought to do better than one with 25.
c) no allowance is made for clear-cut winners who have achieved success in level-weight or weight-for-age races, or when carrying top weight in a handicap. These horses rank pari passu with those whose earnings are entirely due to a weight advantage, and so reflect less credit on themselves or their sires.

This last-named point is a very relative matter which I fear does not always receive the consideration that it deserves. In order to reach sound conclusions from any Turf statistics, it is essential to differentiate between these two categories.

Due to some extent to the inclusion of winners with a weight advantage coupled with the operation of the seasonable distribution of winnings, a first-class sire may rarely, or never be champion sire. For instance, the only time Hurry On was top of the tree was in 1926, when his stock won £17,000 more than his nearest rival. The next year he was second, only £3,800 behind Buchan. Both therefore held the coveted honour of being champion sires but their general merit as stallions was widely different. In his 18 years at stud Hurry On-sired stock won some £326,000 in stakes among them, or an average of £18,000 per crop. Buchan had 16 batches of foals who secured about £166,000, or an average of £10,300. The Hurry Ons won seven Classic races and one Ascot Gold Cup, whilst Buchan's sole Classic winner was Booklaw (St. Leger). It is manifest that these two champion sires were of different calibre.

Another misleading result worth mentioning is that of Isonomy (1875). Except for Bayardo's wartime successes, he was the only sire to get two Triple Crown winners—Common and Isinglass—yet he was never champion sire, in spite of the fact that Isinglass won more in stakes (£57,455) on English racecourses than any other horse until the arrival of Tulyar (£76,417) in the 1950s. Another aspect of the matter is that there have been not a few useful stallions who from time to time have worked their way into the upper ten leading sires of winners through the achievements of their numerous handicapper offspring, whose victories were due to a weight advantage. The names of some of these stallions find no place in the pedigrees of high-class parental stock.

Another point of interest is the number of foals a stallion sires compared to the number of those who appear on a racecourse. So far as I am aware there are no statistics of this published, and the information can only be extracted from the General Stud Book and volumes of Races Past after laborious research. For instance, if a horse from four seasons at stud has 120 foals of racing age and only 30 of these have run, one is inclined to be suspicious. What has happened to the others? Is he getting stock who are unsound or who suffer from some defect which prevents them from racing?

In the case of Irish-based stallions commanding moderate fees, it is very likely that many of his progeny are being kept as "stores" for National Hunt racing, but amongst others any marked discrepancy between the numbers of live foals and the number of runners calls for some satisfactory explanation. In these days much publicity is given to individual stallion's fertility returns but, after all, if their stock cannot, or will not race, a high fertility rate is worse than useless.

ASSESSMENT BY POINTS SYSTEM

In an endeavour to place stallions with few and with many runners on a level basis of calculation for their position in the list of sires of winners, a points system was developed in America. It works as follows:

First the total amount of stake money won on all courses during the Flat-race season is added up. For the season 1970, for example, this came to £3,985,868. Then the total number of horses who faced the starter once or more is ascertained. In 1970 these came to 7,677. The next step is to divide the stake money (£3,985,868) by the number of starters (7,677). The result comes to £519.2. So if every starter won in turn, and took exactly the same amount out of the stakes pool, they would credit their sires with £519.2. A stallion with ten runners would thus be the sire of winners of £5,192 in stakes for the season, whilst one with 20 runners would be the sire of winners of £10,384, and so on. Therefore a stallion whose runners can show an average winning figure of over £519.2 is an above-par stallion, whilst one whose stock average under £519.2 is below par.

It is now necessary to align the mythical figure of earnings explained above, with the actual earnings. The method of doing this will probably be easier to understand if I take a concrete example. I will cite Hethersett, who in 1970 had 29 runners (winners and non-winners) who between them won £46,525. If each of these had won exactly their share of the stakes pool (£519.2), they would have credited Hethersett with siring the winners of £15,056.8 for the year. But they did not, and it is necessary to ascertain to what degree they earned more or less than a par stallion. This is done by dividing the actual amount won (£46,525) by the entitlement figure of £15,056.8 given above. This comes to 3.09, and is termed the average earning index figure for Hethersett. It compares with the leading 1970 figure of 68.09 for Northern Dancer, whose seven runners included the Triple Crown winner Nijinsky, hence the imbalance with the rest of the list, headed by Charlottesville on 5.11.

The formula for the system may be summarized as follows:

The actual amount won by sire's progeny
divided by
the number of starters multiplied by their
average entitlement
gives his average earning index figure

In Hethersett's case this is:

Average
earning index
figure

$$\frac{\text{Actual winnings £46,525}}{29 \text{ (starters)} \times £519.2 \text{ (average entitlement of each starter)}} = \frac{£46,525}{£15,056.8 \text{ (average entitlement for all 29 starters)}} = 3.09$$

As I have mentioned, a mythical "par" stallion with ten runners would have accumulated £5,192 to his credit in the sires' list and his figures in conformity to the above would be:

Average
earning index
figure

$$\frac{\text{Actual winnings £5,192}}{10 \text{ (starters)} \times £519.2 \text{ (average entitlement of each starter)}} = \frac{£5,192}{£5,192 \text{ (average entitlement for all 10 starters)}} = 1$$

I am indebted to the *Bloodstock Breeders' Review* 1971 for the figures quoted above, and for their explanation of the system, which was devised by the American expert Mr. J. A. Estes.

It seems to me to be an interesting method of grading stallions, especially if the indices are compiled for the complete stud careers of all the horses whom it is desired to compare, and without doubt it allows those with few runners an equal chance of holding a high position in the list of successful sires as those with many starters.

However, a single season's figures may give most misleading information, particularly regarding those sires with only one or two runners, as in the case of some American-located stallions. In 1953, for instance, Princequillo had one runner (Blue Prince II) in England who won one race value £2,049, and as a result gave his sire an average earning index of 8.06. And as has been shown, 17 years later Northern Dancer was ascribed a figure of 68.09, being loaded greatly by the exploits of Nijinsky, winner of over £218,000 in the year. These figures compare with Hyperion's 2.93—which is plainly misleading.

But the great weakness in the system is that it does not overcome the difficulty that winners with a weight advantage and winners in level-weight or weight-for-age company are mixed up willy nilly. The same is true in the case of the English method of compilation, even with the revised system of including place money. But the more detailed English system giving the name, sex, age, winnings etc., of each winner, number of runners etc., provides the best insight to a stallion's merit of any plan that I am aware of, especially if his whole stud career is examined. I have already outlined, relative to Blast, the considerable amount of information and probably inclinations which can be garnered from the bare statistics.

The merit of a stallion—or for that matter any horse—cannot be reduced to a figure, or worked out in a formula. His whole history must be considered and his achievements mentally balanced against his opportunities.

THE JERSEY ACT

No book on modern bloodstock breeding would be complete without mention of that unfortunate subject, the so-called Jersey Act. This controversy raged around the conditions under which horses were accepted for registration in the English Stud Book. Up to the early days of this century, broadly speaking, they were confined to two categories. First, horses whose pedigrees could be traced without flaw to progenitors entered in previous editions of the book, and second, to foreign horses who were registered in the Thoroughbred Stud Books of their country of origin.

As the French, American, Australian, Russian and one or two more Stud Books permitted entry on much looser terms, the acceptance of their horses sometimes allowed the admittance to the English Stud Book of parental stock of impure blood. The chief protagonists in this wordy war were the Americans and the English. The English and the Irish use the same Stud Book so they were allied. The American team consisted of themselves, the French and, to a lesser extent, the Australians. The Russians were deflected from the conflict by a mere political revolution.

Between about 1900 and 1913 a wave of anti-gambling laws were introduced in many states in America causing racing to be outlawed. As a result of this there were a horde of American horses surplus to requirements. A considerable number of these were shipped to England and Ireland. As there was no apparent likelihood of the revival of American racing it seemed plain that they would remain and some would go to stud here.

Earlier there had been a limited number of importations but, as is the case today, they were mostly high-grade stock who, not infrequently, returned to their native land after their racing days were over. Amongst the large influx in the early days of the century there were a number of horses (and mares) of impure blood who were classified as thoroughbreds by the American Stud Book Rules which were different from ours. As they were in the American Book, however, in accordance with the reciprocal agreements then in force, they were entitled to entry in the English Book. The English breeder became alarmed and feared the results of the wholesale introduction of impure blood into our studs. Before going further I will touch on how these so-called "half-bred" strains came into existence and I will also ask readers to view the position from that of breeders in, say, 1905 and not in the 1970s.

ORIGIN OF IMPURE AMERICAN STRAINS

The first complete edition of the American Stud Book was not published until 1868 so at the relative time was not 50 years old. Fifty years in a horse's genealogical tree is a short span which sometimes only covers his grandparents. The condition of acceptance for entry in this 1868 Book was that the animal must have an uncontaminated pedigree for five generations. Supposing the word thoroughbred is substituted for uncontaminated, even then there is a considerable difference between the background of a horse eligible for entry in the American Book and one normally acceptable in the English Book. Except in the case of a very few foreign imports, entry in the latter volume was confined to horses whose pedigrees were without flaw since the inauguration of the breed. But the word uncontaminated does not appear to have been synonymous with thoroughbred.

The horse who was almost the centre-piece in this affair was Lexington, who was foaled in 1850 and was at stud at the very time the original American Stud Book was published. A descendant of Diomed in the male line, Lexington was a fine racehorse, a great stayer and one of the greatest stallions the world has ever known. He was by Boston, a fine racer got in old age by Timoleon, himself a splendid performer who won 13 consecutive races, usually three- or four-mile heats, in 1816 and 1817 before retiring to stud in Alabama the following year.

Lexington was from an established and accepted sire line, but elements on his dam's side were open to the gravest suspicion. The result was that in the English terminology Lexington could only be described as a half-bred. If such a great and recent pillar of the American Stud Book as Lexington was of impure blood, is it to be wondered that the English breeders of 1905 were doubtful about other entries? Furthermore, in the American book there was a considerable number of other animals who traced in tail female to mares of undisputed non-thoroughbred origin. In fairness it is only right to make it perfectly clear that also there were many American horses of flawless pedigrees.

STATUS OF THE ENGLISH STUD BOOK

The English Stud Book is maintained and published by Messrs. Weatherby. It is not the property of the Jockey Club; nor has the Government anything to do with it, as is the case in France. I mention the latter point because I have even encountered people who were under the impression that the Jersey Act was an Act of Parliament! Apart from any other aspect of the matter, few will dispute Weatherbys' absolute right to enter in their own book such names as they think fit. However, at the turn of the 20th Century the Stud Book authorities began to have doubts about the international expansion of bloodstock trade, and they approached the Stewards of the Jockey Club, pointing out the discrepancies between American and English pedigrees and asking for guidance.

The outcome was the insertion of a qualifying clause, restricting entry in the General Stud Book to horses "able to prove satisfactorily some eight or nine crosses of pure blood, to trace back for at least a century, and to show such performances of its immediate family on the Turf as to warrant the belief in the purity of its blood". This was followed in 1909 by the ruling that "no horse or mare can be admitted unless it can be traced to a strain already accepted in the earlier volumes of the Book".

However, the fears were not allayed, and in 1913 Lord Jersey introduced a motion at a meeting of his fellow members of the Jockey Club, to the effect that they should recommend Weatherbys to confine all future entries in the Stud Book to horses, whether home-bred or foreign, whose pedigrees could be traced without flaw to progenitors entered in previous volumes. This resolution was passed as the Jersey Act, and Weatherbys acted accordingly. Thus, before entry, foreign-bred horses had to conform to exactly the same standards as the English- and Irish-bred ones. At the time this ruling came into force there were certain short pedigree American and other horses already entered in the English Stud Book. These and their descendants were not eliminated but further entries of the same kind were debarred. This produced some paradoxical cases as I will explain in a minute.

The new regulations were quite understandable and, I think, reasonable from the English side's point of view. But the Americans and their friends took the greatest exception to them and, from their standpoint, not without reason. By 1913 Lexington's blood was as widely distributed throughout American blood-stock as St. Simon's is amongst English pedigrees of today—furthermore it was of similar beneficial effect. Supposing for some purely American reason that they in the 1970s suddenly declared that in future all horses with St. Simon blood would be regarded by them as "half-breds" would not we be highly indignant? I have no shadow of doubt that we would.

EFFECTS OF AMERICAN RACING BOOM

As time went on the diverse interests became less acute. A law action in New York ruled that oral betting on the racecourse was not illegal after all, and by 1920 racing in America was booming to a degree never previously attained. All possibility of their unwanted horses being dumped in England, Ireland or elsewhere disappeared and, in fact, they had insufficient to meet the needs of the many new racecourses which sprang up all over America. So only their very best came to Europe to race, and most of these returned to the land of their birth for breeding purposes.

Then, the conditions of entry for the later editions of the American Stud Book were on very similar lines to our own—namely that horses' pedigrees had to trace without flaw to progenitors already entered in previous volumes. Admittedly many of the original entries were not thoroughbred but as the years rolled by the impure strains became more and more remote until they had no significant influence. In the early part of the century this defect may have been present in a great-grandparent but 40 years later it could not be closer than in the seventh or eighth remove—and almost certainly further back in the pedigree. So that by 1949 the anomalies and contradictions of the Act, as revealed in practice, made it clear that the time to conclude the long, drawn-out war was at hand, and a compromise was effected. In that year rules for entry in the English Stud Book were amended to read:

"Any animal claiming admission to the General Stud Book must be able

(1) To be traced at all points in its pedigree to strains already appearing in pedigrees in earlier volumes of the General Stud Book

or

(2) To prove satisfactorily some eight or nine crosses of pure blood, to trace back for at least a century, and to show such performances of its immediate family on the Turf as to warrant belief in the purity of its blood.

The publishers reserve to themselves the sole right to decide what horses or mares can, under the above qualifications, be admitted to or excluded from the Book."

These regulations covered nearly all the disputed strains, but a similar enactment in 1913 would have been of little or no avail, as then many of the American horses could neither be proved to have eight or nine crosses of pure blood, nor yet be traced back for 100 years.

A further revision was carried out in 1969, so that the qualifying clauses now read:

"Any animal claiming admission to the General Stud Book must be able

(1) To be traced at all points of its pedigree to strains already appearing in pedigrees in earlier volumes of the General Stud Book, these strains to be designated 'thoroughbred'

or

(2) To prove satisfactorily eight 'thoroughbred' crosses consecutively including the cross of which it is the progeny and to show such performances on the Turf in all sections of its pedigree as to warrant its assimilation with 'Thoroughbreds' ''.

These alterations, which introduced a definition for the term "thoroughbred" for the first time, further rationalized entry to the English Stud Book, but by now the decision to repeal the Jersey Act had already been proved sound. Had the Act not been amended, the Derby winners Galcador (in 1950), Larkspur (1962) and Relko (1963) would have been excluded from the Book, along with the St. Leger winners Aurelius (1961), Hethersett (1962) and Ragusa (1963), as well as Never Say Die, winner of both Classics in 1954.

DIFFERING VIEWS ON A STUD BOOK

I think the whole crux of the dispute which resulted in the Jersey Act rested very largely on the different viewpoints of the Americans and ourselves regarding the purport of the Stud Book. Many Americans considered that a horse (or mare) with a fine racing record had some claim to registration even although his pedigree was blemished or unproved. We regard the Stud Book purely as a genealogical record of the breed.

As things panned out it may have been that had the Jersey Act never been passed, no great harm would have resulted. The undesirable "half-breds" would have been eliminated in the process of stud competition, and it is unlikely that the Americans would have bred more of this calibre especially for export. Their influx was solely due to the collapse of American racing 70 years ago. On the other hand the anxiety of stud masters of the time to keep their heritage of a pure breed intact is understandable.

I have remarked that the ruling that impure strains already registered should remain but that no new ones would be accepted produced some curious results. As an example I set out below the pedigree of the 1925 Goodwood Stakes winner Diapason (1921).

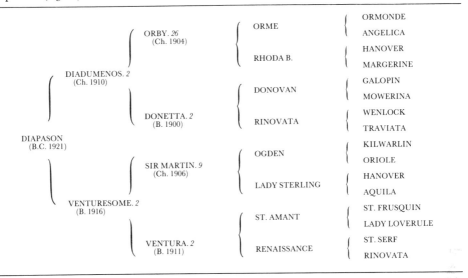

It will be seen that Diapason carried Hanover's blood both on the sire's side and on the distaff side of his pedigree. Hanover, an American horse, was by Hindoo, and the latter's dam was a Lexington mare. Hanover thus had an impure strain. Hanover's daughter Rhoda B was imported to England and registered before the Jersey Act came into force, and so her descendants, including Diadumenos, were regarded as thoroughbred. But Hanover's grandson Sir Martin, who was also bred in America, went to stud in England after the introduction of the new rules and so was debarred from the English Stud Book and classified as "half-bred". Thus although Diapason had Hanover's blood in exactly the same relationship on both sides of his pedigree, his sire was thoroughbred but his dam "half-bred". He of course was also "half-bred". With the sudden changing of the Stud Book conditions, such cases were bound to occur. It was hardly a practical proposition to eliminate the names of all the tainted strains already at stud. It was no use closing the stable door after the horse had bolted.

Amongst the American-bred stock with faulty pedigrees who were registered in the English Stud Book before the Jersey Act were:

1) Americus (1892). Imported as a three-year-old, he sired Americus Girl, ancestress of a line of top horses including Mumtaz Mahal, Fair Trial, Mahmoud, Nasrullah, Tudor Minstrel, etc.

2) Rhoda B (1895). Imported as a yearling, she became the dam of the Derby winner and important sire Orby.

3) Sibola (1896). Winner of the English 1,000 Guineas and second in the Oaks, she became the grandam of Nogara, who in turn produced Nearco, Niccolo dell'Arca and others of note.

Faulty pedigree or not, these played a most important part in the build-up of modern British bloodstock, and their names are found in the family trees of thoroughbreds the world over.

Probably the most important horse excluded through this set of rules was the earliest to be revealed, the 1914 Derby winner Durbar II, who was bred in France by the English-bred but French-based sire Rabelais out of an American-bred mare Armenia, by Meddler out of Urania, by Hanover. The presence of Hanover debarred Durbar II from the English Stud Book—yet it has been seen how the 1907 Derby winner Orby, who was out of the Hanover mare Rhoda B, managed to beat the ban, since his dam was imported before the Jersey Act.

Durbar II sired Durban, on whom M. Marcel Boussac's Turf fame was founded and who produced the fine stallion Tourbillon (1928). Following the 1949 Stud Book amendments, Tourbillon's position and that of his descendants was regularized, but he himself ranked as a "half-bred" until the last few years of his life, and this classification operated against the interests of English and Irish bloodstock. If he had been a thoroughbred, I feel pretty sure that there would have been more of his invaluable blood in our studs today. For example, his French Derby-winning son Cillas (1935) was at stud in Ireland and failed to attract patronage at a fee of only £25 a mare. Then, the mare Loika (1926) was entered for the Newmarket December Sales, covered by Tourbillon, but failed to change hands. The produce was Djebel (1937), who later proved a first-class racer and sire of international fame. I often wonder to what extent the non-sale was due to the fact that Tourbillon was not then in our Stud Book.

Indirectly however, it was the success of Tourbillon which helped bring about the repeal of the Jersey Act. In 1940 Djebel won the 2,000 Guineas, and eight years later Djebel's son My Babu (1945) took the same race. In 1948 a second English Classic, the St. Leger, was won by a horse barred from the Stud Book, Black Tarquin, whose fourth dam Frizette was by the ineligible American sire Hambourg; and so the defences of the Jersey Act crumbled in the face of these "half-bred" successes.

THE AID OF THE BIOLOGISTS

When I study works on biology I find great difficulty in aligning the scientists' view with practical horse breeding. They commonly devote considerable space to the explanation of Mendelism, almost invariably illustrated through the medium of the complications produced by mating Albino Himalayan and whole coloured rabbits. Having waded through much of this I come to the conclusion that it is time for a brown sherry and to go out to look around the greys, bays, browns and chesnuts in the paddocks and to see something practical. On my return I again apply myself to the work—but, joking apart, I have yet to come across a book of this nature which is helpful to the normal breeder.

I remember some years ago reading a scientific book largely devoted to horse breeding. It was written in a way which induced me to believe that at long last I was about to learn something new and of practical value. After many pages of learned discourse the writer disclosed that he had discovered that certain tail female thoroughbred lines produce Flat-racers and others 'chasers. Well, I thought to myself, if he had only asked a normally intelligent stud hand with a couple of years experience he would have had that point made clear in a twinkling and saved himself much laborious research—and his readers much unprofitable reading.

Then, there was the gentleman who, in a well-known book, advanced the theory that staying ability was ruled by a horse's lung capacity. His idea was that a horse filled himself up with air at the start of a race and did not breathe again during the contest. If his air supply gave out before he reached the winning post, then that horse could not stay the distance of that race. The principle was somewhat similar to an underwater swimmer taking a deep breath before his dive. But the author did not disclose how to account for the noise made by whistlers. I do not wish to be indelicate, but most of us are of the opinion that this is due to his respiration, but if the above premises are true I can only conclude that it comes out of some other part of his anatomy.

It would ill become me to ridicule or to disparage works written by biologists whose knowledge of the subject is far in excess of mine. But I greatly wish that they would place their knowledge at the disposal of breeders in a practical, easily understood manner relative to the thoroughbred horse of today. At present this is almost invariably confined to an exposition of the principles of Mendelism—a subject which is commonly familiar to the breeders of all pedigree stock. I have little doubt that a scientist's trained brain and perception could garner many practical and useful lessons from a research of thoroughbred records which are not apparent to the untrained mind. Moreover, the practical stud master has his stud to look after and can rarely spare the necessary time for extensive research.

CHAPTER 7

Stallions

A TABLE SHOWING the stallions who headed the list of sires of winners in England and Ireland for the past 120 years (1856–1976) is shown on the next two pages. They have been divided into groups according to their own successes on the racecourse.

There is such a variety of circumstances to be considered that I do not feel that any useful purpose would be served by splitting hairs between the successes achieved by stallions in the various groups of Classic, etc., winners. For instance, there is little doubt that the Ascot Gold Cup winners St. Simon and Cyllene would have scored in one or more of the three-year-old Classics if they had been started, and how is it possible to differentiate between the merits of a Derby and St. Leger winner as opposed to a Derby and Ascot Gold Cup winner? I will confine myself to drawing attention to the fact that horses who have themselves won a Classic, the King George VI and Queen Elizabeth Stakes, Arc de Triomphe, Ascot Gold Cup or Grand Prix de Paris have headed the list of sires of winners twice as frequently as those who did not score in any of these races.

In these days, and for many years past, foreign horses have figured prominently amongst our leading sires. If any comparison is made between the stud achievements of Classic and non-Classic winners these must be taken into consideration. I will therefore divide stallions into three classes, based on their racing merit, as follows: (a) those who have won an English, French or American Classic, the King George VI and Queen Elizabeth Stakes, Arc de Triomphe, Ascot Gold Cup or Grand Prix de Paris; (b) those who have been placed in any of these 13 races, and (c) those who have failed to gain distinction in any of these races.

The 1976 edition of *Register of Thoroughbred Stallions* contains details of almost 1,300 stallions, all but a few of them based in England and Ireland and including those used entirely for "half-breds". Excluding the latter group, about half of the stallions each cover ten mares or fewer in a breeding season, so can be entirely ignored. Of the rest, the "Register" gives clear indication of the upper-crust of the stallion population by allotting them a separate page containing tabulated pedigree and racing performance. This select band numbers 182 sires, and taking it fine and large, they are the ones with the most reasonable chance of making good at stud.

CLASSIC ETC. WINNERS

	WINNERS OF	NAME OF CHAMPION	NUMBER OF CHAMPION-SHIPS OF EACH SIRE	NUMBER OF CHAMPION-SHIPS OF EACH GROUP
1	Triple Crown & war substitute Gold Cup	Gainsborough (1915)	2	2
2	Derby (Italian), Arc, King George VI & Queen Elizabeth	Ribot (1952)	2	2
3	Leger, Arc, King George VI and Queen Elizabeth	Ballymoss (1954)	1	1
4	Derby, Leger, Ascot Gold Cup	Persimmon (1893)	4	4
5	Derby, Leger	Never Say Die (1951)	1	
		Hyperion (1930)	6	11
		Blair Athol (1851)	4	
6	Kentucky Derby, Preakness	Northern Dancer (1961)	1	1
7	Derby, Gold Cup	Thormanby (1857)	1	1
8	Leger, Gold Cup	Solario (1922)	1	
		Bayardo (1906)	2	3
9	Leger, 2,000 Guineas	Stockwell (1849)	7	7
10	Derby only	Psidium (1958)	1	
		Lemberg (1907)	1	
		Galopin (1872)	3	13
		Hermit (1864)	7	
		Orlando (1841)	1	
11	Leger only	Hethersett (1959)	1	
		Chamossaire (1942)	1	
		Tehran (1941)	1	
		Fairway (1925)	4	12
		Hurry On (1913)	1	
		Swynford (1907)	1	
		Lord Clifden (1860)	1	
		Newminster (1848)	2	
12	2,000 Guineas only	Court Martial (1942)	2	
		Big Game (1939)	1	6
		Tetratema (1917)	1	
		St. Frusquin (1893)	2	
13	Arc only	Vaguely Noble (1965)	2	2
14	King George VI & Queen Elizabeth only	Aureole (1950)	2	2
15	Gold Cup only	Alycidon (1945)	1	
		Cyllene (1895)	2	12
		St. Simon (1881)	9	
16	Grand Prix de Paris only	Nearco (1935)	2	2
		Total		81

NON-CLASSIC WINNERS

WINNERS OF	NAME OF CHAMPION	NUMBER OF CHAMPIONSHIPS OF EACH SIRE	NUMBER OF CHAMPIONSHIPS OF EACH GROUP
17 Horses placed in 2,000 Guineas, Derby, Leger, Arc, King George VI & Queen Elizabeth, Ascot Gold Cup, or Grand Prix de Paris	Great Nephew (1963)	1	
	Never Bend (1960)*	1	
	Court Harwell (1954)	1	
	Chanteur II (1942)	1	
	Nasrullah (1940)	1	
	Pharos (1920)	1	16
	Buchan (1916)	1	
(* placed in Kentucky Derby)	Polymelus (1902)	5	
	Flageolet (1870)	1	
	Speculum (1865)	1	
	King Tom (1851)	2	
18 Others	Queen's Hussar (1960)	1	
	Mossborough (1947)	1	
	Petition (1944)	1	
	Fair Trial (1932)	1	
	Blandford (1919)	3	
	Phalaris (1913)	2	
	The Tetrarch (1911)	1	
	Son-in-Law (1911)	2	
	Sundridge (1898)	1	
	Desmond (1896)	1	23
	Orme (1889)	1	
	Gallinule (1884)	2	
	Kendal (1883)	1	
	Hampton (1872)	1	
	Adventurer (1859)	1	
	Buccaneer (1857)	1	
	Melbourne (1834)	1	
	Birdcatcher (1833)	1	

Total sire championships won by non-Classic winners = 39 (32.5%)
Total sire championships won by Classic, etc. winners = 81 (67.5%)

120

Since there can be no more than 13 winners of the major European races in a year (discounting dead heats) and not all of them go to stud over here, sheer weight of numbers should enable those stallions which were not Classic winners to register more successes than the winners. But in general results prove that the opposite is the case . . . with one proviso that will be examined later. I detail below the percentages of achievements of each of the three classes of sires amongst (1) Champion sires 1856–1975, (2) Sires of the 77 Derby winners of this century, and (3) the ten leading sires of winners for each of the past ten years (1967 to 1976). Groups 1 and 3 will thus cover 100 stallions and Group 2 will embrace 77 sires.

PERCENTAGES OF SUCCESS

(handwritten margin note: Pick Classic Winners)

(handwritten margin note: XXX 6)

		A CLASSIC, ETC. WINNERS	B HORSES PLACED IN CLASSICS ETC.	C OTHERS
1	Amongst the champion sires 1856–1975	67.5%	13.3%	19.2%
2	Amongst the sires of Derby winners of this century 1900–76	61.0%	16.9%	22.1%
3	Amongst the 10 leading sires 1967–76	39%	15%	46%

The greatly superior results obtained by Classic etc. winners as champion sires and sires of Derby winners is valid proof of the value of these races for pin-pointing potential stallions of merit. However, another angle can be observed from reference to the third part of the above table, where horses which have neither won nor been placed in the Classic etc. races are in the majority. This is a new feature and is worth investigating.

Twenty years ago a study of the ten leading sires for the ten years 1946 to 1955 showed that 60 per cent had been Classic etc. winners, 9 per cent had been placed in such races, and 31 per cent had gained neither distinction. The current study for the ten years 1967 to 1976 has enlarged the number of events termed as Classic etc. races but still shows a dramatic swing away from domination by this category, the percentage attributable to them being 39 per cent, compared with 46 per cent for sires which neither won nor were placed in Classic etc. races. Taking the five years 1972 to 1976 alone, the trend is even more illuminating, since Group A accounts for 28 per cent, Group B 24 per cent and Group C 48 per cent.

The conclusion must be that Classic etc. winners no longer enjoy an overall advantage in the season's list of leading sires, and for this a number of reasons may be advanced.

The incidence of American-bred and owned Classic etc. winners has increased in recent years, but comparatively few of these horses have been left in Europe to make their mark at stud. What representatives they have had have been imported. Thus the pool of available stallions, and the resulting numbers of their offspring, has been depleted, leaving room for others to figure high in the lists. The emergence of Royal Palace and Blakeney may help to redress the balance at least as far as English Derby winners are concerned.

As the shift of stallion "fashion" has grown away from stayers, so the emergence of sires whose best form was shown at a mile or a mile and a quarter has become apparent. Thus Wolver Hollow and Habitat have shown up prominently in the lists for 1974 to 1976, and Derring-Do was notable for his appearances in the top ten sires between 1971 and 1973. Coupled with this feature is the fact that races other than those used for identifying Classic etc. winners in the above tables have grown in importance, with the Eclipse Stakes and Champion Stakes taking pride of place. Handsome increases in prize money for these and similar races have no doubt played their part in the process.

It is possible that the classification of races for the upper-crust stallions will have to be revised in the light of these new trends, although a sample comprising five years alone is not sufficient to be dogmatic. However, the theory still holds good that the higher racecourse merit, the more likely is a horse to be a successful sire.

REASONS FOR SUCCESS AT STUD

The stud successes of the Classic etc. winners are due to two causes. First, they are superior racehorses and in a position to transmit their qualities to their offspring, whilst second, individually they are probably mated to better class mares than most non-Classic winners at the outset of their stud careers. I do not propose here to go into which of these elements is probably paramount (if either), but I will touch on the subject when I review the sires of Derby winners of this century (see page 478). Neither do I propose to enlarge on the reason why a horse who is beaten a short head in a Classic, etc., is not uncommonly a stud failure at home unless he has made a name for himself in other races. I will only observe that to all intents and purposes he is within a tiny fraction of a second the equal of the winner. But figuratively speaking if his stock on an average show the same narrow inferiority, the winner will sire many winners and be thought a good sire, whilst the second will sire many placed horses and be deemed an indifferent sire.

In these cases however I have not the slightest doubt that the quality of their mates is the relative cause of the difference which they, not infrequently, record at stud. I stress that I am referring here to horses whose major achievement has been a place in a Classic or the equivalent and not to those who have won other important races. There have been any number of the latter who have been great sires including ten who have headed the list of sires of winners during the past 120 years.

There have been many outstanding racehorses who have proved to be most unsatisfactory stallions. The Triple Crown winners Pommern (1912) and Gay Crusader (1914), the double Ascot Gold Cup victors The White Knight (1903), Prince Palatine (1908), Invershin (1922) and Trimdon (1926) who were all stud failures are but a few examples of this. But there have been exceedingly few bad racehorses who have achieved great fame as sires at home although a few have done well after export. I will refer to these later.

Amongst the stallions who have headed the list of sires of winners during the past 120 years two were racers of limited worth. These were (1) Gallinule, who was a useful two-year-old and then became a bleeder, he raced for three more seasons but was of no account after his affliction became manifest, and (2) Desmond, who won three good races as a two-year-old and then became a rogue who refused to try. On a slightly lower plane there have been a few sires who have left an important mark on the Stud Book who were either unraced or bad racers.

The chief amongst these are: (1) Wisdom (1873) by Blinkhoolie out of Aline by Stockwell who ran a dozen times but never won and at one time was sold for £50. He went to stud as a six-year-old and is the only horse for over a hundred years who was a non-winner but a sire of outstanding overall merit at home. His get included the Derby winner Sir Hugo (1889), the 2,000 Guineas winner Surefoot

(1887), the Oaks winner La Sagesse (1892) and the Ascot Gold Cup winner Love Wisely (1893). On nine occasions his name appears amongst the ten leading sires of winners, his best year being 1890 when he was second to St. Simon.

(2) Next in merit comes Tredennis (1898) by Kendal out of the 1,000 Guineas winner St. Marguerite by Hermit. He also failed to win a race of any kind and was sold for stud work for £100. He commenced duties as an Irish country sire, covering "half-breds" and all sorts of mares, but ere he died in his 29th year he had got the winners of 480 races worth £147,479 in all. Wonderful as his achievements were considering his humble start and background, his stock were not quite of the same class as Wisdom's get. He sired mostly handicappers and none scored in a Classic, although his son Golden Myth (1918) won the Ascot Gold Cup. Further details concerning him will be found on page 206.

(3) Cherry Tree (1891) by Hampton out of Cherry by Sterling never raced and his sole but important contribution to the build-up of modern pedigrees was his daughter Cherimoya (1908). This mare won her only race—the Oaks—and became the grandam of the 2,000 Guineas and Derby winner Cameronian (1928) and the ancestress of other important horses. Further particulars about Cherry Tree are recorded on page 222.

(4) Captivation (1902) by Cyllene out of Charm by St. Simon only ran once, when he was unplaced. Although on a minor scale, his stud career was something on the lines of that of Tredennis. He started as an Irish country sire as a four-year-old at a fee of £5 a mare. He spent 23 seasons at stud and got a great number of winners of small races. A few of his daughters were broodmares of the highest class, whilst his son Kircubbin (1918) was a great stud success in France whose tail male descendants include Pinza (1950), Chanteur II (1942), The Phœnix (1940), etc. I will refer again to Captivation's history (see page 340).

(5) St. Florian (1891) by St. Simon out of Palmflower by The Palmer and so half-brother to the Oaks winner Musa (1896). He won the Duke of York Handicap at Kempton Park carrying seven stone which can be taken as a fair guide to his racing ability. He sired one good horse only and that was the Derby and Eclipse Stakes winner Ard Patrick (1899).

(6) Janissary (1887) by Isonomy out of the Oaks and St. Leger winner Janette by Lord Clifden was nearly useless for racing but sired the very moderate Derby winner Jeddah (1895). Apart from his one success he was a most indifferent stallion, whilst his Derby-winning son was of no account at stud.

There have been a few other sires of the same kidney, but so far as I can recollect none who has had so much influence on modern Anglo-Irish pedigrees as those cited—and for that matter the names of St. Florian and Janissary are rarely encountered in these days. Perhaps I ought to mention Loved One (1883) by See Saw out of Pilgrimage (who won both the 2,000 Guineas and the 1,000 Guineas). He was a poor racehorse who won three races worth £1,847 and cut no ice as a sire of winners, but he got Gondolette (the great grandam of Hyperion) and Doris (the dam of Sunstar). Further details regarding these two mares will be found on pages 201 and 177 respectively. So whatever shortcomings Loved One may have appeared to suffer from during his lifetime he, in fact, made a most valuable contribution to modern blood lines.

The most notable addition in recent times was the unraced Never Say Die stallion Immortality (1951), out of La Troienne's daughter Belle of Troy, who before his export to Argentina sired the 1967 1,000 Guineas winner Fleet.

In order to discover the bad racehorses who have achieved some distinction at stud I have gone back some 100 years. Since those far-off days there have been literally thousands with the same defect tried at stud at home with almost universally unsatisfactory results. After export there have been a few who have produced outstanding results such as: (A) Alibhai (1938) by Hyperion out of Teresina, who was placed in both the Oaks and St. Leger. He was sold to U.S.A. as a yearling and broke down on both forelegs before he even saw a racecourse. At stud he proved a stallion of the highest class.

(B) Beau Père (1927) by Son-in-Law out of the 1,000 Guineas winner Cinna won three small races worth £974 in all. He went to stud at Newmarket for one season at a fee of nine guineas a mare but failed to attract patronage. He was then sold for £100 for export to New Zealand. He was champion sire there twice and in Australia three times. Later he was transferred to U.S.A. at a big price, where he sired the winners of over a million and a quarter dollars in stakes during the few remaining years of his life.

(C) Grafton (1894) by Galopin out of Maid Marian and so half-brother to Polymelus. This horse was wrong in his wind and never won. He also went to the Antipodes where he proved an outstanding stallion.

(D) Limond (1913) by Desmond out of Lindal won one race value £731. He spent three seasons at stud at Newmarket at a fee of 18 guineas with reduced terms for approved mares. He was sparsely used and I doubt if he got one single winner on the Flat before export to New Zealand. In his new home he sired stock who won nearly a quarter of a million pounds sterling.

(E) Selim Hassan (1939) by Hyperion out of Blanc Mange (half-sister to Blandford) won two races worth £586 together. He was sent to Argentina and there became champion sire.

(F) Balloch (1939) by Obliterate out of Cinna and so half-brother to Beau Père mentioned above. He never ran and followed Beau Père to New Zealand where he headed their list of sires of winners more than once.

(G) Mr. Standfast (1931) by Son-in-Law out of Cinna. He won one race worth £100 and then spent two or three seasons at stud in Ireland. There he received scant patronage and produced unsatisfactory results. On export to the Antipodes he proved to be a stallion of equal stature to his full brother Beau Père.

(H) Afghan (1939) by Mahmoud out of Coronal managed to win one race worth £162. He was a leading sire in Chile.

(I) Polystome (1912) by Polymelus out of Battels won three small races worth £571. Sent to South Africa he was champion sire there no less than 11 times.

(J) Grassgreen (1936) by Fairway out of Lucerne only ran once (unplaced) before retiring to stud in England for a season or two at a charge of nine guineas a mare. He was hardly used even at this nominal fee and was a complete "write-off".

He also was sent to South Africa where he turned out excellently but unfortunately died at a comparatively early age.

The racing records of these horses were so poor (or non-existent) that breeders at home refused to use them. On transfer abroad they were generally mated to an ample quantity of good-class mares and produced wonderful results. This brings up the question as to whether we, in England and Ireland, do not allow a good deal of invaluable blood to slip through our fingers by our insistence on racing merit in sires. The above cases prove that occasionally we are guilty of this. But whether we are at fault is another matter.

To discover by a process of trial and error a bad racer who transmits the necessary qualities to maintain and enhance the standard of our bloodstock is like looking for a needle in a haystack. Hundreds, if not thousands, of horses of this class could be tried and found wanting before one of value came to light. In the process hordes of useless daughters of the failures would be born who would be sent to stud in due course with the most retrograde results. Moreover there is no guarantee that because a sire is a great success in a country where standards are a trifle (or a great deal) lower than ours that he would make the grade in competition with the great stallions who are usually available to breeders in England and Ireland.

No—we are right. These occasional leakages do more good than harm as they encourage the export trade. So far as the development of genetics has gone it is impossible to say which stallion will be an asset and which will be an encumbrance (or worse) to the breed until their stock have had an opportunity of proving their abilities on the racecourse. But Turf records prove that bad racers usually make very bad sires and that some—not all—good racers make good sires. Beyond this it is impossible to go, so we are wise to accept the proved facts and to pin our faith to stallions with good racecourse records.

ISOLATED SUCCESSES—AND THE REASON

Amongst the bad racers I have mentioned who gained some important results at stud it will be noticed that three of them—Cherry Tree, St. Florian and Janissary—each sired one good horse and one only. There are countless cases of the same sort recorded in Turf annals. Irish Elegance (1915), who was an exceptionally fine performer, was the only good winner sired by Sir Archibald; the dual Ascot Gold Cup winner Invershin (1922) was by the otherwise poor stallion Invincible; Roi Herode sired nothing in the same street as The Tetrarch (1911); Windy City (1950), who was the outstanding two-year-old of his generation, is by the stud failure Wyndham; Gold Bridge's (1929) sire Gold Boss sired no other racer of distinction and was sold to the American Remount Department. The Derby and Oaks winner Signorinetta (1905) was by the Cesarewitch and Manchester November Handicap winner Chaleureux who got nothing else of worth, etc.

In the great majority of these cases the stallion who sires one good horse only is not a popular stud proposition and is not mated to a full complement of good-class mares. When his good winner appears breeders flock to him hoping for a repetition. Not infrequently if this does not occur it is alleged that the horse is either over used or that he is an animal of sub-normal vitality capable of giving his best only when mated to a very limited number of mares. Whilst each breeder must draw his own conclusions according to the specific case, I think as a general rule the cause lies in a different direction. The fact that prior to the appearance of his good winner the sire is not much used is, generally speaking, an indication that he has some defect which renders him unlikely to be a good sire in the opinion of breeders.

For example in the cases I have mentioned, and which I have taken at random, Cherry Tree, St. Florian, Janissary, Invincible were all bad racers, Roi Herode was a useful staying handicapper and no more, whilst Wyndham, Golden Boss, Chaleureux and Sir Archibald were all indifferently bred. In the normal course of events they could hardly be expected to be great overall stud successes, but all stallions carry somewhere in their make-up the blood of illustrious ancestors. Their one great success is due to a lucky combination of their best genes happening to be transmitted on one particular occasion, but the odds against that occurring again are enormous.

Early in this book I compared the chances of inheritance to the chances of selecting a card at random from a pack of various denominations (see page 6). The more high cards there are in the pack the greater is the chance of picking, unseen, a high one. Similarly in the case of the stallion who sires one good horse: his "pack" is composed of low cards with one or two high ones amongst them. His success is represented by the merest chance of a high card (or good horse) coming out of the pack, but it is possible to try and try again before the same thing is repeated. The whole thing is in accordance with the normal experience of breeding. From time to time an unusual result will crop up but this cannot be reasonably relied upon to happen again.

I do not think that vitality is usually a relative matter nor can I accept the theory that some horses show better results during a period that they are mated to a few under-bred mares than when they are put to good ones. This latter idea seems to me to be contrary to both common sense and the accepted principles of breeding. In the majority of these alleged cases the stallion concerned was either poorly bred or deficient in racing worth and by the merest chance happened to score one or two bull's eyes. Even with better class mares continued repetition could not reasonably be expected.

There have been a number of sires who have started at stud in a lowly capacity and worked their way to higher and yet higher planes as the quality of their mates improved. Wisdom (1873), Tredennis (1898), Captivation (1902), Sundridge (1898), Bay Ronald (1893), Roi Herode (1904) and Marcovil (1903) all started at stud at a fee of 25 guineas or less and reacted to their improved chances as time went on. But for every one of those who made good there have been hundreds who have failed to do so. If this theory of improved results from bad mares holds good, a high-grade horse who at first fails as a stallion and then comes down in the world should show improvement when he is mated to lower-class mares. But it is hard to call to mind a single one who has done so.

There are some stallions who have produced their best get in their old age.

Admiral Drake's (1931) best was Phil Drake (1952) who was 21 years younger than his sire. Moreover until he was 14 years old Admiral Drake got only one good horse—namely the French 2,000 Guineas winner Mistral (1943). Vedette (1854) was 18 years old when his best son, Galopin (1872), appeared; Musket (1867) was the same age when Carbine (1885) was born, and Carbine's best son, Spearmint (1903), was also 18 years younger than his sire. Santoi (1897) was 23 years old when he sired the Ascot Gold Cup winner Santorb (1921). Bubbles (1925) got none better than the full brothers the French 2,000 Guineas winner Guersant (1949) and unbeaten Ocarina (1947) who were 24 and 22 years respectively his juniors—and so on. All these stallions although they produced their best in their old age had, however, met with a fair measure of success throughout their stud careers and could be expected at any time to sire a top-grade horse. I will refer again to this subject. (See page 312.)

INTERMITTENT SUCCESS BY STALLIONS

It is impossible to know at what age a stallion will be most successful. Hurry On (1913) sired Derby winners at the age of 5, 9 and 10, 1,000 Guineas winners at 7 and 10, Oaks winners at 11 and 12, and then, although he got any number of good winners, nothing of outstanding ability appeared until he was 20 years old and Precipitation (1933) first saw the light of day. Orme (1889) sired the Triple Crown winner Flying Fox (1896) when he was six, and then no other Classic winners until Orby (Derby) and Witch Elm (1,000 Guineas), who were both 15 years his juniors. Hyperion (1930) had a wonderful overall stud career, during which he sired horses of the highest class, but his biggest stakes winner, Aureole (1950), came when he was 20 years old. And at the age of 22 Chamossaire (1942) distinguished himself as leading sire for the only time, and Red God (1954) was represented by the brilliant two-year-old Blushing Groom.

On the other hand Tetratema (1917) sired his only Classic winners, Mr Jinks (2,000 Guineas) and Four Course (1,000 Guineas) during his fourth and sixth seasons at stud respectively, and a few years later faded from the scene as a first-class sire. John O'Gaunt (1901) sired the St. Leger winner Swynford (1907) at five and the 2,000 Guineas winner Kennymore (1911) at nine, but after that nothing except horses of the most moderate class and ere he closed his long stud career his stud fee had been for some years reduced to 25 guineas a mare.

I think that there is an increasing inclination in these days to discard too quickly parental stock who do not show immediate results. This is due in some degree to the necessity of sales' yearlings being either from parents with good winner producing records, or from young parents untried at stud but with good racing records. But it is beyond dispute that some of the greatest horses of all time were the produce of an old parent as the following examples will show.

HORSE'S NAME AND YEAR OF BIRTH	CLAIM TO FAME	SIRE	SIRE'S AGE AT THE TIME OF FAMOUS HORSE'S BIRTH
1. Pantaloon (1824)	Tail male ancestor of the Roi Herode sire line	Castrel	23
2. Windhound (1847)	Tail male ancestor of the Roi Herode sire line. The hereditary aspect will be noticed here and in 4, 5, 6, 7, 8, 9 and 10 below	Pantaloon (See 1 above)	23
3. Beeswing (1833)	Ran 64 races. Won 51 including the Ascot Gold Cup, the Doncaster Cup (four times), etc. Dam of Newminster (St. Leger and champion sire) and Nunnykirk (2,000 Guineas). One of the most important tap root mares in the Stud Book.	Dr. Syntax	22
4. Voltigeur (1847)	Won Derby and St. Leger and a great sire. Founded both the St. Simon and Speculum sire lines	Voltaire	21
5. Galopin (1872)	Won Derby. Three times champion sire. Sire of St. Simon	Vedette (son of Voltigeur. See 4 above)	18
6. Galicia (1898)	Dam of the Derby winner Lemberg, the St. Leger winner Bayardo, and Kwang-Su (beaten a neck for the Derby)	Galopin (See 5 above)	26
7. Chaucer (1900)	Sire of the dams of Fairway, Pharos, Hyperion, etc.	St. Simon (son of Galopin. See 5 above)	19
8. Oxford (1857)	Tail male ancestor of both the Blandford and Gallinule sire lines	Birdcatcher	24

HORSE'S NAME AND YEAR OF BIRTH	CLAIM TO FAME	SIRE	SIRE'S AGE AT THE TIME OF FAMOUS HORSE'S BIRTH
9. Doncaster (1870)	Won Derby, Ascot Gold Cup, etc. Tail male ancestor of the following sire lines: (A) Phalaris (B) Teddy (C) Gold Bridge and Panorama	Stockwell (grandson of Birdcatcher. See 8 above)	21
10. Iroquois (1878)	The only American-bred horse to win the Derby and St. Leger prior to Never Say Die	Leamington (by Faugh-a-Ballagh, a full brother to Birdcatcher. See 8 above)	25
11. Emperor of Norfolk (1885)	One of the greatest racers ever bred in U.S.A. Sire of Americus who has widespread influence through his daughter Americus Girl (ancestress of Mahmoud, Fair Trial, Nasrullah etc.)	Norfolk	24
12. Heaume (1887)	Won French Derby. Founder in France of the Hermit sire line which includes Guersant and Ocarina. (See page 253)	Hermit	23

Let us not forget The Little Wonder (1837) who stood but 14.3 and was the smallest Derby winner in the long history of that race. His sire was 27 and his dam 21 years old when he was born.

It might be suggested that the above are freak cases and cannot be accepted as a guide to reasonable probabilities, but it must not be overlooked that the horses cited were of quite outstanding merit whose influence on the breed has been of an exceptionally high order. I most emphatically do not advocate the continued use of old discredited sires, but I feel today there is a tendency to chop and change stallions too hastily. If a horse has the right genes and transmits them he might sire progeny of outstanding value at any time up to his death, but a horse with a defective racing record and pedigree seldom does much good at stud.

CHAPTER 8

More About Stallions

Colts superior to fillies as racers: mating outstanding colts to non-winner fillies: a sire's average progeny inferior to himself: rise and fall of sire lines: dams who produce good (or bad) sires: stallions with hard racing careers: duration of the racing careers of various sires compared

THE AMOUNT A horse, or group of horses, win in stake money cannot be accepted as an indisputable criterion of racing merit on account of victories secured by weight advantages and other outside factors, but it will serve as a medium for pointing out the wide discrepancy in racing ability between the average stallion and the average broodmare.

Every year up to 5,000 thoroughbred foals are born—roughly half are males and half females. Also yearly there are about 2,500 two-year-olds put into training. Again the sexes are equally divided. The remaining 2,500 foals are not up to Flat-racing standard, are exported, kept for 'chasing, meet with accidents or in other ways become casualties. For the purposes of the following calculation they can be ignored, together with the very few who are not put into work for the Flat until they are three-year-olds. The total amount of stakes distributed under Jockey Club Rules was almost £7,000,000 in 1976. The 2,500 two-year-olds put into training is the annual intake of recruits into the ranks of Flat racehorses. If each of these won exactly his or her share of the stakes pool, and no more, he would win £2,800. This may be termed the mean winning expectation of every colt and filly who races on the Flat. If a stallion was put to stud who had won but the bare mean expectation stakes of £2,800 he could be regarded as an average specimen of his breed, but breeders of good-class stock would fight shy of him on account of lack of racing merit.

STALLIONS USUALLY SUPERIOR TO THEIR MATES

In order to have any prospect of attracting patronage at stud a horse must have won five, ten or more times the mean expectation. In other words he is not a true representative of the average ability of his breed. He is a super specimen. On the other hand, there are countless mares, even in the best studs, who have won below their mean expectation and a great many who failed to win anything at all. This is unavoidable as the number of fillies who win £2,800 or over is limited. A stallion normally covers 40 mares each season but never gets a complement of mates who all have secured £2,800 in stakes. He is put to some good winners, some small winners and some non-winners and if the combined racecourse earnings of his 40 mates amounts to £112,000 he is receiving excellent opportunities. Even then individually they are streets below his standard, as he himself has probably won £50,000 or more. It is not open to doubt that the average stallion at stud is of incomparably better racing ability than his average mate.

Amongst the human race and certain animals who practice monogamy, the average parents are drawn from the ranks of the mean of their races or species. In the case of these wild animals (or birds) parentage is reserved for the strong after the elimination of the weaklings in the battle for existence. In bloodstock breeding, on an average, the super specimen (the stallion) is mated to the mean or subnormal (broodmares). Some people may query whether this is a wise policy and whether it would not be better to breed from parents who do not possess this wide discrepancy in their merits.

CONFORMING TO NATURE'S WAY

I have no doubts but that the present policy is right for the following reasons. First it has stood the test of time and produced the best results. Second, as I have already shown, when stallions of mean ability (i.e., winners of about £2,800) have been tried at stud, they have proved failures in the overwhelming majority of cases. Third, it is in accordance with the practice of Nature amongst wild equines and some other species which are polygamous.

Under natural conditions, unhampered by man's interference, the stallion runs loose with a herd of mares. His sons are born, grow to maturity and in common with other male intruders, are chased away from the mares by the king of the harem, who thus shows his superiority to the mean of his sex. As time goes on a colt even better and stronger than he turns up and displaces him. So the sire is always a super specimen.

It may be that amongst thoroughbreds the difference between the male and the female parent is more accentuated than amongst other animals. Moreover, not only is the stallion greatly superior to the mare, but the same disparity is present in nearly every link in the ancestral chains of both parents. In fact it is in the blood. This accounts for the enormous difference in racing merit between the various produce of the same sires and the same mares. Nature has the choice of genes of widely different calibres to select from for the production of a foal. A horse may sire a youngster who turns out to be the best racer in Europe yet from his very next covering, as likely as not, gets a foal incapable of winning a race of any kind, even although both dams are of equal status. This is accepted amongst bloodstock breeders as a natural phenomenon yet amongst other animals the unevenness of their produce is not so apparent.

STALLION'S AVERAGE GET INFERIOR TO HIMSELF

The breeder of pure bred domestic fowls of proved egg-laying strains does not find that some of the hens lay 260 eggs a year and others 50. The general output is more or less level. I do not for one moment suggest that bloodstock breeders should eschew the use of the marvellous super specimens available to them. Such a course could only result in the general lowering of the standard of the breed, but I point out the disparity between parents' racing merit as a matter of general interest and to explain the reason for the vast difference in the worth of their offspring. It also accounts for the fact that the average produce of a stallion has considerably lower racing ability than he himself showed—except, of course, in the case of the non-winner, or very small winner, used as a stallion whose get cannot be worse than himself.

Some stallions sire a few progeny who are their equals or betters—just as the horse in the wild state sometimes has to surrender his harem to his own son. But however successful a bloodstock sire may be, he also gets a quota of non-winners, with the result that his average output are his inferiors. To illustrate this I will cite St. Simon as, although never beaten, he was not a big stakes winner through force of circumstances which I will explain later (see page 182). Moreover, he was the greatest sire of winners for the past 120 years. In all he got 423 live foals who between them won 571 races of £553,158, or an average of £1,300; whereas he himself won £4,675 in spite of his very limited opportunities. The total stakes winnings of £553,158 by St. Simon's progeny was enormous, as in their day the total prize money distributed during a Flat race season was about £480,000, compared to 14 times that sum in 1976. To achieve the same standard of success it would be necessary for a great modern stallion to sire stock to win over six and a half million pounds sterling!

There have been many discussions about the rise and fall of certain sire lines. Even the once all-powerful St. Simon line suffered practical extinction in England and Ireland between the Wars, but with the aid of foreign racers like Nearco and Ribot it has been revived. The Gallinule and Santoi stirps are dead. The Hermit line failed totally at home for generations. The Carbines and the Springfields are no more, and the Wild Dayrell and Sweetmeat lines succumbed generations ago. I do not think that these failures can be regarded as abnormal or attributable to any errors of judgement made by breeders. Basically these are not separate sire lines, but branches of the great Eclipse, Matchem and Herod lines.

It is common for a tail female line to achieve great successes at one time, then to fade into obscurity only to reappear as a first-class stirp after many generations. I have never heard a satisfactory explanation to this phenomenon.

What is sauce for the goose is sauce for the gander. The branch sire lines produce excellent results and then fade, but in their case there is little or no opportunity of retiring out of sight for some generations and then coming to life again. Once their scions have proved to be stallions not up to standard they are not used and so the line dies out—either totally or until revived by an importation from another country who have been lucky enough to possess its sole (or practically sole) worthy representative. In my opinion this is all there is to the matter so far as the present knowledge of genetics has developed. In the process of competition some branch sire lines fall by the wayside, then a great stallion appears from the main stem and forms a new patriarchy. I have no doubt that in 50 or 100 years time some of our present day lines will be extinct and other stirps fan out into further branches of the paternal tree.

It will be remembered that I mentioned that Bruce Lowe maintained that certain tail female lines carried factors which produced or helped to produce good stallions (see page 41). He pin-pointed families 3, 8, 11, 12 and 14, as the ones concerned. Experience has proved that this theory is quite untenable when applied to such a wide group as a whole family. But whether certain individual mares have a tendency to produce good (or bad) stallions is, I think, at least debatable. I will now set out a few examples of both successes and failures as illustrations.

I
MARES WHO PRODUCED GOOD SIRES

A MARE	B SONS AT STUD	C RACING RECORDS OF SONS	D STUD RECORDS OF SONS
(1) Pocahontas (1837) by Glencoe	(A) Stockwell (1849) by The Baron	Winner of the St. Leger and 2,000 Guineas, etc.	Champion sire seven times.
	(B) Rataplan (1850) by The Baron	Winner of the Doncaster Cup, Ascot Gold Vase	Sire of Classic and other good winners.
	(C) King Tom (1851) by Harkaway	Second in the Derby and a good stayer	Champion sire twice.
	(D) Knight of Kars (1855) by Nutwith		A useful stallion who had a particularly strong influence on the breeding of 'chasers.
	(E) Knight of St. Patrick (1858) by Knight of St. George		Sire of Classic and other good winners.
(2) Perdita II (1881) by Hampton	(A) Florizel II (1891) by St. Simon	Won £7,858. Not a Classic horse	Sired three Classic and many other good winners.
	(B) Persimmon (1895) by St. Simon	Won the Derby, St. Leger, Ascot Gold Cup, etc.	Champion sire four times.
	(C) Sandringham (1896) by St. Simon	Never ran	Sent to U.S.A. where he sired some winners but an indifferent stallion.
	(D) Diamond Jubilee (1897) by St. Simon	Won 2000 Guineas, Derby, St. Leger, etc.	Very successful during his four seasons at stud at home and several times champion sire in Argentina.

MARES WHO PRODUCED GOOD SIRES (*contd.*)

A MARE	B SONS AT STUD	C RACING RECORDS	D STUD RECORDS
(3) Plucky Liége (1913) by Spearmint	(A) Sir Gallahad III (1920) by Teddy	Won French 2,000 Guineas, Lincolnshire Handicap, etc.	Champion sire U.S.A. four times.
	(B) Bull Dog (1927) by Teddy	Won moderate races in France	Champion sire
	(C) Quatre Bras II (1928) by Teddy	A useful but not top-class racer in France and U.S.A.	Moderately successful in U.S.A.
	(D) Admiral Drake (1931) by Craig-an-Eran	Won Grand Prix de Paris	Sire of Classic winners in France and England.
	(E) Bel-Aethel (1933) by Athelstan (by Teddy)	Won four small races in France and England	Made no mark at stud in U.S.A.
	(F) Bois Roussel (1935) by Vatout	Won English Derby	Sire of Classic and other good winners in England.
(4) Selene (1919) by Chaucer (See page 232 for her son Moonlight Run	(A) Sickle (1924) by Phalaris	Third in 2,000 Guineas and a good two-year-old winner	A leading sire in U.S.A.
	(B) Pharamond (1925) by Phalaris	A good two-year-old winner in England	A leading sire in U.S.A.
	(C) Hunter's Moon (1926) by Hurry On	Won Newmarket Stakes	A leading sire in Argentina
	(D) Salamis (1927) by Phalaris	Useless as a racehorse	During a short stud career in Ireland and England met with limited success.
	(E) Hyperion (1930) by Gainsborough	Winner of Derby, St. Leger, etc.	Champion sire six times.
(5) Nogara (1928) by Havresac II	(A) Nearco (1935) by Pharos	Never beaten. Winner of Grand Prix de Paris, Italian Derby, etc.	Champion sire in England twice, and great modern influence on pedigrees.
	(B) Niccolo dell 'Arca (1938) by Coronach	Won Italian 2,000 Guineas and Derby, etc.	Champion sire in Italy, and sire of many winners in England.

MARES WHO PRODUCED GOOD SIRES (*contd.*)

A MARE	B SONS AT STUD	C RACING RECORDS	D STUD RECORDS
	(C) Nicolaus (1939) by Solario	Won nine races in Italy but not a top-class racer	Sire of Classic winners in Italy, and of NH winners in England, including Nicolaus Silver (Grand National).
	(D) Nakamuro (1940) by Cameronian	Second in Italian Derby and won four races there	Sire of Classic winners in Italy. Moderate success in six years at stud in Ireland before returning to Italy.
	(E) Niccola d'Arezzo (1943) by Ortello	Won three small races in Italy	Indifferent sire in U.S.A.
	(F) Naucide (1945) by Bellini	Useful 2-y-o in Italy, and third in Italian Derby	Moderate sire in England, best winner being Seascape (useful 2-y-o and second in Cesarewitch).
	(G) King of Tara (1947) by El Greco or Torbido	Useless as a racehorse	Exported to South Africa.

Others are (6) Guiccioli (1825) the dam of Birdcatcher (champion sire) and Faugh-a-Ballagh (sire of Classic winners). (7) Canterbury Pilgrim (1893), the dam of Swynford (champion sire), Chaucer (twice head of the list of maternal grandsires) and Harry of Hereford (a non-winner whose name sometimes appears in French and American pedigrees). (8) Galicia (1898), the dam of Bayardo and Lemberg (both champion sires) and also of Kwang-su (spent most of his stud life as an Irish country sire with no chances). (9) Scapa Flow (1914), the dam of Fairway and Pharos (both champion sires) and also St. Andrews (a non-winner who had a short and unsuccessful stud career). (10) Double Life (1926), the dam of Precipitation and Persian Gulf (both sires of Classic winners) and also Casanova and Holywell (who both achieved some sucesses with limited opportunities), etc. (11) Cinna (1917) whose sons Beau Père, Mr. Standfast and Balloch were all champion sires in Australasia and whose other sons Gay Shield and Dink reached near the top of the tree in the same part of the world, but Hot Haste failed him.

I would not for one minute suggest that mares who produce good stallions are other than extremely rare, but I think the above examples give some substance to the claim that they have existed, and doubtlessly others of the same kidney will appear from time to time. One who has emerged in more recent years is Relance

III (1952), by Relic out of Polaire II, by Le Volcan out of Stella Polaris. Relance's son Relko (1960) gave the Sceptre family its first Derby winner, and within two years the mare had provided the family with its first French Derby winner, Reliance II (1962). Relance was also responsible for Match III (1958), and between them this trio won seven English and French Classics before going on to enjoy considerable success at stud.

To turn to the other side of the picture:

II
MARES WHO PRODUCED BAD SIRES

A MARE	B SONS AT STUD	C RACING RECORDS OF SONS	D STUD RECORDS OF SIRES
(1) Pilgrimage (1875) by The Earl or The Palmer	(A) Loved One (1883) by See Saw	Won Wokingham Stakes (Ascot) and two other small races	Sired the dam of Sunstar and the great brood mare Gondolette. Otherwise a failure.
	(B) Lourdes (1884) by Sefton	Won £1,957	Failure at home.
	(C) Knight of Malta (1886) by Galopin	A non-winner	Failure at home.
	(D) Pilgrim's Progress (1889) by Isonomy	Won one race in England value £1,000	A fairly successful sire in Australia.
	(E) Jeddah (1895) by Janissary	Won Derby	Failure at home.
(2) La Vierge (1890) by Hampton	(A) Sir Geoffrey (1895) by St. Angelo	A moderate racehorse	Failure at home.
	(B) Innocence (1896) by Simonian	Third in Derby	An indifferent sire in Argentina
	(C) Prince William (1903) by Bill of Portland	A moderate racehorse	Failure in France.
	(D) Lycaon (1908) by Cyllene	Second in St. Leger and third in 2,000 Guineas. Also a useful winner	A failure at home and sent to Germany.
	(E) White Magic (1910) by Sundridge	Second in St. Leger and a useful winner	A failure at home and in U.S.A.
(3) Be Sure (1898) by Surefoot	(A) Cocksure II (1905) by Count Schomberg	A good middle distance handicapper	Failure at home.
	(B) Decision (1907) by Count Schomberg	A good middle distance handicapper	Failure at home.

MARES WHO PRODUCED BAD SIRES (*contd.*)

A MARE	B SONS AT STUD	C RACING RECORDS OF SONS	D STUD RECORDS OF SIRES
	(C) Stedfast (1908) by Chaucer	Second in Derby and 2,000 Guineas. Won £26,479. An excellent racer	Sired the Oaks winner Brownhylda. Otherwise a most disappointing stallion.
(4) Countess Zia (1910) by Gallinule	(A) The Panther (1916) by Tracery	Won 2,000 Guineas, etc.	Failure at home, and in Argentina.
	(B) Eaglehawk (1918) by Spearmint	A good handicapper. Won Liverpool Cup, etc.	Failure at home.
	(C) Bhuidhaonach (1919) by Royal Realm	A good handicapper. Won Queen's Prize, Manchester Cup, etc.	Failure at home. Descended to the ranks of hunter sires.
(5) Tillywhim (1913) by Minoru	(A) Monk's Way (1923) by Friar Marcus	One of the best two-year-olds of his generation in England	Failure in U.S.A.
	(B) Tommy Atkins (1924) by Spion Kop	One of the best sprinters of his time	Failure at home.
	(C) Santilio (1928) by Sansovino	Won Red Rose Stakes Manchester	Failure at home.
	(D) Figaro (1930) by Colorado	One of the best sprinters of his time	Failure at home.

Also in this category are such as (6) Mowerina (1876) the dam of Donovan (Derby and St. Leger winner) and Raeburn (third in the Derby and the only horse who ever beat Isinglass). (7) Ornament (1887) full sister to unbeaten Ormonde, the dam of the great filly Sceptre and of the bad sires Collar, Star Ruby (U.S.A.) and Naledi. Also grandam of the failure St. Denis (third in the Derby and a good winner). (8) St. Louvaine (1898) the dam of Louvois (2,000 Guineas winner) and Louviers (second in the Derby; third in 2,000 Guineas; stud failure at home: sent to Russia). (9) Dame D'Or, the dam of Diamond Stud, All Gold (U.S.A.), Sir Eager and Bay D'Or. (10) Donnetta (1900), the dam of Diadumenos (best sprinter of his time) and Diophon (2,000 Guineas winner). Also grandam of Galvani (by Prince Galahad) and On Parade. (11) Tagale (1900) dam of the 1,000 Guineas and Derby winner Tagalie and of the bad sires Poltava and Montdidier. Also grandam of the failures Allenby (second in 2,000 Guineas) and Tagrag. (12) Gay Laura (1909) the dam of Gay Crusader (winner of the Triple Crown and an exceptional racehorse), Manilardo and Leonardo. (13) She (1909) the dam of He

(Coronation Cup, etc.), Yutoi (Cesarewitch) and Roi Hero. (14) Salamandra (1913) the dam of Salmon-Trout (St. Leger), Sagacity, St. George and Star of Destiny, and so on *ad infinitum*.

HEREDITARY TENDENCIES

Some of the junior members of the matriarchies mentioned above had insufficient racing merit to warrant the belief that they were likely to make great stud successes but they all failed even at the level where they could have been expected to do reasonably well. It is true that there are a large number of stallions tried at stud and that only a few of these make good, but the number of failures amongst full or half brothers provides food for thought. In some cases the tendency appears to be hereditary and in others very much the reverse. Great broodmare that Gondolette (1902) was, her sons were unsatisfactory stallions. The eldest Great Sport (1910) by Gallinule was a racehorse of class who was third in the Derby. He went to stud at the late Lord Wavertree's famous establishment (later the National Stud) and had the advantage of being mated to mares of the very highest class, yet he was so unsuccessful that as a 13-year-old he was weeded out and sent to Belgium.

Next came Let Fly (1912) by White Eagle who also was a good racer and was second in the Derby before being exported to Argentina where he met with limited success as a stallion. He was followed by the Derby winner Sansovino (1921) by Swynford whose stud career was spent in the late Lord Derby's stud where he was put to a number of the best mares in the world. He sired produce who won £113,126 in stakes, including the St. Leger winner Sandwich but considering his opportunities Sansovino cannot be regarded as other than a very disappointing stallion. His best season was in 1931 when he was 50th on the list of sires of winners but even in that year his stock won over £20,000 less than the progeny of the champion sire (Pharos). Neither have his daughters come up to expectations as broodmares. So bearing in mind their high merit as racehorses and wonderful chances at stud it must be conceded that Gondolette's sons were all indifferent stallions. She also produced Serenissima whose two sons Bosworth (Ascot Gold Cup) and Schiavoni were equally disappointing at stud. Serenissima was additionally the dam of Selene and here came a complete change over the line's record, for she was the dam of Hyperion, Pharamond, Sickle and Hunter's Moon, all of whose stud records must satisfy the most capricious critic.

Sansovino was incestuously inbred at the third remove to Pilgrimage whose sons were all bad stallions but this circumstance is not sufficient to account for his failure as his half-brothers Great Sport and Let Fly were free from it. Moreover Selene also was inbred to Pilgrimage but at a remoter incidence. Although Pilgrimage's sons were no good as sires, her daughter Canterbury Pilgrim produced Swynford and Chaucer who were very much the reverse. To take another case, as we have seen, Perdita II's sons were first-class stallions but her half-sister Dorothy Draggletail was the great grandam of Mésange (1903) who produced Jaegar (second in the Derby), Argos (Middle Park Plate), Lanius (Australia), Meleager, and the dam of Spearwort who all made little or no headway as stallions.

On the other hand the action of heredity seems operative in other cases. Stallions tracing to Bridget (1888) by Master Kildare have excellent records and on the whole better than their racing records would warrant expectation. These include Bridge of Earn, Bridge of Canny (U.K. and Argentina), Legionaire (Australia), Knockando (Venezuela), Light Brigade (U.S.A.), Oliver Goldsmith (South Africa), Gulf Stream (Argentina), Shannon II (bred in Australia and later in U.S.A.), Menow (U.S.A.), Heliopolis (U.S.A.), Manitoba (Australia), etc. In fact it is hard to call to mind any racehorse of class from this tap root who was an unsatisfactory stallion except the Derby winner Mid-day Sun who had very limited chances at home for reasons outlined on page 247. Bridget was a member of No. 8 Family which can claim other important sires such as Newminster, Marske, Humphrey Clinker, Orville at home, Sun Briar, Blue Larkspur, Sunreigh and Sweep in U.S.A., Perth in France, Valais and Linacre in Australia, etc., but some of its other scions were shocking bad stallions. The latter group include Royal Lancer (St. Leger), Sleive Gallion (2,000 Guineas), The Panther (2,000 Guineas), Dark Lantern II (French 2,000 Guineas) and Invershin (Ascot Gold Cup, twice). These were all racehorses of high class who could reasonably be expected to make good at stud. They prove that it is quite impossible to substantiate any claim that a whole family has a tendency to produce good sires. The most that can be said is that certain branches of a family for a limited number of generations may perhaps show an inclination in this direction.

The descendants of Concussion (1885) by Reverberation appear to have a tendency the reverse way. Many of them have had excellent racing records but most indifferent stud careers. The following examples show this (1) Llangibby (£13,907 including the Eclipse Stakes). (2) Comrade (Grand Prix de Paris but died young). (3) Cellini (National Breeders' Produce Stakes). (4) Orpheus (£11,971). (5) Spion Kop (Derby winner). (6) Craigangower (good winner: third in the Derby: sent to U.S.A. and after failing at stud there, to Canada). (7) Salmon-Trout (St. Leger), and (8) Epigram (£7,445, including the Goodwood and Doncaster Cups). It is difficult to pin-point a single horse tracing to Concussion who has lived up to expectations as a stallion although the line has produced a number of first-class racers and broodmares.

Whilst it is totally incorrect to say that some whole families breed good stallions, it is not inaccurate to claim that some of the smaller and less important ones cannot boast a sire of merit for a hundred years past—and in some cases have never produced a good one.

These include Families 15 (except for Hanover in America), 25, 29, 30, 32, 33, 34, 35, 36, 39, 40, 41, 42 and 43. Each of these stirps is numerically weak, as instanced by reference to the *Register of Thoroughbred Stallions* 1976, which denotes 182 of the most important stallions in tabulated pedigree form, and includes only Scottish Rifle (Family 15) and Ragapan (42) from the above list. Some of these families have almost, if not totally died out, and their failure to produce good stallions is due to this circumstance, combined with the fact that they can claim very few good racehorses. They are, in fact, the submerged section of their breed.

RACING CAREERS OF SIRES COMPARED

It is often said that the modern stallion of class is lightly-raced and hustled off to stud before his ability—or defects—are fully exposed, and the oncoming of greater syndication of stallions for ever-larger sums has merely accentuated the theory. Yet life was ever thus. With a few exceptions the average leading stallion of days gone by was not subject to appreciably harder tests than his modern counterpart. Indeed, it is possible to state that the top-class modern stallion is slightly more overworked than his forbears. I set out overleaf for comparison the details of the ten leading sires of 1974 and ten outstanding sires of the 19th Century, showing for each (A) the number of seasons they raced, and (B) the number of times they ran.

It will be seen that for both categories the average number of seasons' racing was approximately the same, but that the average number of races run was slightly higher for the top ten sires of 1974, and thus the average number of times they started in each season is slightly more for modern stallions. The details seem to suggest that the outstanding sires of the 19th Century were certainly not subject to more racing than their present-day successors. On the other hand, in the former period—and particularly in the early part of it—there were probably a considerably higher percentage of horses at stud who raced until well on in years, than is the case today. Amongst them were some who rendered yeoman service to their breed, but it must not be overlooked that in those days some animals did not commence their Flat-racing careers until four or five years old.

DOES HARD RACING JEOPARDIZE A STUD CAREER?

Some breeders maintain that a great deal of work on the course undermines a horse's constitution to such a degree as to jeopardize his stud career. At first sight colour is lent to this theory by the fact that none of the horses mentioned overleaf was raced to excess. Can it therefore be assumed that the comparatively lightly-raced stallion is likely to meet with greater success at stud than one who has been more severely tested? I think not. It will be remembered that I examined the frequency of running of 1,000 horses of both sexes, taken at random, for the years 1911 and 1975, and I found that on an average the figure for each horse had altered only slightly from 4.6 races per season to 4.8 (see page 16). These figures correspond with only slight variation to average annual frequency of running of the sires in the above tables, namely 4.1 for 19th Century sires, and 5.7 for 1974 sires. Moreover, the average Flat-racing life of every thoroughbred at home is about two seasons, which again approximately tallies with these stallions' periods of racing.

So it can be said that they were actually neither lightly nor excessively raced but were subject to average racing careers. I therefore conclude that the stud achievements of the so-called lightly raced horses is not due to the curtailment of their racing careers but merely to the fact that those in this category vastly outnumber the ones who have had a hard time on the Turf. In other words, they are the natural results of numerical superiority. A contributory element is that the horses who have long and hard racing careers, besides being comparatively few (except for geldings), are, in this country, mostly lacking in class and exploited in

1974 SIRES			19TH CENTURY SIRES		
NAME AND YEAR FOALED	NUMBER OF SEASONS RACED A	NUMBER OF RACES RUN B	NAME AND YEAR FOALED	NUMBER OF SEASONS RACED A	NUMBER OF RACES RUN B
Vaguely Noble (1965)	2	9	Touchstone (1831)	5	15
Petingo (1965)	2	9	*Melbourne (1834)	3	14
Busted (1963)	3	13	Stockwell (1849)	3	14
*Firestreak (1956)	3	15	Lord Clifden (1860)	2	7
Habitat (1966)	1	8	*Blair Athol (1861)	1	4
Reform (1964)	2	14	*Hermit (1864)	2	11
Faberge II (1961)	2	11	Galopin (1872)	2	10
Sing Sing (1957)	2	9	St. Simon (1881)	2	9
Great Nephew (1963)	3	23	Persimmon (1893)	3	11
Queen's Hussar (1960)	3	20	Cyllene (1895)	3	11
10	23	131	10	26	106
Average degree of testing	2.3	13.1		2.6	10.6

1974 SIRES

Average number of races per season
(23 divided into 131) = 5.7

19th CENTURY SIRES

Average number of races per season
(26 divided into 106) = 4.1

*Indicates did not run as a two-year-old.

lowly company. Their whole background is not of sufficient status to warrant the expectation that they will make good stallions. A potential high-class sire is not exposed to the risks of racing longer than necessary.

HARD RACING OF SUCCESSFUL U.S.A. SIRES

Reference to the racing careers of important American sires proves that a more strenuous life than is normal in England and Ireland in no way harms a horse's stud career. The facts that Bold Ruler (23 wins) and Princequillo (12 wins) each ran 33 times, and Round Table (43 wins) raced 66 times have not prevented their having enormous modern influence at stud.

Opposite are a dozen examples of stallions who raced in America and who have been responsible for prominent European Classic winners since 1970.

Comparison with the table above for the leading sires in Britain in 1974 shows that while the average degree of testing was similar for the number of seasons raced, the American sires took part in over two races more each year.

However, an observation must be made with regard to figures published in an earlier edition of this book. A list of American-raced sires whose influence had been felt in Britain up to 1956 showed that the average number of races per season was 12.8. The stallions were Equipoise (1928), Discovery (1931), War Admiral (1934), Challedon (1936), Eight Thirty (1936), Alsab (1939), Devil Diver (1939), Stymie (1941), Polynesian (1942) and Citation (1945), who together raced 539 times. In the following 20 years the average number of races per season has dropped to 8.3.

NAME AND YEAR FOALED	NUMBER OF SEASONS RACED	NUMBER OF RACES RUN	STUD RECORD, ETC.
	A	B	
Northern Dancer (1961)	2	18 (won 14)	Leading sire in Britain 1970. Sire of Nijinsky (Triple Crown)
Sadair (1962)	1 (as 2yo)	12 (won 8)	Sire of Pampered Miss (French 1,000 Guineas, 1970)
Never Bend (1960)	2	23 (won 13)	Leading sire in Britain 1971. Sire of Riverman (French 2,000 Guineas, 1972), Mill Reef (Epsom Derby, Arc de Triomphe, 1971)
Traffic (1961)	3	22 (won 4)	Sire of Rheffic (French Derby, 1971)
*Gun Bow (1960)	3	42 (won 17)	Sire of Pistol Packer (French Oaks, 1971)
Bold Lad (U.S.A.) (1962)	2	19 (won 14)	Sire of Bold Fascinator (French 1,000 Guineas, 1971)
Hail to Reason (1958)	1 (as 2yo)	18 (won 9)	Sire of Roberto (Epsom Derby 1972)
Bald Eagle (1955) (including 6 races in England)	4	29 (won 12)	Sire of San-San (Arc de Triomphe 1972)
Graustark (1963)	2	8 (won 7)	Sire of Caracolero (French Derby 1974)
Nearctic (1954)	4	47 (won 21)	Sire of Nonoalco (2,000 Guineas 1974)
Ack-Ack (1966)	4	27 (won 19)	Sire of Youth (French Derby 1976)
Lucky Debonair (1962)	3	16 (won 9)	Sire of Malacate (Irish Sweeps Derby 1976)
12	34	281 (won 147)	
Average degree of testing	2.8	23.4 (won 12.2)	

Average number of races per season
(34 divided into 281) = 8.3

*Indicates did not run as a two-year-old

It is impossible to draw exact conclusions, since the sample taken is not large enough, and different figures may have been obtained from a different survey. But it does suggest that perhaps the modern Americans are not quite so tough on their horses as their forebears. Their success in breeding top-class horses, however, goes on undiminished.

CONSTITUTIONAL STRAIN OF RACING

The weakest part of the racehorse is his forelegs and injury to them is the most frequent cause of a horse's early forced retirement from the course. As American horses are generally raced more heavily than ours I assume that their race tracks are easier on an animal's legs than our courses, although the fact that some of them are possibly inclined to be of somewhat smaller and lighter build than many of ours may be relative. But in any case I cannot accept any suggestion that racing on American tracks is a materially smaller strain on a horse's constitution than racing on ours. Excluding (1) exceptional underfoot conditions (when our high-class horses are not normally started) (2) races in which a horse is "eased", either because he is greatly superior to his rivals or because he has no chance of success and (3) horses who refuse to exert themselves, all races entail approximately the same strain on a participant's constitution and physical powers. Regardless of the track and regardless of the class of race, all horses, assisted by their jockeys, put forth their maximum efforts in an endeavour to win.

Apart from his great numerical inferiority and from the fact that at home he is usually a very moderate-class animal, the hard-raced horse suffers another material disadvantage compared with his rivals, if put to stud comparatively late in life. If he starts to hold court as a seven-year-old, other sires of his generation will have runners available to make their reputations for them, before he has any foals at all. His very first crop will not be of racing age until he is a ten-year-old and unless they achieve immediate outstanding successes, breeders will regard him as an old horse, and refuse patronage, before he has had a reasonable chance of making good. Moreover this type of horse rarely sires early maturing stock, so by the time there is an opportunity for them to show their ability in their own *métier* the odds are that the stallion himself will be regarded as a stud failure.

His failure is not due to his hard racing career but to the general set-up of the bloodstock world. His potentialities as a parent are exactly the same as if he had been retired to stud as a four-year-old, and furthermore he has proved that he possesses exceptional qualities of soundness, hardiness and gameness which he may transmit—and these are very valuable attributes. It should be remembered that Hampton, Chaucer and Sundridge all started their stud careers at seven years old and all sired important winners amongst their first few crops of foals.

I believe American horses are not given so much fast work on the training grounds as ours. The many rich prizes there are an inducement to race horses often, whilst here it pays better to enhance their stud values by successes in a few kudos events.

CHAPTER 9

Broodmares

Important races and fillies' records: vital importance of tail female lines: some great tap roots: racing merit of broodmares: effect of hard racing: some great broodmares with hard racing careers: important broodmares who achieved success only in their old age: successful first foals

I HAVE PREVIOUSLY referred to the fact that the male thoroughbred is, on average, an incomparably better racehorse than the female, and I make no apology for returning to the subject, as I have no doubt whatsoever that wrapped up in this circumstance there are considerations of the greatest importance to the breeder.

For the benefit of those who are not familiar with English racing conditions, I will record that there are three Classic races and three other events of approximately corresponding importance open to horses of both sexes contested each year. These are:

1 The 2,000 Guineas over one mile at Newmarket. Run at the end of April or early in May;
2 The Derby over 1½ miles at Epsom. Run at the end of May or early in June;
3 The St. Leger over 1¾ miles 127 yards at Doncaster. Run early in September;
4 The Eclipse Stakes over 1¼ miles at Sandown. Run at the beginning of July;
5 The King George VI and Queen Elizabeth Stakes over 1½ miles at Ascot. Run in the second half of July;
6 The Champion Stakes over 1¼ miles at Newmarket. Run in the middle of October.

The first three are termed Classics—the other two Classics, the 1,000 Guineas (one mile at Newmarket) and the Oaks (1½ miles at Epsom), are for fillies only. All the Classics are confined to three-year-olds each carrying nine stone, except in the 2,000 Guineas and Derby where fillies shoulder 8st 9lb, and the St. Leger, where fillies receive 3lb from the colts. The Eclipse Stakes, King George VI and Queen Elizabeth Stakes and Champion Stakes are open to horses of three years old and upwards. In the first two, three-year-olds carry 8st 8lb and older horses 9st 7lb, and in each case fillies are allowed 3lb. In the Champion Stakes three-year-olds carry 8st 10lb and older horses 9st 3lb, and in each case fillies are allowed 3lb.

Both the Eclipse Stakes and King George VI and Queen Elizabeth Stakes attracted sponsorship from commercial organizations in the 1970's, and their titles have been slightly amended to accommodate the outside support, so they are now referred to as the Coral Eclipse Stakes and King George VI and Queen Elizabeth Diamond Stakes. For the sake of simplicity, however, I have retained the old names in the text.

Another race of comparable value and importance is also sponsored, the Benson & Hedges Gold Cup, run over 1¼ miles at York in August, but it cannot yet be added to this list because of its short history. Inaugurated in 1972, it quickly developed notoriety for its shock results, which included the only defeat for Brigadier Gerard in 18 races. It will, however, remain a most significant race as long as rich prize money is offered, and the provisional figure for 1978, £70,000 added to stakes, was the same as that for the St. Leger and more than for the Champion Stakes (£60,000), Eclipse Stakes (£45,000) and Coronation Cup at Epsom (£40,000).

It was inexcusable not so long ago to exclude the Ascot Gold Cup, over 2½ miles, from this list of important races, but times have changed, and the Gold Cup no longer carries the same weight of prestige as it did before the Second World War. Breeders appear to fight shy of Ascot Gold Cup winners as stallions (as I shall explain in the next chapter), and it is of further significance that only two fillies have won the race this century—Quashed (in 1936) and Gladness (in 1958).

The Ascot Gold Cup has thus been eliminated from the list and replaced by the Eclipse Stakes and Champion Stakes, which due to increased prize money and the shift of emphasis to middle-distance races, play a more important role in the modern pattern of racing. Whether the decline of the Ascot Gold Cup is a benefit to thoroughbred breeding must be doubted, but it has also to be accepted as a matter of fact.

These races confer on the owner, trainer, breeder, jockey and parents of the winner great kudos, whilst the winner himself (or herself) gains a stud value of special significance.

Naturally all connected with the Turf leave no stone unturned to succeed in one or more of them. Moreover they are run at level weights or weight for age, with no handicaps or allowances, except that fillies always have a weight advantage. They therefore afford a useful comparison of merit between the two sexes and I set out details of results during the past 100 years.

The most interesting figures as they affect fillies are those for the Eclipse Stakes and Champion Stakes, races open to three-year-olds and upwards and run over the same distance but the first in midsummer and the second at the end of the season. Considering that fillies are generally reckoned to train off towards the end of the season, it is interesting to note that they have not yet won the Eclipse Stakes but provided 16 Champion Stakes winners (including a dead-heater) in the years 1876 to 1976, eight coming since 1957. If anything, these figures prove that not every filly trains off towards the end of the season!

Readers may have noticed that I usually pay considerable importance to results compared to opportunity, and to provide a more balanced picture in this instance

SUCCESSFUL FILLIES

2,000 GUINEAS		DERBY	
1876	Pilgrimage	1882	Shotover
1882	Shotover	1908	Signorinetta
1902	Sceptre	1913	Tagalie
1944	Garden Path	1916	Fifinella
	Total = 4		Total = 4

ST. LEGER

1878	Janette	1919	Keysoe
1882	Dutch Oven	1923	Tranquil
1888	Seabreeze	1927	Book Law
1890	Memoir	1942	Sun Chariot
1892	La Fleche	1943	Herringbone
1894	Throstle	1955	Meld
1902	Sceptre	1959	Cantelo
1904	Pretty Polly		Total = 15

ECLIPSE STAKES	KING GEORGE VI & QUEEN ELIZABETH STAKES		CHAMPION STAKES	
NIL	1966	Aunt Edith	1878	Janette
	1969	Park Top	1894	La Fleche
	1973	Dahlia	1903	Sceptre
	1974	Dahlia	1905	Pretty Polly
	1976	Pawneese	1918	My Dear
		Total = 5	1933	Chatelaine (dead heat)
			1938	Rockfel
			1944	Hycilla
			1957	Rose Royale II
			1958	Bella Paola
			1959	Petite Etoile
			1960	Marguerite Vernaut
			1963	Hula Dancer
			1969	Flossy
			1973	Hurry Harriet
			1975	Rose Bowl
				Total = 15 and one dead heat

	SUCCESSFUL COLTS	SUCCESSFUL FILLIES
2,000 Guineas	96	4
Derby	95	4
St. Leger	84	15
Eclipse Stakes	80	0
King George VI & Queen Elizabeth Stakes	20	5
Champion Stakes	84	16
Total	459 (91%)	44 (9%)

NB. The St. Leger was not run in 1939. The Eclipse Stakes was inaugurated in 1866, but was not run in 1887, 1890, 1915–1918 and 1940–1945, and resulted in a dead heat in 1910. The Champion Stakes was not run in 1939, and resulted in dead heats in 1884 and 1933.

I detail below the opportunities which fillies have been given to run in these races in the 30 years 1947 to 1976 inclusive. The races are tabulated in chronological order through the year.

EARLIER in YEAR ↓

LATER in YEAR

RACE	TOTAL RUNNERS	NUMBER OF FILLIES	WINNING FILLIES	FILLIES IN 1ST 3
2,000 Guineas	534	2 (0.4%)	0	1
Derby	691	1 (0.1%)	0	1
Eclipse Stakes	230	9 (3.9%)	0	2
King George VI & Queen Elizabeth Stakes	242	22 (9.1%)	5	11
St. Leger	350	21 (6.0%)	2	7
Champion Stakes	251	39 (15.5%)	8	19
Total	2,298	94 (4.1%)	15	41

A number of impressions can be gained from these figures, not least that considering the very small percentage of fillies (4.1 per cent of the total runners) which has contested these races, they have a highly creditable record. Almost 44 per cent of the fillies which have run, have gained a place in the first three, and 16 per cent have been successful.

It is also obvious that the earlier in the year the less chance there is of fillies taking on the colts, despite the fact that of three fillies which have run in either the 2,000 Guineas or Derby, two finished second. The other, incidentally, finished last of 17 in the 2,000 Guineas.

This latter seasonal distribution is explained away by reference to a clash of interests in the first half of the season. There are comparable races for fillies only—the 1,000 Guineas and Oaks for instance—which are not matched for prestige or prize money in the second half. Yet this argument itself draws me into the belief that the incomparably better results achieved by the colts speak for themselves, despite the apparent lack of opportunity which fillies have had, for I am convinced that more fillies do not contend these races because their connections do not consider they are up to the required standard. The rewards for success against the colts in the six races set out above are too glittering to be cast aside, even if there is only an outside chance of victory.

Our trainers of top-grade horses are men of the highest competence, and it would be an impertinence to suggest that they neglect to run fillies who have a winning chance in these races. Very occasionally, for one reason or another, a very high class one is not entered, but then some of the fillies in the above list owe their successes to the absence of greatly superior colts of their generations. The incidence of opportunity therefore is not synonymous with the number of starters of each sex for these races, but is dependent on the judgement of widely experienced trainers as to which animals are up to the necessary standard to compete. Research into Turf history shows how extremely accurately the judgement has been exercised down the years, and I have every confidence in recording that, taking it fine and large, the results of these six races afford a fair comparison between the abilities of male and female thoroughbreds.

Similarly to every other chain the strength of an ancestral chain is dependent on its weakest links. As I have explained at some length, the weakest links in the thoroughbred's make-up lie in his various female ancestors. An examination of the pedigree of the average horse (or mare) shows that usually his sire is a great deal better racer than his dam, his maternal grandsire a better performer than his grandam, and so on all through the pedigree. In order to take matters a step further it is necessary to consider the ancestry of these various sires, i.e. the sire's ancestry, the dam's sire's ancestry, the grandam's sire's ancestry, etc.

Assuming that the animal under consideration is reasonably well bred it can be said that these sires were racers or stallions of merit. Therefore, in a rough and ready way, it can be inferred that their progenitors, both male and female, have proved their worth by being contributing factors in the production of these various sires. Whatever their shortcomings may have been either on the racecourse or at stud they have at least vindicated themselves by their part in the build-up of at least one horse of standing, i.e. the sires. Broadly speaking this covers all the various sires and their ancestors in the pedigree but leaves untouched the tail female line, i.e. the dam, grandam, etc., of the animal under investigation. If they fail to come up to approximately the same standard as the other parental stock in the pedigree they will constitute the weak links in the whole structure. Here, I think, lies the key to the generally accepted principle, amongst breeders, that the tail female line is of the utmost importance.

Every tail female line in the Stud Book can claim winners of one sort or another but the abilities of these are widely separated. Some lines have a fine record for producing Classic class stock, whilst others throw up in profusion winners of small handicaps, generation after generation, without bearing hardly a single high-grade performer. Then there are the lines which specialize in jumpers. These are largely located in Ireland and it is through them that the Irish 'chaser has won his great reputation.

There are a considerable number of families who have produced horses of the highest grade for two or three generations and who have then faded from the scene. In addition there are a few lines who have continued their successes for periods of 50 years or more. The mare Chelandry (1894), whose descendants include the 1944 and 1954 Derby winners Ocean Swell and Never Say Die, and the 1962 St. Leger winner Provoke, traces in tail female to Prunella (1778), whose son Pope won the 1809 Derby and whose daughter Pelisse won the 1804 Oaks. It is not suggested that Provoke was directly influenced by the merit of Prunella, but the example does show how the chain of success in the top races can extend over many decades.

I do not propose to attempt to make a complete list of these great long-term Classic and near-Classic winner-producing families, but below will be found a number of examples, with some of their more important descendants:

A   CHELANDRY (1894), by Goldfinch. 1,000 Guineas winner
FAMILY NUMBER: 1

Neil Gow (1907) 2,000 Guineas, Eclipse. Sire.

Wrack (1909) 10 Flat and 6 hurdle races at home. Sire in U.S.A.

Magpie (1914) 2nd in 2,000 Guineas, winner in Australia of Caulfield and Melbourne Cups. Champion sire.

Pogrom (1919) Oaks etc.

Heroic (1921) Won £38,062 in Australia and champion sire seven times.

Saucy Sue (1922) 1,000 Guineas, Oaks.

Book Law (1924) St. Leger.

Pay Up (1933) 2,000 Guineas.

Rhodes Scholar (1933) Eclipse Stakes.

Galatea II (1936) 1,000 Guineas, Oaks.

Ocean Swell (1941) Derby, Ascot Gold Cup. Sire.

Iona (1943) 2nd in Oaks, 3rd in 1,000 Guineas.

All Aboard (1947) Dam of stakes winners Copenhagen, Temptress, Captain Kidd, Smuggler's Joy, The Bo'sun.

Never Say Die (1951) Derby, St. Leger. Sire.

Tomy Lee (1956) Kentucky Derby.

Tudor Melody (1956) Leading two-year-old of his year. Sire.

Trelawny (1956) Smart stayer.

Cynara (1958) Queen Mary Stakes. Dam of Stintino, Ormindo.

Fighting Ship (1960) 3rd in St. Leger

Provoke (1962) St. Leger.

Ardent Dancer (1962) Irish 1,000 Guineas.

Rarity (1967) 2nd in Champion Stakes.

B   CONJURE (1895), by Juggler
FAMILY NUMBER: 1

Winkipop (1907) 1,000 Guineas.

Blink (1915) 2nd in Derby.

Short Story (1923) Oaks.

Pennycomequick (1926) Oaks.

Sunny Devon (1928) Coronation Stakes and £7,600.

Volume (1928) Park Hill Stakes.

Mannamead (1929) Unbeaten.

Pink Flower (1940) 2nd in 2,000 Guineas.

Pensive (1941) Kentucky Derby, Preakness Stakes.

High Stakes (1942) £21,702.

Court Martial (1942) 2,000 Guineas.

Ambiguity (1950) Oaks, etc.

Hornbeam (1953) 2nd in St. Leger.

Almiranta (1959) Park Hill Stakes.

Craighouse (1962) Irish St. Leger.

Sodium (1963) St. Leger, Irish Sweeps Derby.

Paveh (1963) Irish 2,000 Guineas.

Mount Hagen (1971) Sussex Stakes.

Lady Singer (1973) 3rd in Irish 1,000 Guineas.

C   MISS GUNNING (1897), by Carbine
FAMILY NUMBER: 3

Silonyx (1916) Ascot Gold Vase.

Uganda (1921) French Oaks and St. Leger.

Ukrania (1926) French Oaks.

Ut Majeur (1927) Cesarewitch under 8st 3lb, record for a 3-y-o.

Udaipur (1929) Oaks.

Umidwar (1931) Jockey Club Stakes, Champion Stakes.

Airway (1937) Sire.

Umiddad (1940) Wartime substitute Gold

Cup (Newmarket).

Claro (1943) Irish 2,000 Guineas. Sire in Argentina.

Hindostan (1946) Irish Derby.

Palestine (1947) 2,000 Guineas and £38,515. Sire.

Neemah (1950) Royal Lodge Stakes.

Fiorentina (1956) Irish 1,000 Guineas.

Prominent (1967) Smart handicapper.

D   SILVER FOWL (1904), by Wildfowler
FAMILY NUMBER: 3

Silver Tag (1912) 2nd in 1,000 Guineas.
Fifinella (1913) Derby, Oaks.
Silvern (1917) 2nd in St. Leger. Won £6,277.
Shrove (1920) 2nd in Oaks, 3rd in 1,000 Guineas.
Press Gang (1927) £10,580.
Shred (1929) 2nd in French Derby.

Tai Yang (1930) Unbeaten.
Pasch (1935) 2,000 Guineas, Eclipse.
Chateau Larose (1938) 2nd in St. Leger.
Pinza (1950) Derby and £47,401.
Wallaby (1955) French St. Leger, Ascot Gold Cup.
Le Bavard (1971) French Gold Cup.

E   GONDOLETTE (1902), by Loved One
FAMILY NUMBER: 6

Ferry (1915) 1,000 Guineas.
Selene (1919) £14,386.
Tranquil (1920) 1,000 Guineas, St. Leger.
Sansovino (1921) Derby and £17,732.
Sickle (1924) 3rd in 2,000 Guineas. Champion sire in U.S.A.
Bosworth (1926) Ascot Gold Cup. 2nd in St. Leger.
Hunter's Moon (1926) Leading sire in Argentina.
Hyperion (1930) Derby, St. Leger, and £29,509. Leading sire 6 times.
Myrobella (1930) 3rd in 1,000 Guineas, and won £16,143.

Big Game (1939) 2,000 Guineas. Leading sire.
Chamossaire (1942) St. Leger. Leading sire.
Mossborough (1947) Leading sire.
Thunderhead II (1949) 2,000 Guineas.
Raise You Ten (1960) Ascot Gold Cup.
Double Jump (1962) Leading two-year-old of his year.
Hopeful Venture (1964) Hardwicke Stakes, Grand Prix de Saint-Cloud. 2nd in St. Leger.
Ardale (1968) Italian Derby.
Snow Knight (1971) Derby.

F   AMERICUS GIRL (1905), by Americus
FAMILY NUMBER: 9

Lady Juror (1919) Jockey Club Stakes and £8,057.
Mumtaz Mahal (1921) £13,933 and 2nd in 1,000 Guineas. One of the fastest fillies of all time.
Fair Trial (1932) £5,100. Leading sire.
Mahmoud (1933) Derby in record time and £14,426. Leading sire in U.S.A.
Nasrullah (1940) 3rd in Derby. Won £3,348. Leading sire in U.S.A. and Britain.
Tudor Minstrel (1944) 2,000 Guineas and £24,629. Sire.
Combat (1944) Unbeaten in 9 races.
Migoli (1944) Eclipse, Arc de Triomphe. 2nd in Derby, 3rd in St. Leger.
Abernant (1946) Leading sprinter, won

£26,394. Sire.
Diableretta (1947) £14,492.
Palariva (1953) Smart sprinter.
Ommeyad (1954) Irish St. Leger.
Prince Taj (1954) 3rd in French 2,000 Guineas. Sire.
Scot (1954) French St. Leger.
Ginetta (1956) Irish 1,000 Guineas.
Petite Etoile (1956) Oaks, 1,000 Guineas, Champion Stakes, Coronation Cup (twice) and £67,786.
Kashmir II (1963) 2,000 Guineas.
Celina (1965) Irish Oaks.
Erimo Hawk (1968) Ascot Gold Cup.
Kalamoun (1970) French 2,000 Guineas.
Habat (1971) Won 5 races and £49,636.

G   DAME MASHAM (1889), by Galliard
FAMILY NUMBER: 9

St. Astra (1904) French Oaks.

Fair Play (1905) Leading racehorse and sire in U.S.A.

Friar Rock (1913) Belmont Stakes. Leading sire in U.S.A.

Asterus (1923) French 2,000 Guineas. Leading sire in France.

Bateau (1925) Top-class filly in U.S.A.

Corrida (1932) Grand Prix de Saint-Cloud, Arc de Triomphe (twice).

Marsyas II (1940) French Gold Cup (twice).

Pensbury (1940) Grand Prix de Paris.

Caracalla (1942) Unbeaten. Grand Prix de Paris, French St. Leger, Ascot Gold Cup, Arc de Triomphe.

Arbar (1944) Ascot Gold Cup.

Galgala (1946) French 1,000 Guineas.

Galcador (1947) Derby.

Asmena (1947) Oaks.

Philius II (1953) French Derby.

Talgo (1953) Irish Derby.

Fidalgo (1956) Irish Derby.

Ancasta (1961) Irish Oaks.

Crepellana (1966) French Oaks.

Northern Treasure (1973) Irish 2,000 Guineas.

N.B.—Dame Masham was bred in England, but it is curious that her line has achieved greater success in France and America than at home, though there is a high degree of success in Ireland in more recent years. Her success abroad appears to be largely due to the export of her daughter Fairy Gold (1896), by Bend Or, to America, from where some of her descendants went to France.

H   FRIZETTE (1905), by Hamburg. Bred in U.S.A. and exported to France.
FAMILY NUMBER: 13

Banshee (1910) French 1,000 Guineas.

Durban (1918) Foundation mare of M. Boussac's stud.

Tourbillon (1928) French Derby. Leading sire in France.

Cillas (1935) French Derby.

Caravelle (1940) French 1,000 Guineas and Oaks.

Jet Pilot (1944) Kentucky Derby. Leading sire in U.S.A.

Djeddah (1945) Eclipse Stakes, Champion Stakes.

Black Tarquin (1945) St. Leger.

Corejada (1947) French 1,000 Guineas, Irish Oaks.

Djelfa (1948) French 1,000 Guineas.

Auriban (1949) French Derby.

Cote d'Or II (1951) French 2,000 Guineas.

Apollonia (1953) French 1,000 Guineas and Oaks.

Sing Sing (1957) Leading sprinter.

Baldric II (1961) 2,000 Guineas, Champion Stakes.

Roan Rocket (1961) Sussex Stakes.

Burglar (1966) Smart sprinter.

Dahlia (1970) Irish Oaks, King George VI and Queen Elizabeth Stakes (twice), Washington International and European record of £488,545 (win only).

I   ADMIRATION (1892), by Saraband
FAMILY NUMBER: 14

Pretty Polly (1901) 1,000 Guineas, Oaks, St. Leger and £33,297.

Craganour (1910) £10,990. Won Derby but disqualified. Leading sire in Argentina.

King John (1915) Irish Derby.

Spike Island (1919) Irish 2,000 Guineas.

Zodiac (1921) Irish Derby and St. Leger.

Cresta Run (1924) 1,000 Guineas.

Kopi (1926) Irish Derby.

Foxbridge (1930) £2,520 in England. Leading sire 11 times in New Zealand.

Cappiello (1930) Grand Prix de Paris.

Donatello II (1934) Italian Derby. Leading sire.

Tehran (1941) St. Leger. Leading sire.

Daumier (1948) Italian Derby and St. Leger.

Supreme Court (1948) £36,949.

Guersant (1949) French 2,000 Guineas.

I ADMIRATION (*contd.*)

Premonition (1950) St. Leger.
Nearctic (1954) Stakes winner in U.S.A. Sire.
Arctic Explorer (1954) Eclipse Stakes.
Marguerite Vernaut (1957) Champion Stakes.
St. Paddy (1957) Derby, St. Leger, Eclipse Stakes.
Psidium (1958) Derby.

Arctic Vale (1959) Irish St. Leger.
Only for Life (1960) 2,000 Guineas.
Dolina (1964) Italian 1,000 Guineas and Oaks.
Mistigo (1965) Irish 2,000 Guineas.
Brigadier Gerard (1968) 2,000 Guineas, Champion Stakes and £243,924.
Hard to Beat (1969) French Derby.
Flying Water (1973) 1,000 Guineas.

N.B.—This is one of the most prolific families of the 20th Century, and its other important winners include High Treason, Gratitude, Court Harwell, Even Money, Welsh Abbot, Carnoustie, Vienna, Entanglement, Pardao, Whistling Wind, Great Nephew, Felicio II, Lucyrowe, Huntercombe, Parmelia, Magic Flute and Patch.

There is evidence to suggest that the lumping together of the various branches into one family is now too general a view. For example, there are five generations between Pretty Polly and Brigadier Gerard, and using Galton's Law, by which the figure 4,096 represents the progeny, the contribution of Brigadier Gerard's parents would be 1,024 each, while Admiration in the sixth generation would contribute 1. Even the naming of Admiration as the ultimate matriarch can now be regarded as too distant, and future generations may be distinguished further, particularly in regard to Pretty Polly's daughters Molly Desmond (from whom Brigadier Gerard and St. Paddy are descended), Dutch Mary (Donatello II and Psidium), Baby Polly (Vienna) and Polly Flinders (Supreme Court, Only for Life).

J SCEPTRE (1899), by Persimmon. Winner of 2,000 Guineas, 1,000 Guineas, Oaks, St. Leger and £38,225.
FAMILY NUMBER: 16

Sunny Jane (1914) Oaks. 2nd in 1,000 Guineas.
Buchan (1916) Eclipse Stakes and £16,658. Leading sire.
Craig an Eran (1918) 2,000 Guineas, Eclipse Stakes.
St. Germans (1921) £7,905. Leading sire in U.S.A.
Tiberius (1931) Ascot Gold Cup. Leading National Hunt sire.
Flyon (1936) Ascot Gold Cup.
Petition (1944) Eclipse Stakes and £18,028.
Noor (1945) 356,940 dollars in U.S.A. 3rd in Derby.

Zucchero (1948) £14,837.
Northern Light (1950) Grand Prix de Paris.
Counsel (1952) Sire.
Taboun (1956) 2,000 Guineas.
Match III (1958) French St. Leger, King George VI & Queen Elizabeth Stakes, Grand Prix de Saint-Cloud.
Relko (1960) Derby, French 2,000 Guineas and St. Leger, Coronation Cup.
Reliance II (1962) French Derby and St. Leger, Grand Prix de Paris.
Full Dress II (1966) 1,000 Guineas.
Weimar (1968) Italian St. Leger and 12 other races from 16 runs.

N.B.—This family supplied four Derby seconds between 1919 and 1924—Buchan, Craig an Eran, Tamar and St. Germans.

I have tabulated the above mares in sequence of family numbers and not in order of merit. I have confined this list to mares who have been at stud in the 20th Century and whose descendants are achieving important results today.

There are, however, other families that have emerged through more recent success, and though they cannot perhaps yet lay claim to the long-standing importance of the ten mares named above, they may eventually prove of equal worth. Representatives of these are:

 LA TROIENNE (1926), by Teddy out of Helen de Troie
FAMILY NUMBER: 1

Black Helen (1932) CC American Oaks, American Derby.

Bimelech (1937) Preakness Stakes, Belmont Stakes.

Busher (1942) Champion filly in U.S.A. 13 races, including Hollywood Derby, and 334,035 dollars.

Hall of Fame (1948) American Derby and 8 other races, and 234,430 dollars.

Jet Action (1951) 308,225 dollars.

Searching (1952) 25 races, and 327,381 dollars.

Cohoes (1954) 13 races and 210,850 dollars. Sire.

Harmonizing (1954) Man O'War Stakes.

Immortality (1956) Unraced. Sire.

Hitting Away (1958) 309,079 dollars.

Royal Record II (1958) Sire.

The Axe II (1958) Imperial Stakes in England, and 13 races and 393,391 dollars in U.S.A.

Affectionately (1960) 28 races and 546,660 dollars.

Malicious (1961) 14 races and 317,237 dollars.

Buckpasser (1963) 25 races and 1.4 million dollars.

Priceless Gem (1963) Frizette Stakes and 6 other races in U.S.A.

Personality (1967) Preakness Stakes, Horse of the Year in U.S.A.

Boucher (1969) St. Leger.

Allez France (1970) French 1,000 Guineas and Oaks, Arc de Triomphe, and £463,557.

MARCHETTA (1907), by Marco out of Hettie Sorrel
FAMILY NUMBER: 1

from Rose Red (1924)
(Swynford x Marchetta)

Borealis (1941) Coronation Cup. 2nd in St. Leger.

Alycidon (1945) Ascot Gold Cup.

Festoon (1951) 1,000 Guineas.

Acropolis (1952) 3rd in Derby, 2nd in King George VI & Queen Elizabeth Stakes.

Even Star (1954) Irish 1,000 Guineas.

Celtic Ash (1957) Belmont Stakes.

Larkspur (1959) Derby.

Altesse Royale (1968) 1,000 Guineas, Oaks.

Bustino (1971) St. Leger, 2nd in King George VI & Queen Elizabeth Stakes.

Imperial Prince (1971) 2nd in Derby and Irish Sweeps Derby.

from Sweet Lavender (1923)
(Swynford x Marchetta)

Sayani (1943) Cambridgeshire as 3-y-o under 9st 4lb.

My Babu (1945) 2,000 Guineas, Sussex Stakes.

Ambiorix (1946) Leading 2-y-o in France. Leading sire in U.S.A.

Cagire II (1947) King George VI Stakes.

Turn-To (1951) 6 out of 8 races and 280,032 dollars in U.S.A. Sire.

Klairon (1952) French 2,000 Guineas.

Die Hard (1957) Ebor Handicap. 2nd in St. Leger.

Pushful (1961) Timeform Gold Cup.

Atilla (1961) Vaux Gold Tankard.

Haltilala (1963) Prix Vermeille.

Dan Kano (1964) Irish St. Leger.

English Prince (1971) Irish Sweeps Derby.

N.B.—The branch from Sweet Lavender is now further distinguished by its branch from her granddaughter Perfume II (1938), by Badruddin out of Lavendula II, by Pharos.

ALOE (1926), by Son-in-Law out of Alope.
FAMILY NUMBER: 2

Hyperion (1943) 1,000 Guineas.

Sideral (1948) 9 races. Leading sire and broodmare sire three times each in Argentina.

Aureole (1950) King George VI & Queen Elizabeth Stakes, Hardwicke Stakes. 2nd in Derby, 3rd in St. Leger.

Doutelle (1954) Ormonde Stakes, John Porter Stakes.

Round Table (1954) American Derby. Leading sire in U.S.A.

Alcide (1955) St. Leger, King George VI & Queen Elizabeth Stakes.

Above Suspicion (1956) St. James's Palace Stakes.

Parthia (1956) Derby.

Atan (1961) Sire.

Ben Marshall (1961) Italian St. Leger.

Ravageur (1962) 6 steeplechases and over £50,000 in France.

Cadmus (1963) Prix d'Harcourt and 4 other races in France.

Tammuz (1968) Schweppes Gold Trophy (hurdle).

Lassalle (1969) Ascot Gold Cup, French Gold Cup.

Highclere (1971) 1,000 Guineas, French Oaks.

Import (1971) Wokingham Stakes, Goodwood Stewards' Cup.

SIMON'S SHOES (1914), by Simon Square out of Goody Two-Shoes
FAMILY NUMBER: 5

from Carpet Slipper (1930) (Phalaris x Simon's Shoes)

Godiva (1937) 1,000 Guineas, Oaks.

Windsor Slipper (1939) Irish Triple Crown.

Silken Glider (1954) Irish Oaks, 2nd in Oaks.

Time Greine (1958) 3rd in 2,000 Guineas.

Val de Loir (1959) French Derby. Leading sire in France.

Alciglide (1963) Queen Alexandra Stakes,
Prix Gladiateur.

Valoris (1963) Oaks, Irish Oaks.

Reform (1964) Champion Stakes, Sussex Stakes.

Allangrange (1967) Irish St. Leger.

Vincennes (1968) 2nd in Irish Oaks.

Roi Lear (1970) French Derby.

Val's Girl (1972) 2nd in Oaks.

from Dalmary (1931)
(Blandford x Simon's Shoes)

Le Sage (1948) Sussex Stakes.

Tudor Era (1953) 28 races, including Washington International, and 477,386 dollars in U.S.A.

Sarcelle (1954) Leading 2-y-o.

Barclay (1956) Irish St. Leger.

Light Year (1958) Irish 2,000 Guineas.

Christmas Island (1960) Irish St. Leger.

Sammy Davis (1960) Leading sprinter.

Super Sam (1962) News of the World Handicap and 6 other races.

Lorenzaccio (1965) Champion Stakes.

Marquis de Sade (1973) King Edward VI Stakes.

from Rough Shod (1944)
(Gold Bridge x Dalmary)

Ridan (1959) 13 races and 635,074 dollars. 2nd in Preakness Stakes.

Lt Stevens (1961) 9 races and 240,949 dollars.

Moccasin (1963) Champion filly in U.S.A. 11 races and 388,075 dollars.

Gamely (1964) 16 races and 574,961 dollars.

Drumtop (1966) 17 races and 493,738 dollars.

Thatch (1970) Sussex Stakes.

Cellini (1971) Dewhurst Stakes.

Apalachee (1971) Observer Gold Cup.

King Pellinore (1972) 2nd in St. Leger and Irish Sweeps Derby.

Take Your Place (1973) Observer Gold Cup.

ORLASS (1914), by Orby out of Simon Lass
FAMILY NUMBER: 21

Netherton Maid (1944) Princess Elizabeth Stakes. 2nd in Oaks.
Neasham Belle (1948) Oaks.
Gay Time (1949) 2nd in Derby.
Elopement (1951) 2nd in St. Leger.
Narrator (1951) Champion Stakes, Coronation Cup.
Bride Elect (1952) Queen Mary Stakes.
Nucleus (1952) King Edward VII Stakes. 2nd in St. Leger.
None Nicer (1955) Ribblesdale Stakes.
Sanctum (1956) Victoria Cup.
Hethersett (1959) St. Leger.
Cursorial (1961) Park Hill Stakes.
Derring-Do (1961) Queen Elizabeth II Stakes.
Bamboozle (1964) Princess Royal Stakes.
Dart Board (1964) 3rd in Derby.
Flossy (1966) Champion Stakes.

Saraca (1966) Prix Vermeille.
Humble Duty (1967) 1,000 Guineas.
Droll Role (1968) Washington International.
Coup de Feu (1969) Eclipse Stakes.
Jacinth (1970) Cheveley Park Stakes, Coronation Stakes.
Peleid (1970) St. Leger.
Simbir (1970) Criterium de Saint-Cloud.
Tudenham (1970) Middle Park Stakes.
Dakota (1971) Ebor Handicap.
Dumka (1971) French 1,000 Guineas.
Northern Gem (1971) Princess Royal Stakes, 2nd in Irish 1,000 Guineas and Champion Stakes. Sold for 108,000 gns December 1974.
Rubric (1972) 2nd in Spanish Derby.
Hittite Glory (1973) Middle Park Stakes.

Two families whose top-class representatives are not so great in numbers but which nevertheless are influential are those of Democratie and Schiaparelli. The exploits of Democratie are interesting because this was an undistinguished family until the emergence of her daughter Fair Freedom (1945), by Fair Trial, and granddaughter Zanzara (1951), by Fairey Fulmar out of Sunright, by Solario.

DEMOCRATIE (1927), by Epinard out of Queenly
FAMILY NUMBER: 6

from Fair Freedom

Marshal Ney (1951) Jersey Stakes.
Liberal Lady (1955) Lowther Stakes.
Be Careful (1956) Gimcrack Stakes, Champagne Stakes.
Spree (1960) 2nd in 1,000 Guineas and Oaks.
Juliette Marny (1972) Oaks.

from Zanzara

Matatina (1960) Leading sprinter.
Showdown (1961) Coventry Stakes, Middle Park Stakes. Leading sire in Australia.
Enrico (1962) Victoria Cup.
New Chapter (1966) Lincoln Handicap.
Farfalla (1967) Queen Mary Stakes.

SCHIAPARELLI (1935), by Schiavoni out of Aileen
FAMILY NUMBER: 8

Herringbone (1940) St. Leger, 1,000 Guineas.
Swallow Tail (1946) 3rd in Derby.
Entente Cordiale (1951) Doncaster Cup.
Shantung (1956) 2nd in Derby. Sire.
Lynchris (1957) Irish Oaks and St. Leger.
Roi Dagobert (1964) Leading 3-y-o in France.
Djakao (1966) Grand Prix de Deauville. 3rd in French Derby.

Sassafras (1967) French Derby and St. Leger, Arc de Triomphe.
Irish Ball (1968) Irish Sweeps Derby.
Coral Beach (1969) Cheshire Oaks.
Paysanne (1969) Prix Vermeille.
Ragstone (1970) Ascot Gold Cup.
Cloonlara (1974) Leading 2-y-o in Ireland.
Conglomerat (1974) Criterium de Saint-Cloud.

LASTING DOMINANCE OF A FEW MARES

In the early part of this century there were nearly 6,000 thoroughbred mares at stud, and it is most remarkable that a small band of about a couple of dozen have secured greatly superior lasting results to the other 5,976. I think this is primarily due to their overall super excellence as matrons, but also the number of fillies produced in each family is relative. However successful a mare may be as a dam of winners, the fewer fillies she produces the smaller will be her chance of founding a great family.

A case in point is Nogara (1928), the dam of unbeaten Nearco, the Italian Derby and 2,000 Guineas winner Niccolo dell'Arca and other winners of high class, but who had only one daughter during her lengthy stud career. This filly, Nervesa (1941, by Ortello), won three races including the Italian Oaks and finished second in the Italian 1,000 Guineas and fourth in the 2,000. She had daughters at stud before her death in 1957, but her tail female line hangs by a thread. An early success came through her daughter Natalina da Murano (1949, by Orsenigo), whose son Pinturischio was ante-post favourite for the 1961 Derby but missed the race in mysterious circumstances. The most significant contribution by Nervesa has been through her granddaughter Ranavalo III (by Relic), the dam of stallion Fortino II (1959), whose most successful produce before his export to Japan included Fortissimo, Fine Blade, No Mercy, Knockroe, Pidget and Flintham in Britain, and Caro in France.

On the other hand, Montem (1901) produced no less than 12 fillies, and through these became a notable tap root mare. Her chief successes are detailed below:

MONTEM (1901), by Ladas out of Kermesse
FAMILY NUMBER: 11

Beta (1918) Royal Lodge Stakes.
Ellangowan (1920) 2,000 Guineas, Champion Stakes.
Charley's Mount (1921) Park Hill Stakes.
Plack (1921) 1,000 Guineas.
Colombo (1931) 2,000 Guineas.
Patriot King (1931) Irish Derby.
Afterthought (1939) 2nd in Oaks.
Suntop (1940) Irish 1,000 Guineas and Oaks.

Midas (1942) 2nd in Derby.
Sweet Solera (1958) 1,000 Guineas and Oaks.
Aunt Edith (1962) King George VI & Queen Elizabeth Stakes.
Our Mirage (1969) Jockey Club Stakes. 2nd in English and Irish St. Legers.

However, time appears to be catching up with the Montem branch, and only three of the winners named above have been recorded since 1942—the fillies Sweet Solera and Aunt Edith, neither of which has been a notable success at stud, and the colt Our Mirage, who was just below top class.

Whether the Montem branch can be revived remains to be seen, but for the moment she has been eclipsed by another member of the No. 11 family, Felucca (1941), by Nearco out of Felsetta. Felucca also has a good record for throwing fillies, and the notable winners from her family are as follows:

FELUCCA (1941), by Nearco out of Felsetta.
FAMILY NUMBER: 11

Ark Royal (1952) Ribblesdale Stakes, York-shire Oaks, Park Hill Stakes.

Kyak (1953) Park Hill Stakes.

Cutter (1955) Park Hill Stakes, Yorkshire Cup.

Ocean (1961) Coronation Stakes.

Hermes (1963) Great Voltigeur Stakes.

Sloop (1964) Craven Stakes.

Mariner (1964) King Edward VII Stakes.

Torpid (1965) John Porter Stakes, 2nd in French St. Leger and Ascot Gold Cup.

Anchor (1966) Nell Gwyn Stakes.

Furibondo (1966) 10 races in Italy.

Raft (1966) Princess Elizabeth Stakes.

Fluke (1967) Jersey Stakes.

Buoy (1970) Coronation Cup, Yorkshire Cup, 2nd in St. Leger, 3rd in Irish Derby.

Sharp Edge (1970) Irish 2,000 Guineas. 3rd in 2,000 Guineas.

Sea Anchor (1972) Doncaster Cup.

Sky Ship (1974) July Stakes.

Though all the above are winners of high-class races, not all were noted for their reliability, and several did best when equipped with blinkers. It is of further interest to note the amount of overseas' interest in the colts as stallions, with Hermes being exported to New Zealand, Sloop to Argentina, Mariner to Australia and Torpid to Poland.

Montem and Felucca had an advantage in the number of fillies they foaled, but there are some comparatively small families which have met with wonderful success in spite of their paucity of numbers, such as Stella (1890) by Necromancer; and Snoot (1906) by Perigord.

STELLA (1890), by Necromancer out of Holly Leaf
FAMILY NUMBER: 22

Blakestown (1902) Irish Oaks.

Athgreany (1910) Irish Oaks.

Flying Orb (1911) Sire.

Trigo (1926) Derby, St. Leger, Irish St. Leger.

Mate (1928) Preakness Stakes.

Loaningdale (1929) Eclipse Stakes.

Harinero (1930) Irish Derby and St. Leger.

Primero (1931) Irish Derby and St. Leger.

Sun Castle (1938) St. Leger.

Ciel Etoile (1946) French St. Leger.

Tulyar (1949) Derby, St. Leger, King George VI & Queen Elizabeth Stakes.

Le Geographe (1951) Grand Criterium.

Tropique (1952) Eclipse Stakes.

Milesian (1953) Imperial Produce Stakes.

Nagami (1955) Coronation Cup. 3rd in 2,000 Guineas, Derby, St. Leger.

Saint Crespin III (1956) Eclipse Stakes, Arc de Triomphe. 3rd in Derby.

Ninabella (1958) Italian Oaks.

Anticlea (1960) Italian 1,000 Guineas and Oaks.

Soleil (1963) French 2,000 Guineas, Grand Criterium.

Young Emperor (1963) Leading 2-y-o of his year.

Mistigri (1971) Irish St. Leger.

Dona Barod (1972) 6 races in France.

Centrocon (1973) Lancashire Oaks.

Cramond (1974) Queen Mary Stakes.

SNOOT (1906), by Perigord out of NRA
FAMILY NUMBER: 7

Snow Maiden (1916) Irish Oaks.

Caligula (1917) St. Leger.

American Flag (1922) Belmont Stakes.

Noble Star (1927) Jockey Club Cup, Cesarewitch.

Sandjar (1944) French Derby.

Do Well (1948) Irish St. Leger.

Nashua (1949) Irish 1,000 Guineas.

La Sorellina (1950) French Oaks, Arc de Triomphe.

SNOOT (*contd.*)

Taittinger (1954) 4th in Oaks.

Supreme Sovereign (1964) Lockinge Stakes.

Frascati (1968) St. Simon Stakes. 2nd in Hardwicke Stakes.

The Admiral (1969) Henry II Stakes.

Chili Girl (1971) 2nd in Queen Mary Stakes.

Steel Heart (1972) Gimcrack Stakes.

Ampulla (1974) Cherry Hinton Stakes.

The Snoot family, which went into the doldrums for some time following the racing career of the smart French-trained filly La Sorellina, has undergone a comparative resurgence since 1964, mainly through the produce of the minor two-year-old winner Asti Spumante (1947), by Dante out of Blanco, though it has yet to return to the Classic-winning status of earlier years. A family of comparatively small numbers which has re-discovered its former glories, after an unfortunate setback, is that of Black Cherry (1892), by Bendigo.

Black Cherry's family suffered severely through the confiscation by the Germans of some of its important members from H. H. the Aga Khan's stud in France. They were never recovered after the War, but with the help of astute selection of stallions the family has got back on its feet and resumed Classic status in 1963 through the Derby winner Santa Claus. Sister Clara, grandam of Santa Claus, never ran and in her first three years at stud was presented with modest stallions. However, as a result of the exploits of her half-sister Sun Chariot, Sister Clara was afforded better sires and to a mating with the Derby winner Arctic Explorer she foaled Aunt Clara, dam of Santa Claus. This branch of the Black Cherry family, tracing from Clarence (1934), by Diligence out of Nun's Veil, by Friar Marcus out of Blanche, by White Eagle out of Black Cherry, has since produced Ragtime, Matahawk and Star Bird. The wider importance of the family can be seen below:

BLACK CHERRY (1892), by Bendigo out of Black Duchess
FAMILY NUMBER: 3

Cherry Lass (1902) 1,000 Guineas, Oaks.

Night Hawk (1910) St. Leger.

Blandford (1919) Leading sire.

Selim Hassan (1938) Leading sire in Argentina.

Sun Chariot (1939) 1,000 Guineas, Oaks, St. Leger.

Grand Corniche (1941) 2nd in 1,000 Guineas.

Blue Train (1944) Unbeaten.

Gigantic (1946) Leading sire in Australia.

Krakatao (1946) Sussex Stakes.

Bebe Grande (1950) Gimcrack Stakes, Champagne Stakes, Cheveley Park Stakes. 2nd in 2,000 Guineas, 3rd in 1,000 Guineas.

Skyraider (1950) King Edward VII Stakes.

Landau (1951) Sussex Stakes.

Martine (1952) 2nd in Irish 1,000 Guineas.

Carrozza (1954) Oaks.

Amante (1955) Irish Oaks. 2nd in 1,000 Guineas.

Pindari (1956) King Edward VII Stakes, Great Voltigeur Stakes.

Opaline II (1958) Cheveley Park Stakes, Leading 2-y-o.

Sherluck (1958) Belmont Stakes.

Merchant Venturer (1960) Dante Stakes. 2nd in Derby.

Santa Claus (1961) Derby, Irish Sweeps Derby and 2,000 Guineas.

Ragtime (1962) July Stakes and five other races at two years.

Pieces of Eight (1963) Eclipse Stakes, Champion Stakes.

Caro (1967) French 2,000 Guineas.

Bonne Noel (1969) Ebor Handicap.

Matahawk (1972) Grand Prix de Paris.

Star Bird (1973) Challenge Stakes.

The small families cannot be expected to equal the great ones detailed in my original list in the multiplicity of Classic winners, but although I have no figures to quote, from the examples given above, I would hazard that in relation to their numbers at stud they can show very similar results. Broodmares tracing to the great tap roots are fairly numerous, but they are vastly outnumbered by the great pool of other mares at stud. Apart from their capacity for throwing up high-grade horses, a remarkable thing about them is how very few of their representatives fail to produce winners of one sort or another.

A glance down the list of leading tap root mares which I have given shows that seven out of the ten were by stallions who were but moderately successful as sires of winners, i.e. Chelàndry, Conjure, Miss Gunning II, Silver Fowl, Gondolette, Americus Girl and Admiration. Other great matriarchs who had the same peculiarity were Altoviscar (1902) by Donovan, Concussion (1885) by Reverberation, Black Cherry (1892) by Bendigo, Eryholme (1898) by Hazelhatch, Canterbury Pilgrim (1893) by Tristan, Grand Duchess (1871) by Lozenge, Marchetta (1907) by Marco, Petit Bleu (1902) by Eager, Rinovata (1887) by Wenlock, Snoot (1906) by Perigord, and Stella (1890) by Necromancer. This is a formidable list and covers some 60 per cent of the mares with claims to be considered in this class. I can offer no satisfactory explanation of this phenomenon unless it is purely due to the weight of numbers. For every mare whose sire is a stud success, there are many who are got by horses with disappointing stud records.

Taking it fine and large, down the years the leading sires of winners of one period have been the leading maternal grandsires a few years later. This is attributable not only to their ability to transmit genes of superior quality but also to the fact that usually they are mated to the best mares in the country. The result is that their daughters have highly successful strains on both sides of their pedigrees.

THE TAIL FEMALE WINNING CHAIN

It must be conceded that a mare from a good winner-producing tail female line is more likely to do well at stud than one with a poor maternal record. It is not possible for everyone to obtain descendants from the great long-term Classic-winning tap roots, but in default of these nearly any mare who is by a stallion of standing and who can claim close relationship in tail female with a performer of merit is worthy of consideration. But once the winner-producing chain has been lost for a couple of generations the situation alters. A mare who is a sister or a half-sister to a good winner or one whose dam is similarly related to a horse of standing is usually a sound stud proposition. But one whose dam and grandam have produced no worthwhile racers must be regarded with reserve. When expressing this opinion I am referring to the normal average results to be expected. Some very great broodmares have come "out of the blue" with poor immediate backgrounds. Admiration and Gondolette, whose vast successes I have already mentioned, both came from families who had produced in the direct line no good-class horses for 50 years prior to the dates of their respective births. It is beyond the wit of man to say why or when these phenomenal mares will make their abilities manifest until their progeny prove their worth on the racecourse. It is, however, certain that for every one of this class of mare who makes good there are literally tens of thousands who fail.

SELECTION OF POTENTIAL BROODMARES

The selection of a potential broodmare is best done by balancing pedigree, performance, temperament, conformation, constitution, soundness, etc., against each other and then choosing the animal with the fewest defects. No mare is perfect.

Whether a non-winner from a great female line is likely to make a better broodmare than a winner from a poor family is one of those things which could only be conclusively determined by a vast research into the histories of the Stud Book and Racing Calendars. I have already drawn attention to the necessity of racing merit in sires and it is only reasonable to assume that the same attribute is of an advantage to broodmares. It is, however, impossible to draw a true comparison between the two, as the non-winning stallion is put to very poor quality partners whilst the non-winning mare, if well enough bred, commonly goes to good-class sires. When considering the winning element the class factor is very relative. An indifferently bred filly who has won a small race is patently a poorer stud proposition than one who is beautifully bred and has finished second in the Oaks but failed to earn brackets.

I think that sometimes there may be an inclination to give too much consideration to the fact that a mare is a winner. There is no doubt that the first few foals out of a winner are better sales' propositions than those out of a non-winner, and the dam's success looks well in the sales' catalogue. But not infrequently a thorough examination of her racing record discloses that her abilities were of the humblest order and that her one or two victories were achieved not through any merit, but by means of weight advantages or against moderate rivals. In these days there are beautifully bred fillies running in the lowest-class races, worth £600 or less to the winner, purely for the purpose of earning a bracket and making their early yearlings more saleable. I do not think for one moment that racing successes of this calibre give the slightest guidance as to whether the filly is likely to breed winners or not.

MISLEADING STATISTICS

From time to time I have come across statistics showing the stud achievements of mares who won compared to non-winners, but all these have been unsatisfactory. Not infrequently winners who secured clear-cut victories under top, or equal top weight are jumbled up with those who owe their successes to the handicap element. A further point which is not provided for—and a most vital one—is that winners are commonly mated to better-class sires than non-winners. Let me make it quite clear that I do not question for a moment the wisdom of breeding from winning fillies, but the mere fact of securing success in a minor event should not be allowed to warp other considerations.

Some breeders aver that they would prefer to breed from a sister or a half-sister to a good winner rather than from the winner herself. It is easy to quote any number of cases of half-sisters proving better broodmares than their more accomplished relatives, but most good winners have several half-sisters, so the weight of numbers becomes operative. The winner's maximum annual output is one foal, but her various sisters or half-sisters may between them produce six or more times that number. I do not think that the strain of racing successfully necessarily jeopardizes a mare's stud career, and I will refer to this again.

Meanwhile, I will touch on the well-known fact that in the history of the Turf the mating of an Epsom Derby winner with an Epsom Oaks winner has never produced an English Derby winner—though Charlottown all but caused the record books to be re-written after his success in 1966, for he was by the French Derby winner Charlottesville out of the Epsom Oaks winner Meld.

Perhaps in these days of international breeding Charlottown deserves a record of his own; but for the moment the fact stands in relation to the premier English Classic. This is not due to any mysterious incapacity on the part of the parents, but simply to the heavy odds against success. At any given period in racing annals there have never been more than about ten former Oaks winners at stud who have been in competition with several thousand other mares. A great number of the latter are of a class which renders them unlikely to breed a winner of the Derby, but, then, Oaks winners are not tied to Derby winners as mates. I estimate that at least half of them have been annually put to sires who have won other important races including foreign Classics.

The pool of sires and dams who have a reasonable chance of producing a Derby winner is comparatively large, and the very few matings of former Derby winners to former Oaks winners is but a tiny drop in it. I think probably in the course of time this specific union will be rewarded with success, but the operation of chance, dictated by numbers, must be allowed full play and patience exercised. Oaks winners have produced Derby winners to other sires, i.e. Humorist (1918) by Polymelus out of Jest (Oaks); Gainsborough (1915) by Bayardo (St. Leger) out of Rosedrop (Oaks); Blair Athol (1861) by Stockwell (2,000 Guineas, St. Leger, etc.) out of Blink Bonny (Derby and Oaks), etc. Charlottown (1963) is the most recent.

DOES HARD RACING HARM A MARE'S STUD CAREER?

Now to turn to the question of whether hard racing usually has any harmful results on a mare's breeding capacity. This is yet another question on which a lengthy scientific research into racing and breeding history would probably shed useful light. I should like to see a statement covering some 50 years showing, by categories, the percentage of mares who have produced winners (excluding those who won through a weight advantage), who themselves had raced for 0, 1, 2, 3, 4, etc., seasons, the average value of the races won by the progeny—as a guide to class—the average number of winners produced by the mares of each group and similar details concerning mares who have run 0, 5, 10, 15, 20, etc., races. May be with the aid of a computer someone will find the time to carry out such a study.

In the absence of specific information of this nature the only thing a breeder can do is to form his own opinion according to his own experience and observation. For my part I should expect a statement on the lines indicated would show that the percentage of successes in each class would more or less correspond, thereby proving that normally the exertion of racing is not harmful. But that is purely a personal opinion. It is possible to quote a number of incidences of hard-raced mares who have turned out disappointments as matrons, but also there are any quantity of lightly or unraced mares who have been stud failures. One of the differences between the two is that the former's frequent appearances on the racecourse focusses public attention and memory on them, and if they fail at stud they are held up by the world and his wife as examples of the dangers of hard racing. On the other hand, nothing much is expected of the latter, except by their owners and a limited circle, so that their shortcomings escape general comment.

It must be remembered that for every filly who races on the Flat up to the age of five or beyond there are hundreds who retire from the fray earlier, or who never race, so the operation of numbers is strongly against those with a long racing career. Moreover, the filly subject to a lengthy sojourn on the course is not uncommonly one whose breeding is not of the highest standard. The better bred ones are usually retired early. In my view each case must be judged separately, dependent on the individual's constitution, temperament, etc., but there is no possible doubt that some hard-raced mares have contributed more than their fair share to the make-up of their breed. I quote the following examples of this, and there are many more for those who care to investigate the matter in greater detail.

GREAT BROODMARES WITH HARD RACING CAREERS

MARE, YEAR FOALED AND SIRE	BRIEF DETAILS OF RACING AND STUD RECORD
1. Beeswing (1833) by Dr. Syntax	Raced for eight seasons. Won 52 races, including the Ascot Gold Cup and the Doncaster Cup (four times). Then used for a year as a park hack. Dam of Newminster (2,000 Guineas and St. Leger) and Nunnykirk (2,000 Guineas). A great tap root mare.
2. Alice Hawthorn (1838) by Muley Moloch	Raced for seven seasons. Ran 71 times and won 52 races including the Goodwood and Doncaster Cups. Dam of Thormanby (Derby and Ascot Gold Cup). A great tap root mare.
3. Haricot (1847) by Mango or Lanercost	Raced until she was a six-year-old. Produced her first foal at eight and last at 25. Dam of Caller Ou (St. Leger) and at the age of 21 Lady Langden. The last named produced Sir Bevys (Derby), Hampton (Goodwood and Doncaster Cups and an exceptionally influential sire).
4. Little Lady (1856) by Orlando	Won the last race (¼ mile) for yearlings run under Jockey Club Rules. Ran between 30 and 40 times. Winner of 16 races in all from one- to five-year-old. At the age of 16 produced the 2,000 Guineas winner Camballo.

GREAT BROODMARES WITH HARD RACING CAREERS (*contd.*)

MARE, YEAR FOALED AND SIRE	BRIEF DETAILS OF RACING AND STUD RECORD
5. Moorhen (1873) by Hermit	Ran 77 times under both Rules. Raced when in foal and after foaling was returned later for further racing. Finally quitted the course as a nine-year-old. Dam of Gallinule (champion sire).
6. Kincsem (1874) by Cambuscan	Bred in Hungary. Possibly the greatest race filly of all times. Ran for four seasons winning all of her 54 races, including the Austrian Derby, Hungarian Oaks and St. Leger and the Goodwood Cup. Dam of Buda Gyonge (German Derby) and None Such (Hungarian St. Leger). A great continental tap root mare. Her family sustained very heavy losses in both World Wars, but there are a few members left.
7. Mowerina (1876) by Scottish Chief	Bred in Denmark. Raced for four seasons in England, starting 37 times, winning 16 events. Raced when in foal. Dam of Donovan (Derby and St. Leger), Semolina (1,000 Guineas) and Raeburn (third in the Derby and a good winner). A foundation mare of the late Duke of Portland's stud. Lived until she was 30 years old.
8. Atalanta (1878) by Galopin	Ran 37 times in four seasons racing. Raced when in foal. Dam of Ayrshire (2,000 Guineas and Derby). Another foundation mare of the late Duke of Portland's stud.
9. Wavelet (1880) by Paul Jones	Ran 75 races under both Rules, winning 23 of them. She was an indifferently bred mare who could not be expected normally to be a great broodmare, but she produced Wavelet's Pride who won the Doncaster Cup, Great Metropolitan, etc., and was a kind of Brown Jack of his time. He lived until he was 24 and was a good NH sire.
10. Broad Corrie (1889) by Hampton	Raced until she was a five-year-old, winning ten times. A foundation mare in the late Lord Derby's stud. Ancestress of Canyon (1,000 Guineas), Toboggan (Oaks), Colorado (2,000 Guineas), Citation (U.S.A.), etc.
11. Admiration (1892) by Saraband	Ran for four seasons on the flat and also ran a few times under Irish National Hunt Rules. Bred 13 foals in 13 years. Dam of Pretty Polly (1,000 Guineas, Oaks and St. Leger). One of the great tap root mares of modern bloodstock. (See page 118, Item I).
12. Sceptre (1899) by Persimmon	Ran for four seasons winning the 2,000 Guineas, 1,000 Guineas, Oaks and St. Leger. Lived until she was 27 years old. Another great tap root mare. (See page 119, Item J.)
13. Donnetta (1900) by Donovan	Raced on the Flat until eight years old. Dam of Diophon (2,000 Guineas). Diadem (1,000 Guineas) and Diadumenos (Champion sprinter). She was 21 years old when Diophon was born and lived until she was 25.

GREAT BROODMARES WITH HARD RACING CAREERS (*contd.*)

MARE, YEAR FOALED AND SIRE	BRIEF DETAILS OF RACING AND STUD RECORD
14. Hammerkop (1900) by Gallinule	Raced until she was a six-year-old winning the Cesarewitch, Queen Alexandra Stakes and other long-distance races. At the age of 17 produced the Derby winner Spion Kop.
15. Pretty Polly (1901) by Gallinule	Raced for four seasons. Won 22 out of 24 races, including 1,000 Guineas, Oaks, St. Leger, Coronation Cup and Champion Stakes. Bred only four winners but became a first-class tap root mare whose descendants include Donatello II, Supreme Court, Premonition, St. Paddy, Psidium, Brigadier Gerard, etc.
16. Escarpolette (1917) by Fitzhardinge	Raced in U.S.A. Half-sister to the dam of the great American racer Equipoise. Ran 261 races between the ages of two and nine. Won 32 and was placed in 81 others. Her first five foals were all winners.
17. Athasi (1917) by Farasi	Raced under both Rules until six years old. Dam of Trigo (Derby and St. Leger), and Harinero and Primero (both won Irish Derby and St. Leger). Ancestress of Tulyar (Derby etc., and £76,417), Nagami (Coronation Cup), Saint Crespin III (Eclipse Stakes, Arc de Triomphe), Young Emperor (leading 2-y-o, 1965), Anticlea (Italian 1,000 Guineas and Oaks), Mistigri (Irish St. Leger).
18. Selene (1919) by Chaucer	This filly did not have an exceptionally long racing career but certainly did not eat the bread of idleness for in two years on the course she won 16 races. Dam of Hyperion (Derby, St. Leger and six times champion sire), Sickle and Pharamond (both leading sires in U.S.A.)
19. Corrida (1932) by Coronach	Bred in France. Hard raced until the age of five, winning high-class races in France, Belgium, England and Germany. One of the best race fillies in Europe of the between the Wars period. Dam of Coaraze (French Derby) and then taken by the Germans during the war and never recovered.

I think the above list shows that down the ages some-hard-raced fillies have proved abnormally good matrons, and I emphasize that the percentage of fillies who are subject to such extensive racing as those I have quoted is extremely small, compared to the ones who have a fairly easy time or are not trained at all.

Some of the animals in the list are from a bygone age, but I do not consider the argument that nowadays competition is keener, to be relative. A recent example, the filly Dahlia (1970), will illustrate the point. She raced for four seasons, and won 13 of her 35 races, in most of which she took on the highest class of rival and had to travel to five countries and two continents to achieve a European stakes-winning record of £488,545 in win money alone. Yet Dahlia was merely continuing her family's record for toughness, since her dam Charming Alibi won

16 of her 71 races in four seasons, while the latter had a half-brother, Mighty Fine, and a half-sister, Adormie, who ran 167 and 114 times respectively. Since Adormie has already produced winners, Dahlia's record at stud is awaited with interest!

Provided a filly runs her races out truly—and those who do not are usually retired early—the physical strain occasioned is the same whether the races are staged in the 1800s, the late 1900s, or at any other period. Moreover, the advances of training and feeding methods, veterinary science, etc., all tend to put the animal in a better condition to withstand this strain.

A relative matter when comparing results in the various epochs is that in former days a far higher percentage of fillies were severely tested than is the case today. It is often averred that a weakness in our present-day breeding methods lies in the comparatively light racing some sires are subject to, thus exposing the breed to the risk of the introduction of blood of doubtful tenacity. But why should the sires be blamed if their mates are raced even more lightly, or not at all?

WEEDING OUT UNSATISFACTORY MARES

I will now turn to the age element. I am inclined to think that nowadays there may be a tendency to discard too hastily mares who have failed to produce useful winners early in life. This is quite understandable as yearlings out of oldish mares who have foaled nothing worthwhile are poor sales propositions, more or less regardless of their dams' racing record and breeding. This may cause a considerable wastage of good blood. Mares of impeccable antecedents are quite commonly weeded out of the best studs if they have not made good as dams of winners in their early days. They have little or no second chance, as after sale not infrequently they are mated to inferior stallions. This weeding out is far from a new practice, but owing to modern economic considerations, and the reduced dimensions of many non-commercial studs, it seems to be more ruthlessly adhered to than formerly. Amongst the mares who have had a poor stud record until getting on in years are:

MARE, YEAR FOALED, ETC.	NOTES ON MARE'S STUD CAREER
A. Octaviana (1815) by Octavian Ancestress of the French Derby and Grand Prix de Paris winner Ajax (1901) and other important horses.	Bred nothing of note until 22 years old when she foaled the 2,000 Guineas, 1,000 Guineas and Oaks winner Crucifix (dam of the Derby and St. Leger winner Surplice).
B. Mare (1816) by Comus out of a mare by Delpini	Bred nothing of note until 22 years old when she produced Miss Lydia ancestress of Fair Play (U.S.A.), Star Shoot (U.K. and U.S.A.), Cyllene and Astérus (France) who were all great sires. The only good racer this mare foaled was Nutwith (St. Leger) when she was 24.

MARE, YEAR FOALED, ETC.	NOTES ON MARE'S STUD CAREER

C. Lacerta (1816)
by Zodiac
Ancestress of St. Simon
and Orme

Her only good winner The Little Wonder (Derby) was foaled when she was 21 years old.

D. Flying Duchess (1853)
by
The Flying Dutchman.
Half-sister to the dam of
Moslem (2,000 Guineas)

Bred nothing of merit until 19 years old. At this age she was sold with her foal at foot for £100. The foal was the Derby winner and champion sire Galopin.

E. Anonyma (1859)
by
Stockwell

Produced the good broodmare Stock Dove (dam of the excellent Australian racer The Australian Peer) at 14, but no racer of worth until she was 23 when she foaled Lonely (Oaks).

F. St. Angela (1865)
by
King Tom
Her grandam was a half-
sister to The Little Wonder
(Derby)

At 14 produced Angelica (the dam of Orme) but no good racer until she was 16 when she foaled St. Simon—possibly the greatest horse of all time. Her bad stud record was the probable reason that St. Simon was not entered for the Derby or St. Leger.

G. Kizil Kourgan (1893)
by
Omnium II
A great race filly who won the
French 1,000 Guineas, Oaks
and Grand Prix de Paris.

Produced some quite useful stock but nothing comparable to Ksar (French Derby and a first-class stallion) who appeared when she was 19 years old.

H. Hammerkop (1900)
by
Gallinule
A good race filly who won the
Cesarewitch and Queen Alex-
andra Stakes (twice). Sister to
the dam of Electra (1,000
Guineas).

The one and only winner she produced was Spion Kop (Derby) and she was then 17 years old.

I. Pearl Cap (1928)
by
Le Capucin
Another great race filly who won
the French 1,000 Guineas, Oaks
and Prix de L'Arc de Triomphe.
Half-sister to Pearlweed (French
Derby) and Bipearl (French
1,000 Guineas).

Producing nothing of note until 16 years old when the English Derby winner Pearl Diver appeared.

J. Mercia (1931)
by
Teddy
Full sister to the grandam of Phil
Drake (1955 English Derby and
Grand Prix de Paris).

Produced nothing above selling-plate class until she was 20 years old when she foaled Le Petit France (French Derby, 1954).

MARE, YEAR FOALED, ETC.	NOTES ON MARE'S STUD CAREER
K. Barclays (1930) by Son-in-Law Her dam was sister in blood to Stedfast who was second in both the 2,000 Guineas and the Derby and won £26,479.	Produced nothing of note until 21 years old when she foaled Popof (Grand Prix de Paris, 1954).

This list is by no means complete, and the greater emphasis nowadays on weeding out older mares, as noted earlier, is probably the reason for there being few comparative examples in the 20 years. But the list is there, and it shows that through Turf history mares with useless stud records have produced Classic winners in their old age. It would be idle to claim that this is an everyday occurrence. At a rough guess I should estimate that possibly three per cent of the dams of Classic winners have been in this category, whilst I think probably less than 0.1 per cent of all mares who have been at stud at home in the past hundred years have produced a Classic winner at one time or another during their lives. Now that the Stud Book has expanded greatly not more than 0.04 per cent can expect to achieve this great distinction.

ACUTE COMPETITION AMONG MARES

I sometimes hear people say "Such-and-such a mare produced the Derby winner So-and-so but nothing else"—thereby implying that the production of Classic winners or high-class horses is something akin to sausages being rolled out of a machine. The dam of a Classic winner is a wonderful animal. She is in competition with several thousand other mares and has only one foal each year, whereas a stallion has only several hundred rivals for the honours and may sire 30 to 40 foals in a year.

St. Simon, Lady Langden (dam of Hampton) and Blanche (dam of Blandford), to name but a few, were all out of old mares, and the Stud Books of the world today would make very different reading if there had never been a St. Simon, a Hampton or a Blandford.

The mare who is the highly prized recruit to the best stud on coming out of training carries the same potentialities for breeding high-class winners in her more mature age that she possessed earlier in life. Provided that there is a goodly supply of silver pieces in the Christmas pudding there is always a chance of encountering one until the very last bite, but if they have been sparsely used or forgotten, luck will probably be out. I do not for one moment advocate retaining old mares who have proved their incapacity indefinitely, but I think that the yearling sales element is inclined to impose the tendency of going from pillar to post always looking for mares whose progeny are more and yet more likely to attract good prices.

FIRST FOALS

Tulyar, Mahmoud, Son-in-Law, Hampton and many other famous horses were the first produce of their dams, but, so far as I am aware, no statistics have ever

been worked out showing the percentage of successes amongst first foals compared to second, third, fourth, and later issue. Obviously there are and always
have been far more firstlings in the world than later produce, as every mare who
breeds must have a first foal but some have no more.

In the annals of the Stud Book there have been a few cases of fillies being mated
as two-year-olds. This is a most unusual—and I might term it freak—procedure,
but in two cases it yielded good results. Monstrosity (1938) as a three-year-old
produced The Ugly Buck (2,000 Guineas) and an unnamed mare (1840) by
Plenipotentiary at the same age foaled the great tap root mare Queen Mary (dam
of the Derby and Oaks winner Blink Bonny and ancestress of many good horses).
Eye Pleaser (1876), the grandam of Admiration, and Poet's Star (1917), the
grandam of the 1941 Derby winner Pont L'Eveque, were also covered as
two-year-olds, but in neither case was the produce (if any) of any account.
Breeding from immature animals is repugnant to many of us and there are no
grounds for believing that it yields good results.

CHAPTER 10

Stallion Syndication and Fees

Syndication explained and discussed: large capital outlay: drawbacks to syndication: syndication of stayers to preserve staying blood: no foal no fee, live foal, and other terms discussed: payment of fees

SINCE THE SECOND World War the introduction of the syndicated stallion system has effected a considerable change in the general set-up of high-class bloodstock breeding arrangements in England and Ireland.

For generations past it has been the general practice to limit stallions of standing to 40 mates per season. There is no authority which enforces this limitation, but by mutual agreement between stallion owners, broodmare owners and all concerned it has long been the order of the day.

Prior to the War, when stallions were generally owned by private individuals, a mare owner paid a fee for the right to use a stallion for one season only. The following year he repeated the transaction or transferred his allegiance elsewhere. These arrangements, which amounted to a year-by-year hiring system, worked satisfactorily from time immemorial, but after the War events brought about a fundamental change.

Firstly, as a succession of various government taxes were raised to their present high levels, the situation arose whereby the State took the major portion of the stallion's annual earnings, leaving his owner with little worthwhile return for his capital outlay. Secondly, and directly coupled with the first point, as demand for British bloodstock grew in foreign countries such as America and Japan, individual British stallion owners found it difficult to repel lucrative offers from these wealthier quarters. The position was further worsened when the value of sterling against foreign currencies fell in the 1970s.

THE SYNDICATION SYSTEM

To overcome these difficulties, the syndication system was evolved, whereby capitalization of the stallion into 40 shares takes place. Ownership of a share entitles the individual to a free nomination to the stallion annually as long as the horse is at stud. The general management of the horse is usually vested in a small committee of shareholders, and the member of the syndicate can normally use his annual nomination in one of two ways—either he can take up the option for one of his own mares, or he can dispose of it either privately or at one of the public auctions which have sprung up in recent years. Members of the syndicate may be called upon to pay their share of the expense of keeping the horse, or the syndicate committee will have the power to sell one or more extra nominations each year to cover running costs.

By this arrangement the original owner obtains a capital appreciation when he sells his horse to the syndicate, but he may also retain a number of shares for his own use. Shares retained by the original owner can thus be taken up for his own mares, exchanged for nominations to other stallions, or sold annually for profit.

Since international bloodstock values have risen enormously since the Second World War, trade in syndication has reached proportions that could not have been foreseen in 1950. Whereas a Classic-class winner of impeccable breeding would have been sold to 40 shareholders at a rate of £2,500 per share in 1950, nowadays the general figure is more likely to be in the range of £25,000 to £50,000. A figure from America is worth quoting in this context, following the syndication early in 1976 of What a Pleasure. Leading American sire of winners in 1975, What a Pleasure had a capital value of eight million dollars, which under the usual American quota of 32 shares, and using the Jockey Club's prevailing rate of exchange, represented a value per share of £123,762!

As well as enabling bloodstock breeders to band together so as to be in a better position to ward off huge bids from overseas interests, syndication has clearly worked to the advantage of owners of high-class horses, to which the system is largely confined. The more modest animal seldom is involved, largely because shares in him normally are not a readily saleable commodity.

If the syndicated stallion turns out well at stud and is a longlived horse, the buyers have made an excellent bargain, as instead of paying £3,000 to £4,000 a year for 12 to 15 years, they have a free nomination in exchange for their capital investment. Moreover, whatever heights of success the horse reaches, they are certain that this nomination will be at their disposal.

If the horse dies young, they will probably recover most of their investment through insurance, but if he is a stud failure and his stock are unsaleable, things are not so satisfactory. However, even in these circumstances, shareholders may have recuperated a considerable proportion of their capital outlay by the medium of sale of the first few crops of his yearlings before his inability to sire good winners becomes manifest. Yearlings by a horse of a capitalized value of £500,000 or more make high figures in the yearling market until the racecourse test proves that they are of no use.

The mere fact that there are 40 breeders of experience and judgment willing to invest a considerable sum in the horse, shows that there is an appreciable weight of opinion that he is likely to do well at stud, and it does not take 40 determined bidders at the ringside to ensure a good price at the yearling sales. On the other hand, the shareholder has to face the risk that the first mares he sends to the horse may be barren or their foals unsaleable from accident, disease, etc., so that he has no yearlings to sell until the horse's reputation is tarnished. His position then is bad, as he has little or no prospect of a return on his investment. He is tied to the horse; his share is unsaleable and can hardly be given away. Moreover, if he avails himself of his rights in the hopes that the horse will show better results later, he must bear in mind the fact that if these hopes are not fulfilled, he is not only producing an unsaleable yearling but also wasting a year of his mare's life.

The latter point is particularly relative in the case of a young mare who has yet to make her reputation as a winner-producer. If she was mated to a more successful stallion, her chances of breeding a useful winner would be enhanced, with the result that her subsequent stock would be more saleable. Perhaps the most disastrous prospect a shareholder has to reckon with is the chance of the horse being sub-normally fertile or even becoming sterile early in life. The 1976 Irish Sweeps Derby winner Malacate has recently proved an important example. He did not come up to scratch in his first season at stud, and after insurance details had been confirmed, he was returned to racing in the hope that time would solve his problems. If all fails, there is nothing for it but to accept the investment as a dead loss, whereas if the breeder had merely taken an annual nomination to a stallion with the same failing, he is in a position to transfer his patronage elsewhere.

LARGE CAPITAL OUTLAY INVOLVED

As a result of the upsurge in syndication, and as a safeguard against having to use not-so-good stallions, shares in syndicates with the certainty of nominations to top-grade horses are almost a necessity for a breeder who is aspiring to produce the highest-class stock. Moreover, if he is breeding on a comparatively large scale, he will require a share in half a dozen or more syndicates. This entails a considerable speculative capital outlay.

Owing to the operation of competition, even amongst horses of impeccable breeding and racecourse record, only a few of these will reach the top of the tree as stallions with the certainty of their stock being in heavy demand over a long period. Thus at the outset the mare owner has to accept the fact that at least some of his shares will prove to be unremunerative investments.

Unfortunately, the increase in syndication appears to have been accompanied by a failure on the part of some shareholders to accept this point. It seems to me that there is a growing tendency to overlook the profits made on the horse's early stock and to expect him to be a kind of inexhaustible gold mine for the rest of his days; or to give a stallion only limited opportunity to prove himself before he is packed off abroad.

No horse can make headway at stud unless he is mated to an adequate number of good-class mares. Many horses do not sire their best in the first few seasons, and it is not uncommon to see some of them sold for export and proving great successes in their new countries. The deflection of their services cannot be other than a loss to home breeders. It is true that some of them were of a class which rendered them unlikely to achieve success at home, but I feel there is a tendency today continually to flit from pillar to post, always searching for some new super sire. This results in an excessively wide distribution of our broodmare pool, and condemns to failure the stallion who makes a tardy start at success at stud.

Nor has it been unknown for a stallion to produce his best racecourse representatives only a short time after he has been exported. Tesco Boy (1963), for example, stood one season in Ireland before being sent to Japan, where he had been resident for three years when his single crop, racing in Britain as three-year-olds in 1971, produced the laudable statistics of ten winners of 15 races worth £17,869 from only 14 individual runners. Tesco Boy has since been an outstanding success in Japan. On a different plane, Saint Crespin III (1956) went also to Japan in 1970, and the following year his daughter Altesse Royale won the 1,000 Guineas and both the English and Irish Oaks.

DRAWBACKS TO SYNDICATION

Whether the advent of large-scale syndication has been to the advantage of the British thoroughbred is a much-debated question, and the answer usually depends on which side of the bloodstock fence one sits. Those who have successfully engaged in syndication will doubtless point out its advantages; those who have been unable to join in will clearly have a different view.

It cannot be avoided, however, that syndication places extra restrictions on the use of a stallion, for whom the choice of mare is limited by the identity and circumstances of the shareholders, who may use the nomination even if their mare is unsuitable. There is a danger that economics will become a more important factor in drawing up a list of matings, rather than the pedigree, confirmation or temperament of the two animals in question. However, I will record my opinion that the complaint sometimes heard that syndicated sires attract bad mares (not to be confused with unsuitable ones) is based on false premises. It is unreasonable to anticipate that mare owners who have paid between £25,000 and £50,000 for a share in a young stallion will deliberately ruin their chances of a sizeable return by sending him indifferent mates.

A comparatively recent trend has been the arrangement for syndication of a horse whilst he is still in training, which situation has been accused of shaping or curtailing a career unnecessarily or even unreasonably. A syndicate may stipulate which subsequent races their acquisition must, or must not contest, and when his career should end. Those who are against this trend argue that it may prevent a horse from being tested to the full in his racing career. Shareholders, however, will retort that they are protecting their investment, which might lose its value if the horse were to be beaten in a succession of races or be kept in training for another season with little reward. It is again a matter of which side of the fence one sits.

Whatever the answer it is interesting to note that of the 15 Derby winners between 1962 and 1976, eight were retired at the end of their three-year-old seasons, and the seven who ran as four-year-olds all won at least once. Of the seven who stayed in training Relko, Charlottown, Royal Palace, Mill Reef and Roberto all won the Coronation Cup at Epsom, while Blakeney won the Ormonde Stakes at Chester and finished second in the King George VI and Queen Elizabeth Stakes, and Snow Knight won in North America. It clearly did the reputation of these horses no harm by their being trained as four-year-olds. But who is to say the likes of Sea-Bird II, Sir Ivor, Nijinsky and Grundy would not have suffered from similar treatment?

On the subject of Derby winners and syndication, it is an impressive indication of how values have risen with inflation to compare Santa Claus, whose capital value on retirement in 1964 was £400,000, with Nijinsky, who six years later was syndicated for £2,250,000.

SYNDICATION OF STAYERS

It is as pertinent today as it was 20 years ago to say that if we are to maintain our position in the world bloodstock market, staying blood is needed, but its actual preservation is the problem.

There is little incentive to breed a slow-maturing yearling, by a long-distance sire, as he is a poor sale proposition. Nor has the position been made any better by the comparative failure to keep prize money for staying races in step with those for sprints or middle distance, and by the direct policy in America and France of downgrading long-distance races, either by shortening the distance or decreasing the prize.

The situation is hardly a creation of the 1970s, for the state of the poll was well illustrated at the record-breaking 1954 Newmarket December Sales, where two young Ascot Gold Cup winners were sold for low prices and both to go abroad—namely Pan II, seven years old for £12,000, and Souepi, six years old for only £2,100. Winners of the Ascot Gold Cup have fared little better since then. Of the ten individual winners trained in England between 1965 and 1976, Fighting Charlie died young after standing at a stud fee of £248; Parbury was exported to Chile; Precipice Wood stands in Cheshire at a fee of £300, and Random Shot in Ireland at £150; Erimo Hawk was exported to Japan; and Ragstone was partially syndicated to stand in Sussex with nominations available at £1,000 no foal no fee. Even Rock Roi, who gained distinction by being twice disqualified in the Ascot Gold Cup, was exported to Australia. It is a record which hardly suggests that winning the Ascot Gold Cup is an automatic entry to popularity at stud.

Perhaps the purchase in 1977 by the National Stud of Sagaro, who ranks as one of the outstanding stayers of all time after his achievement in winning a record three Ascot Gold Cups, will help to start redressing the balance. And I should particularly like to see more stayers syndicated, preferably amongst shareholders who own good-class mares.

Today the general demand is for early-maturing, fast horses, but the prestige of Anglo-Irish bloodstock rests to a marked degree on its possession of a leavening of stamina. America, for instance, has a wealth of fast blood but interbreeding to this element can, in the end, result only in shorter and shorter runners. While English breeders can benefit from using some of the best American elements of speed, it is equally true that judging by their breeders' purchases in England, the Americans appear to be seeking parental stock with an appreciable dosage of stamina in their pedigrees. The same tendency is clear among other overseas buyers.

Thus staying blood is needed in England. But a situation has arisen which is something akin to passing the buck. Many people are anxious to breed from broodmares with an element of stamina in their veins if someone else will produce them. I think that this difficulty would be eased if there were more good-class stayers syndicated.

No stallion owner at home is keen to have a long-distance horse at stud with the prospect of poor patronage and consequent failure, but the ownership of a one-fortieth share in one is a different proposition. The operation of the syndication system would more or less ensure a good supply of mares, and with these the horse might well turn out satisfactorily. If this were so, his yearlings would be far more saleable than the produce of many staying sires who are so sparsely patronized that they cannot build up a reputation. But the objective I have in mind is primarily the production of broodmares of the future rather than of early two-year-old winners.

If the horse failed after patient backing to produce satisfactory results as a sire of slow-maturing winners, the only thing to do would be to drop him, but the loss to each of the 40 shareholders would be small. Naturally no-one wishes to breed from a broodmare who is by a total stud failure, but I am convinced that horses of the type to which I am referring are frequently condemned to ignominy by the shortage and poor quality of their mates.

The number of stayers who could be usefully treated in this way would necessarily be limited, firstly through the difficulty of finding suitable horses and secondly of enrolling sufficient shareholders with the required patience and with mares of sufficient class to portend success. But every little helps, and if nothing is done, the shortage of staying strains in our racehorses is sure to become more and more manifest as the years roll by. Whether or not the shareholders would in the long run receive an adequate return would rest in the lap of the gods. There is an element of gambling in all bloodstock breeding—especially stallion ownership—and I think their chances of coming out to advantage would be reasonably good; moreover, their commitments would not be heavy. But I am confident that the syndication and patient backing of more staying stallions would be in the best interests of the bloodstock industry. I also hold the opinion that it might be possible to encourage the breeding of stayers by means of premiums for the breeders of winners in this class of race, but that is another story, to which I shall refer later (see pages 194 and 306).

SERVICE FEES

The terms of service fees for the use of unsyndicated stallions vary considerably but usually come under the following headings: (A) A fixed fee which is payable whether or not the mare proves in foal, (B) No foal no fee. In this case the fee is payable in the autumn after service unless a veterinary surgeon's certificate is produced certifying that the mare is barren. (C) Live foal terms. Under these conditions there is no charge unless the mare produces a live foal. (D) A fixed fee with a free return clause. This allows a mare owner a free nomination the following year if the original services are ineffective.

To weigh up the advantages of these different conditions it is necessary to consider various points. First it must be clearly understood that a stallion owner sells his horse's blood at the time of service in exactly the same way as the mare owner sells his foal as a weanling, yearling or later in life. He cannot normally be either morally or actually responsible if the services are unproductive any more than the mare owner can be implicated if the yearling he sells fails to win races. I strongly deprecate the attitude of the mare owner who considers that he is under no obligation to pay the stud fee if his mare proves barren unless the service conditions specifically provides for this arrangement. To take matters logically a step further, if this line of ethics is correct the buyer of a yearling, etc., who fails to give satisfaction has an equal right to demand a refund of the purchase price!— which on the face of it is absurd. With but the rarest exceptions it can be said that all stallion owners do their utmost to get every single mare sent to their horse pregnant, for the simple reason that every time a mare is sent home barren the horse has lost an opportunity of siring a winner—and a stallion's reputation rests on the number of winners he gets.

It has been my personal luck in life to have a considerably wide practical experience amongst stallions and mares. I do not mean, shall I say, in an advisory or superficial capacity but by actual manual contact. As a result of this and the most careful study of the animals I am absolutely sure that the cause of infertility lies very much more frequently with the mare than with the horse, and after all it is only reasonable to expect such to be the case. A sire who is a poor foal-getter is soon forced into retirement through lack of patronage but there are innumerable mares with bad foaling records at stud, for year after year.

The main reason some horses show a higher fertility return than others is that their owners or managers are in a position to pick and choose their horses' mates and to refuse to accept mares with doubtful foaling records. In proof of my contention I draw attention to the case of the Premium Stallions who under the conditions of their contract are obliged to cover any "half-bred" mare presented to them—unless of course she is obviously diseased. Over a period of years the fertility percentages of all these sires are approximately the same. In one specific year a particular horse may—and does—show above-normal or sub-normal results but, generally speaking over a period of say five years the average return for each of them shows remarkable uniformity. After all, if a sire can get in foal the majority of his mares why should he fail with the others, considering that if a mare does not conceive to first service he has innumerable shots at the target? On the other hand there are a few definitely infertile or sub-normally fertile sires. These do not come within the scope of normal everyday stud contracts. Their incapacity becomes well known amongst breeders almost as soon as the failing has made itself apparent and special arrangements are introduced if they continue at stud. If this is not done the wise mare owner declines to have anything to do with them.

INSURANCE AGAINST BARRENNESS

The arrangement involving straightforward terms of the stud fee being payable whether or not the mare conceives has been in vogue ever since the dawn of bloodstock breeding, has stood the test of time, and in my view is usually the best and fairest for all concerned. If a breeder using a horse under these conditions wishes to protect himself against financial loss, provided his mare is a normal risk, he can generally without trouble insure against barrenness with a reputable insurance company.

A stallion owner has to cut his coat according to the cloth. To make ends meet he must budget for a reasonable return to cover his heavy capital outlay, running expenses, etc. He is aware that approximately 45 per cent of the total number of thoroughbred mares covered annually do not produce live foals, so if he offers no foal no fee, or live foal terms, he must, and does, raise the charge for his horse's service to meet this contingency. This increase need not necessarily be the full 45 per cent. If he can with confidence anticipate that he will be able to fill his horse with mares with good foaling records, he may consider that a small increase will meet the case. If this is not so, he is forced to provide against the full incidence of probable disappointments.

Therefore taking things find and large it can be said that the basic fees for horses who stand under these terms are considerably higher than for horses of the same class who are at stud under the more usual "pay and take your chance" conditions. Naturally a breeder must expect to pay more for the extra accommodation of no foal no fee or live foal arrangement and must make up his mind whether the extra expense is sound business. I am inclined to think, in some cases, he would be wise to consider whether it would not be a sounder proposition to lay out more on veterinary surgeons' fees and other costs, in an endeavour to present his mare for service in the peak of breeding condition to a sire standing under the normal terms, rather than paying the higher no foal no fee, etc., rates.

NO FOAL NO FEE TERMS

I have found that there is often a genuine misconception regarding the no foal no fee arrangement. This does not provide for the non-payment of the service fee if the mare fails to produce a foal. It only allows for the fee being waived if she is sent home barren. If this is the case, in the autumn after service, the mare owners sends the stallion owner a veterinary surgeon's certificate to that effect. In default of this the stud fee is payable.

I am sure that amongst people who have not actually worked on a stud farm the frequency of slipping foal or aborting amongst thoroughbreds is not fully realized. This accident is not uncommon especially during the first few months of pregnancy. I often hear breeders express astonishment when their mare has been certified in foal by a veterinary surgeon and later fails to produce. They frequently say that they are certain she did not slip foal as no foetus or other signs of the accident had been detected. In the case of an early miscarriage the foetus may vary in size between that of a half-cigar and a rabbit. It is quickly devoured by rats, etc., and often there is little or nothing to show that the mishap has occurred.

For evidence that the mare was in foal at one time the veterinary surgeon's report must be accepted, coupled often with the very closest observation of the mare's behaviour. For instance it is not unusual for a mare after mating to go for several weeks without coming into use and then to do so unexpectedly and at a time not within her normal oestrus cycle. Very often this is attributable to an early "slip".

A stallion owner who offers no foal no fee terms cannot reasonably be expected to forego his fee if a mare suffers this mishap after she has been sent home and, in consequence, is not under his control at the relative time.

The live foal conditions are almost universal in America but not so common in Europe. They constitute one of the reasons why stud fees are on a much higher scale there than closer home. Furthermore under the American arrangements stud fees are often payable in advance and sometimes at the time the nomination is booked. If the mare fails to produce a live foal the fee is returned in full.

The remarks I have made regarding the no foal no fee terms are also applicable to the free return basis. The stallion owner of necessity must enhance his horse's fee to cover the contingency of unproductive services.

In order to tidy things up there is one minor point which perhaps I should make clear. About one-third of the thoroughbred mares covered annually are barren. This is a rough average figure worked out on the basis that mares who are known to have aborted or slipped foal are not barren. Somewhere about two per cent of in-foal mares leave visible proof that they have suffered this accident and so came within this category. In addition there are an unknown proportion of undetected slips. Therefore the number of mares who conceive must certainly be higher than those who produce a live foal.

It is, of course, obvious that a stallion owner offering no foal no fee terms cannot adjust his fee relative to this unknown factor and must work approximately on the incidence of failure to foal shown in the Stud Book.

FEES AND FOAL REGISTRATION

In France the Stud Book is maintained by a branch of the Ministry of Agriculture who will not register the birth of a foal until the breeder produces a clearance certificate from the relative stallion owner acknowledging that the stud fee has been paid. If this is not forthcoming by certain dates the mare owner has to pay increasing fines for non-compliance until the time comes when the foal is debarred from registration. I sometimes wonder if something on the same lines would not be in the best interests of the Anglo-Irish bloodstock industry—or even the American system of advance payment.

All good-class studs at home pay their bills on presentation, but I am afraid that amongst small breeders this is not always the case. The owner of a stallion who is not in a position to pick and choose often has to wait ages for payment—and in fact sometimes never receives settlement. To allow for these bad (or delayed) debts he must charge a higher service fee than if he was certain of punctual payment. It must be remembered that some of the most influential sires in the build-up of our bloodstock have sprung from small beginnings and it does not seem right to leave any stone unturned which will encourage owners of this class of horse to give them every chance at stud. This is particularly so in these days when many of them have difficulty in making ends meet. Race entrance fees and jockeys' fees are collected centrally, so it should not be difficult to make similar arrangements in respect to stud fees.

Those engaged in the bloodstock industry are mostly of a conservative outlook, and it is not the custom at home to sue at law for outstanding stud bills, neither is it usual to demand payment for service fees in advance. Any stallion owner who was rash enough to pursue either of these courses would in the end, in all probability, wish he had not done so as, for a certainty, he would lose general goodwill and custom. I most emphatically do not wish to imply that the non-payment or late payment of stud bills is a common occurrence and probably the general consensus of opinion amongst the bloodstock industry is that, on the whole, the present arrangements work satisfactorily. Nevertheless as a purely personal view it seems to me if the French and Americans can conduct their affairs on a more business-like footing that we might be wise to consider doing the same. Similarly, as in other trade transactions, long term credit increases the price of a commodity, and a more rigid system of payment should be reflected in a tendency to scale down stud fees.

Before concluding this chapter I will add, lest I have given an impression that there is a kind of cold war continually being waged between stallion owners and mare owners regarding stud fees, etc., that nothing could be further from the truth. The wonderful world-wide results produced by the industry is ample proof that the machinery works with great efficiency and these achievements could not have been attained without a very high degree of co-operation between all concerned. My objective in recording both sides of the various stud terms is merely to place the facts, as I see them, before readers.

CHAPTER 11

Stud Management

Stallion management: Stallion temperament: fertility of thoroughbreds: broodmare management: shy breeders: case sheets for mares: some causes of barrenness

A STUD MANAGER requires the very maximum knowledge of practically everything connected with horses—particularly their sex behaviour and sex ailments. He must be conversant with the lessons of the annual volumes of Races Past and Stud Book. He must understand the principles of breeding and the various theories and practices appertaining thereto. He must be familiar with the fortunes of the various blood lines in all parts of the world. He must also be a good administrator and general farmer besides possessing a working knowledge of a host of other matters.

Last but not least he should be a good salesman. This quality is one of some significance in these days of commercial stud farming. An expert will suitably dispose of a dozen nominations to the stud's stallion whilst a poor salesman is selling a couple, and the value of his services at the time of the yearling sales is great. It will be obvious that to cover even a fraction of the attributes needed by that paragon, a good stud manager, would require many volumes written by many technical experts. My remarks on the subject will therefore be confined to a very few points.

First I will record my opinion that no man is qualified to take charge of a stud farm, either as a stud manager or as a stud groom, unless he has spent some years on an establishment where a stallion holds court. During five years so employed he will see upwards to a couple of hundred different mares—or proportionately more if there are several sires kept. I stress the point that they are different mares as it is from their variety that comparatively wide experience can be gained within a few years.

The opportunity will be afforded to study their idiosyncrasies and especially their behaviour when foaling and under sexual excitement. A considerable chance of acquiring a knowledge of the various ailments they suffer from will be at hand. Facilities will be on offer to peep behind the scenes of other studs and to garner from the circumstances of visiting mares, the manner in which other people combat the numerous difficulties which are inseparable from broodmare management, etc. A man so engaged will quickly realize that a broodmare or a stallion is not a kind of penny in the slot machine who functions in a purely mechanical manner. Anyone who has spent a lifetime on a stud without a sire will never have the same opportunities, and will rarely be able fully to appreciate all the practical troubles often encountered of mating two somewhat highly strung and unpredictable animals.

THOROUGHBRED STALLION'S EXCITABILITY

I think that in recent years the general standard of stallion management has shown a marked improvement, but still there often is room for further advances. To understand the difficulties concerning stallions it is necessary to consider various points. To begin with there is the question of temperament. It is common knowledge that the average thoroughbred stallion is far more excitable than the average thoroughbred mare. Why should this be so? Every cab driver from Tunis to Tokyo drives stallions in his gharry without trouble, and not infrequently yoked in the vehicle with a mare. A man is no more excitable than a woman, a dog than a bitch, a cock than a hen, etc. A bull, a ram, or a boar do not usually show the signs of being so overwrought as is commonly noticeable amongst thoroughbred stallions. Observation of a herd of Exmoor or Dartmoor ponies running out in wild surroundings shows that great temperamental disparity amongst the equine sexes is not natural.

EFFECTS OF ARTIFICIAL CONDITIONS

There are various causes which so commonly produce irritability in the thoroughbred entire. He often leads an isolated, lone life, is under-exercised and often overfed. In the years gone by it was the common practice to keep a stallion cooped up in a box with only a couple of hours daily walking exercise, often less. He was scarcely allowed to see a mare except for covering purposes, and during the breeding season was taken out of his box more frequently for this purpose than for any other. In fact he was nearly trained to show sexual excitement whenever he saw a mare and could almost anticipate that that was expected of him whenever he was let out of his prison. Very often his quarters were located far from any other horses or mares. Is it to be wondered at that he became overwrought? If a man were similarly treated for many years on end he would assuredly end up in the lunatic asylum.

This old-fashioned régime is unhappily still far too frequent, but the practice of turning out stallions in reasonably sized paddocks where they can see mares and sometimes other horses is on the increase. There they can graze, exercise themselves and obtain to some degree the benefits of sun, air, and natural surroundings which are denied them by the artificial conditions imposed by the necessities of their vocation.

A horse in training does a considerable amount of fast or fairly fast work which is sufficient to provide him with all the exercise he requires. When he goes to stud he becomes gross, so that his body and organs are not in a condition to withstand this treatment. He then used to be walked out for a couple of hours daily during which he covered a bare eight miles. This is far below his natural requirements, for if horses or mares left to themselves in the paddocks are watched it will be noticed that they are everlastingly on the move. It is true that at times when grazing their rate of progress is slow but it is continuous. As they eat they are always moving. After feeding in one corner of their field for a short time they will walk over to another spot. If they are out day and night they graze and move about for some 20 hours out of the 24, and during this time probably cover 20 or more miles, so an eight-mile walk offers a poor substitute. Furthermore the risks for an excitable animal of road exercise with slippery surfaces, motor and lorry traffic are large. A stallion loose in a paddock exercises himself for the period he is at liberty.

The usually quiet demeanour of the average travelling Premium Stallion affords a useful lesson. The necessity of his function ensures that he has a fair measure of exercise, being led from place to place daily. He sees everything that is going on in the world. He constantly meets and passes other horses and mares during his perambulations—and in fact leads a reasonably interesting life. As a result he is generally of an equable disposition.

ACCIDENTS TO STALLIONS AT GRASS

When comparing the advantages of allowing a stallion liberty in a paddock with those ensuing from boxed isolation and walking exercise, the question of the risk of accidents needs consideration. Some horses when loose are much more inclined to gallop about and thus expose themselves to harm than others. I think to some extent this tendency can be controlled by careful management. If, on coming out of training and before going to stud, a horse is given good bouts of exercise and then turned out when he is tired he will be inclined to settle down quietly more quickly than if he is simply let loose without any preliminary work. This treatment can be repeated and as time goes on his walking exercise can be curtailed and his period of freedom increased. He will thus become accustomed to being free. It is a mistake to get alarmed because he kicks and bucks about when first turned out and to hastily bring him in. He must be given a reasonable chance to get used to his freedom. It is also an error to bring him in in the early stages of this education, just as he has settled down.

Although each animal must be treated individually according to his temperament I do not think that a higher percentage of accidents is suffered amongst stallions at liberty than amongst those who lead a wholly artificial life. The latter in their overstrung—often neurotic—condition, kick and dance about the place, constantly endangering themselves, and if they slip up it is on a hard road or stable yard. Hard as their leader tries to control them he can do little; often he unwittingly does more harm than good.

To a marked extent a horse, left to himself, will steer clear of trouble, but when man tries to aid him it often ends in disaster. For example a horse jumping fences completely free, with no one on his back, rarely falls, but if he is ridden, until he is a fully experienced 'chaser or hunter, spills are common. The position in regard to the excitable led stallion is at times something akin to a man standing on one leg who can be readily knocked down by a light blow. So also is the horse standing on his hind legs easily overbalanced by a slight but painful chug under the chin from the leading rein—and with the uncertainty of his movements no man can be sure of avoiding this. Except as the result of kicks and other extraneous impact a horse hardly ever suffers fatal injury unless he falls. Naturally a stallion who has been isolated and hand exercised for years is not a good medium for the running loose policy.

Whichever method is followed I am sure that his temperament is influenced by his treatment immediately after he has covered his first few mares. He should then at once continue his normal routine as if nothing had happened. It is fatal to his nervous system to make any change just as he has discovered his new powers. On no account should he be shut up to wait until the next mare is ready for his attentions.

If it is possible it is probably in a horse's best interests for him to be looked after during his early stud career by the same man as tended him when in training, provided of course that man and horse are on friendly terms. A stallion is an individualistic creature who undoubtedly shows partiality for the person who looks after him. The two know each other's little tricks and foibles so it imposes an avoidable nervous strain on the horse if he is subject at the same time to a new outlook on life and to a strange attendant.

Some horses have their tempers more or less permanently ruined by inconsiderate treatment during their racing days. The bad stable lad continually rebukes his charge to show how clever he is in overcoming the animal's alleged wickedness. This is also true of some stallion men. If a horse stands on one side of his box they lack the patience to wait a few seconds until he is ready to move and give him a dig in the ribs to hasten him. If he is brought out for inspection he is nagged first to put one leg forward and then the other until he stands like the lions in Trafalgar Square. This is totally unnecessary and harmful.

HORSES FEAR QUICK MOVEMENTS

It has been my experience when dealing with stallions that tact rather than force produces the best results. A point I have noticed is that until complete confidence has been established between man and horse all movements should be slow—almost lazy. A hasty grab at a head collar or any jerky action is inclined to produce nervousness. The fear of quick movement is an inborne impulse in horses. All are familiar with the animal who will face a steady stream of traffic rolling by on the road but who loses his nerve when a single vehicle travelling at 60 miles per hour approaches on a lonely stretch. The only antidote for frayed nerves is the establishment of complete confidence between man and horse. The former must be absolutely sure the horse is not going to do him any harm and the latter must be totally free from any fear of his man. When this state has been reached I have found that a stallion is usually a most obedient, kindly, pleasant animal. For an example of mutual trust see the Frontispiece.

EXCITABILITY NO BAR TO FERTILITY

To turn to another aspect of the matter, there is no evidence to prove that a highly excitable stallion is less fertile or shows a lower ability to sire winners than one of equable temperament. The drawback to his idiosyncrasies is more or less limited to the increased chance of accident he is liable to occasion. The defect is transmittable to a horse's progeny on the same lines as any other factor in his make-up. Some will pass it on with comparative regularity, some practically never, and some spasmodically.

When considering in advance probabilities, it is necessary to give thought to the reason why the particular horse is overwrought. If he was born that way and always suffered from nervous disability it is only reasonable to anticipate that he may be liable to pass on the trouble. If he has been sympathetically handled and met with no outside accidents to account for his state it can be assumed that he is an animal of sub-normal structure who has proved his inability to withstand the racket of normal life. He therefore has a weakness in his make-up which he may transmit. But if his unequable temperament can be traced to an accident or maltreatment I do not think that transmission need be feared.

The expression "a temperamental horse" covers a variety of conditions. One may be merely nervous, a second highly excitable, a third self-assertive, and a fourth show inclinations to attack with teeth and forelegs all and sundry. In a rough and ready way it can be said that these are all closely allied and only divided by degree. As the animal's nerves get more and more the better of him his behaviour deteriorates from the nervous to the dangerous—unless he falls into hands capable of arresting or reversing the process.

A horse who will not do his best on the racecourse is generally an entirely different proposition. His trouble may be due to sheer softness and lack of guts, or possibly just as, if not more often to some hidden bodily or organic defect which causes him pain when under pressure. In either of these circumstances he must be looked upon with suspicion as a stallion, as undoubtedly he suffers from a weakness of mind or body—and weaknesses are heritable.

INDIVIDUAL TREATMENT ESSENTIAL

I have dealt at some length with some aspects of stallion management, but I have no wish to lay down the law. The only indisputable principle in dealing with them, that I am aware of, is that each horse must be subject to different treatment according to his disposition, constitution and general make-up. In some cases the use of the stick is the only remedy for wrong-doing, whilst in others it would be asking for trouble to allow a horse to roam a paddock on his own.

The number of mares a stallion can usefully cover is a debatable point. Does the unusually free use of him or a stringent conservation of his powers have any effect on his fertility or the quality of his stock? This is yet another matter which could only be satisfactorily settled after a most comprehensive scientific research into the lessons of the Stud Book and the annual volumes of Races Past. On the face of things a marked degree of conservation probably means fewer mates and fewer mates results in fewer foals. Therefore a horse treated in this way is handicapped from being a stud success at the outset owing to the small number of his stock to appear on the racecourse. But provided he has a full complement of mares who readily conceive to him, there is no point in using him more often than necessary.

I have previously remarked (see page 92) that a stallion who attracts only a few mares and does well with them so that he rises to popularity, only to fail when mated to a full quota is, more often than not, an indifferent racer or poorly bred animal whose failure is not due to increased use but to defective hereditable elements. I will also later (see page 274) refer to 'chaser sires who have covered 60 or more mares, shown normal fertility and produced excellent results.

It should not escape notice that in order to gain honours under National Hunt Rules their stock must possess a high degree of robustness. Moreover a 'chaser sire is often used with considerable liberality and the order of the day is to cover most of their mates twice or more every time the latter are in use. Incidentally the greatest Flat-race sire of modern times—St. Simon—was mated to 46 or 47 mares per season for years on end. In these days most of the leading Flat-race sires in Europe are put to 40 mares per season, so it is safe to say that a horse can produce his best with this number. It is impossible to determine if there would be a drop in fertility or quality with an additional ten or a dozen.

From present knowledge of the subject I think that the situation may be summed up as follows: (A) A normal stallion can with success cover at least 40 mares—possibly a good many more. (B) The essential objective is to get these mares in foal. (C) The fewer services given to obtain this objective the better, but the objective must have over-riding consideration.

Perhaps the greatest bugbear breeders of thoroughbreds have to put up with is the losses sustained by infertility. I however consider the fact that its incidence is not higher is a wonderful tribute to all connected with the bloodstock industry. The set-up of breeding racehorses imposes several circumstances which are fundamentally opposed to Nature and directly likely to produce disappointments from this cause. In order to meet the exigencies of racing conditions the universal practice is to carry out matings at a time of the year which is at variance with natural conditions. In her wisdom Nature sets apart the spring and summer for this purpose when the parents are enjoying the benefits of new grass, ample sun and warm conditions. Her wishes are clearly shown by the comparative ease with which many mares who failed to do so in the early part of the thoroughbred breeding season will conceive in May or June.

SCIENCE VERSUS NATURE IN BREEDING

Amongst Dartmoor, Exmoor and other wild ponies, running loose with the sexes together, January and February foals are extremely rare. Their average fertility is considerably higher than amongst thoroughbreds, as in addition to this advantage the stallion from natural instinct covers his mares at the most likely period of their œstruses to produce foals. Observation shows that under these conditions and free from man's promptings a stallion does not cover to excess. For a long time past, and especially during recent years, the veterinary profession and stud managers have given great attention to determining the best psychological moment for mating, but for my part I doubt if science is equally efficient to Nature in deciding the correct time for this purely natural act. When comparing the overall fertility between thoroughbreds and other breeds the most modern scientific methods are only relative to a limited degree, as their use is confined to but a minority of the great bulk of broodmares at stud and for all practical considerations they are not used amongst the others.

A further point which ensures greater fertility amongst the wild breeds is that when one of their mares comes into use, however short the period, the stallion does not miss his opportunities. Amongst thoroughbreds the detection of this state rests in man's hands aided by a teaser. I have had ample personal proof that sometimes the œstrus period only extends for such a short time as 24 hours and occasionally even less.

In these cases, however carefully a stud is managed, failure to recognize the mare's condition cannot always be avoided, especially if it is only present in full for a short period during the hours of darkness. The result is that the delinquent is returned to her owner genuinely believed in foal, unless veterinary examination shows that this is not so. These cases are exceptional, but a condition on somewhat similar lines is quite common. It is not unusual to encounter mares who rarely breed two foals in succeeding years. Very often these are matrons who are exceptionally fond of their offspring and are so wrapped up in them that they do not show in use in the normal way. With care it is usually possible to get them covered on or about the ninth day after foaling, but for man to "catch" them if they "break" is a different story. Here again under natural conditions the stallion is quickly aware of the state of affairs.

SHY BREEDERS

A very serious handicap to a higher overall fertility percentage amongst thoroughbreds is the retention at stud of many mares with previous poor foaling records. The probably high value of their produce makes it an economic proposition to persevere with them and to accept their failing, in the hopes of the little gold mines they will (or may) eventually produce. The numbers of these are large and their retention probably is in the best interests of their breed. Many of the great pillars of the Stud Book have been sons or daughters of shy breeders. They, however, act as an effective brake on the fertility ratio of bloodstock whilst their counterparts amongst other breeds are quickly weeded out.

Another adverse factor which operates against higher fecundity is that thoroughbred mares are usually confined to small paddocks rendering the transmission of disease amongst them easy. Even those owned by small breeders, who keep only two or three broodmares, are normally sent during the breeding season to a stud where this practice is in vogue. A number of the maladies which cause barrenness are traceable to germs which may gain entrance to the system through the mouth whilst the mare is grazing. Try as one will with all the modern scientific methods of keeping paddocks clean, it is not possible to prevent a mare, unsuspected of infection, discharging small quantities of germs to the danger of her companions. Pony breeds and "half-breds" are not exposed to this risk. The ponies roam over an almost unlimited acreage of fresh clean ground, whilst the "half-breds" have their being on farms in ones and twos, and are not sent to a central stud for mating.

Thus amongst bloodstock the following conditions militate against fertility: (A) the unnatural time of the year of the breeding season. (B) The difficulty of mating at the psychological moment and of detecting short periods or sub-normal signs of œstrus. These are common to other stock artificially bred. (C) The presence of a considerable number of shy breeders at stud. (D) The danger of the transmission of disease from mare to mare.

Owing to the methods of their production it is not possible to obtain a completely accurate overall figure of the fertility percentages amongst Exmoors, Dartmoors and other wild ponies, but an estimate of about 90 per cent, running up in some cases to close on 100 per cent, would be about correct. Amongst thoroughbreds the average is some 60 per cent and amongst "half-breds", covered by Premium Stallions, about 54 per cent. The "half-breds" have few of the handicaps of the thoroughbreds so the greater fecundity of the latter is a wonderful tribute to the efficiency of the methods employed by bloodstock breeders and to members of the veterinary profession engaged in this work. Further, amongst a limited band of the better-class thoroughbreds, where mares with poor foaling records are eliminated and the most modern scientific methods used, fertility incidences of 80 per cent, 90 per cent or even more are the order of the day.

DECLINE OF FERTILITY

Unhappily, however, the overall fertility amongst both thoroughbreds and "half-breds" is lower now than in the days gone by. The earliest statistics published in the General Stud Book refer to 1846 and show that 76.5 per cent of the mares then covered proved in foal. Since then there has been a gradual but steady decline. The oldest records of the Premium Stallions I have available are for 1895 when their fecundity was 61 per cent. It is hard to account for these drops, especially amongst bloodstock, some of whom have the advantages of more advanced scientific treatment.

Figures vary from year to year, but those for 1971 will serve as a general indication of modern trends. The Return of Mares for that year showed that 14,555 mares were recorded, and 13,468 were covered in 1970. The total number of mares who proved barren or for whom no returns were made was 4,639 (or 31.9 per cent), while the total number of mares who produced live foals was 7,486 (52 per cent). The percentage of live foals to the total number of mares, 52 per cent, corresponds roughly to the average for the previous 14 years, during which time the total number of mares accounted for rose steadily from 7,826 to 14,555. Thus it can be seen that there has been no improvement in the number of live foals produced; it was 54 per cent in 1957, and 52 per cent in 1971.

As time goes on, one looks for improvement, not stagnation or retrogression. The possible effect of shortage of trained personnel with a lifetime experience of breeding is not an acceptable answer to the conundrum, as the steepest declines in both breeds occurred in the first quarter of this century when an ample supply of knowledgeable staff was available. Neither is any theory tenable which is based on the supposed over-racing of early two-year-olds as the "half-breds", who do not work until fully matured, show a corresponding fall in their fertility incidence to the thoroughbreds. Moreover, some of the Premium Stallions, who sire the "half-breds", never raced at all or only very lightly. Furthermore, so far as I am aware—but I am open to correction on this point—higher fertility amongst thoroughbreds is not registered in countries such as France, South Africa, Australia, etc., where two-year-old racing is not greatly in vogue, than at home. Any satisfactory explanation to the phenomenon must cover the conditions of the lives and breeding arrangements of both thoroughbreds and "half-breds", and for my part I have never heard one.

I am, however, absolutely convinced from personal experience that increased fertility would result if stallion managers, veterinary surgeons and stud grooms were in a position to know more about the individual mares visiting their sires. Broodmares vary enormously in their behaviour and particularly in their sex reactions. One mare will habitually be in use for ten or 12 days whilst another's œstrus period will be confined to a couple, or less. One mare will show readily in use to a teaser and another indicate her condition only to her own sex. One mare will be highly excitable and even almost dangerous at the time of mating and another totally placid. One mare when first turned out after foaling will gallop about regardless of her unfortunate foal's safety and another will take the greatest care of her newborn's interests. There is no predicting what a strange mare will do.

CASE SHEETS FOR MARES

When mares are sent to a stallion at public stud and are there only a few weeks, the staff have no chance to ascertain in such a short time the peculiarities of each individual. This is particularly relative when they come from long distances and it is not practical to get into touch with their owners at short notice. It is hard to see how this difficulty can be overcome but I think that something might be done in the way of passing on from stud to stud a record of a mare's behaviour somewhat on the lines of a soldier's conduct sheet. I myself keep the particulars of mares visiting my sires on the pro-forma shown overleaf.

These are kept in a loose-leaf binder. At first sight the pro-forma appears a formidable document, but in every well-managed stud practically all the detail recorded thereon is kept somewhere or other in the stud's books. I have used it for up to 80 visiting mares and have found it a labour-saving device to have all particulars of each mare entered on one piece of paper, instead of rummaging about in several books. It will be seen not only are full details of the mare's stud behaviour provided for, but also that accounts can readily be made up from the form. At the conclusion of the breeding season a copy of it can be sent to the mare owner, who is thus in a position to give the stud manager of the stallion the mare visits the following season a comprehensive picture of her previous record. If something on these lines could be generally adopted and passed from stud to stud a complete case history of each mare would be available. The result could only be increased fertility from greater knowledge of the individual, a great saving of time and the reduction of the risk of mishaps now caused by the discovery of peculiarities by a process of trial and error.

I am confident that all stallion managers and veterinary surgeons will agree with me that their work would be much simplified if they knew more about some of the animals they have to deal with. Moreover the plan would be an incentive to better stud management. No stud would like their misdeeds—such as failure to try a mare at the appropriate time—exposed to public record. There might occasionally be a stud manager or groom tempted to "cook" the detail, but these would be extremely rare and it is not easily possible to guard totally against crooks.

THE FOLLY OF SECRETIVENESS

In the bad old days there was an air of secretiveness about visiting mares' behaviour in some poorly-managed studs. There was a feeling if things did not turn out satisfactorily and full details of their actions were known, that the stud would be blamed for not taking some alternative steps to put matters right. But that outlook is passing and today there is greater co-operation. It is realized that difficulties are bound to occur and that mutual help to avoid them—or to avoid their repetition—is in the best interests of all.

If a system on the lines I have outlined came into general use it would be advisable, if possible, to secure some general agreement on the exact detail of the case sheet and its size, in order to facilitate filing. In the layout I use there is certain embroidery which is not relative, such as details of a mare's racing record and family history. I record these as I find the stud hands work better if they know something about the mares they are handling instead of identifying them by such terms as "the bald-faced bay with the moth-eaten tail".

To turn to other matters. I have formed the opinion that the life-long segregation of mares often produces bad effects on their nervous systems and sex impulses. This is shown occasionally in their somewhat stallion-like behaviour, and quite commonly in their habit of showing in use to other mares when they will not do so to a teaser. This does not necessarily decrease their chances of conceiving except from the fact that it is often hard to detect whether or not they are in use, and so the moment for successful mating may be missed. To alleviate this trouble it is a good policy to turn out a quiet gelding with the barren mares for a few months occasionally during the non-breeding season. Foaling mares are safest in one another's company.

ISOLATION OF VISITING MARES

If practical I consider all newcomers to a stud should be isolated for ten or 14 days after arrival, especially if they have travelled by other than the mare owner's own transport. I am well aware that private companies have rigid regulations about disinfecting horses boxes after use, but I am not convinced that these measures are totally effective. However thoroughly they are carried out it is not easy for the disinfectant to penetrate into every nook and corner of the vehicle so, as a precautionary step, the isolation of mares after arrival is a wise one. From time to time mysterious outbreaks of contagious or infectious disease break out in studs when all visiting mares come from homes beyond suspicion. I often wonder if these are due to infection picked up in transit. From personal experience I know

FRONT OF PRO-FORMA

Mare's Name   Mated to   [Stallion's name]

Owner's name, address, and phone number

Date and method of arrival.   Date and method of return.

Kit on arrival.

Due to foal (date).   Foaled (date, sex, colour)   by

In use	Covered	Off		Oestrus Cycle	
				Number of days in use	Number of days off
			1st heat		
			2nd "		
			3rd "		
			4th "		
			5th "		

Tried	
February	
March	
April	
May	
June	

BACK OF PRO-FORMA

Mare's Name and Markings
_____

Racing record, breeding and family history
_____
_____
_____
_____

Mare's past foaling record	19....	19....
	19....	19....
	19....	19....

Foal's Markings
_____
_____
_____

Veterinary treatment of mare and foal whilst at stud
_____
_____
_____
_____
_____
_____

Date and result of pregnancy test
_____

Mare's disposition at teasing bar, covering time and other notes
_____
_____
_____
_____
_____

Shoeing
_____

of more than one case when all evidence pointed to this being so, but luckily through isolation the trouble was nipped in the bud before spreading.

VULVA STITCHING

In recent years the practice of vulva stitching has become very prevalent as an antidote for sucking in wind, a habit often known as "fluting". It has undoubtedly resulted in many mares becoming pregnant who could not have been without this little operation. But I am inclined to think that sometimes it is resorted to too readily. The mares who are afflicted with this trouble usually have their anuses and vulvas in an almost concave position. To use a simple term, that part of their anatomy is dropped in. In perfect health these parts are convex and sticking out.

The concave condition is not infrequently due to general debility rather than any other cause. I therefore think that the best course to take is to get the animal into prime bodily health. Her lack of it may be due to worms, bad teeth, indigestion or a dozen other things, and if these are tackled the complaint often rights itself and moreover all the other organs of the body are toned up so as to bring her into first-rate breeding trim. However when a mare is sent to a horse in a debilitated state the time available for her to build up her frame is insufficient for the purpose and stitching is the only course to pursue.

Sucking in wind seems much more prevalent today than it appeared to be some years ago, and I think this may be due to the over-use of the speculum in some studs. The continual stretching of the parts to permit the introduction of this instrument produces a condition which is conducive to the trouble. Too frequent rectal examinations may perhaps have the same inclination.

BARREN MARES NEED CARE

When a mare comes back barren from a stallion I am afraid that there is sometimes a tendency to turn her out in the paddocks and to do little or nothing for her until the next breeding season is at hand. This is a grave mistake. If a mare fails to breed she must be looked upon with suspicion. It is not by any means always possible to detect or to cure the cause of her failure. She should be most carefully watched, her œstrus cycle recorded, her tail daily inspected for the very slightest sign of a discharge, etc., and a veterinary surgeon called in. Very often the most careful observation will fail to disclose any reason known to man for her deflection from her natural duty, but no stone should be left unturned to trace possibilities whilst there is sufficient time to effect a cure if necessary.

It is unreasonable, and often too late, to place a mare suffering from some disability under the veterinary surgeon's care only a short time before she is again due to go to horse. Often up to six months elapse before any concrete steps are taken to put matters right. This is obviously wrong. The time of action is directly her failure to breed is known, and care should not be relaxed until she goes to stud in the following year. If treatment is necessary and effects a cure her owner must realize she may suffer a relapse or develop some other malady. A veterinary surgeon can say that an animal is in a certain condition only at the time he sees her. He cannot be a sort of fairy godmother to guard her against all ills for the rest of her days.

I am absolutely confident that if greater care was taken of barren mares the incidence of fertility throughout the breed would increase appreciably. Incidentally I often hear disappointed mare owners say that they cannot understand a specific mare being barren as the veterinary surgeon said she was in perfect breeding condition before being sent to horse. I then wonder if the veterinary surgeon did, in fact, make such a sweeping statement. If he did he must have been somewhat rash, to say the least of it. From a short inspection how is he to know that the mare's œstrus cycle and various other matters are normal and right?

It has been my experience that mares with irregular cycles of heat are extremely hard to get in foal. Moreover to the best of my belief there is no effective cure for this condition available at present. I have tried many but none has been satisfactory. In some cases the fault rights itself late in the season when the mare has had the benefits of the spring grass, the sun and warm weather, but most breeders want an early foal and so patience does not altogether meet their needs. When mares with this trouble have been successfully treated the veterinary surgeon is often proud of his cure, but it should not be overlooked that there is no certainty that his actions had anything to do with recovery. As likely as not this was due to the natural factors which I have mentioned.

To be accepted as satisfactory any artificial method of overcoming the malady must show a very high percentage of immediate successes early in the breeding season. I had better add that when I use the term an irregular cycle I mean what I say and not an abnormal cycle. In the former case a mare will come into use after intervals of 9, 12, 15, 23 days or at any other time and her period of heat will vary at each cycle. In the latter circumstance her periods will be regular but abnormal, i.e. instead of "coming on" every 21 days she will do so every 19th or 23rd day with precision. Generally no harm results from abnormal cycles.

THE TEASER

There is a humble but important member of all establishments where a stallion holds court, namely the teaser. He is the animal who will tell us whether the mares are in use or not, and therefore bears considerable responsibility for the incidence of fertility. He is far too often chosen for his job because he happens to be a comfortable hack for the stud manager or because he is a reliable slave in the shafts of the manure cart. Very little attention is paid to, shall I say, his technical skill in carrying out his duties. If a stallion and a mare in use, completely free from man's interference and in natural surroundings, are watched it will be noticed that the stallion is usually extremely gentle and quiet with his mate. Naturally the severity of his attentions depends on the individual mare, but in the great majority of cases the absence of shrieking or biting teeth, etc., are most noteworthy. The little nips he gives her are barely enough to hurt a man's hand. As the teaser is enacting the part of the stallion under wild conditions he must behave in the same way. Under the acutely artificial and provocative circumstances in which he is forced to carry out his duties it is not easy to find a horse who will do so. The teaser who rushes at the mares and chews lumps out of them will not only put many mares off showing that they are in use, but also may directly cause an early slipped fœtus when trying a mare who is a few weeks in foal. His demeanour must be extremely circumspect but he must also be willing to show robust attentions on occasions. A good teaser is almost worth his weight in gold.

PART TWO

DERBY WINNERS 1900–1976
AND THEIR RELATIONS

CHAPTER 12

1900–1914
The Great Days of Racing

1900

1　HRH the Prince of Wales's b c Diamond Jubilee (St. Simon – Perdita II)
2　Duke of Portland's b or br c Simondale (St. Simon – Ismay)
3　J. B. Keene's br c Disguise II (Domino – Bonnie Gal)
　　14 ran.　Time: 2 min 42 sec

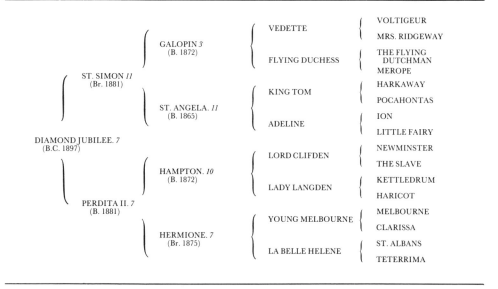

THE NAME OF Diamond Jubilee will be found amongst the few horses who have won the Triple Crown but it is extremely doubtful if he was as good a racer as his elder brother Persimmon, who won the Derby, St. Leger and Ascot Gold Cup, whilst another brother Florizel II was a Goodwood Cup winner. Their dam Perdita II started her racing career as a selling plater but later proved quite a good performer, winning seven races during her four years in training. Her last owner during her racing career (Mr. A. Falconer) considered her a terrible jade and gladly accepted an offer of £900 for her made on behalf of the Prince of Wales. Jade or no jade, she proved herself to be one of the most influential mares ever bred, as all her three sons did well at stud.

Diamond Jubilee himself was a very highly strung and queer tempered horse. This trait was so bad that the Royal jockeys of the time could do nothing with him and in his races he was ridden by his stable lad, Herbert Jones, who later made a great reputation as a jockey. The horse's behaviour was so bad as a two-year-old that orders were issued for him to be castrated. When the veterinary surgeon arrived to carry out the operation, it was found that he was a one-sided rig and nothing was done.

Later in life he became a normal stallion. He spent a few seasons at stud in England and the list of sires of winners for 1907 shows him as seventh from the top—one place above his brother Persimmon who had far more opportunities. Diamond Jubilee was sold to Argentina in 1906 for 30,000 guineas and he was an outstanding success before dying at the age of 26.

1901

1 W. C. Whitney's b or br c Volodyovski (Florizel II – La Reine).
2 Duke of Portland's b c William the Third (St. Simon – Gravity).
3 Douglas Baird's ch c Veronese (Donovan – Maize).
    25 ran. Time: 2 min 40.8 sec

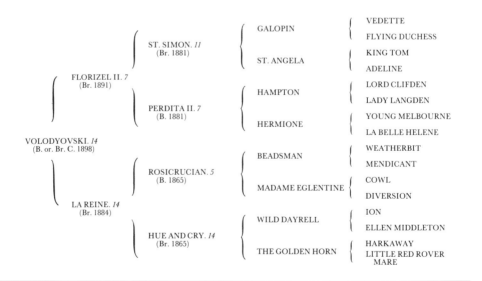

This was the first Derby to be started with a starting-gate—previously a flag had been used—and it drew the largest field since 1867. It is also the last Derby which included a gelding amongst the runners, although they were not excluded by the conditions of the race until 1906.

Volodyovski's sire Florizel II was a brother to Diamond Jubilee, who had won the great event the previous year. Volodyovski was the best two-year-old of his generation when leased to Lord William Beresford who died in December 1900. The question then arose as to whether the lease was valid to Lord William's executors and the Law Courts decided it was not, with the result that the horse was re-leased by his breeder Lady Meux to the American William Cornelius Whitney.

After his Derby victory Volodyovski gradually deteriorated and finished his racing career by running unplaced in the Manchester November Handicap as a four-year-old. He retired to stud at the modest fee of £48 per mare, but proved a failure during the few seasons he held court in England. In contrast, the runner-up William the Third made up into an excellent stayer the following year and at stud sired the Gold Cup winner Willonyx, the 1,000 Guineas winner Winkipop (later grandam of the Oaks winner Pennycomequick), and Roseland, who appears in the pedigree of Bahram as the sire of his grandam Garron Lass.

1902

1 J. Gubbin's br c Ard Patrick (St. Florian – Morganette).
2 Colonel H. McCalmont's ch c Rising Glass (Isinglass – Hautesse).
3 Duke of Portland's br c Friar Tuck (Friar's Balsam – Galopin Mare).
   18 ran. Time 2 min 42.2 sec

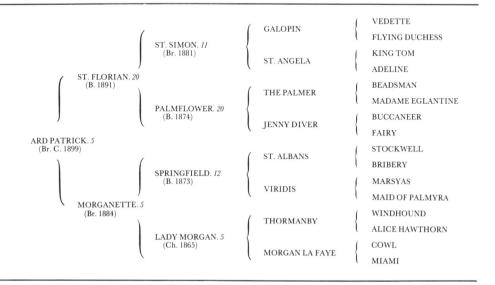

Amongst the "also rans" for this race was the great Sceptre, who won all the other Classics of the season and beat Ard Patrick in the 2,000 Guineas. The colt could not run in the St. Leger on account of leg trouble but he again beat the filly in the Eclipse the following year. Ard Patrick was a half-brother to Galtee More, who won the Triple Crown in 1897, and to the Irish Derby winner Blairfinde. Their dam Morganette never advanced beyond selling plates during her racing career and was a roarer. On retiring to stud her first mate was a "half-bred" horse. She was, however, well bred, as her dam was a half-sister to the Oaks and St. Leger winner Marie Stuart, whilst her great grandam was the Oaks winner Miami. It will be noticed that she was 15 years old when she produced Ard Patrick.

Ard Patrick's sire St. Florian was not much of a racehorse himself and except for his solitary Derby winner made no mark at stud, although he sired the dam of the 2,000 Guineas winner Slieve Gallion. St. Florian was nicely bred, being a half-brother to the Oaks winner Musa, whilst their dam was a sister to another Oaks winner Jenny Howlet. Ard Patrick was sold to the German Goverment for £21,000 as a four-year-old, and though not an outstanding success, he did sire the German Derby winner Ariel.

1903

1   Sir J. Miller's br c Rock Sand (Sainfoin – Roquebrune).
2   Edmond Blanc's b c Vinicius (Masque – Wandora).
3   Sir Daniel Cooper's b c Flotsam (St. Frusquin – Float).
    7 ran. Time: 2 min 42.8 sec

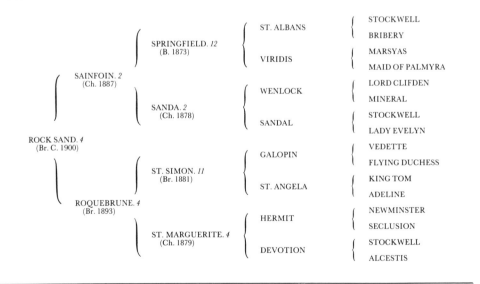

This was the smallest field for the Derby in this century. Amongst the "also rans" was Rabelais, who later made a great name for himself as a stallion in France after a deal which would have taken him to Russia fell through.

Rock Sand, a small horse and a first foal, also won the 2,000 Guineas and St. Leger. I have heard but cannot vouch for the truth of the story, that after he was weaned and until he went into training, he lived an isolated life with no companion in his paddock, and thus did not have the normal opportunities of developing his muscles and stride offered by the gambols of two youngsters together. Be that as it may, he was a first-class racer who was sold for £25,000 on the death of his owner. After a brief period at stud in England he was shipped to America and later went to France, where he died in 1914. Amongst his get in America were Mahubah (dam of the great racer Man O'War) and Tracery, who came to England to win the 1912 St. Leger and remained to sire the Derby winner Papyrus, the 2,000 Guineas winner The Panther, and other horses of merit including one of the Aga Khan's foundation mares Teresina.

Rock Sand's sire Sainfoin was bred by Queen Victoria and won the Derby of his year in Sir James Miller's colours, but apart from Rock Sand he was not a particularly good sire of winners. Rock Sand's dam Roquebrune won two races and was a half-sister to the Oaks and St. Leger winner Seabreeze and other horses of note, including the successful sire but bad racer Tredennis. Their dam St. Marguerite won the 1,000 Guineas, as also did her sister Thebais, whilst the latter in addition won the Oaks. Rock Sand therefore was not only a very good racehorse but also he was a very well-bred one. It will be noted he had only 14 great-great-grandparents—Stockwell's name appearing three times in his pedigree at the fourth remove.

1904

1   Leopold de Rothschild's b c St. Amant (St. Frusquin–Lady Loverule).
2   Sir J. Thursby's b c John O'Gaunt (Isinglass – La Flêche).
3   S. B. Joel's b c St. Denis (St. Simon – Brooch).
        8 ran. Time: 2 min 45.4 secs

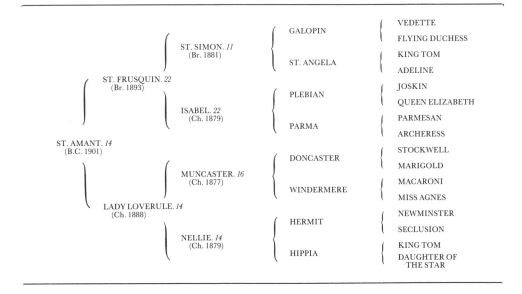

St. Amant, a contemporary of the brilliant filly Pretty Polly, also won the 2,000 Guineas, but after his Derby victory trained off badly and finished his racing career as a five-year-old when unplaced four times in handicaps. He was an unreliable and bad-tempered horse who made no great mark at stud before he died at the age of 23. His sire St. Frusquin finished second in the Derby but won the 2,000 Guineas and other important races before proving a great stud success.

St. Amant's dam Lady Loverule never ran but her own dam Nellie won seven races value £4,224 and her grandam Hippia scored on 16 occasions, including in the Oaks.

An unusual feature of St. Amant's Derby was that the second, John O'Gaunt, was ridden by the amateur George Thursby, half-brother to the owner to whose baronetage he later succeeded. Mr. Thursby was also second in the 2,000 Guineas on John O'Gaunt and second in the 1906 Derby on Picton. John O'Gaunt, who was by the Triple Crown winner Isinglass out of the 1,000 Guineas, Oaks and St. Leger winner La Flèche, ensured himself permanent record of his name by siring the St. Leger winner and important stallion Swynford in his first season and later the 2,000 Guineas winner Kennymore (who died after one season at stud), though his capacity to get good winners dwindled and he ended his career at an advertised fee of 25 guineas a mare.

1905

1   Lord Rosebery's ch c Cicero (Cyllene – Gas).
2   Edmond Blanc's b c Jardy (Flying Fox – Airs and Graces).
3   Chev. E. Ginistrelli's b c Signorino (Best Man – Signorina).
      9 ran. Time: 2 min 39.6 sec

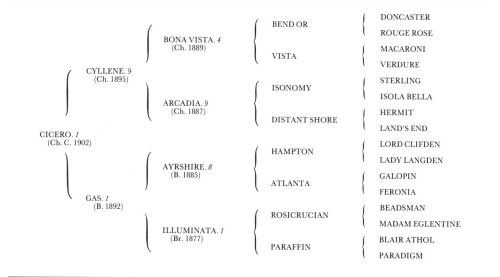

	BONA VISTA. *4* (Ch. 1889)	BEND OR	DONCASTER / ROUGE ROSE
CYLLENE. *9* (Ch. 1895)		VISTA	MACARONI / VERDURE
	ARCADIA. *9* (Ch. 1887)	ISONOMY	STERLING / ISOLA BELLA
CICERO. *1* (Ch. C. 1902)		DISTANT SHORE	HERMIT / LAND'S END
	AYRSHIRE. *8* (B. 1885)	HAMPTON	LORD CLIFDEN / LADY LANGDEN
GAS. *1* (B. 1892)		ATLANTA	GALOPIN / FERONIA
	ILLUMINATA. *1* (Br. 1877)	ROSICRUCIAN	BEADSMAN / MADAM EGLENTINE
		PARAFFIN	BLAIR ATHOL / PARADIGM

Cicero was unbeaten as a two-year-old and did not run in either the 2,000 Guineas or the St. Leger. In the latter race it is interesting to note that the great stallion Polymelus, who is sometimes referred to as a non-stayer, finished second. Cicero was a fairly good but by no means top-class stallion and his name lives on mainly through his son Friar Marcus, who sired the 1,000 Guineas winner Brown Betty, and also Friar's Daughter, dam of Bahram.

Cicero's sire Cyllene, foaled on May 28, was so small as a yearling that he was not entered for the Classics. Later he grew to full size and after winning the Ascot Gold Cup, etc., turned out to be a remarkably successful stallion, getting four Derby winners (Cicero, Minoru, Lemberg and Tagalie) before he was sold to Argentina as a 13-year-old for £25,000. There his blood proved equally valuable before he died aged 30.

Cicero's dam Gas was closely related to the 2,000 Guineas and Derby winner Ladas (by Hampton out of Illuminata), and half-sister to the 1,000 Guineas winner Chelandry.

Third in Cicero's Derby was Signorino, whose half-sister Signorinetta was to win the race three years later and who himself became several times champion sire in Italy.

1906

1 Major Eustace Loder's b c Spearmint (Carbine – Maid of the Mint).
2 J. L. Dugdale's ch c Picton (Orvieto – Hecuba).
3 Duke of Westminster's b c Troutbeck (Ladas – Rydal Mount).
   22 ran. Time: 2 min 36.8 sec

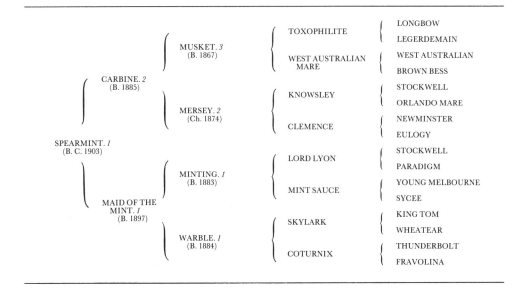

This was a vintage year and amongst the "also rans" were such as Radium, Beppo, Gorgos, Lally and The White Knight (Ascot Gold Cup twice) who all enjoyed considerable reputations on the course.

Spearmint was bred at the famous Sledmere Stud and realized but 300 guineas as a yearling at auction. He was an exceptionally good racer whose Turf career was curtailed by his bad forelegs. He only appeared on a racecourse five times during his one-and-a-half seasons in training. Eleven days after winning the Derby he scored in the Grand Prix de Paris, but could not withstand further work and spent his four-year-old days in healthy exercise before taking up stud duties in 1908 at a fee of 250 guineas. He did fairly well but never got anything as good as himself, his best being the Derby winner Spion Kop and the rather moderate St. Leger winner Royal Lancer. Spearmint also sired Plucky Liege, dam of Bois Roussel (Derby), Admiral Drake (Grand Prix) and Sir Gallahad III, and is further notable as having started one of the only two three-generation dynasties of Derby winners in the 20th Century, since his son Spion Kop sired the 1928 winner Felstead. Gainsborough founded the other such example which was completed by Hyperion and Owen Tudor.

Spearmint's sire Carbine was bred in New Zealand and was a great racer there and in Australia, winning no less than 33 races during his four years at work, including the Melbourne Cup in record time, under 10st 5lb and from 38 rivals. He spent four years at stud there with fairly satisfactory results and in 1895 was bought for £13,000 by the Duke of Portland. Truth to tell that although he sired plenty of winners, he was not an unqualified success in England in spite of having the best of opportunities. This probably accounts for Spearmint's very low valuation as a yearling, coupled with his bad legs. Carbine was a very tough horse as in spite of his long and meritorious racing career he lived until he was 29 years old.

Spearmint's dam Maid of the Mint, who never ran, was a half-sister to the Cesarewitch winner Wargrave, who was also by Carbine.

1907
1   R. Croker's ch c Orby (Orme – Rhoda B).
2   Colonel E. W. Baird's b c Wool Winder (Martagon – St. Windeline).
3   Captain Greer's bl c Slieve Gallion (Gallinule – Reclusion).
    9 ran. Time: 2 min 44 sec

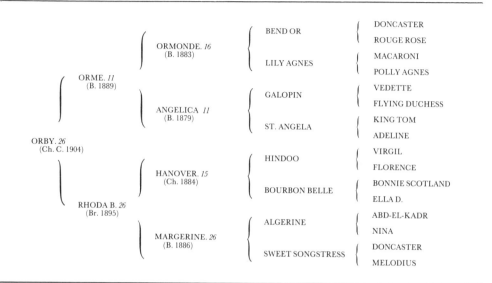

ORBY. *26* (Ch. C. 1904)	ORME. *11* (B. 1889)	ORMONDE. *16* (B. 1883)	BEND OR { DONCASTER / ROUGE ROSE
			LILY AGNES { MACARONI / POLLY AGNES
		ANGELICA *11* (B. 1879)	GALOPIN { VEDETTE / FLYING DUCHESS
			ST. ANGELA { KING TOM / ADELINE
	RHODA B. *26* (Br. 1895)	HANOVER. *15* (Ch. 1884)	HINDOO { VIRGIL / FLORENCE
			BOURBON BELLE { BONNIE SCOTLAND / ELLA D.
		MARGERINE. *26* (B. 1886)	ALGERINE { ABD-EL-KADR / NINA
			SWEET SONGSTRESS { DONCASTER / MELODIUS

A remarkable thing about this Derby was that the winner was trained by a doctor (Dr. McCabe) who had suspended his medical interests in order to become private trainer to Mr. Richard "Boss" Croker.

The merit of the field was in no way comparable to that of the previous year. Orby scored four times as a three-year-old, including both the English and Irish Derbys, before breaking down in his St. Leger preparation. At stud he sired the Derby winner Grand Parade, the 1,000 Guineas winner Diadem and a number of other useful horses who were chiefly endowed with speed rather than stamina. The vicissitudes of bloodstock breeding are brought out by the fact that while the sire line of the greatly superior Spearmint is now dead, that of Orby still exists, though it has gone into serious decline in more recent years, and is in danger of collapsing completely. The line has been carried on not through Grand Parade, but through an inferior son The Boss, whose six victories netted £1,584 and included the Gosforth Park Cup.

After starting at stud at a modest 15 guineas per mare, The Boss sired several speedy horses, and Golden Boss and Sir Cosmo founded the two main branches of the Orby line, which can be seen today in Be Friendly, Track Spare, Roi Soleil and Sweet Revenge.

Golden Boss appeared destined to be a stud failure and after his fee dwindled to nine guineas a mare, he was sold as an 11-year-old to the American Government as a remount sire, leaving behind in France a two-year-old called Gold Bridge.

Foaled very late, on May 26, Gold Bridge later came to England, and after showing he was the best sprinter of his time, proved an excellent sire, with Golden Cloud (sire of Matador and Ennis), Denturius and Vilmorin his most influential sons.

Sir Cosmo, a headstrong racehorse, got Panorama, who in turn sired Whistler and has become an important sire of broodmares, including the dams of the Irish Derby winners Chamier (by Chamossaire) and Panaslipper (by Solar Slipper).

Orby's sire Orme was a very good racer but could not run in either the 2,000 Guineas or Derby owing to poisoning—and was not, as the eminent breeder Federico Tesio has written, "unable to stay the distance of the Derby". Whether this poison was maliciously administered or the result of dental infection was never proved. In the St. Leger he was beaten out of a place by La Flêche but took his revenge on the filly in the Eclipse the following year. During his second year at stud he got Flying Fox, who was both a better racer and stallion than his sire, and it is curious that eight years elapsed before Orme got another Classic winner, and in that year besides Orby he was responsible for the 1,000 Guineas winner Witch Elm.

Orby's dam Rhoda B was bred in America and imported as a yearling. She also produced the 1,000 Guineas winner Rhodora, who died foaling to her half-brother Orby.

1908

1 Chev. E. Ginistrelli's b or br f Signorinetta (Chaleureux – Signorina).
2 Duke of Portland's b c Primer (St. Simon – Breviary).
3 Barclay Walker's b c Llangwm (Missel Thrush – Llangarren Lass).
    18 ran. Time: 2 min 39.8 sec

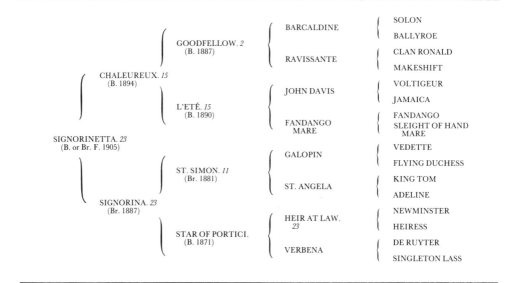

It is sometimes represented that this race produced a freak result, since the winner started at 100-1, but I cannot altogether agree. Some 25 years previously Chev. Ginistrelli had settled in Newmarket, where he bred, trained and raced his horses. Amongst these was Signorinetta's dam Signorina, who was the best two-year-old of her year, was beaten less than a length in the Oaks and in all won 11 races value the remarkable sum of £20,902 during her three seasons in training. Her son, as I have mentioned, was second in the 2,000 Guineas and third in the Derby of 1905. It cannot therefore be truthfully said that Chev. Ginistrelli lacked experience in racing and bloodstock matters. In fact, considering the small string he maintained, he seems to have been a man of unusual acumen.

Neither can I regard Signorinetta as a sub-standard Derby winner. Her preparation was probably handicapped by the lack of good galloping companions and trial horses, yet she won the race easily by two lengths in 2 minutes 39.8 seconds which was considerably faster than Diamond Jubilee, Ard Patrick and Rock Sand, who are generally looked upon as good Derby winners. Two days later she scored in the Oaks with ease in 2 minutes 42.4 seconds, and thus became the third filly in the history of the Turf able to accomplish the double, her predecessors being Eleanor in 1801 and Blink Bonny in 1857. Fifinella added her name to the list in 1916, but her victories were in war-time substitute races at Newmarket.

After her Epsom triumphs Signorinetta failed to win in three attempts and following her purchase by Lord Rosebery retired to the paddocks. At stud she produced six winners of races value £5,021, and her impact rests on her daughter Erycina, who bred a Scottish Grand National winner, and her son The Winter King, sire of the Grand Prix winner Barneveldt, who in turn sired Pont l'Eveque.

It will be noticed that Signorina was 18 years old when her Derby-winning daughter was born and thereby hangs a tale of remarkable perseverance on the part of Chev. Ginistrelli, as for the first ten seasons of her stud career this brilliant racer was barren—her first foal being Signorino and her second Signorinetta.

Signorinetta's sire Chaleureux started life as a selling plater but later became a useful handicapper, his victories including the Cesarewitch, the Manchester November Handicap and the Chesterfield Cup, Goodwood. He was a badly bred horse and breeders looked askance at him. Although he sired nothing else of note, he must be given credit for getting a Derby and Oaks winner from so few opportunities.

1909

1 His Majesty's br c Minoru (Cyllene – Mother Siegel).
2 W. Raphael's b c Louviers (Isinglass – St. Louvaine).
3 Lord Michelham's ch c William the Fourth (William the Third – Lady Sevington).

   15 ran. Time: 2 min 42.4 sec

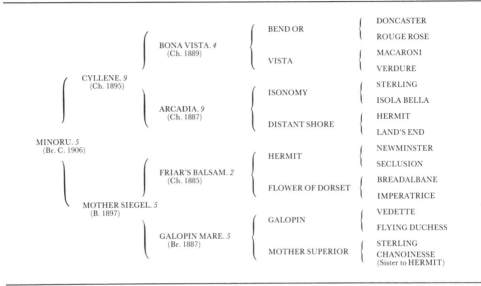

MINORU. *5*
(Br. C. 1906)

CYLLENE. *9*
(Ch. 1895)

BONA VISTA. *4*
(Ch. 1889)

BEND OR
{ DONCASTER
  ROUGE ROSE

VISTA
{ MACARONI
  VERDURE

ARCADIA. *9*
(Ch. 1887)

ISONOMY
{ STERLING
  ISOLA BELLA

DISTANT SHORE
{ HERMIT
  LAND'S END

MOTHER SIEGEL.. *5*
(B. 1897)

FRIAR'S BALSAM. *2*
(Ch. 1885)

HERMIT
{ NEWMINSTER
  SECLUSION

FLOWER OF DORSET
{ BREADALBANE
  IMPERATRICE

GALOPIN MARE. *5*
(Br. 1887)

GALOPIN
{ VEDETTE
  FLYING DUCHESS

MOTHER SUPERIOR
{ STERLING
  CHANOINESSE
  (Sister to HERMIT)

This is the only Derby to date to be won by a horse carrying the colours of the reigning monarch and the victory of Minoru was probably the most popular in the history of the race. King Edward VII was not only an exceptionally popular king but he also obviously greatly enjoyed racing, the sport of his people.

Minoru was bred by Colonel Hall-Walker (later Lord Wavertree) and leased to His Majesty for his racing career. He won the 2,000 Guineas and other important races before developing eye trouble and being retired to stud. A fortnight after his retirement, King Edward VII died, and Minoru went to Col. Hall-Walker's stud in Ireland, where for two years he stood at a fee of 98 guineas, before being exported to Russia, where the Derby runner-up Louviers also ended his days.

During his short spell in Ireland Minoru made an indelible mark on the Stud Book, for he sired Serenissima, whose produce included the 1,000 Guineas and St. Leger winner Tranquil, the Ascot Gold Cup winner Bosworth, and the famous mare Selene. The last-named became the dam of Hyperion, Sickle and Pharamond (who were influential sires in America), and All Moonshine, dam of champion sire Mossborough.

Minoru disappeared in the 1917 Russian Revolution. A story was current that he and Aboyeur, who won the 1913 Derby before export to the same destination, turned up with some White Russian refugees at the end of the 1914–18 war at Constantinople, where British troops were in occupation. Although two thoroughbred stallions undoubtedly arrived there, substantiation of their identity is doubtful.

A feature of Minoru's pedigree is his inbreeding to Hermit, whose name occurs at the third and fourth remove and whose sister Chamoinesse was Minoru's great-great-grandam.

1910

1 Mr. "Fairie's" b c Lemberg (Cyllene – Galicia).
2 Lord Villier's b c Greenback (St. Frusquin – Evergreen).
3 A. P. Cunliffe's b c Charles O'Malley (Desmond – Goody Two Shoes).
   15 ran. Time: 2 mins 35.2 secs

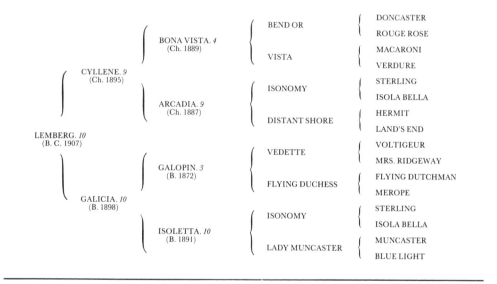

Owned by an Australian, Mr. A. W. Cox, who raced as "Mr. Fairie", Lemberg was also second in the 2,000 Guineas and third in the St. Leger. He can be regarded as well up to the average of a Derby winner but was probably not the best of his generation. This honour must be given to the St. Leger winner Swynford, who had not reached his peak at the time of the Derby.

At stud Lemberg retired at a fee of 250 guineas and did satisfactorily, being champion sire in 1922. His best get were the Grand Prix de Paris winner Lemonora, and the 2,000 Guineas winner Ellangowan, both of whom were moderate stallions, and the 1,000 Guineas winner Taj Mah.

Lemberg's dam Galicia was no less than 26 years younger than her sire Galopin but this did not prevent her from being one of the greatest broodmares in the history of the Stud Book. To her mating the year before Lemberg, she produced Bayardo (St. Leger, and sire of Gainsborough etc.) and also bred Kwang-Su (second in the 2,000 Guineas and Derby). Galicia also produced Silesia, dam of the Oaks winner My Dear and grandam of the very brilliant Picaroon, who won

£13,034 but unfortunately died in training. Galopin's advanced age at the time he sired his distinguished daugher is but one instance of many which shows that the age element does not necessarily preclude a stallion getting outstanding stock. Lemberg had only seven great-grandparents, his sire's dam and his dam's dam both being by Isonomy.

1911

1   J. B. Joel's br c Sunstar (Sundridge – Doris).
2   Lord Derby's ch c Stedfast (Chaucer – Be Sure).
3   Captain F. Forester's b c Royal Tender (Persimmon – Tender and True).
     26 ran. Time: 2 min 36.8 sec

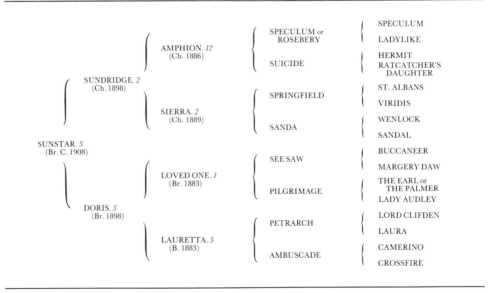

Sunstar, the product of parents who began their careers in selling plates, won the 2,000 Guineas and was an outstandingly game horse, for he broke down in the Derby some 300 yards from the winning post and battled on most gallantly. At stud, although he never headed the list of sires of winners, he was twice leading sire of broodmares, and his stock won collectively 440 races value £229,712. These figures compare satisfactorily with the returns of a great sire like Swynford, whose get scored in 344 races to the tune of £231,722. Sunstar's best produce included the very good racer and sire Buchan (1916), whose record I will deal with later.

Sunstar also got the 2,000 Guineas winner Craig an Eran, who in turn sired April the Fifth (Derby), Mon Talisman (French Derby) and Admiral Drake (Grand Prix de Paris). Others by Sunstar were Galloper Light (Grand Prix de Paris), Saltash (£11,113 and sent to Australia), Sunny Jane (Oaks and sent to America) and Alan Breck (leading sire in Argentina in 1932).

During his stud career Sunstar was mated with an unusually large number of mares and was very fertile. His overall stud record shows that the prolific use of a stallion does not necessarily portend poor quality stock. He died in his 19th year.

Sunstar's sire Sundridge, who made his racing debut in a selling plate, was unsound in his wind but a hardy racer, who withstood five years in training and won 16 short-distance events worth £6,716. He went to stud at the humble fee of nine guineas a mare but lived to head the list both of sires of winners and of maternal grandsires. Apart from Sunstar he got the 1,000 Guineas and Oaks winner Jest (dam of the Derby winner Humorist), Sun Worship (dam of the St. Leger and Ascot Gold Cup winner Solario) and others of merit, including the brothers Sun Briar and Sunreigh, who both exercised influence on American bloodlines.

Sunstar's dam Doris was also a selling plater but in her case she never advanced to superior status. At stud, however, she produced ten winners who between them secured 32 races value £38,460, and these included the Classic winners Sunstar and Princess Dorrie (1,000 Guineas and Oaks).

1912

1 W. Raphael's gr f Tagalie (Cyllene – Tagale).
2 L. Neumann's b c Jaeger (Eager – Mésange).
3 A. Belmont's br c Tracery (Rock Sand – Topiary).
    20 ran. Time: 2 min 38.8 sec

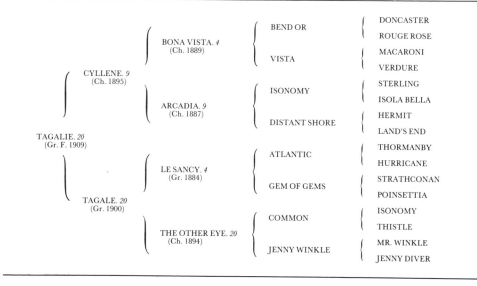

Apart from the fact that Tagalie was only the fifth filly to win the Derby since the race's inception, her grey colour caused much comment. She was only the second successful grey, of either sex, following Gustavus in 1821, and only two more—Mahmoud (described as grey or roan) in 1936 and Airborne in 1946—

have won since. At the time of Tagalie greys were extremely rare on English racecourses and were looked upon as something almost comic. A great number of them owe their coat colour to Roi Herode (a grandson of Le Sancy) who went to stud in Ireland in 1910, and whose most famous grey offspring was the brilliantly-fast The Tetrarch, nicknamed the Rocking Horse.

Tagalie also derived her hue from Le Sancy through her dam Tagale, and there is documentation of the family's colouring back to 1722. In addition to the Derby Tagalie won the 1,000 Guineas (at 20-1), but was unplaced in the Oaks and St. Leger. She was without doubt a very good racer, although it must be allowed that except for Tracery her generation was below par. On the small side, she had bad forelegs—up on her joints and back at the knees—faults often observed in her descendants. At stud she produced four winners, easily the best of whom was the Newmarket Stakes winner and 2,000 Guineas second Allenby. He suffered from the same defects as his dam and moreover transmitted them with such regularity that he had very few representatives able to stand racing. In consequence he was a complete failure at stud, and Tagalie's influence in this country is long defunct, though she does have descendants in Chile and America.

Tagalie concluded Cyllene's wonderful record for siring English Derby winners as he left for Argentina the year before Tagalie was foaled, but his prowess as a sire of Derby winners was not over, and he was responsible for three winners of the Argentine equivalent.

Tagalie's dam Tagale (1900) won two small races in France before being imported to England, where she produced seven winners in all, but none of the others calls for special mention. Tagale was a sister to Cypriote, whose descendants include Fairy Legend (1924), Feerie (1935) and Mary Legend (1925), who all won the French Oaks whilst the two first-named also secured the French 1,000 Guineas. Tagale's grandam Jenny Winkle was a half-sister to Jenny Howlet, who won the English Oaks of 1880.

1913

1 A. P. Cunliffe's b c Aboyeur (Desmond – Pawky).
2 W. Raphael's b c Louvois (Isinglass – St. Louvaine).
3 Colonel W. Hall-Walker's b c Great Sport (Gallinule – Gondolette).
    15 ran. Time: 2 min 37.6 sec

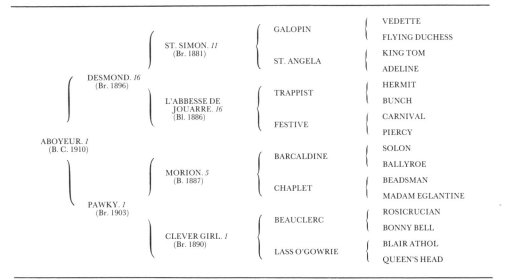

		GALOPIN	VEDETTE
	ST. SIMON. *11*		FLYING DUCHESS
	(Br. 1881)	ST. ANGELA	KING TOM
DESMOND. *16*			ADELINE
(Br. 1896)		TRAPPIST	HERMIT
	L'ABBESSE DE		BUNCH
	JOUARRE. *16*	FESTIVE	CARNIVAL
	(Bl. 1886)		PIERCY
ABOYEUR. *1*		BARCALDINE	SOLON
(B. C. 1910)			BALLYROE
	MORION. *5*		BEADSMAN
	(B. 1887)	CHAPLET	MADAM EGLANTINE
PAWKY. *1*			ROSICRUCIAN
(Br. 1903)		BEAUCLERC	BONNY BELL
	CLEVER GIRL. *1*		BLAIR ATHOL
	(Br. 1890)	LASS O'GOWRIE	QUEEN'S HEAD

This was perhaps the only bad Derby in the history of the race as it was marred by two regrettable incidents. In those days women had no parliamentary vote and the suffragette movement sprung up to draw attention to the injustice. In order to demonstrate their political worthiness, women engaged in all kinds of petty disorders. A favourite feature of their militant campaign was to chain themselves to the railings of government offices and whilst attempts were made to release them, to wield their umbrellas in no uncertain fashion on the persons of the unfortunate police. They also showed a partiality for activity at mere males' sporting events. In the 1913 Derby, as the field was rounding Tattenham Corner, one of the suffragettes' most militant members, Miss Emily Davison, ducked under the rails, ran on the course and seized the bridle of one of the passing horses. Horse, jockey and suffragette crashed to the ground. Miss Davison died a few days later from her injuries but neither the jockey nor the horse suffered any permanent ill effects. As luck would have it the horse involved in this outrage was His Majesty's Anmer.

The second episode which marred this Derby was that a considerable amount of rough riding, bumping and boring occurred as the field ran down the straight. Craganour (Desmond – Veneration II) passed the winning post first, a head in front of Aboyeur. His number was hoisted as the winner and he was just being led out of the unsaddling enclosure when a messenger rushed out to say that the Stewards on their own initiative had raised an objection. The upshot was that Craganour was disqualified and the race awarded to the 100-1 chance Aboyeur.

Never before or since have the Stewards disqualified a horse in the Derby without an objection from an owner of one of the other runners.

Craganour was sold shortly afterwards for £30,000 to Argentina where he proved an outstanding stud success, with the winners of 616 races worth £295,327 from 1917 to his death. There is little doubt that he was the best racer of his generation in England. His dam Veneration II by Laveno was a half-sister to the great Pretty Polly. She also produced Nassovian (placed in the 2,000 Guineas and Derby and sent to U.S.A., but an indifferent stallion).

Desmond, by coincidence sire of both Craganour and Aboyeur, was a good two-year-old who won the Coventry Stakes (Ascot) and the July Stakes (Newmarket), but then lost his zest for racing—a defect he did not transmit to his stock. His son the dual Ascot Gold Cup winner The White Knight (1903) was an exceptionally game racer, but a very bad stallion. Desmond himself was a good sire whose name is familiar mostly through his Ascot Gold Vase winning son Charles O'Malley, who got Blenheim's dam Malva. Desmond's dam L'Abbesse de Jouarre won the Oaks and nine other races.

Aboyeur's dam Pawky (1903) was a non-winner and produced nothing else of note. Her sire Morion (1887) won both a Royal Hunt Cup and an Ascot Gold Cup, but he was an evil-tempered animal and poor stallion.

After the Derby Aboyeur never won again and was sold to Russia for 13,000 guineas. His ultimate fate I have mentioned when referring to Minoru.

1914

1 H. B. Duryea's b c Durbar II (Rabelais – Armenia).
2 Sir E. Cassel's br c Hapsburg (Desmond – Altesse).
3 H. J. King's ch c Peter the Hermit (St. Petersburg – Blare).
   30 ran. Time: 2 min 38.4 sec

		GALOPIN	VEDETTE
	ST. SIMON. *11*		FLYING DUCHESS
	(Br. 1881)	ST. ANGELA	KING TOM
RABELAIS. *14*			ADELINE
(B. 1900)		SATIETY	ISONOMY
	SATIRICAL. *14*		WIFEY
	(Ch. 1891)	CHAFF	WILD OATS
DURBAR II. *X*			CELERRIMA
(B. C. 1911)		ST. GATIEN	ROTHERHILL or THE ROVER
	MEDDLER. *1*		ST. EDITHA
	(B. 1890)	BUSYBODY	PETRARCH
ARMENIA. *X*			SPINAWAY
(Ch. 1901)		HANOVER	HINDOO
	URANIA. *X*		BOURBON BELLE
	(Ch. 1892)	WANDA	MORTEMER
			MINNIE MINOR

(Durbar II came from an American tail female line with no Bruce Lowe number).

The vast concourse which assembled on Epsom Downs on May 27, 1914, had no inkling they were witnessing the last Derby held in the traditional carefree atmosphere of generations. In ten weeks' time the First World War was to break out.

The field for the race itself was not a particularly good one. The outstanding horse of the generation, The Tetrarch, concluded his unbeaten career as a two-year-old owing to a rapped joint. Durbar II, of English blood but bred and trained in France and American owned, had made no particular mark on the racecourse of his native land. He was unplaced in the French 2,000 Guineas and later only fourth in the French Derby and third in the Grand Prix de Paris. His stud career was complicated by the fact he was not a thoroughbred horse according to Messrs. Weatherbys' classification, although he qualified as such in both France and America. Bearing this in mind, he can be said to have done well as a stallion, his best son being the hardy Scaramouche, who won eight races value 393,310 francs. Scaramouche sired Pantaloon, who in turn got Talon, a big winner in Argentina and America. Durbar II's position, and that of his descendants, was put right by changes to the rules in the 1949 Stud Book. His most effective contribution to the bloodstock of the world was when he sired Durban, who after being the best two-year-old of her generation in France became one of the pillars of M. Boussac's famous stud, her produce including Tourbillon. Durbar II held court in France for ten seasons and in his old age went back to America, where he died in his 21st year.

Durbar II's sire Rabelais was a small horse who won races value £10,049 besides being third in the 2,000 Guineas and fourth in the Derby. His abortive sale to Russia was referred to in discussing the 1903 Derby, and after being resold to France for £900 he turned out to be a sire of the highest order.

Durbar II's grandam Urania was bred in America, where she raced no less than 87 times, winning 35 and being placed in a further 34—a wonderful record of consistency.

I have already referred to the non-thoroughbred strain in this pedigree in my notes on the so-called Jersey Act (see page 83).

Rabelais was 19 years younger than his sire St. Simon (1881), who can lay claim to being the most influential horse in the world foaled in the past 100 years. His name is familiar to all who have even the slightest knowledge of the thoroughbred horse and is common in pedigrees in all parts of the globe. He was by Galopin (see page 313) out of St Angela, who was 16 years old when her great son was born and who previously had produced nothing of racing merit. But she was well enough bred, her sire being the one time champion stallion King Tom (1851) and her grandam a half-sister to the Derby winner The Little Wonder (1837) and her daughter Angelica (sister to St. Simon) became the dam of dual Eclipse Stakes winner Orme. It was probably owing to St. Angela's poor record as a dam of winners that St. Simon was not entered for the Derby or St. Leger, but only for the 2,000 Guineas. He could not fulfil this engagement as his owner died when the colt was a two-year-old and in those days entries became void under such circumstances. St. Simon was sold by auction to the late Duke of Portland for 1,600 guineas, which was an extremely lucky purchase. The Duke had attended the sale with the intention of buying the three-year-old Fulmen, who had shown himself to be the best juvenile the previous season, but Fulmen fetched £500 more than His Grace was prepared to give, so he bought instead St. Simon, the next lot to enter the sale ring.

St. Simon raced only as a two- and three-year-old, winning all his ten races, value collectively £4,675, including the Ascot Gold Cup by the margin of 20 lengths. In this race he was so full of running on completion of the 2½-mile course that his jockey could not pull him up for a further mile. He had not been entered for any of the important four-year-old races so, although absolutely sound, did not run that season. He went to stud as a five-year-old at a fee of 50 guineas a mare, raised to 100 guineas the following year. From then on the charge for his services gradually increased until by the time he was 18 years old it was 500 guineas and remained at this figure until his death at the age of 27 in 1908.

The hereditary longevity of the line is worth noting. His sire Galopin lived for 28 years and his son Rabelais for 29.

During his long stud career St. Simon covered 775 mares of whom 554 proved in foal and 423 delivered live foals. All his foals were bays or browns except the very last, who was a grey. Between them they won 571 races and £553,158 in stakes. This was a gigantic sum in those days when the total prize money distributed annually was about £480,000 as compared to more than £7,000,000 in 1976. St. Simon headed the list of sires of winners nine times and of maternal grandsires on six occasions—both these are records unequalled in the history of the Turf. He and his sons Persimmon, St. Frusquin and Desmond held the sire of winners championship no less than 16 times between them in the years 1890 to 1913. His get included Persimmon (Derby, St. Leger and Ascot Gold Cup); St. Frusquin (2,000 Guineas); Semolina (1,000 Guineas); La Flêche (1,000 Guineas, Oaks, St. Leger and Ascot Gold Cup); Amiable (1,000 Guineas and Oaks); Memoir (Oaks and St. Leger); Mrs Butterwick (Oaks); Diamond Jubilee (2,000 Guineas, Derby and St. Leger); Winifreda (1,000 Guineas); La Roche (Oaks); William the Third (Ascot Gold Cup).

He had ten individual winners of 17 Classics, and reached his peak in 1900, when his progeny won all five Classics.

Persimmon, who died at the age of 15, was St. Simon's best son at stud, being leading sire four times, and his produce included Sceptre (1,000 Guineas, 2,000 Guineas, Oaks and St. Leger) and Prince Palatine (Ascot Gold Cup twice). Except for America, to where no good-class St. Simon stallion was imported, his sons and grandsons have been leading sires in practically every corner of the globe where bloodstock is bred, whilst his daughters as broodmares have had a tremendous world-wide influence.

CHAPTER 13

1915–1919
Racing During the First World War

1915

1   S. B. Joel's b c Pommern (Polymelus – Merry Agnes).
2   Colonel W Hall-Walker's b c Let Fly (White Eagle – Gondolette).
3   Sir John Thursby's bl c Rossendale (St. Frusquin – Menda).
       17 ran. Time: 2 min 32.6 sec   Run at Newmarket.

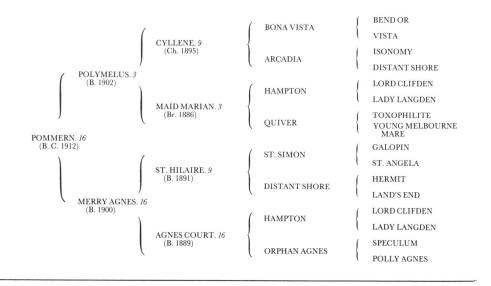

THE BACKGROUND UNDER which this race was run was very different from that of any previous Derby. The First World War had produced repercussions which were little dreamt of at its outbreak. The Army in France was hanging on against desperate odds, and although in previous wars our forces had met with reverses, never before or since have the scale of casualties been in any way comparable. The

183

First World War cost 812,317 killed from the British Isles alone as opposed to 244,723 in the longer Second World War.

However, racing went on, and since the Epsom course had been requisitioned by the Army, the Derbys of 1915 to 1918 were run over 1½ miles at Newmarket. The 1915 winner Pommern was by far the best of his generation and also won the substitute 2,000 Guineas and St. Leger. He retired to stud at a fee of 300 guineas, later raised to 400 guineas, but was a disappointing sire, his best get being the 2,000 Guineas winner Adam's Apple, who when exported to Argentina made no great show at stud. Pommern's only other significant contribution was through his daughter Vesper Bell, dam of the 1,000 Guineas winner Campanula.

Pommern's sire Polymelus won races value £16,725 during his four years in training mostly over middle distances but, as I have already mentioned in notes on the 1905 Derby winner Cicero, he could stay well enough to be second in the St. Leger. At stud Polymelus was an outstanding success, being head of the list of sires of winners five times—a record only surpassed in the 20th Century by Hyperion's total of six. Polymelus's dam was a half-sister to the great race fillies Le Flêche and Memoir. He owes his tremendous influence on modern pedigrees not to his Classic winners but to the handicapper Phalaris, whose achievement at stud I have referred to at length in Chapter 5.

Pommern's dam Merry Agnes won two small races and produced two other winners, but of very lowly class. He himself was an inbred horse, as not only were his sire's dam and his dam's dam both by Hampton, but his dam's sire (St. Hilaire) was a half-brother to Cyllene's dam.

1916

1   E. Hulton's ch f Fifinella (Polymelus – Silver Fowl).
2   Mr. "Fairie's" b c Kwang-Su (Cicero – Galicia).
3   J. Sanford's b c Nassovian (William the Third – Veneration II).
        10 ran. Time: 2 min 36.6 sec.   Run at Newmarket.

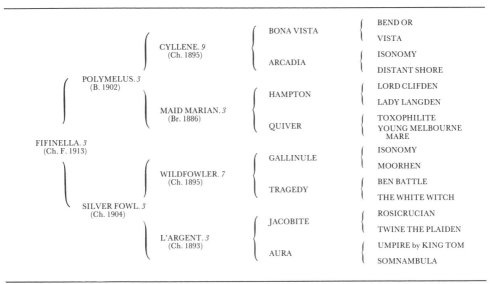

Fifinella, emulating the feat of Signorinetta eight years previously, won both the Derby and Oaks. She was highly temperamental and although an exceptionally good performer for one of her sex, was in no way comparable to unbeaten Hurry On, who won the 1916 St. Leger. At stud she produced eight winners who collectively won 15½ races value £13,898 including Middle Park Stakes winner Press Gang, who contributed £10,580 to this total, and Fifine, the dam of Portofino (£5,397 and sent to Australia). The tail female line is one that had a remarkable record over a long period of producing high-class stock. Fifinella's half-sister Soubriquet won five races value £4,539, finished second in both the 1,000 Guineas and Oaks, and produced unbeaten but unsound Tai-Yang, and Pasca (dam of the 2,000 Guineas winner Pasch, who died as a four-year-old; Chateau Larose, the St. Leger second who went to Argentina; and Pasqua, dam of the Derby winner Pinza). Fifinella's brother Silvern won nine races value £6,277 and was second in the St. Leger before taking up stud duties at the National Stud, where he did not do well, whilst a half-sister, Silver Tag, won the Cambridgeshire and nine other races to the tune of £7,288 besides being second in the 1,000 Guineas. In more recent years the family has distinguished itself in France through the stayers Wallaby (French St. Leger and Ascot Gold Cup) and Le Bavard (French Gold Cup).

Fifinella's dam Silver Fowl won £1,176 and bred 11 winners of over £27,000, whilst the latter's sire Wildfowler was a St. Leger winner but an indifferent sire. Wildfowler was sold to France for 5,000 guineas as a 13-year-old but when he came into the auction ring eight years later was knocked down for the equivalent of £10.

Here it may be opportune to draw attention to the remarkable success achieved in the early Derbys of this century by brothers, half-brothers etc., which I recapitulate below:

1 Kwang-Su (second in 1916) and Lemberg (winner in 1910) were both out of Galicia, while Kwang-Su's sire Cicero was by Lemberg's sire Cyllene.
2 Nassovian (third in 1916) and Craganour (disqualified in 1913) were by William the Third and Desmond respectively, out of Veneration II.
3 Let Fly (second in 1915) and Great Sport (third in 1913) were both out of Gondolette, while Let Fly's sire White Eagle was by Great Sport's sire Gallinule. Gondolette was also responsible for Sansovino (by Swynford), winner of the 1924 Derby.
4 Louvois (second in 1913) and Louviers (second in 1909) were brothers, by Isinglass out of St. Louvaine.
5 Signorinetta (winner in 1908) and Signorino (third in 1905) were half-brother and sister, out of Signorina.

Bearing in mind that during the 17 years reviewed there were literally thousands of mares at stud, the fact that only five of these produced over 17½ per cent of the winners and placed horses effectively shows the extraordinary dominance of a few over their rivals.

1917

1 Mr. "Fairie's" b c Gay Crusader (Bayardo – Gay Laura).
2 Sir Hedworth Meux's br c Dansellon (Chaucer – Tortor).
3 Sir W. Cooke's br c Dark Legend (Dark Ronald – Golden Legend).
    12 ran. Time: 2 min 40.6 sec.   Run at Newmarket.

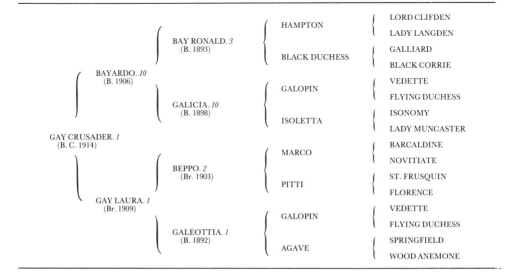

Gay Crusader, a first foal, added his name to the list of Triple Crown winners, and even in that class must be regarded as above average. He also won as a three-year-old a substitute Ascot Gold Cup, run at Newmarket, but which produced only three starters. That very experienced jockey Steve Donoghue, who steered six Derby winners to victory, used to say he was the best horse he ever rode. Though kept in training as a four-year-old, he never ran that year owing to a sprung tendon. He retired to stud at a fee of 400 guineas, and sired a large number of winners but few of special note. Although a lion-hearted horse himself, he gained a reputation for getting ungenuine stock. Long after his death his worth as a sire was resuscitated through his daughters Loika (dam of Djebel), Indolence (dam of Prince Rose, probably the best racer in Europe of his time and sire of Prince Chevalier, Prince Bio and Princequillo) and Hellespont (grandam of the Derby and St. Leger winner Airborne).

Gay Crusader's sire Bayardo, a half-brother to Lemberg and Kwang-Su as previously mentioned, won 22 of his 25 races, including the St. Leger and Ascot Gold Cup, before retiring to stud where he was an unqualified success during the few years he held court. His death from thrombosis as an 11-year-old was a grievous loss to breeders. In addition to Gay Crusader he got the Triple Crown winner of the following year Gainsborough (sire of Hyperion) and the Oaks winner Bayuda. He was champion sire in 1917 and 1918.

I have already mentioned that Bayardo's dam Galicia was 26 years younger than her sire Galopin, and I now record that his sire Bay Ronald was 21 years younger than his paternal grandsire Hampton. There is no escaping the fact that old parents have had a trememdous influence on the build-up of bloodlines, in spite of their numerical inferiority compared to young and middle-aged progenitors.

Gay Crusader's dam Gay Laura won once as a two-year-old and bred five other winners of 14 races value £9,906 between them, including Sea Rover, who made quite a name for himself under National Hunt rules. Gay Laura's dam Galeottia was a 1,000 Guineas winner.

Dark Legend, who was third in the 1917 Derby is worthy of note. He scored three times as a three-year-old in races of small value, before being sent to India, where he showed himself to be one of the best racers in their history. He returned to England as a six-year-old and went to stud in France, where he did very well. Amongst his get was Rosy Legend (dam of the Derby winner Dante and the St. Leger winner Sayajirao). It is unusual for a horse who has raced in India to be a good sire but not unique, as Fastnet (1933, by Pharos) had a similar record.

1918

1 Lady James Douglas's b c Gainsborough (Bayardo – Rosedrop).
2 Major W. Astor's br c Blink (Sunstar – Winkipop).
3 Sir W. J. Tatem's ch c Treclare (Tredennis – Clare).
    13 ran. Time: 2 min 33.2 sec.   Run at Newmarket.

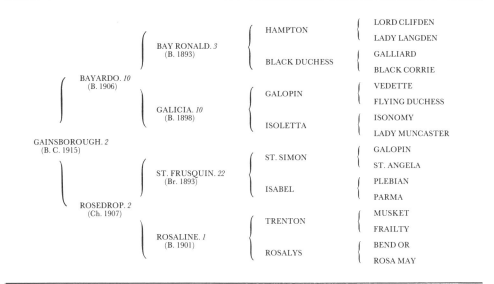

Gainsborough, the first Derby winner owned by a woman, was a Triple Crown winner, an exceptionally good racehorse and a quite exceptional stallion. I do not think the fact that Pommern, Gay Crusader and Gainsborough all won the Triple Crown in wartime substitute races at Newmarket, instead of over the usual courses, in any way detracts from their achievements. They were all easily the best of their respective generations and except for some untoward accident would have asserted their superiority on any course.

It is a curious fact that about this period a number of outstanding horses were foaled whose influence on modern pedigrees has been quite phenomenal. The year 1911 saw the birth of The Tetrarch and Son-in-Law; 1913 Hurry On and Phalaris; 1915 Gainsborough; then after four years break Blandford appeared in 1919. Without these names the Stud Books of the world would make very different reading. It is impossible to ascribe any reason for Nature's bounty of such wealth of blood within a few years. It just happened, and will doubtless happen again.

An early foal (January 24), Gainsborough, who also won a substitute race for the Ascot Gold Cup at Newmarket as a three-year-old, was an individual of almost faultless conformation and most charming temperament. He rested as a four-year-old, started at stud in 1920 and died in his 31st year. His success at stud is almost legendary. His colts were decidedly better than his fillies—he never sired the winner of either the 1,000 Guineas or Oaks—and his St. Leger winners Solario,

Singapore and Hyperion, after their distinguished racing careers, which included a Derby for Hyperion and an Ascot Gold Cup for Solario, all sired Classic winners. His 2,000 Guineas winner Orwell was not a success as a stallion.

Gainsborough's fillies turned out most valuable broodmares, and in 1932 he was both leading broodmare sire and leading sire for winners, which latter position he repeated the following year. His most important mares were Una Cameron, dam of the 1931 Derby winner Cameronian; and Mah Mahal, dam of the 1936 Derby winner Mahmoud. Gainsborough was also responsible for Imagery, dam of the Irish Derby winners Museum and Phideas.

When referring to Gay Crusader, I mentioned that great racer and stallion Bayardo, and it may be of interest to say something about his sire Bay Ronald, as although his name is familiar, his history is not so well known.

Bay Ronald's sire Hampton was a small horse, barely 15.2, who commenced his racing career in selling plates but showed vast improvement later. Withstanding training for five seasons, he won 20 races including the Goodwood Cup, Doncaster Cup and Northumberland Plate, before taking up stud duties at a fee of only 30 guineas a mare. He soon made a great name for himself, and before he died at the age of 25 he had sired the Derby winners Merry Hampton, Ayrshire and Ladas, and many other good winners, in addition to establishing a sire line (Hyperion's) which holds good today.

Bay Ronald's dam Black Duchess, who won one small race, also produced Black Cherry, dam of Cherry Lass (1,000 Guineas and Oaks), Blanche (dam of Blandford) and Jean's Folly (dam of the St. Leger winner Night Hawk). So no exception could be taken to Bay Ronald's breeding. As an individual he was markedly over at the knees. His racing career was that of a useful handicapper and his victory under 8st as a five-year-old in the City and Suburban gives a fair picture of his merit. He started at stud in 1899 at a modest 25 guineas a mare whilst dams of winners were accepted free. Even at these terms he received poor patronage and after the 1905 season was sold for 5,000 guineas to France, where he died in 1907. In spite of the adverse circumstances he made great use of his few opportunities and in addition to Bayardo sired Rondeau (dam of the great French racer and stallion Teddy), Macdonald II (French St. Leger and a useful sire) and Dark Ronald. The last named before export to Germany got that great stayer and sire of stayers Son-in-Law and also Dark Legend who I have already mentioned (see page 187). In Germany Dark Ronald is regarded as probably the best sire in their bloodstock annals. Amongst his get there was Prunus, whose son Oleander got Pink Flower, sire of that versatile horse Wilwyn (1948).

Gainsborough's dam Rosedrop won the Oaks whilst her sire St. Frusquin netted nearly £30,000 in stakes including the 2,000 Guineas before twice heading the list of sires of winners. Gainsborough's grandam Rosaline, once sold for 25 guineas, was by the New Zealand-bred Trenton, who started 13 times during his four years in training in New Zealand and Australia, winning on eight occasions and being placed in four other races. He spent nine seasons at stud in Australia with excellent results, for his stock won 404 races value £101,933—and in those days stakes in the Antipodes were on a much lower scale than is the case today. Trenton was brought to England as a 15-year-old when it was too late to expect him to make any great mark as a sire of winners, but his contribution to the make-up of both Gainsborough and Buchan amply justified his importation.

Somewhat similarly to Gay Crusader, Gainsborough was inbred to Galopin but in the latter case at the third and fourth remove with a subsidiary line through the 2,000 Guineas winner Galliard.

1919

1   Lord Glanely's bl c Grand Parade (Orby – Grand Geraldine).
2   Major W. Astor's b c Buchan (Sunstar – Hamoaze).
3   Sir Walter Gilbey's b or br c Paper Money (Greenback – Epping Rose).
        13 ran. Time: 2 min 35.8 sec

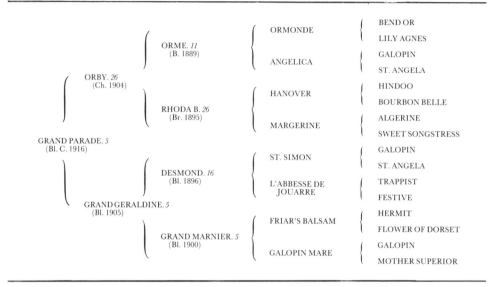

With the war over, the Derby returned to Epsom. It was a period of inflation and false prosperity. Large fortunes had been made during the war and the demobilized personnel were spending their war gratuities, which were on a far more generous scale than those granted for the Second World War. But industrial troubles were looming ahead; Ireland, still governed from Westminster, was in turmoil, and income tax had risen to six shillings in the pound—its highest peak, in peace or war, until the coming of the Second World War.

Favourite for the race was the 2,000 Guineas winner The Panther (Tracery – Countess Zia) who I mention as he was the best mover I have ever seen. He became over-excited at Epsom and gave a poor display. He went to stud in Argentina, where he sired the winners of some £68,000 in stakes before repatriation in 1929, but died two years later. It was discovered he had a heart disease.

The Derby winner Grand Parade was bred in Ireland by Mr. R. Croker who sold him as a foal to Lord Glanely for only 470 guineas. He was only the second horse black in colour to win the Derby, his predecessor in this peculiarity being

Smolensko in 1813. There has not been one since, though so strong is the prejudice against this coat colour that it is not uncommon to find blacks registered as browns. Grand Parade was not a particularly good Derby winner but he managed to win five of his six races in England and Ireland as a two-year-old, and after his Epsom triumph scored in the St. James's Palace Stakes at Ascot. He was then taken out of training and went to stud as a four-year-old at a fee of 400 guineas a mare. He sired a large number of moderate winners but his only Classic winner was the Aga Khan's 2,000 Guineas victor Diophon, a moderate stallion who was eventually disposed of to Greece, but before departing got Diolite who, like his sire, won the 2,000 Guineas. Diolite also was a stud disappointment and was sent to Japan.

I have already referred to Grand Parade's sire Orby (see page 171) and his maternal grandsire Desmond (see page 180). His dam Grand Geraldine never raced, produced nothing else of note, and is rumoured to have spent part of her life between the shafts of a cart. Her dam Grand Marnier also never raced and had no winners to her credit, but she was sister to Mother Siegel, the dam of the Derby and 2,000 Guineas winner Minoru (see page 174). Their sire Friar's Balsam was an exceptionally brilliant two-year-old who did quite well at stud without ever reaching the top class. He was 21 years younger than his sire Hermit.

Grand Parade has only 13 instead of 16 great-great-grandparents, Galopin's name appearing three times in his pedigree at that remove, and St. Angela's twice.

Grand Parade later reappeared in the pedigrees of a top-class winner through My Love (1948 Derby) and Ambiguity (1953 Oaks), both of whose great-grandams were by Grand Parade. However, it was the slow-starting 1919 Derby runner-up Buchan who had the better stud record, being leading sire in 1927 and responsible for the St. Leger winner Book Law and the Oaks winner Short Story, as well as the dams of Rhodes Scholar, Pay Up and Airborne. I will refer later to other Derby seconds who were successful at stud, including two of Buchan's half-brothers.

CHAPTER 14

1920 – 1925
Some Famous Mares of
Humble Origins

1920

1   Major G. Loder's b c Spion Kop (Spearmint – Hammerkop).
2   Lord Derby's ch c Archaic (Polymelus – Keystone II).
3   Sir H. Cunliffe-Owen's b c Orpheus (Orby – Electra).
      19 ran. Time: 2 min 34.8 sec

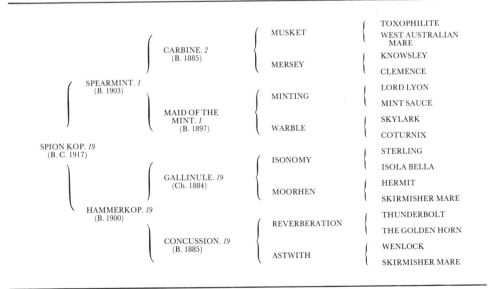

APART FROM THE Derby only one other race (value £420) was won by Spion Kop during his three seasons in training and he owes his Derby success to superior staying capacity compared to his rivals. The field embraced some horses who were both good racers and influential sires but pace was their forte. Amongst them were such as the 2,000 Guineas winner and champion sire Tetratema (The Tetrarch – Scotch Gift), Abbots Trace (six races value £5,433 and a useful sire. By Tracery – Abbots Anne), Sarchedon (a brilliant but totally undependable racer who was sent to Australia. By The Tetrarch – Perfect Peach), the bad-tempered Orpheus (11 races value £11,971 and sire of many moderate winners. By Orby – Electra), and so on.

Spion Kop, who died a few weeks short of his 29th year, sired a great number of useful winners, but only one of Classic class—the 1928 Derby winner Felstead. His Irish Derby-winning son Kopi did quite well at stud in France whilst his Yorkshire Cup winner The Bastard (renamed The Buzzard) proved a most valuable sire in Australia. His daughter Kopje produced the Grand Prix de Paris victor Cappiello.

Spion Kop's sire, the 1906 Derby winner Spearmint, I have already mentioned (see page 169). His dam Hammerkop was a great race mare who stood up to training for no less than five years, winning 11 times including the Cesarewitch, Queen Alexandra Stakes and July Stakes. She was a very difficult mare to get in foal and only had three other produce, none of whom won. Her sire Gallinule won three races value £1,985 as a two-year-old. He then became a pronounced bleeder and failed to score as a three-, four- or five-year-old. Retiring to stud, he made a great name for himself, as his get included Pretty Polly (1,000 Guineas, Oaks, St. Leger, etc.), Slieve Gallion (2,000 Guineas), Night Hawk (St. Leger), Wildfowler (St. Leger) and others of high class.

Gallinule's dam Moorhen had a most remarkable career. She commenced racing as a three-year-old, when from 24 starts she won six selling plates. She was mated to Vulcan as a four-year-old but continued racing, running nine times on the Flat and five times over hurdles without success. At five years old she produced a colt foal and was covered again. At six years old she slipped foal but her owner did not like her to eat the bread of idleness and promptly put her back into training. At six, seven, eight and nine, she ran in all sorts of races, selling plates on the Flat, hurdles, 'chases, and hunter Flat races, winning 20 of them. But during her last year's exertions she was covered again and produced a foal as a ten-year-old. She was then mated to the dual Ascot Gold Cup winner and most successful sire Isonomy and at 11 years old produced Gallinule.

Hammerkop was a sister to Sirenia who produced the 1,000 Guineas winner Electra (dam of Orpheus, third in this Derby and grandam of the St. Leger winner Salmon Trout); Siberia (dam of the Oaks winner Snow Marten); and Sourabaya (by Spearmint and dam of the Grand Prix de Paris winner Comrade). Hammerkop and Sirenia were incestuously-bred, their sire's dam Moorhen and their dam's dam Astwith being half-sisters.

1921

1   J. B. Joel's ch c Humorist (Polymelus – Jest).
2   Lord Astor's b c Craig an Eran (Sunstar – Maid of the Mist).
3   J. Watson's ch c Lemonora (Lemberg – Honora).
      23 ran. Time: 2 min 36.2 sec

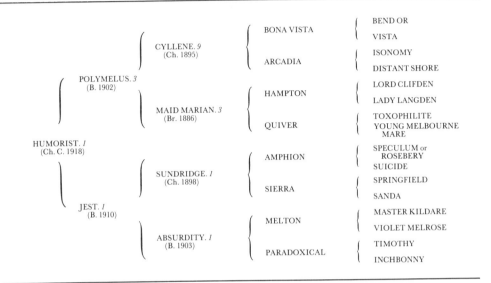

An innovation in this Derby was that £500 out of the stakes went to the breeder of the winner. This practice was continued up to 1949, except for the war years, but then was dropped. In 1919 and 1920 a similar sum had been set aside for the nominator of the winner. These may be different people. Mr. Croker bred Grand Parade and sold him as a foal to Lord Glanely who entered him for the Derby. Thus Lord Glanely received the £500 nominator's money and Mr. Croker got nothing.

Breeders' prizes are prevalent in France, where their scope has been gradually enlarged so that in 1977 they were paid to breeders of French-bred horses that finished in the first four in all races at the rate of 25 per cent of the nominal prize money in major races and 15 per cent in others. In addition, owners' premiums, as these extra payments have become known, have been paid in respect of French-bred winners in all French races for several years, and in 1977 these amounted to 30 per cent of the nominal value of the race to the winner, and 30 per cent of the nominal amounts for second, third and fourth places. Owners' prizes were even higher for the biggest races, with the result it was possible for the owner-breeder of a French-bred Classic winner to receive an additional 75 per cent incentive premium.

I am strongly in favour of the plan. After all, one of the basic considerations in horse racing is to maintain and improve the standard of the horse, and how can this be done better than by rewarding the breeder when he produces a successful specimen? It may be argued that if a breeder sells his produce, he receives his full entitlement at the time of sale, and that stakes, often largely subscribed by owners, should not in part be deviated to his interests. That is all very well in theory, but a great many horses are sold at prices which bring little grist to their breeders and a great deal to their buyers.

Although it may be said in a rough and ready way that a vendor receives the market price for his horse, the sale of yearlings and foals is such a vast gamble that there are countless exceptions. It always seems to me that it would be more equable for the breeder to receive a tiny fraction less at the time of sale and an honorarium when his product proves its worth by winning.

A number of schemes have been put forward in recent years to bring widescale breeders' premiums to Britain, but all had been turned down for lack of finance.

However, following encouragement from the Thoroughbred Breeders Association, a pilot scheme for fillies' premiums was introduced in 1978, whereby the Horserace Betting Levy Board set aside £169,050 to be paid to the owners of British-bred winners of selected two-year-old fillies' races at a rate of 50 per cent of the added money to the winner. This does not go to the breeder, but official thinking is that such a premium will be a direct incentive to purchasers at the sales to buy British-bred fillies, and thus will benefit breeders in this way. Until (if ever, that is) breeders' prizes are introduced, this new scheme is a welcome innovation.

Returning to Humorist, his breeder Mr. J. B. Joel was the first to pick up a breeders' prize in the Derby, though in this case he was also the owner. Humorist must be regarded as probably the most gallant Derby winner in the history of the race. His racing and training career were marked by most uneven performances. He would be going great guns at one moment and suddenly shut up to nothing the next. The cause was unknown until 18 days after his Derby triumph, when Humorist was found dead in his box. The post-mortem disclosed he had tuberculosis of the lungs which brought on a fatal hæmorrhage.

I have already mentioned his sire Polymelus when referring to the 1915 Derby winner Pommern, and his maternal grandsire Sundridge in my notes on the 1911 Derby winner Sunstar.

Humorist's dam Jest won the 1,000 Guineas and Oaks but bred no other winners, although her son Chief Ruler (by The Tetrarch) did all right as a sire in New Zealand. She was a half-sister to the St. Leger winner Black Jester and a sister to the great New Zealand-based sire Absurd.

The 1921 runner-up Craig an Eran was disappointing in this country, though he got the Derby winner April the Fifth, but he did better in France, with Admiral Drake (winner of the Grand Prix and sire of the Derby winner Phil Drake) and Mon Talisman (winner of the French Derby and sire of the Grand Prix winner Clairvoyant).

1922

1 Lord Woolavington's ch c Captain Cuttle (Hurry On – Bellavista).
2 Lord Astor's b c Tamar (Tracery – Hamoaze).
3 Barclay Walker's ch c Craigangower (Polymelus – Fortuna).
  30 ran. Time: 2 min 34.6 sec

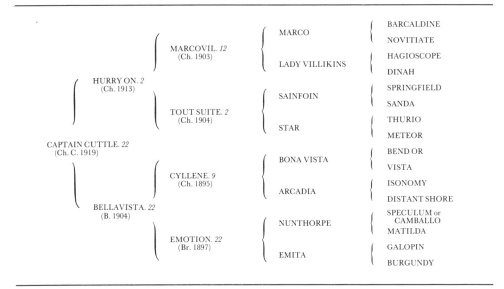

				MARCO		BARCALDINE

Captain Cuttle was a magnificent specimen of his breed and an above-average Derby winner. Owing to his size he could not be trained seriously as a two-year-old. The following year he was beaten only once, in the 2,000 Guineas when he was third and suffering from the effects of indigestion. He missed the St. Leger because of leg trouble, and as a four-year-old won his only race before being taken out of training as he could not withstand further work. In his first season at stud, 1924, he sired the 1,000 Guineas winner and Oaks second Scuttle, whilst his next crop included the Derby second Walter Gay. He also sired Glenabatrick, dam of the Gold Cup winner and significant National Hunt stallion Tiberius. But before his progeny could gain further Classic honours, Captain Cuttle was sold at the end of the 1927 breeding season for £50,000 to Italy. Incidentally, during the period the Italians were building up their studs from somewhat moderate class to their more recent status, he was the only young stallion of the highest credentials they bought. He did well in Italy but unfortunately when covering a mare in 1932 fell off her and was killed. At the time Lord Woolavington sold him, his sire Hurry On was a comparatively young horse and an established stud success, whilst Coronach, also by Hurry On, had recently scored in the Derby, St. Leger, etc., and was due to commence stud work the following season.

Captain Cuttle's massive frame was rather too heavy for his legs and, oddly enough, in the same generation there was another horse, Blandford, with the same complaint. The two never met on the racecourse, so their relative merits are pure conjecture, but I understand that very experienced trainer Richard Dawson considered his horse at least a match for Captain Cuttle.

Captain Cuttle's sire Hurry On, who was not foaled until May 7, was one of the great horses of modern times. A chesnut of outstanding physique, he stood 17 hands, had a girth of 82½ inches and 9½ inches of bone below the knee. On the racecourse he never knew defeat. At stud he headed both the list of sires of winners and sires of winning broodmares. Moreover, Hurry On maintained and revived the Matchem line which was almost dead at the time he went to stud. What more could a horse do?

Hurry On's get included three Derby winners, Captain Cuttle, Coronach and Call Boy; one St. Leger winner, Coronach; two Oaks winners, Toboggan and Pennycomequick; two 1,000 Guineas winners, Plack and Cresta Run; numerous other winners of class, not forgetting his Ascot Gold Cup winner Precipitation; and his daughter Instantaneous bred Court Martial, winner of the 2,000 Guineas and twice champion sire before his export to America.

Hurry On's colts and fillies were equally good, whilst his stock were both good two-year-olds and trained on. The very first foal he sired was Captain Cuttle, born on January 11, 1919, and it is a remarkable tribute to Hurry On's virility that his influential son Precipitation came along in 1933.

Precipitation, who did not race until he was three, proved that Ascot Gold Cup winners can be successful at stud by producing English Classic winners in Airborne (Derby and St. Leger), Chamossaire (St. Leger and leading sire), Why Hurry (Oaks) and Premonition (St. Leger), as well as Supreme Court (King George VI and Queen Elizabeth Stakes) and Sheshoon (Ascot Gold Cup and important European sire).

Since Captain Cuttle's best produce was the 1,000 Guineas winner Scuttle, and Coronach's was the dual Prix de l'Arc de Triomphe winning filly Corrida, it has fallen to Precipitation to carry on the Hurry On sire line, and though there have been disappointments, it is managing to survive. Airborne was a failure; Premonition was disappointing, and Chamossaire's chief sons Santa Claus and Cambremer have not been noticeably successful. But Sheshoon, leading European sire in 1970, has provided new hope through his son Sassafras, winner of the Prix de l'Arc de Triomphe, French Derby and French St. Leger.

I have previously remarked that generally speaking it is dangerous to make excuses for horses. Taking it fine and large a horse's merit either as a racer or as a stud proposition shows itself in the results he achieves.

There is no doubt that Hurry On's sire Marcovil was a far from first-class performer but he may have been a trifle unlucky in his stud career. I say "may have been" not "was". His racing life was greatly hampered by bad forelegs. He was kept in training for four years but his appearances on the racecourse were few. He won two small races as a three-year-old and the Cambridgeshire carrying the comparatively light weight of 7st 11lb as a five-year-old. This gives a fair picture of his ability.

In 1909 he went to stud at a fee of 25 guineas a mare and was located in Sussex, which was not a great bloodstock breeding county at the time. He soon made headway as is evinced by the fact that his fee had gone up to 250 guineas a mare by

1915—before Hurry On had appeared on a racecourse. He died of paralysis before the 1917 breeding season, just as he was getting into his stride as a stallion. Considering his modest start at stud and not very attractive background, I feel he did very well and that his death was probably a great loss to breeders. In addition to Hurry On he left behind him My Prince (a very useful racer and later one of the greatest sires of 'chasers of all time); the hardy Milton (nine races value £6,267 and sire of Penny Royal, Millrock, etc.); Golden Orb (whose daugher Orbella produced Bellacose); Miss Matty (dam of the Derby winner Papyrus); Marcarême (dam of Pinxit, who won £5,302, and grandam of Panipat, French 2,000 Guineas) and others of merit.

Whichever way one looks at his whole career there is no possible doubt that his beneficial impact on the Stud Book has been far greater than could reasonably be expected from his racing career. He was an inbred horse, his sire and his dam both being out of Hermit mares. I have mentioned earlier other cases where the same incidence of inbreeding produced excellent results (see page 23).

Hurry On's dam Tout Suite was an undersized mare who never went into training and at stud produced no other important winners, whilst his grandam Star won one race and was a half-sister to Light of Other Days and to Radiancy (by Tibthorpe), who both produced useful winners.

Captain Cuttle's dam Bellavista was a good racer who won three races value £2,421 and at stud produced eight other winners, including the good but unreliable Tom Pinch (£5,275), a brother to Captain Cuttle.

1923

1   B. Irish's br c Papyrus (Tracery – Miss Matty).
2   Lord Derby's b c Pharos (Phalaris – Scapa Flow).
3   M. Goculdas's b c Parth (Polymelus – Willia).
     19 ran. Time: 2 min 28 sec

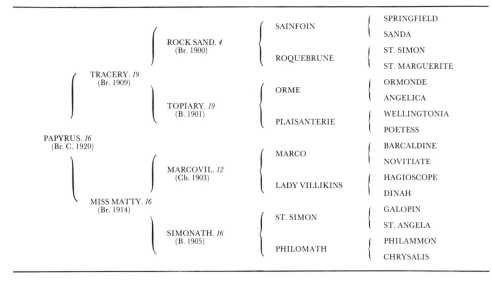

PAPYRUS. *16* (Br. C. 1920)	TRACERY. *19* (Br. 1909)	ROCK SAND. *4* (Br. 1900)	SAINFOIN	SPRINGFIELD
				SANDA
			ROQUEBRUNE	ST. SIMON
				ST. MARGUERITE
		TOPIARY. *19* (B. 1901)	ORME	ORMONDE
				ANGELICA
			PLAISANTERIE	WELLINGTONIA
				POETESS
	MISS MATTY. *16* (Br. 1914)	MARCOVIL. *12* (Ch. 1903)	MARCO	BARCALDINE
				NOVITIATE
			LADY VILLIKINS	HAGIOSCOPE
				DINAH
		SIMONATH. *16* (B. 1905)	ST. SIMON	GALOPIN
				ST. ANGELA
			PHILOMATH	PHILAMMON
				CHRYSALIS

Papyrus, who was a trifle on the small side, won six races as a two-year-old and two more in addition to the Derby the following year. He was then sent to America for his famous match with Zev where he failed badly. But, truth to tell, the match was a farce. The course was more or less under water, and Papyrus was both unacclimatized and unfamiliar with American racing conditions. Moreover a horse who had been wound up for the St. Leger on September 12 could not reasonably be expected to be racing fit again by October 20 when the match was run, after a comparatively long sea voyage. The fact that horses these days switch continents with no apparent loss of form means nothing in the context of the 1920s.

At the end of his three-year-old days Papyrus was sold to Mr. J. P. Hornung for £25,000 but never won for his new owner, though he was second in the Eclipse Stakes. He went to stud at a fee of 300 guineas, and though his career was not outstanding, he got the winners of 282 races worth £112,149 without siring anything of special merit before he died at the age of 23. His most successful mare was Honey Buzzard, who to a mating with Fairway (a brother to the 1923 Derby runner-up Pharos) produced the top-class sprinter and sire Honeyway.

Tracery, the sire of Papyrus, was bred in America and imported as a yearling to England, where he proved a first-class racehorse, winning the St. Leger, Eclipse Stakes, Champion Stakes etc., and running third in the 1912 Derby. His effort to win an Ascot Gold Cup was defeated by a madman who ran on to the course and brought him down. He went to stud in 1914 in England, where he sired in addition to Papyrus, the 2,000 Guineas winner The Panther, Transvaal (Grand Prix de Paris), Obliterate (sire of Quashed), Flamboyant (who reappears later) and the 'chaser sire Cottage.

In 1920 Tracery was sold for 53,000 guineas to Argentina. He remained there only a short time, as after Papyrus had won the Derby in 1923 he was brought back by an English syndicate, but died the year after. I have already mentioned Tracery's sire, the Triple Crown winner Rock Sand (see page 166). His dam Topiary was a winner herself and half-sister to the Cesarewitch winner and useful sire Childwick. Their dam, the French-bred Plaisanterie, was a great racemare who won 16 races value £11,295 and had the rare distinction of scoring as a three-year-old in both the Cesarewitch and Cambridgeshire.

Miss Matty, the dam of Papyrus, did not win but in addition to her Derby-winning son produced eight other winners, including Bold Archer, who sired any number of small winners; the hardy Paddington (11½ races value £2,212), and also Great Star and Comus, who both went to Australia as stallions. Her last winner Miss Matilda appeared when she was 22 years old. Miss Matty was a half-sister to the Cesarewitch winner Bracket (dam of Parenthesis, second in the St. Leger, winner of £5,885 and later sent to Australia) and to Flamboyant, who won the Doncaster and Goodwood Cups and was second in the Ascot Gold Cup. Flamboyant spent four seasons at stud in England before being sold to Germany, but during his time at home he sired a good number of useful winners, including the 2,000 Guineas Flamingo. The latter was a disappointing stallion but got the well-known 'chaser sire Flamenco (1931) during his first season at stud and four years later the Ascot Gold Cup winner but very bad sire Flyon.

Incidentally, Flamingo's dam Lady Peregrine was a half-sister to the Kentucky

Derby winner Omar Khayyam (1914), who was bred in England and sent to America as a yearling.

The 1923 Derby runner-up Pharos went on to run for two more seasons and retired with a record of 14 wins from his 30 races. He might not have been a great racehorse, but he did become a most influential sire, far outstripping the horse that beat him at Epsom. Reference will be made to Pharos later, but it is worth pointing out that like Buchan before him and Tehran, Aureole and Ballymoss after, he was a Derby second who at least once headed the list of leading sires.

1924

1   Lord Derby's b c Sansovino (Swynford – Gondolette).
2   Lord Astor's b c St. Germans (Swynford – Hamoaze).
3   S. Tattersall's b c Hurstwood (Gay Crusader – Bleasdale).
       27 ran. Time: 2 min 46.6 sec

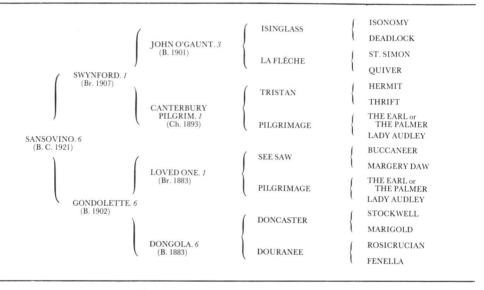

The Derby Stakes, first run in 1780, was named after the 12th Earl of Derby who won the race in 1787 with Sir Peter Teazle. In spite of lavish support of the race from successive holders of the earldom from that day until 1924 victory had not come their way. Sansovino's owner had had his share of bad luck, for his nominations were second in 1911 (Stedfast), 1920 (Archaic) and 1923 (Pharos). Another interested party in this success who had previously been unlucky in the race was Sansovino's dam Gondolette, for she had produced the third in the 1913 Derby (Great Sport) and second in 1915 (Let Fly). It was ironic that when Lord Derby and Gondolette did get their break through Sansovino, the runner-up should be Lord Astor's St. Germans, whose dam Hamoaze was also responsible for the Derby runners-up Buchan in 1919 and Tamar in 1922, while Lord Astor was also second with Craig an Eran in 1921.

Sansovino won his only two starts as a two-year-old, three times from six efforts as a three-year-old and one small race as a four-year-old. He did not run for the 2,000 Guineas and was unplaced in the St. Leger. The value of his six wins came to £17,732. He can be described as a Derby winner of average standard although his subsequent stud career was disappointing. Retiring in 1926 at a fee of 200 guineas, he got the winners of 205 races worth £113,126 before he died 14 years later. This result is not good considering that, amongst others, he had as mates a number of Lord Derby's peerless collection of mares. His best get were the 1931 St. Leger winner Sandwich (who during his short stud career made no show as a stallion); Jacopo (best two-year-old of 1930 and later a top-grade sire in America); St. Magnus (£3,045 and a leading sire in Australia); and Sansonnet (£2,875 and dam of several winners of class including Tudor Minstrel, who won the 2,000 Guineas and £24,629, and sired the fast two-year-olds and subsequently successful sires Sing Sing and Tudor Melody before being exported to America, where he did fairly well).

Sansovino's sire Swynford was an extremely lucky horse ever to go to stud as, when a four-year-old, he broke a fetlock bone—an accident which sometimes means immediate destruction. However, the skill and patience of the veterinary surgeons resulted in his being saved for the stud, where he was able to commence work 18 months after the accident. Swynford, who did not come to his best until towards the middle of his three-year-old days, did not run for the 2,000 Guineas and was unplaced in the Derby but won the St. Leger. In all, in his comparatively short racing life, he won eight races value £25,508 and his victory in the Eclipse gave abundant proof that he was no mere plodder. At stud he was champion sire in 1923 and head of the list of maternal grandsires in 1932. He died at the age of 24. His Classic winners were Sansovino (Derby), Tranquil (1,000 Guineas and St. Leger), Keysoe (St. Leger), Saucy Sue (1,000 Guineas and Oaks), Ferry (1,000 Guineas—full sister to Sansovino) and Bettina (1,000 Guineas), but his most influential get was his son Blandford whom I will refer to later (see page 212).

Swynford's sire John O'Gaunt has been referred to in detail earlier. His dam Canterbury Pilgrim won the Oaks, Jockey Club Cup, etc., and produced a number of other good winners including the famous sire Chaucer. She was by the bad-tempered but game and hardy Tristan, who won 25 races during his five years in training including the Ascot Gold Cup. Canterbury Pilgrim's dam Pilgrimage was a rare good mare who won six races including the 2,000 Guineas and 1,000 Guineas and foaled amongst others the 1898 Derby winner Jeddah and Loved One (sire of the dams of Sansovino and Sunstar).

Sansovino's dam Gondolette was one of the best broodmares of modern times. Sold as a yearling for 75 guineas to Mr. George Edwardes of Gaiety Theatre fame, for whom she was a small winner, she was re-sold as a three-year-old for 360 guineas to Colonel Hall-Walker (later Lord Wavertree). Whilst she was in his Tully Stud (later the National Stud), she bred the 1913 Derby third Great Sport and the 1915 Derby second Let Fly. He sold her in 1912 for 1,550 guineas to the late Lord Derby, for whom she produced Serenissima, Ferry (1,000 Guineas) and Sansovino. Prior to this sale she gave birth to Dolabella, whom Lord Wavertree included in the gift of his stud to the nation. Dolabella produced the wonderfully fast filly Myrobella (£16,143, dam of the 2,000 Guineas winner and champion sire Big Game, grandam of the St. Leger winner Chamossaire, and great-grandam of Hardwicke Stakes winner Hopeful Venture). More recently Myrobella appears as the fourth dam of the 1973 Derby winner Snow Knight.

The first foal Gondolette produced for Lord Derby was Serenissima—a winner herself and dam of eight winners who between them scored in 40 races value £47,850. These included the Ascot Gold Cup winner Bosworth (sire of the St. Leger winner Boswell, and Plassy, a useful stallion in France); Tranquil (1,000 Guineas and St. Leger); and Selene (winner of £14,386; dam of ten winners of 30 races value £47,345, amongst whom were Hyperion, Sickle, Hunter's Moon, Pharamond and Moonlight Run; grandam of leading sire Mossborough; and great-grandam of Ascot Gold Cup winner Raise You Ten).

I have previously drawn attention to the fact that whilst Gondolette's sons Sansovino, Let Fly and Great Sport were all disappointing stallions, Selene's sons Hyperion, Sickle, Hunter's Moon and Pharamond were all extremely successful in that capacity (see page 105). Furthermore Serenissima's sons Bosworth and Schiavoni only achieved limited stud successes.

Following Serenissima, Gondolette produced the 1,000 Guineas winner and Oaks second Ferry (1915), who was an unsatisfactory broodmare, being eventually sold for export for 55 guineas at the December Sales of 1930. Then came the small winner Casa d'Oro by Chaucer, followed by Sansovino, Domenico (£162) and Piazetta (£1,549 and dam of winners).

Gondolette's sire Loved One was a moderate racer, winning three races value £1,847 before retiring to stud in 1891. He made no show as a sire of winners and his name only lives through his two daughters Gondolette and Doris (dam of Sunstar). He was an old horse when he got both these, Doris being 15 years and Gondolette 19 years his juniors. Loved One's sire See Saw won the Royal Hunt Cup, Cambridgeshire, etc., and did well as a stallion without ever reaching the top rank. The line traces back to in tail male to King Herod. Loved One together with Grey Leg (see page 296) were near the last influential sires of this once powerful stirp until the importation of Roi Herode (see page 245) some years later. Loved One and Grey Leg were members of the Highflyer branch of the family which is now extinct, but the Roi Herodes owe their origin to Woodpecker.

Sansovino's dam's sire (Loved One) and his sire's dam (Canterbury Pilgrim) were half-brother and sister, so he was a closely inbred horse.

1925

1   H. E. Morriss's b c Manna (Phalaris – Waffles).
2   H.H. Aga Khan's b c Zionist (Spearmint – Judea).
3   A. K. Macomber's br c The Sirdar (McKinley – Gibbs).
      27 ran. Time: 2 min 40.6 sec

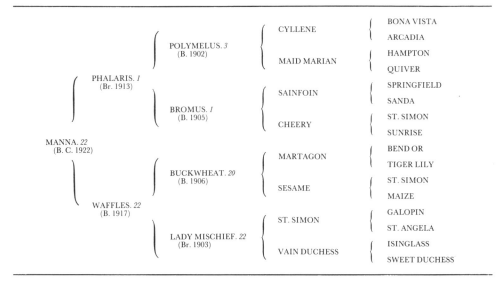

Solario, who lost several lengths at the start before finishing fourth in the Derby, beat Manna a fortnight later in the Ascot Derby, won the St. Leger and Ascot Gold Cup, and at stud bred two Derby winners in Mid-day Sun (1937) and Straight Deal (1943). But it is Manna's name which is recorded as the 1925 Derby winner, following his success in the 2,000 Guineas. He broke down badly in the St. Leger, was retired to stud as a four-year-old and died at the age of 18. Without producing any startling results he did well as a sire, his winners of 237½ races including Colombo, who won nine of his 11 starts including the 2,000 Guineas and never ran unplaced, netting £26,228 for his owner. Colombo had the rather unusual distinction of starting odds-on for all his races except one. At stud he sired the 2,000 Guineas winner Happy Knight and the 1,000 Guineas winner Dancing Time.

Others by Manna were unbeaten Mannamead, who after a few seasons at stud in England went to Hungary, where he died soon afterwards; Miracle, £14,600 but a bad sire; Manitoba, £4,082 and champion sire when exported to Australia; and Pasca, dam of the 2,000 Guineas winner Pasch, who died as a four-year-old, and grandam of the 1953 Derby winner Pinza.

The stud career of Manna's sire Phalaris I have commented on in some detail in Chapter 5, but it is worth reiterating that during his 12 years at stud he achieved astonishing success and effected an influence second to none on modern pedigrees the world over, most notably through the brothers Pharos and Fairway, though among his best winners was the filly Fair Isle (1,000 Guineas).

Pharos, who spent his stud life between France and England, amongst others got Pharis II (champion sire in France) and Nearco. The latter in turn sired Nasrullah, whose enormous stud influence I will refer to later; Dante, 1945 Derby winner; Royal Charger, a most successful sire; Nimbus, 1949 Derby and St. Leger; Sayajirao, 1947 St. Leger; and others of the highest class.

Fairway was four times champion sire in England. His sons included the Derby winners Blue Peter and Watling Street, and the champion sire Fair Trial, who in turn got the 2,000 Guineas winners Court Martial, Palestine and Lambert Simnel, in addition to that very good racer and sire Petition. Court Martial followed the family tradition and was champion sire in 1956 and 1957. Incidentally, Fairway has since been accorded the status of being the last champion sire whose pedigree comprises only stallions and mares bred in the British Isles, such has been the advance of international patterns in bloodstock.

I have referred to Polymelus, sire of Phalaris, on page 184. Bromus, dam of Phalaris, won one small race and produced nothing else of note, unless Hainault, who was a moderate racer but whose name is sometimes encountered in modern pedigrees, can be so classified. Cheery, the dam of Bromus, was a non-winner and foaled no other winners. In her old age she was sold to Germany, and thus was one of the very few St. Simon mares to leave her native shores. Cheery's dam Sunrise was a good winner of £3,500 and dam of the good performers Dunure and Greenan, who both did fairly well as sires in Austria-Hungary and America respectively.

Manna's dam Waffles, who stood barely 15 hands, was never broken. She produced five winners, all of some interest. First came Bunworry, who won four two-year-old races and was then sent to India, where she did not score. After Manna's Derby victory, Bunworry was retrieved and found her way into Senator Tesio's famous Italian stud where she produced Bernina, winner of the Italian 2,000 Guineas, 1,000 Guineas and Oaks, and grandam of Botticelli, who won the 1955 Ascot Gold Cup; Bozzetto, three times champion sire in Italy; Brueghel, by Pharos, three wins in Italy and later a leading sire in Australia; Bernina, Italian Oaks and 1,000 Guineas; and Saucy Silver, dam of the American Grand National winner Burma Road. It is most unusual for a filly who has been to India to turn out a top-grade broodmare.

Waffles' next foal after Bunworry was Manna, followed by his brother Parwiz (£3,897 and sent to Argentina). Then came the St. Leger winner Sandwich, who proved a bad sire during his nine seasons at the stud, and finally the notorious Tuppence, the top-priced yearling at auction in 1931 when bought by Miss Dorothy Paget. Although he never won a race, the popular Press and the public combined to make him a 10-1 chance for Hyperion's Derby. He was a thoroughly bad horse and paid for his sins by being sent to Russia.

Waffles' sire Buckwheat (£4,245) was a useful handicapper whose general merit as a racehorse and sire is indicated by his stud fee of £48. He was later exported to Australia. Waffles was an incestuously-bred mare, her dam and her sire's dam both being by St. Simon—so the mating was one of half-niece to half-uncle. Manna carried further St. Simon blood through his sire.

CHAPTER 15

1926 – 1929
Influence of Coronach and Blandford

1926

1 Lord Woolavington's ch c Coronach (Hurry On – Wet Kiss).
2 W. M. G. Singer's b c Lancegaye (Swynford – Flying Spear).
3 Lord Derby's br c Colorado (Phalaris – Canyon).
   19 ran. Time: 2 min 47.8 sec

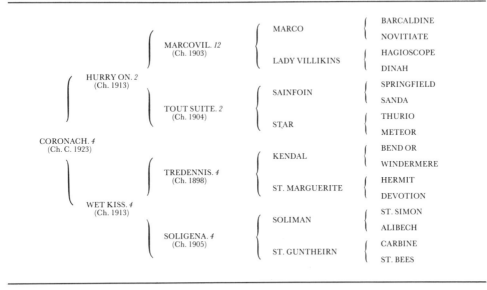

			BARCALDINE
MARCOVIL. 12	MARCO		NOVITIATE
(Ch. 1903)			HAGIOSCOPE
	LADY VILLIKINS		DINAH
HURRY ON. 2			SPRINGFIELD
(Ch. 1913)	SAINFOIN		SANDA
TOUT SUITE. 2			THURIO
(Ch. 1904)	STAR		METEOR
CORONACH. 4			BEND OR
(Ch. C. 1923)	KENDAL		WINDERMERE
TREDENNIS. 4			HERMIT
(Ch. 1898)	ST. MARGUERITE		DEVOTION
WET KISS. 4			ST. SIMON
(Ch. 1913)	SOLIMAN		ALIBECH
SOLIGENA. 4			CARBINE
(Ch. 1905)	ST. GUNTHEIRN		ST. BEES

A TYPICAL SON of Hurry On (see page 197) was Coronach—a great upstanding, strongly-built chesnut with plenty of quality standing 16.2 and girthing 81½ inches. He ran 14 times, winning on ten occasions and being placed in his remaining ventures. He won the Derby in a canter by six lengths, and the St. Leger by an easy two lengths. In all he won £48,224 in stakes. He had two drawbacks: he was unsound in his wind, and was a disappointing sire. He retired to stud in England in 1928 and was given away to New Zealand 12 years later. He was the first Derby winner to go to stud in New Zealand, and before his death in 1949 had considerable Classic influence there.

His best performers on the English Turf were Montrose (£10,866, see page 11) and the gelding Highlander (21 wins value £6,123). Curiously enough, whilst he was in England he sired two or three horses of outstanding merit on the Continent. The best of these was Corrida, one of the best fillies in Europe of her time, winning the Arc de Triomphe twice and the equivalent to some £47,000 in stakes. At stud she produced the French Derby winner Coaraze, and was then taken by the Germans during the war and never recovered. Coaraze made a good start as a stallion in France, and after being exported to Brazil in 1954 made a substantial name for himself as sire of unbeaten Emerson and Coaralde, as well as La Mirambule, dam of Nasram II and Tambourine. Another outstanding Coronach filly was Jacopo del Sellaio, who had the remarkable record of winning all five Italian Classics.

His Italian-bred son Niccolo dell'Arca won nine races in his native land and Germany, including the Italian 2,000 Guineas and Derby before going to stud. When in Italy he sired the Italian Derby and St. Leger winner Daumier, the Italian 1,000 Guineas, 2,000 Guineas and Oaks winner Astolfina and others of note. On transfer to England in 1948 Niccolo dell'Arca got the brilliant Bebe Grande (1950). Another meritorious son of Coronach is the French-bred Cranach, who was confiscated as a two-year-old by the Germans during the war and won 11 races whilst a prisoner of war. Repatriated at the conclusion of hostilities he did well as a sire in France, his get including Violoncelle (won stakes worth about 13m francs) and Ciel Etoilé (French St. Leger).

It will be seen therefore that, although if judged solely on his English get Coronach must be looked upon as a stud failure, the excellence of a few of his Continental stock puts a different complexion on matters.

Coronach's dam Wet Kiss, bought for £3,000 in 1918, won three races value £1,067 and produced two other winners. She was a full sister to the Irish 2,000 Guineas winner Soldennis who during his five years racing won 24 times to the value of £13,072. Their dam Soligena was a first-class broodmare and others from her were Orpi (£6,217), Irish Kiss (£6,306 in America) and Valiant (£2,306). Wet Kiss's sire Tredennis is one of the very few sires with considerable recent influence who was a thoroughly useless racehorse. He was very well bred, being by Kendal (sometime champion sire) out of the 1,000 Guineas winner St. Marguerite, who during her stud career produced some excellent stock including the Oaks and St. Leger winner Seabreeze, Le Var (£8,995), Roquebrune (dam of Rock Sand, whom I have already discussed).

Tredennis spent three years in training but was incapable of winning a race of any sort. He was then sold for £100, transferred to Ireland and started at stud in the most humble circumstances serving "half-bred" and all sorts of mares. He soon began to get winners from the few thoroughbred mares he was put to, so as time went on the quality of his mates improved. Not unnaturally with better mares he got better winners and before he died in his 29th year he had sired the winners of 480 races value £147,479 in England and Ireland—a most remarkable perform-ance.

His name is familiar largely through his son Bachelor's Double (£10,537) whose daughters include Trustful (dam of St. Leger winner Scottish Union and a significant modern tap root mare); Napoule (dam of Oaks winner Lovely Rosa and grandam of Wilwyn); Double Life (dam of Gold Cup winner Precipitation, Persian Gulf etc., and great grandam of 1,000 Guineas, Oaks and St. Leger winner Meld); Celiba (grandam of 2,000 Guineas and Derby winner Blue Peter); and Miss Bachelor (dam of Grand Prix de Paris winner Fiterari). Tredennis never sired an English Classic winner but his son Golden Myth won both the Ascot Gold Cup and Eclipse Stakes before a most disappointing stud career.

A horse worthy of mention is Colorado, who finished third in the 1926 Derby. His fame lies not only on his racing record but on the wonderful results he got during a very short stud career. On the course he won the 2,000 Guineas and Eclipse—beating Coronach in both—and seven other events value £30,358 in all. He went to stud in 1928 and died the following year. During his two seasons at stud he sired Felicitation (Ascot Gold Cup and sent to Brazil after a short spell at stud in England); Loaningdale (Eclipse Stakes and sent to Uruguay after standing in England); Scarlet Tiger (third in the St. Leger and £5,761, and sent to Argentina after standing in Ireland); Coroado (£5,206, the best sprinter of his time and sent to Brazil after standing in England); Furrokh Siyar (sire of many winners in Ireland); Canon Law (£5,630 and sire of winners); Colorado Kid (£9,462 and sire of winners); Figaro (£4,525 and sire of winners); Colorow (£3,025, died as a three-year-old); Yellowstone (£5,974, sent to Chile); and Riot (£1,190, dam of the Oaks winner Commotion who produced unbeaten Combat, Faux Tirage, £8,744 and successful sire in New Zealand, and Aristophanes, £4,085, sold to Argentina for £22,000 in 1953 and a successful sire).

It is hard to name another horse who has achieved such a widespread influence during a two-year spell at stud as Colorado.

Colorado was by Phalaris out of the 1,000 Guineas winner Canyon, who produced six other winners, including Colorado's brother Caerleon (£11,210). She was a sister in blood to Glacier, who amongst others foaled the Oaks winner Toboggan, who in turn became the grandam of the great American racer Citation. The descendants of Glacier played a big part in the fortunes of Lord Derby's and other important studs throughout the world.

1927

1   Frank Curzon's ch c Call Boy (Hurry On – Comedienne).
2   Sir Victor Sassoon's b c Hot Night (Gay Crusader – Tubbercurry).
3   Major J. S. Courtauld's b c Shian Mor (Buchan – Orlass).
     23 ran. Time: 2 min 34.4 sec

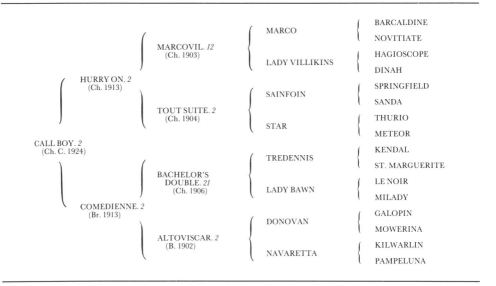

Call Boy, Hurry On's last Derby winner, was probably the best of his generation, which was not a particularly good one. The Derby was his last race as his owner died a month after Epsom and he was taken out of training. He was then sold for £60,000 to Mr. Curzon's brother and sent to stud at Newmarket at a fee of 400 guineas, but was a total write-off in that capacity as he was almost sterile. By the time of his death in 1940 his fee had been reduced to £48.

Call Boy's lack of stud success was emulated by those nearest to him at Epsom—Hot Night, who was a failure as a stallion, and Shian Mor, who was sent to Japan. It was left to Lord Derby's Sickle to perpetuate the 1927 Derby stud annals, since on export to America he reappeared in the pedigree of Native Dancer (1950), one of the great influences of modern times.

I have already referred to Call Boy's sire Hurry On (see page 197) and his maternal grandsire Bachelor's Double (see page 207), but I will just touch on Lady Bawn, the dam of Bachelor's Double. She was a twin and Bachelor's Double, whom she produced as a four-year-old, was her first foal. Then followed Bachelor's Hope (£4,617 and sire of winners); Bachelor's Wedding (Irish Derby and other races, sent to India); Bachelor's Darling and Bachelor's Fare—all winners. Her twin sister Lady Black was the dam of six winners including Melesigenes (£14,924 in Ireland and India). The twins' sire Le Noir (1889) never ran and led the life of an Irish country stallion. He was a member of a branch of No. 29 Family which had failed to produce anything of note for generations and his sole useful contribution to his breed were the twins. Their dam Milady never ran and changed hands for 45 guineas at one of the Dublin Horse Show auctions. She also foaled that very good racer but bad sire Bachelor's Button (by Winkfield), who won 16 races value £16,465 including the Ascot Gold Cup. Milady's dam Alone won one small race but her grandam none, although she was a half-sister to Oaks winner Lonely.

Bachelor's Double therefore was by a non-winning sire, out of a non-winning dam, by a non-winning maternal grandsire, out of a non-winning grandam. The odds against such a combination producing a good horse—let alone a most beneficial sire—must be enormous and his pedigree serves to show how the genes which produce winners can be carried latently in parental stock. As we have seen, Tredennis's sire and dam were both beyond reproach. Le Noir was by the dual Ascot Gold Cup winner Isonomy out of a mare by St. Leger winner Lord Clifden, whilst Milady was by the Derby winner Kisber out of a mare by the Derby winner Hermit. There was thus an abundance of the best blood in the pedigrees which came to the fore when Tredennis was mated to Lady Bawn. Bachelor's Button was a different proposition as he was free from the non-winning strains of Tredennis and Le Noir.

Call Boy's dam Comedienne was bought for 130 guineas as a yearling, won five races and produced Comedy King (£7,442 and a useful sire in France) and Comedy Star (ancestress of 1,000 Guineas winner Dancing Time, and Umberto, who gained eight wins worth £8,327). Her dam Altoviscar won one race, and also was the grandam of the Gold Cup winner Foxlaw, who unfortunately died at the age of 13, but not before he had got the Gold Cup winners Foxhunter and Tiberius. Foxlaw's full sister Aloe did not win but at stud produced several winners, and established an important family based on her daughters Feola and Sweet Aloe, whose influence is set out in Chapter 9.

Altoviscar is also the ancestress of the French 2,000 Guineas winner Drap d'Or, who after six seasons at stud in England went to Kenya. Her sire Donovan (1886) won 18 races value £55,154—an immense sum in the 1880s—including the Derby and St. Leger. He was moderately successful at stud but never got anything nearly as good as himself before meeting with a fatal accident in his 20th year. His dam Mowerina (1876) was a first-class racer, winning 16 races during her five years in training. At stud, in addition to Donovan, she foaled the 1,000 Guineas winner Semolina, Raeburn and other useful stock before dying aged 30. Her career is one example of many which shows that hard racing does not necessarily jeopardize a filly's stud career. Her son Raeburn was one of the few horses capable of beating those two great rivals Isinglass and La Flèche. However, he did not do very well as a stallion in England and was sent to Hungary, where competition was not so keen. Mowerina (by Scottish Chief) was bred in Denmark, a country not usually associated with the production of high-class racehorses. In her case there is an unusual duplication of names in the Stud Book as her great-grandam, also called Mowerina (1843, by Touchstone) produced the Triple Crown winner West Australian.

1928

1 Sir Hugo Cunliffe-Owen's b c Felstead (Spion Kop – Felkington).
2 Sir Laurence Philipps' b c Flamingo (Flamboyant – Lady Peregrine).
3 L. Neumann's b c Black Watch (Black Gauntlet – Punka III).
    19 ran. Time: 2 min 34.8 sec

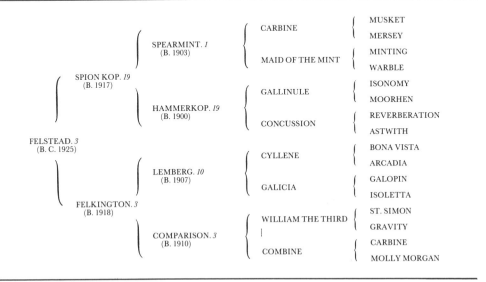

It is usually safest to accept results as they stand and not to make excuses for failures, but it is only fair to mention the unusual circumstances which affected Fairway, hot favourite for the 1928 Derby. As the horses were going to the start the crowd broke through the inadequately policed laneway and mobbed their idol. Some even went so far as to pluck hairs from his tail. Fairway, always rather highly strung, became so upset that his chance of success was ruined before he even reached the starting gate.

Included in the tests which a Derby horse is subject to is the ability to withstand equally the noise and bustle which is part and parcel of Epsom Downs on Derby Day, but mobbing and tail plucking cannot be considered a normal hazard. This was the only race Fairway lost as a three-year-old, whereas Felstead won only two others—a maiden plate value £251 and a handicap value £599.

Be that as it may, Felstead was a very easy Derby winner. Splint trouble prevented further racing, so he retired to stud after a year's idleness. Before he died in his 22nd year he had sired the winners of about £105,000 in stakes. His most famous get was that very game filly Rockfel, winner of £23,431 including the 1,000 Guineas and Oaks before becoming the dam of Rockefella (1941), who although only the winner of three races value £901 was a useful racehorse and did quite well at stud, with the Cambridgeshire winner Richer (sent to Argentina); the Derby second Gay Time; 2,000 Guineas winner Rockavon; Outcrop, the best staying filly of her year; and Italian 1,000 Guineas and Oaks winner Angellai

Rucellai, dam of the Italian Derby winner Appiani II, and Andrea Mantegna, leading sire in Italy.

Felstead also sired the 1946 Oaks winner Steady Aim who was 18 years his junior.

I have already mentioned Felstead's sire Spion Kop (page 193) and his maternal grandsire Lemberg (page 175). His dam Felkington won £1,954 and produced six other winners, amongst them the wartime Gold Cup winner Finis (1935), who went to New Zealand and did well as a sire.

Felkington's dam Comparison won two races and bred Carpathus (£3,592 and sired some winners). She was by the Ascot Gold Cup winner William the Third, who was second on the list of sires of winners twice and champion sire of broodmares once. In all during his 14 seasons at stud he got the winners of 232 races value £112,053. His best were the Ascot Gold Cup winner but very bad sire Willonyx (1907) and the 1,000 Guineas winner Winkipop (1907), who after producing influential stock in England, including the Derby second Blink, went to America. Blink also crossed the Atlantic but died soon after arrival. William the Third's name is more familiar through his daughters, but one son, the rather moderate Roseland, sired Garron Lass, whose descendants include Bahram, Dastur, The Phoenix and Sunny Boy III (France).

Felstead's great-grandam Combine produced the hardy Land League (33 wins including the Cambridgeshire but a bad stallion) and Combination, whose son Collaborator (£3,923) sired a number of winners in France. Combine was a half-sister to Farasi, whose only claim to fame is that his name appears in the pedigree of Derby winners Trigo and Tulyar. I will refer to him later.

Felstead provided a timely revival for the male line of Carbine, to whom he was inbred at the third and fourth removes, and whom he somewhat resembled in appearance.

1929

1   W. Barnett's b c Trigo (Blandford – Athasi).
2   Lord Woolavington's ch c Walter Gay (Captain Cuttle – William's Pride).
3   S. Tattersall's br c Brienz (Blink – Blue Lake).
        26 ran. Time: 2 min 36.4 sec

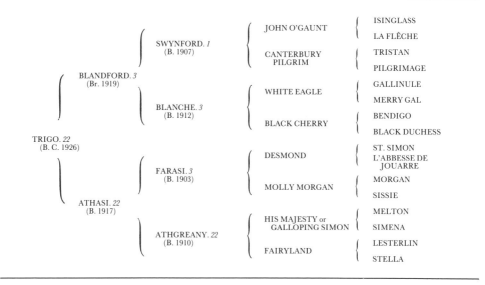

Trigo, who was not foaled until May 23, won two races value £1,652 as a two-year-old in Ireland. On coming to England the following year his only defeat was in the 2,000 Guineas. After winning the English St. Leger he returned to his native land to add the Irish counterpart to his tally and then retired to stud. He cannot be regarded as a success for in 16 seasons he sired the winners of only approximately £46,000 in stakes and very few of distinction, a fact reflected in his stud fee which gradually fell from 400 guineas to £98.

Trigo's sire Blandford was bred at the National Stud shortly after Lord Wavertree had presented it to the public. He had pneumonia as a foal and was extremely lucky to survive. In fact the veterinary surgeon at the time attending the stud told me, many years ago, that Blandford's case during this illness at one period seemed so hopeless that the Stud Director (Sir Harry Greer) suggested that he should take him away as a gift.

Though the horse recovered he was not fit to send to the Sales the following year with the remainder of the stud's yearlings. He was held back until the Newmarket December Sales when he was bought for 720 guineas by Messrs. R. C. and S. C. Dawson. He appeared on a racecourse only four times, winning three value £3,839 and being second, beaten by a short head, in the other. His comparatively small stake winning is but little guide to his racing merit. His legs caused a lot of trouble in training and as I have previously recorded, his very able trainer Mr. R. C. Dawson rated him the equal or superior to the very good Derby winner of his year

Captain Cuttle. He did not run as a four-year-old and started at stud the following year at a fee of £149 a mare. This was later raised to 400 guineas. He died when he was 16 years old, so did not have a long stud career. He sired the winners of 308 races value £327,840 at home, besides some most important ones abroad, and though these figures may not seem outstanding when compared to such as Hyperion, it should be remembered that Blandford had only 9½ crops to represent him. Blandford, twice champion sire, got four Derby winners, Trigo, Blenheim, Windsor Lad and Bahram; two 2,000 Guineas winners, Bahram and Pasch; and the 1,000 Guineas winner Campanula; whilst in France Brantome won the French 2,000 Guineas, St. Leger, Gold Cup, etc.

I will refer later to the stud achievements of his Derby-winning sons but to cover fully the world-wide influence of this great sire would require a book in itself. Blandford was an almost faultless specimen of his breed although he stood over to a marked degree at his knees—not that this is necessarily a fault. At stud I understand he was extremely slow to cover his mares and would sometimes stand behind them for 15 minutes or more before so doing. I mention this as I have often heard breeders express the opinion that a stallion with this idiosyncrasy is unlikely to do much good.

Blandford was by Swynford (see page 201) out of Blanche, one of the National Stud's original mares. Although she was incapable of winning a race herself, she was a half-sister to the 1,000 Guineas and Oaks winner Cherry Lass and others of note as we shall see later. Her other winners were: Silver Hussar (three wins value £4,476—a gelding); Seminole (Cesarewitch and £3,762); Mahoonagh (£506, an Irish country sire); Blanc Mange (three wins for £1,550, dam and grandam of good winners for the Aga Khan); Nun's Veil (three wins value £2,297, grandam of Sun Chariot and great-grandam of unbeaten Blue Train).

Blanche's dam Black Cherry ran only as a two-year-old and scored in a £100 selling plate. On retiring to the paddocks she was mated in her first years with the City and Suburban winner but low-class sire Fullerton. She then passed into Lord Wavertree's possession at a very small figure. She proved a rare bargain as she produced nine winners of 41 races value £28,383. The high average of stakes won over such a large spread is an indication of their class. They included Jean's Folly (£2,588, dam of the St. Leger winner Night Hawk, and grandam of Poisoned Arrow, £7,021); Black Arrow (£7,889 and died before going to stud); Cherry Lass (£15,087, including the 1,000 Guineas and Oaks, died young after producing two winners); and Kingston Black (£262, and second in the St. Leger). Black Cherry was a half-sister to Bay Ronald (see page 189), and her influence to modern times is related fully in Chapter 9.

Blanche's sire White Eagle won eight races and was second in the St. Leger and third in the 2,000 Guineas. His general racing standard was a trifle below that of a Classic winner but at stud, in common with other sons of Gallinule (see page 193), results were disappointing.

As not infrequently happens when a stallion is mated to particularly good mares without siring much in the way of winners, some of the fillies from these unions prove particularly good broodmares, and such was the case with White Eagle. During his whole stud career his name only appears twice among the 20 leading sires of winners, and then in a junior position—11th in 1914 and 20th in 1921. On the other hand it is found ten times among the 20 leading sires of broodmares and not infrequently in a senior position, including third in 1922 and 1932. In addition to Blanche, other famous broodmares by White Eagle were: Dolabella, dam of Myrobella (£16,143) and grandam of the 2,000 Guineas winner and champion sire Big Game; Lady Peregrine, dam of the 2,000 Guineas winner Flamingo and grandam of Honeyway (£10,919); Tudor Honey (£8,114); Welsh Honey (£4,490); Royal Favour, dam of the St. Leger winner and bad sire Royal Lancer, who finished his days in South Africa; Quick Thought, dam of six good winners of £12,810 and grandam of Princequillo; and Erne, great-grandam of Ponder, winner of the Kentucky Derby and 541,275 dollars.

White Eagle's dam Merry Gal was a first-class racer who won the Doncaster Cup, Jockey Club Cup, Princess of Wales's Stakes etc. Her grandam was the Oaks and St. Leger winner Marie Stuart. White Eagle was the result of very close inbreeding, as his sire Gallinule was by Isonomy, whilst his dam Merry Gal was out of an Isonomy mare, Mary Seaton.

Trigo's dam Athasi, bought as a yearling for 270 guineas, won five races value £1,303 between the ages of two and six. The average value of the races she won is a fair indication of her class, i.e. a bottom-grade handicapper just above selling-plate class, and as a six-year-old she was running unsuccessfully in unimportant hurdle races. In 1923 she was covered by the premium stallion All Alone, and in due course produced a small bay filly named Ballygrainey, who never ran and was sent to South America. After Trigo had won the Derby, vain efforts were made to repatriate her but she had been sold and re-sold in her adopted country and it proved impossible to trace her.

When Athasi had Ballygrainey at foot, she was put to Blandford, who was making his debut as a stallion that year. The product, named Athford, won the Doncaster Cup and five other races value £7,973, and after spending one season at stud in England went to Japan. Then followed a long and marvellously successful association between Blandford and Athasi, which produced in all seven winners who won between them £51,337 in stakes. These included Harinero, £6,750, the Irish Derby and St. Leger and a not-too-good sire in Australia; Trigo, English Derby and St. Leger, Irish St. Leger, £27,102; Primero, £4,257, Irish St. Leger and dead heated in the Irish Derby, sent to Japan; Centeno, £961, dam of winners including Artist's Son, who sired the 1955 Grand National Winner Quare Times; Harina, £4,128, grandam of Derby, St. Leger etc., winner Tulyar and Arc de Triomphe winner Saint Crespin III; and Avina, £166, dam of Oatflake, winner of a wartime November Handicap and dam of the smart sprinter Milesian.

Athasi also produced two minor winners by Umidwar, and one by Cygnus. In all she bred ten winners of 28½ races value £51,879 and thus proved herself one of the biggest winner-producers of the period, and all this despite her unpromising start to life.

Athasi's sire Farasi, who cost 35 guineas as a yearling, won £1,014. His best performance on a racecourse was when he beat the following year's Derby winner Spearmint in a nursery at Newmarket. He later trained off and in due course retired to stud in Ireland, where he led the life of an unimportant country stallion until he died in 1922. His valuation can be gauged from the fact he started at a fee of 19 guineas a mare which was reduced to five guineas after a few seasons. He sired a few minor winners and except through Athasi his name has long since been forgotten. He was by Desmond out of the Cambridgeshire winner Molly Morgan, whose other sons Morganatic (by St. Simon) and Morgendale (by Kendal) cut no ice as stallions, but her daughter Combine was the great-grandam of the Derby winner Felstead (see page 210).

Athasi's dam Athgreany won the Irish Oaks and two other races in 17 attempts, and bred four other moderate winners. She was unlucky enough to be continually mated with indifferent sires but nevertheless founded a great winner-producing family.

The status of Athgreany's sire Galloping Simon can be judged by his stud fee of nine guineas, whilst her dam Fairyland was a good two-year-old winner in Ireland and bred three other winners. She was a sister to Zenith, ancestress of the 1941 St. Leger winner Sun Castle and the 1933 Eclipse Stakes winner Loaningdale; to Blakestown, ancestress of the Cesarewitch winner Eagle's Pride; and to Glenesky, who sired a good few winners. She was also half-sister to Flying Orb, who amongst others got the brilliant filly Cos (eight wins value £9,604, dam of four good winners including Rustom Pasha). Flying Orb also sired the grandam of The Bug (£7,415).

An unusual sideline to the 1929 Derby was that the third past the post, Brienz, was later gelded and became one of the best 'chasers in the country, while the fourth horse, Lord Derby's Hunter's Moon, a half-brother to Hyperion, was sent to Argentina and became a leading sire there. Hunter's Moon returned to the Derby story in 1976 through the winner Empery, whose dam Pamplona II, winner of the Triple Crown in Peru, was by Hunter's Moon's son Postin, a top racehorse and six times leading stallion in Peru.

CHAPTER 16

1930 – 1932
Blenheim and Pharos

1930

1 H.H. Aga Khan's br c Blenheim (Blandford – Malva).
2 S. Tattersall's ch c Iliad (Swynford – Pagan Sacrifice).
3 Sir H. Hirst's b c Diolite (Diophon – Needle Rock).
    17 ran. Time: 2 min 38.2 sec

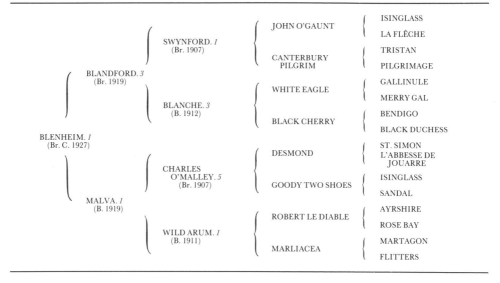

BLENHEIM, BOUGHT AS a yearling for 4,100 guineas, ran seven times as a two-year-old, winning four and being second in the others. The following season he ran once unplaced, was fourth in the 2,000 Guineas and his final race was his Derby success. He then jarred himself and was taken out of training. He started at stud in France as a four-year-old at a fee of some 400 guineas and after six seasons was sold to an American syndicate for £45,000.

His influence as a sire on both sides of the Atlantic has been enormous. Amongst his first crop in France was Mumtaz Begum, dam of Nasrullah and grandam of Royal Charger. Next season came the Derby winner Mahmoud and the brilliant Wyndham (£8,445). The latter, who ran as a two-year-old as the Bossover Colt, was a rather indifferently bred horse who spent ten years at stud in England but sired nothing of note until the last year of his life (1948) when he got Windy City, the best two-year-old of his generation in England and Ireland. Windy City's American-raced son Restless Wind sired the champion two-year-old filly Process Shot; On Your Mark, who showed brilliant speed as a two-year-old when trained in England; and Tumble Wind, who won from 6½ furlongs to a mile and a half in America and was at stud there for four seasons before being exported to Ireland.

In America Blenheim was champion sire, his get including the great Whirlaway (32 wins and 561,161 dollars); Thumbs Up (17 wins and 249,290 dollars); Fervent (17 wins and 347,135 dollars); Jet Pilot (seven wins and 198,740 dollars). His American-bred daughters included Jane Gail (dam of the 1952 Kentucky Derby winner Hill Gail); Easy Lass (dam of Coaltown, who won 415,675 dollars) and Miss Rushin (dam of Kentucky Derby winner Ponder, who won 541,275 dollars).

DONATELLO II (1934)

Amongst Blenheim's 1934 foals were the French 2,000 Guineas winner Drap d'Or and the Italian-bred Donatello II, who won all his races in his native land and suffered his only defeat when second in the Grand Prix de Paris. Donatello II was then bought for £47,500. The original intention was that he should go to stud in France but plans were changed and he commenced his new duties as a four-year-old at Newmarket, where he stayed until his death 17 years later.

He had an erratic stud career—in 1938, his first season, 11 of the 18 mares he covered were barren, and in 1941 his fee dropped from 400 guineas to £198—but he has been a most influential stallion, worthy of close inspection. His best sons were:

Alycidon, the best stayer trained in England since the war, and sire of Alcide (St. Leger, King George VI and Queen Elizabeth Stakes, and important sire), Kalydon (sire of smart racemare Park Top), Twilight Alley (Ascot Gold Cup), Alcibiades (winner in England and America and top-class sire in the States), Preamble (successful sire in South Africa), Alcimedes (Jubilee Stakes twice, and champion sire in Australia), Gloria Nicky (Cheveley Park Stakes, dam of 1,000 Guineas and Oaks winner Never Too Late II, and dam of Without Fear, outstanding first-season sire in Australia), Meld (1,000 Guineas, Oaks and St. Leger, and dam of Derby winner Charlottown), Fair Alycia (unraced, grandam of top-class sire Bold Lad (Ire)), Alconbury (dam of Ascot Gold Cup winner Parbury), Alcazar (winner and dam of top-class winner and stallion Petingo), Alcoa (winner and dam of Oaks winner Lupe), and Homeward Bound (Oaks, and grandam of Grand Criterium winner Super Concorde).

Acropolis, full brother to Alycidon, winner of seven races and second in the King George VI and Queen Elizabeth Stakes, and sire of Espresso (Grosser Preis von Baden twice), and the fillies Fiji II (Coronation Stakes, dam of Ribblesdale

Stakes and Yorkshire Oaks winner Fleet Wahine), Belle of Athens (dam of St. Leger winner Athens Wood) and Fran (dam of smart sprinter Tudor Music).

Supertello, winner of Ascot Gold Cup but a failure at stud whose most notable achievement was through the 66-1 Grand National winner Ayala.

Crepello, bred in the last year of Donatello II's life and winner of the Derby. His stud career will be examined later.

Donatello II, although bred in Italy, was of almost pure Anglo-Irish ancestry. He was also responsible for several influential fillies, among them: Pasqua (1939, dam of the Derby and King George VI and Queen Elizabeth Stakes winner Pinza), Celestial Light (1940, grandam of Sovereign Edition, leading sire in New Zealand), Picture Play (1941, 1,000 Guineas winner and third dam of Derby winner Royal Palace), Donah (1942, grandam of Oaks winner Sleeping Partner), Saracen (1943, dam of Washington DC International winner Wilwyn, a leading sire in South Africa), Valkyrie (1943, dam of Argentine Triple Crown winner Tatan), Angelola (1945, second in the Oaks, dam of tip-top racehorse and stallion Aureole), Belladonna (1952, dam of Italian St. Leger winner Ben Marshall), and Cutter (1955, third in the Oaks and grandam of Irish 2,000 Guineas winner Sharp Edge).

MALVA (1919)

Reverting to Blenheim, he was by Blandford (see page 212) out of Malva, who won three races value £1,052 and produced seven winners, including Blenheim's brother His Grace (£6,556, and died after eight seasons at stud, during which he sired the winners of £46,400 in stakes), and the Eclipse Stakes winner King Salmon, by Salmon-Trout.

King Salmon, who after nine seasons at stud in England was sent to Brazil, sired the filly Herringbone (1,000 Guineas and St. Leger).

Malva's sire Charles O'Malley was third in the 1910 Derby and beaten a neck in the Ascot Gold Cup the following year. However, he won five races value £5,005, including the Ascot Gold Vase, before retiring to stud. Although he cannot be classified as a top-grade sire, his name appeared for some years amongst the leading 20 sires of winners, his best get probably being the 1920 Oaks winner Charlebelle. Unhappily he died at the comparatively early age of 15. He was by Desmond out of Goody Two Shoes, who won one race and although she bred no other foals of special racing merit, was the dam of Simon's Shoes.

Simon's Shoes found the winning of a selling plate beyond her capacity, but at stud bred eight winners of over £10,000 in stakes. Her influence has been narrated in Chapter 9.

Malva's dam Wild Arum won a two-year-old race and bred five other unimportant winners. She was by the Doncaster Cup and City and Suburban winner Robert le Diable, who cannot be accounted as much of a sire but who got the hardy Wrack (ten wins on the Flat value £5,397, plus five hurdle wins value £1,245). It is unusual for a horse who has raced under National Hunt Rules to be a success as a stallion but Wrack, on export to America, most decidedly proved an exception—possibly due to the fact he was a higher-class horse than is normally put to hurdling, his dam being a half-sister to the 2,000 Guineas winner Neil Gow, to the famous Popinjay (ancestress of numerous Classic winners) and to others of great worth. Robert le Diable was also a very well-bred horse, being by the Derby winner Ayrshire out of Rose Bay by the Derby winner Melton out of Rose of Lancaster, sister to the Derby winner Bend Or.

Wild Arum's dam Marliacca won three races and also produced the Ascot Stakes winner Rivoli (£4,914, sent to France after a few seasons at stud in England and got some winners there).

Whether or not Blenheim, who was inbred to Isinglass at the fourth generation, can be regarded as an exceptionally good Derby winner is a debatable point, but it is beyond dispute that he exercised an unusual influence on the Stud Books of the world.

1931

1 J. A. Dewar's b c Cameronian (Pharos – Una Cameron).
2 Sir J. Rutherford's b c Orpen (Solario – Harpy).
3 Lord Rosebery's b c Sandwich (Sansovino – Waffles).
   25 ran. Time: 2 min 36.6 sec

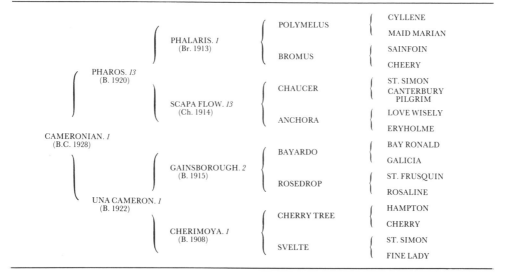

Cameronian was a small bay colt with beautiful quality and action. In addition to the Derby he secured the 2,000 Guineas and three other races, and netted £31,287 in stakes. He started at stud in 1933 but did not prove an overall success, although he sired a host of smallish winners. After eight seasons in England he was exported to Argentina, where he died in 1955. His best get in his native land was Scottish Union, who won six races value £21,587 including the St. Leger and was second in both the Derby and 2,000 Guineas, whilst his daughter Snowberry produced the St. Leger winner Chamossaire. Scottish Union spent 13 years at stud in England and two in Ireland, where he died at the age of 19. His stock won some 260 races and £120,000 in stakes, the best being the gelding Strathspey (11 wins, £12,087 including the Cesarewitch, Goodwood Stakes, etc.).

Cameronian also sired the wartime Gold Cup victor Finis (see page 211) and Nakamuro (four wins in Italy, second in the Italian Derby, brother in blood to Nearco.

PHAROS (1920)

Cameronian's sire Pharos was a dark bay almost brown horse and the second foal of his dam, who later proved her worth as a matron in no uncertain manner.

The beneficial influence of Pharos as a stallion took him above his station as a high-class handicapper rather than a Classic horse. A comparatively long and hard racing career did not adversely affect his capacity as a sire. He started at stud as a six-year-old at Newmarket and after two seasons moved to France, where he remained until he died, aged 17. Cameronian was one of his second crop of foals, whilst next season came the St. Leger winner Firdaussi, who won eight races value £21,550 and was placed in eight more. Firdaussi spent one year at stud at Newmarket, followed by four in France and was then sent to Rumania. Whilst in France he sired the French 2,000 Guineas winner Panipat. Firdaussi's dam was the Oaks winner Brownhylda.

Another son of Pharos who had considerable influence was the Eclipse Stakes winner Rhodes Scholar (three wins value £14,326) who after spending two years at stud in England went to America. He got Black Tarquin, who was imported to England as a yearling and won eight races value £29,483 including the St. Leger.

The year 1935 saw the birth of Pharos's unbeaten son Nearco, to whom I will refer in detail when we come to Dante's Derby victory.

The following year Pharis II, another unbeaten son of Pharos, appeared. He ran only three times, securing both the French Derby and Grand Prix de Paris. He was scheduled to run in the Doncaster St. Leger, where he would probably have crossed swords with that specially good Derby winner Blue Peter, but the race had to be cancelled owing to the outbreak of war. In 1940 Pharis II spent a short season at stud in France and in the summer was sequestered and carted off to Germany. His first foals in 1940 were of sufficient merit to place him at the head of the French list of sires of winners in 1944. He was repatriated at the end of the war and soon regained that position. He died in 1957.

Amongst the get of Pharis II are two French-trained winners of the Doncaster St. Leger, Scratch II (who also won the French Derby) and Talma II. Pharis II also sired Ardan (French Derby, etc., sire of many winners including Hard Sauce); Priam II (ten wins in France and sire of good winners there); Auriban (French Derby), and Pardal (won about £15,000 in France and England, sire of Derby winner Psidium, and Firestreak, sire of Derby winner Snow Knight).

Other outstanding Pharos stock were Mary Tudor II (French 1,000 Guineas and dam of the Epsom Derby and Gold Cup winner Owen Tudor); The Nile (French 1,000 Guineas); En Fraude (French Oaks); Signal Light (£1,885 and sire of many winners); Fastnet (442,912 francs, the Viceroy's Cup in India and a very good stallion); Phideas (Irish 2,000 Guineas and Derby, and sire of the English

Oaks winner Frieze); Bernina (Italian 1,000 Guineas, 2,000 Guineas and Oaks, grandam of top-class Italian racer and sire Botticelli); Bozzetto (123,750 lire, champion sire in Italy) and so on. He was champion sire in England in 1931 and in France in 1935 and 1939.

I have already referred to Phalaris, sire of Pharos, whose dam Scapa Flow was bred by Lord Derby in 1914. At Stockton as a three-year-old Scapa Flow ran second in a selling plate which provided for the winner to be sold for £50. An owner of one of the runners intended to claim her but he thought he could get her a little below the claiming price if he approached Lord Derby's trainer George Lambton. On hearing of this Lambton promptly made arrangements for someone else to put in a friendly claim for the filly, and so she was saved for Lord Derby's stud. She managed to win only three races value £482 but at stud produced eight winners who between them won 63 races value £86,084, as follows:

Spithead (1919) by John O'Gaunt	(£5,641—a gelding)
Pharos (1920) by Phalaris	(£16,594—see above)
Pentland (1923) by Torloisk	(£567—a gelding)
Fairway (1925) by Phalaris	(£42,722—champion sire four times and a great modern influence)
Fair Isle (1927) by Phalaris	(£13,219—including 1,000 Guineas)
Fara (1928) by Phalaris	(£1,950—dam of useful winners)
Highlander (1930) by Coronach	(£6,123—a gelding)
Pharillon (1931) by Phalaris	(£168—a gelding)

It is curious that two mares Scapa Flow and Gondolette (see page 201) who had a tremendous effect not only on Lord Derby's stud, but on the Stud Books of the world were both in humble circumstances at one time during their careers.

Scapa Flow's sire Chaucer by St. Simon was a half-brother to Swynford (see page 201). He won eight races value £5,663—and as their average value (£700 odd) indicates he was a good-class handicapper and not a Classic horse. I have commented extensively on his stud achievements in Chapter 5. Scapa Flow's dam Anchora was bred by Mr. George Edwardes (also one-time owner of Gondolette) and won eight races value £1,966, remaining in training until she was seven years old. This had no adverse effect on her stud career if we are to judge by results as, besides founding a great family through Scapa Flow, her daughter Rothesay Bay was the ancestress of the Eclipse winner Miracle, the Coronation Cup winner and useful French sire Plassy and others of merit. Anchora was by the Ascot Gold Cup winner Love Wisely.

CHERIMOYA (1908)

Cameronian's dam Una Cameron raced for two seasons but failed to win. At stud she also foaled Lovat Scout (£3,990 and a sire in Argentina); Lochiel (£800, a stud failure in Ireland) and Troon (£146 and sent to France where he sired some winners of limited ability). Una Cameron was by Gainsborough (see page 188) out of Cherimoya who won the 1911 Oaks on her only start, and at stud produced three winners including the quite useful The Cheerful Abbot (9½ wins value £4,431 but no use as a sire, possibly through lack of opportunities). The last-named mare was also the grandam of Sunny Trace (11 wins value £7,588 and sire of a good number of small winners) and of The Macnab (four wins value £5,772 who was doing well as a stallion in France up to the war when he was taken off to Germany and never recovered). Cherimoya was the only English Classic winner of modern times whose sire and dam were both non-winners. Her sire Cherry Tree was sent to America as a yearling and raced there under the name Matt Byrne but failed to win. He was then brought back to England and although he lived until he was 30 years of age, never sired anything of note except his single Oaks winner. His stud fee remained at 18 guineas all his life except for a couple of seasons after Cherimoya's success when it was raised to 45 guineas.

Cherimoya's dam Svelte (1899) was by St. Simon out of Fine Lady by Isonomy out of Sonsie Queen by Musket, coming from a good winner-producing line in tail female, so it would appear to be a somewhat unconventional mating to put a young mare of Svelte's status, even if she was not a winner, to a horse of Cherry Tree's limitations—moreover he was 16 years old at the time of the union. The reason probably lies in the fact that Mr. Broderick Cloete who bred Cherimoya also bred and owned her sire and may well have had a feeling of loyalty to his old horse. Whatever the reason, the results were very satisfactory as three years previously the same mating had produced the filly Velvet who won seven races value £2,596—and even at that time Cherry Tree was getting on in life.

Cameronian was inbred to St. Simon at his fourth remove with subsidiary lines of that great horse farther back in his pedigree.

Incidentally, Cameronian's Derby was the first which vindicated the decision to scrap a rule rendering void nominations to big races on the death of an owner. As a result of a friendly court action brought at the suggestion of author and racing enthusiast Edgar Wallace, the Jockey Club changed the rule, just in time to save the Derby nomination of Cameronian, whose owner-breeder Lord Dewar died in April 1930 and left the colt to his nephew J. A. Dewar.

1932

1 T. Walls's br c April the Fifth (Craig an Eran – Sold Again).
2 H.H. Aga Khan's b c Dastur (Solario – Friar's Daughter).
3 Lord Rosebery's b c Miracle (Manna – Brodick Bay).
        21 ran. Time: 2 min 43.2 sec

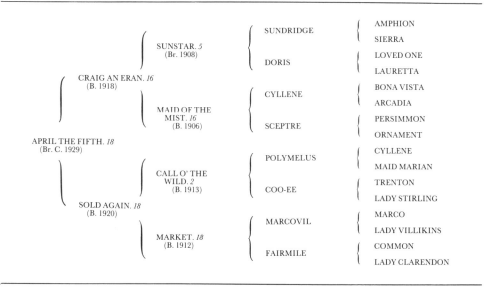

The antecedents of April the Fifth are the most curious of any Derby winner this century. Mr. Walls, whose colours he carried was a well-known manager and comedy actor on the London stage, who as a sideline owned and personally trained a few horses at Epsom. His metier was usually small handicaps and selling plates, where he met with a fair measure of success. April the Fifth was bred in partnership by Mr. Sidney McGregor and Mr. G. S. L. Whitelaw and sold as a yearling for 200 guineas—the purchaser being Mr. McGregor. As the latter received half of the sale price, the net cost to him for the future Derby winner was approximately £100. The yearling was then sent to Mr. Walls to train on the understanding that he should train and race him at his own expense but that any stake money won should be equally divided between trainer and owner. He was to race in Mr. Walls's name and colours. This arrangement continued throughout April the Fifth's racing career and the partnership was continued when the horse went to stud.

April the Fifth failed to win as a two-year-old, ran unplaced in the 2,000 Guineas and an unimportant race, and then won two races before his Derby triumph, after which his only other start was in the St. Leger, where he again failed. It is difficult to assess his merit as he undoubtedly suffered from the lack of galloping companions of class and all the facilities of a big stable. On the other hand he was a horse with bad forelegs and it was a great feather in Walls's cap that he managed to keep him sound enough to win at all. April the Fifth spent some 20 seasons at stud. He was a poor sire of Flat-racers although he got a number of winners under National Hunt Rules, the best being Lord Stalbridge's Red April (£12,900), before he died in 1954.

CRAIG AN ERAN (1918)

His sire Craig an Eran won three races value £15,345, including the 2,000 Guineas and Eclipse Stakes. He was also second in the Derby. He held court for 22 seasons with very uneven results. Except for April the Fifth his get in England were of modest class, but in France his son Mon Talisman won the French Derby etc., and made a good start at stud, siring amongst others the Grand Prix de Paris and French Derby winner Clairvoyant. Both Mon Talisman and Clairvoyant were seized by the Germans early in the war and never recovered. Craig an Eran also sired the Grand Prix de Paris winner Admiral Drake, who did very well at stud in France and had several important sons on both sides of the Channel, including Amour Drake (French 2,000 Guineas, second in the English Derby etc.), Mistral (French 2,000 Guineas etc.), Royal Drake (second in the English Derby and a good winner in France), and Phil Drake (winner of the 1955 Derby and Grand Prix de Paris).

Craig an Eran was by Sunstar out of Maid of the Mist, who after winning three races value £1,850 produced the Oaks winner Sunny Jane (dam of Bright Knight, £6,092, and second in the 2,000 Guineas). Maid of the Mist was also the dam of Hamoaze, who won £1,355 and proved a most influential broodmare. Her first winner was Buchan, who won ten races worth £16,658 and was placed in three Classics. He was a remarkably versatile horse who as a four-year-old achieved the unusual feat of passing the post first for the Ascot Gold Cup (2½ miles) and Eclipse Stakes (1¼ miles) within a month. He was disqualified from the Ascot race for bumping and boring, but only a very good horse could have accommodated himself to the great difference in distances of the two races in such a short time and in competition against high-class opponents. The double has been done before and since—Prince Palatine and Golden Myth are examples—but only by performers of unusual ability, and the feat is rarely attempted nowadays. Buchan was champion sire in 1927, and in 1936 and 1941 was head of the list of maternal grandsires. He spent 16 years at stud and sired the winners of 253 races value £170,230. His best probably was the St. Leger winner Book Law (dam of Rhodes Scholar), whilst daughters of his produced the Derby and St. Leger winner Airborne, and the St. Leger winner Sun Castle.

Hamoaze's second winner Tamar (£1,665 and second in the Derby) had a very brief stud career in England before being sent to Hungary, where he died young. Then came the Eclipse Stakes winner Saltash (£11,113), who went to Australia and did well as a sire.

Hamoaze's fourth and last winner was St. Germans (seven wins value £7,965 and second in the Derby), a most successful sire in America where his best-known get was Twenty Grand, winner of 14 races value 261,790 dollars, including the Kentucky Derby. St. Germans was also the tail male descendant of Bold Venture, the first horse both to win the Kentucky Derby and to sire two winners of the race, namely Assault and Middleground.

Other descendants in tail female of Maid in the Mist are Tiberius (Ascot Gold Cup and best known for his National Hunt produce); Crème Brulée (£9,535; a gelding); Betty (£7,042, dam of winners); and The Chiseller (£4,571, sent to Argentina). Maid in the Mist was by the Ascot Gold Cup winner Cyllene out of that very great racer Sceptre.

SOLD AGAIN (1920)

April the Fifth's dam Sold Again was useless as a racer and given away as a three-year-old to her trainer Sam Pickering. Doubtless sharing Mr. Jorrocks' views on the drawbacks to gifts which eat, Pickering sent her to the December Sales, where she made 20 guineas to the bid of a dealer. Sold Again was then fattened up and sent two years later to one of Messrs. Tattersalls' hunter sales then held in London. She made 230 guineas to Mr. S. McGregor. The mare appeared next season as a hurdler, but after one outing in an event worth £82 to the winner, where she finished among the also-rans, she broke down and was retired to stud.

Her first foal, by the sprinter Vencedor and named Birthday Greetings, went to Australia, where he won two races value £424. She also produced April the Fifth's sister Sybil, who won the Atalanta Stakes at Sandown value £1,620, and his brother Again, who failed to win but sired winners under National Hunt Rules. Sold Again's stud career was somewhat hampered by an unfortunate habit of being frequently barren.

CALL O' THE WILD (1913)

Sold Again's sire Call o' the Wild only ran three times, scoring once. He then broke a leg but was saved for stud. He was not a great success as a sire and after eight seasons at home went to Argentina. Incidentally, I sometimes think we are too ready to run for the gun when a horse sustains a break. There are many cases of the injury mending sufficiently satisfactorily for the animal to carry out stud work without difficulty or pain. Derby winner Mill Reef provides an admirable recent example.

Call o' the Wild was by Polymelus out of Coo-ee by Trenton. Coo-ee also produced Bill and Coo (£1,685, third in both the Ascot Gold Cup and Oaks, and dam of the hardy 'chaser sire Last of the Dandies). She was also the ancestress of Conversation Piece (Irish Oaks, etc.); of Cinq à Sept (Irish Oaks, Ebor and £7,583); of Ballyogan (five wins and sire of the unfruitful 2,000 Guineas winner Ki-Ming among others); and of Society's Way (£2,598).

Sold Again's dam Market won two races value £513 and her only winner was Off the Mark (£386). She was by Marcovil (sire of Hurry On) out of Fairmile, who produced five other winners of selling-plate class. Fairmile was by the Triple Crown winner Common, who was sold for £15,000 to Sir Blundell Maple, but turned out a disappointing stallion, his fee dropping to 19 guineas a mare before he died in his 23rd year.

April the Fifth was a member of the No. 18 Family, which has a very poor record for producing horses of class, but his sixth dam was the Oaks winner Summerside (1856), half-sister to the Derby winner Ellington. April the Fifth was inbred to Cyllene at the third and fourth remove.

CHAPTER 17

1933 – 1935
Influence of Hyperion

1933

1 Lord Derby's ch c Hyperion (Gainsborough – Selene).
2 Sir H. Cunliffe-Owen's b c King Salmon (Salmon-Trout – Malva).
3 Victor Emmanuel's br c Statesman (Blandford – Dail).
   24 ran. Time: 2 min 34 sec

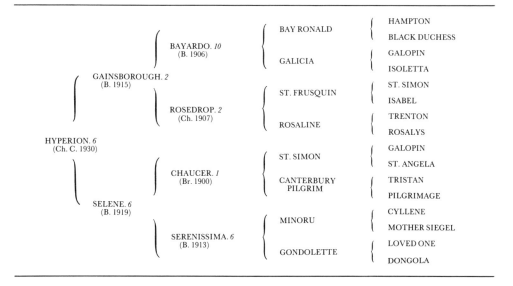

THE 1933 DERBY winner was a totally different proposition from the 1932 winner. Hyperion was the product of careful breeding for generations, the inmate of a large and powerful stable with all training facilities available regardless of expense, and was surrounded by all the benefits a rich owner could bestow on him. What is more, he became a great sire.

Hyperion measured when in training about 15.1½ with a girth of 68 inches and 7¾ inches of bone below the knee. In contrast to this, as I have already mentioned, the great and unbeaten Hurry On stood 17 hands, had a girth of 82½ inches and 9¾ inches of bone. The figures show the tremendous difference in physique amongst thoroughbreds of the highest class.

On the racecourse Hyperion can be regarded as a good two-year-old, receiving 9st in the Free Handicap; a very good three-year-old, when he was unbeaten in the Chester Vase, Derby, Prince of Wales Stakes and St. Leger; and a rather disappointing four-year-old, with two small wins and a modest third in the Ascot Gold Cup from four races, barring him from comparison with smashing performers like St. Simon, Ormonde, Bahram and even Ribot, who showed superiority over their rivals throughout their racing careers.

Hyperion's racing record reads: 13 races, nine wins and three places, for £29,509, an average of £3,278, which compares favourably with some of the great winners of the period such as Coronach (£4,822), Bahram (£4,787), Fairway (£3,560) and Windsor Lad (£3,625).

<h3 style="text-align:center">HYPERION AT STUD</h3>

Commencing stud work as a five-year-old, his success has been phenomenal. With three crops of racing age, 1940 found his name at the head of the list of sires of winners, and in 1954, at the age of 24, he occupied a similar position. By the time of his death, aged 30 in 1960, he had been leading sire six times—1940, 1941, 1942, 1945, 1946, 1954—a record surpassed in the last 120 years only by Hermit (seven), Stockwell (seven) and St. Simon (nine). Hyperion was also second in 1939, 1944, 1950 and 1952, third in 1943, and led the list of sires of winner-producing mares in 1948 and 1957, and was second five times.

Hyperion's English Classic winners were:

2,000 Guineas	1,000 Guineas	Derby	Oaks	St. Leger
Nil	Godiva	Owen Tudor	Godiva	Sun Castle
	Sun Chariot		Sun Chariot	Sun Chariot
	Sun Stream		Hycilla	
	Hypericum		Sun Stream	

It will be seen that his fillies were far more successful—nine to two in the Classics—than his colts, but his biggest stakes-earner was his son Aureole, who after securing £36,225 and finishing second in the Derby went to stud in 1955.

At his death Hyperion had sired the winners of 748 races worth £557,009; he was in the top ten sires 16 times, and outside his Classic winners the most important among his offspring were:

Heliopolis (1936)—brother to Sun Stream; Prince of Wales Stakes; twice leading sire in America.

Rockefella (1941)—successful sire.

High Stakes (1942)—a gelding, won 34 races worth £21,702.

Gulf Stream (1943)—Eclipse Stakes, second in Derby; successful sire in Argentina.

Hyperbole (1945)—Royal Hunt Cup; leading sire in Sweden.

Saturn (1947)—Hardwicke Stakes.

Aureole (1950)—King George VI and Queen Elizabeth Stakes, second in Derby; sire; see below.

Hornbeam (1953)—Great Voltigeur Stakes, second in St. Leger; sire of Intermezzo (St. Leger), Merchant Venturer (second in Derby), Windmill Girl (Oaks, dam of Derby winners Blakeney and Morston); and leading sire in Sweden.

High Hat (1957)—Winston Churchill Stakes; sire of Glad Rags (1,000 Guineas) and White Gloves (Irish St. Leger).

Hyperion's sole Derby winner Owen Tudor sired two very speedy horses in Tudor Minstrel and Abernant, and the French Derby and King George VI and Queen Elizabeth Stakes winner Right Royal V, as well as the Ascot Gold Cup winner Elpenor. Owen Tudor was leading sire in France in 1961, while Tudor Minstrel (sire of Tudor Melody), Abernant and Right Royal V (leading sire in Britain in 1969) all made their mark at stud in America, Britain and France respectively.

It can be seen from the above details that Hyperion's influence stretches the world over, and until Aureole emerged, it was feared he would have a better record outside Europe than closer home. Some of his most important sons abroad have been cited above—Heliopolis, Gulf Stream, Hyperbole and Hornbeam—and others are:

Helios (1937)—leading sire in Australia.

Stardust (1937)—second in 2,000 Guineas and St. Leger; sire of Star Kingdom, seven times leading sire in Australia.

Alibhai (1938)—unraced; 15 times in top 20 American leading sires; sire of Kentucky Derby winner Determine, Santa Anita Derby winner Your Host (sire of five-times Horse of the Year Kelso).

Selim Hassan (1938)—leading sire in Argentina.

Deimos (1940)—twice leading sire in South Africa.

Coastal Traffic (1941)—sire of French St. Leger winner Buisson d'Or.

Pensive (1941)—Kentucky Derby; sire of similar winner Ponder, himself sire of similar winner Needles.

Red Mars (1941)—leading sire in Australasia 1956–7.

Ruthless (1941)—twice leading sire in New Zealand.

Aldis Lamp (1943)—leading sire in Belgium.

Khaled (1943)—sire of the great Swaps (19 races, 848,900 dollars).

Aristophanes (1948)—sire of Forli, winner of Argentinian Triple Crown and sire of top-class winners in Britain.

The list of mares by Hyperion who did well at stud is impressive, both for its quality and range of influence. The most important are:

Dam	Notable produce
Aurora (1936)	Alycidon (Ascot Gold Cup), Acropolis, Borealis, Agricola (champion sire in New Zealand); and grandam of Larkspur (Derby).
Helia (1937)	Supertello (Ascot Gold Cup).
Hydroplane (1938)	Citation (32 races, 1,085,760 dollars, including American Triple Crown, and a leading sire).
Chantress (1939)	Saggy (sire of Kentucky Derby winner Carry Back).
Calash (1944)	Carrozza (Oaks), Snow Cat (top sire in Argentina).
Lady Angela (1944)	Nearctic (21 races, sire of Kentucky Derby winner and leading sire Northern Dancer, and 2,000 Guineas winner Nonoalco).
Fair Edith (1947)	Aunt Edith (King George VI and Queen Elizabeth Stakes).
Jennifer (1948)	Nagami (Coronation Cup and Gran Premio del Jockey Club, Italy); and grandam of leading two-year-old Young Emperor.
Neutron (1948)	Court Harwell (second in St. Leger, leading sire in Britain in 1965, and in Argentina in 1967).
Barley Corn (1950)	Shantung (third in Derby, sire of Classic-winning fillies Ginevra, Full Dress II, Lacquer, Macrina d'Alba, Italian Oaks, and Julie Andrews, South African Oaks); Roi Dagobert (smart middle-distance horse in France).
Lightning (1950)	Parthia (Derby, sire).
Imitation (1951)	Pretense (top-class sire in America).
Libra (1956)	Ribocco and Ribero (brothers who in successive years each won the Irish Sweeps Derby and English St. Leger).
Lakewoods (1958)	Waterloo (1,000 Guineas).

Naturally, Hyperion's direct influence is now receding, as his sons and daughters inevitably decrease in numbers, but his greatness can still be seen in several of the world's bloodstock statistics.

In the British Register of Thoroughbred Stallions 1976, the name of Hyperion can be found at least once in the first five generations of 29 sires from a list of 182, and 15 are in direct male line—Abwah, Gay Fandango, Hill Clown, Home Guard, Jupiter Pluvius, Laser Light, Malicious, Our Mirage, Perdu, Prince Regent, Saintly Song, St. Paddy, Tudor Melody, Welsh Saint and Will Somers.

On the list of the world's all-time leading stakes winners (in American dollars) to the end of 1975, the top 15 included four in which the name of Hyperion is prominent:

Dahlia (1970), 1,356,693—by Vaguely Noble, by Vienna, by Aureole, by Hyperion.

Carry Back (1958), 1,241,165—by Saggy, out of Chantress, by Hyperion.

Forego (1970), 1,163,520—by Forli, by Aristophanes, by Hyperion.

Citation (1945), 1,085,760—by Bull Lea out of Hydroplane, by Hyperion.

And among the world's leading sires, country by country, in 1975 there were four who could attribute a measure, however small, of their pre-eminence to Hyperion:

Australia—Oncidium (1961), by Alcide, by Alycidon, out of Aurora, by Hyperion.

Chile—April Fool (1963), by Court Harwell, out of Neutron, by Hyperion.

Denmark—Andros (1952), by Borealis, out of Aurora, by Hyperion.

Sweden—Hornbeam (1953), by Hyperion.

AUREOLE (1950)

Aureole, bred at the Sandringham Stud by King George VI, was out of the 1,000 Guineas winner Hypericum's half-sister Angelola, winner of the Yorkshire Oaks, Princess Royal Stakes and Newmarket Oaks, and second in the Epsom Oaks. An excitable colt who more than once lost ground at the start, Aureole was not the outstanding horse of his year, being beaten by Pinza in the Derby and King George VI and Queen Elizabeth Stakes. Aureole also failed to stay in the St. Leger, by which time Pinza had been retired. But as a four-year-old Aureole won the Coronation Cup and King George VI and Queen Elizabeth Stakes, and at stud proved a far better stallion than Pinza, or any other colt in the 1950 crop, to such extent he has been among the most successful British sires of the last 30 years.

He got an outstanding colt in each of his first two crops—Saint Crespin III and St. Paddy—and was leading sire in 1960, the first year in which he had three age-groups to represent him, and 1961. Despite his own failure to see out the distance of the St. Leger, he got several top-class horses that stayed extreme distances, including three winners of the longest Classic. He died in the autumn of 1974, the year Vaguely Noble (by Vienna, by Aureole) completed his own double in the list of leading sires, chiefly through the exploits of the filly Dahlia.

Aureole's most important sons were:

Saint Crespin III (1956)—Arc de Triomphe, Eclipse Stakes, sire.
St. Paddy (1957)—Derby, St. Leger, Hardwicke Stakes, Eclipse, sire of Connaught (second in Derby and Eclipse Stakes), Parnell (Irish St. Leger), Patch (second in French Derby), Baccio Bandinelli (Irish St. Leger).
Vienna (1957)—third in St. Leger, sire of Vaguely Noble (Arc de Triomphe, leading sire in 1973 and 1974).
Aurelius (1958)—St. Leger, then gelded and won over hurdles and fences.
Miralgo (1959)—Timeform Gold Cup, Hardwicke Stakes, sent to Japan.
Harvest Gold (1959)—eight wins, £13,126.
Apprentice (1960)—Goodwood and Yorkshire Cups.
Sunseeker (1961)—two wins, £12,478, sent to Argentina.
Provoke (1962)—St. Leger, sent to Russia.
Hermes (1963)—five wins, £13,348, sent to New Zealand.
Hopeful Venture (1964)—Grand Prix de Saint-Cloud, Hardwicke Stakes, sire, sent to Australia.
Laser Light (1966)—smart sprinter.
Saintly Song (1967)—St. James's Palace Stakes.
Orosio (1967)—Cesarewitch.
Buoy (1970)—Yorkshire Cup, second in St. Leger, third in Irish Sweeps Derby.

It is a source of disappointment that Aureole's fillies have not matched the achievements of his colts. His only important winning fillies were Aurabella (1962, Irish Guinness Oaks) and Paysanne (1969, Prix Vermeille), while he shows no signs of breaking through as a sire of broodmares, with only Mary Murphy (1960, dam of Horris Hill Stakes winner Double First), Love-in-the-Mist

(1963, dam of Coventry Stakes winner Perdu) and Aurabella (dam of Queen's Vase winner Royal Aura) rising above the level of ordinary, and even then hardly into the highest bracket.

SELENE'S (1919) SONS

I have already referred in detail to Hyperion's progenitors. Particulars of Gainsborough will be found on page 188 onwards and of Selene on page 202. I will, however, again remind readers that Selene was one of those rare broodmares whose sons have all excellent records as stallions. They were Hyperion (U.K.); Sickle (U.S.A.); Pharamond (U.S.A.); Hunter's Moon (Argentina) and even the non-winning Salamis did well in Ireland during the short time he was at stud. Selene's son Moonlight Run, who won only £231, met with considerable success as a sire in America. Other broodmares with the same facility were Plucky Liége the dam of Bois Roussel (U.K.), Admiral Drake (France), Sir Gallahad III (France and U.S.A.), Bull Dog (U.S.A.), Quatre Bras II (U.S.A.) and Bel Aethel (U.S.A.); Perdita II the dam of Persimmon (U.K.), Diamond Jubilee (U.K. and Argentina) and Florizel II (U.K.); Double Life, the dam of Precipitation (U.K.), Persian Gulf (U.K.) and Casanova (U.K.); and others.

Like many other good horses Hyperion is inbred at the third and fourth remove—and in his case the common ancestor is St. Simon. His dam Selene was also inbred at the same incidence to Pilgrimage (the dam of Canterbury Pilgrim and Loved One).

It is interesting to note that Gainsborough was 14 years old when he sired Hyperion, and Hyperion was 19 when he got Aureole.

1934

1 H.H. Maharaja of Rajpipla's b c Windsor Lad (Blandford – Resplendent).
2 Lord Woolavington's br c Easton (Dark Legend – Phaona).
3 Lord Glanely's b c Colombo (Manna – Lady Nairne).
      19 ran. Time: 2 min 34 sec

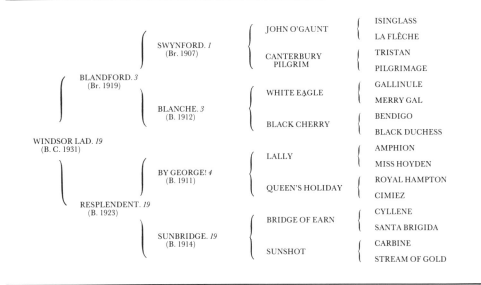

WINDSOR LAD. *19*
(B. C. 1931)

BLANDFORD. *3*
(Br. 1919)

SWYNFORD. *1*
(Br. 1907)

JOHN O'GAUNT
  ISINGLASS
  LA FLÈCHE

CANTERBURY PILGRIM
  TRISTAN
  PILGRIMAGE

BLANCHE. *3*
(B. 1912)

WHITE EAGLE
  GALLINULE
  MERRY GAL

BLACK CHERRY
  BENDIGO
  BLACK DUCHESS

RESPLENDENT. *19*
(B. 1923)

BY GEORGE! *4*
(B. 1911)

LALLY
  AMPHION
  MISS HOYDEN

QUEEN'S HOLIDAY
  ROYAL HAMPTON
  CIMIEZ

SUNBRIDGE. *19*
(B. 1914)

BRIDGE OF EARN
  CYLLENE
  SANTA BRIGIDA

SUNSHOT
  CARBINE
  STREAM OF GOLD

Of his 13 races Windsor Lad won ten, including the Derby and St. Leger, worth a total of £36,257. He was only once beaten as a three-year-old and not at all the next year, when his major objective was the Eclipse, and having secured that, he retired to take up stud duties the following year. An unusual feature of his racing career was that after winning the Derby and before winning the St. Leger he changed hands for £50,000. This rarely happens to a Classic winner.

His stud career was ruined as he developed a brain tumour. The best brains of the veterinary world—augmented by a Harley Street doctor—treated him without success. At one period his owner wished to destroy him on humane grounds but the underwriters would not agree to this. A compromise to this disagreement was effected by the underwriters paying out a certain sum and taking the horse. Thus Windsor Lad finished his stud career as the property of "A Syndicate of Lloyds Underwriters". His first three seasons at stud were normal and he had a good fertility return. He was then withdrawn from active work for a season but on resumption proved a poor foal getter, and was destroyed in 1943. His only get of note were unbeaten Windsor Slipper, who won the Irish Triple Crown, and Phase, one of the king-pins of the stud owned by the late Major L. B. Holliday.

Phase was out of the Yorkshire sportsman's famous mare Lost Soul, and was the dam of Netherton Maid (second in Oaks, Yorkshire Oaks and Park Hill Stakes; dam of Queen Mary Stakes winner Bride Elect, Jubilee Handicap winner and successful sire in New Zealand Chatsworth, St James's Palace Stakes winner and successful sire in New Zealand Pirate King, and sire Pampered King; grandam of St. Leger winner Hethersett); Neasham Belle (won Oaks and £15,630); Narrator (Coronation Cup, Champion Stakes; sire); Kingscavil (grandam of Hevea, winner and dam of Prix Vermeille winner Saraca and useful French two-year-old Simbir); No Pretender (dam of Ebor Handicap winner Proper Pride); Royal Applause (dam of Solario Stakes winner Happy Omen); and None Nicer (Ribblesdale Stakes and Yorkshire Oaks, second in St. Leger; dam of Park Hill Stakes winner Cursorial; grandam of useful two-year-old winner Fine Blade).

WINDSOR SLIPPER (1939)

Exactly how good Windsor Slipper was must remain unproved as, owing to the war, he never ran against top-class horses in England, but the manner in which he won his races in Ireland created the impression that he was a very good horse indeed. He was a half-brother to the English 1,000 Guineas and Oaks winner Godiva who died without produce. At stud Windsor Slipper sired the champion sprinter The Cobbler (£17,011) and the useful Solar Slipper (£5,822, third in St. Leger, and after six years at stud in Ireland was sold to America).

Windsor Lad was by Blandford (see page 212) out of Resplendent who won 2½ races value £3,426, including the Irish 1,000 Guineas and Oaks. She was also second in the English Oaks. At stud in addition to Windsor Lad she produced Lady Gabrial (£1,650, Cheveley Park Stakes and dam of a small winner); Radiant (£1,463, second in the Oaks but a disappointing broodmare); Windsor Lady (a non-winner but dam of three winners) and Isabel De Saye (non-winner but dam of the Irish Cesarewitch winner Winawar).

BY GEORGE! (1911)

The name of Resplendent's sire By George! only occurs in the pedigrees of good-class stock through this mare—I feel largely because he had so very few chances at stud. He was one of the best two-year-olds in 1913, winning the Imperial Produce Stakes and Champagne Stakes (Bibury). He then proved difficult to keep sound and could not reproduce anything like this form in his few starts as a three- and four-year-old. He went to stud in Ireland in 1916 at the rock-bottom fee of nine guineas a mare. Needless to say, with the war situation looking very black, he did not get much patronage but he managed to sire some winners. In 1925—the year before Resplendent showed her merit—he was sold to Canada and so disappeared from the scene.

It is true that his sire Lally (Eclipse and over £20,000 in stakes) was a poor stallion who was exported to Italy after showing his limitations at home, but his dam Queen's Holiday was an excellent mare, winning the Wokingham Stakes at Ascot and other good races before retiring to the paddocks where she foaled five winners. She was by the City and Suburban winner and adequate sire Royal Hampton out of Cimiez, who bred a number of winners and was by St. Simon out of Antibes, sister to the Oaks and St. Leger winner Seabreeze and half-sister to Tredennis. The family history will be found on page 206.

BRIDGE OF EARN (1906)

Windsor Lad's grandam Sunbridge won two small races and produced five winners. In addition to Resplendent these included the Irish 2,000 Guineas winner Soldumeno; the Irish Oaks and 1,000 Guineas winner Sol Speranza, who foaled the Irish St. Leger winner Solferino, sire of Solonaway (Irish 2,000 Guineas, etc.); Ferrybridge (third in the English 1,000 Guineas); and Queen Scotia (ancestress of Eclipse Stakes winner King of the Tudors, and Our Babu, the best two-year-old of 1954 who won the 2,000 Guineas and went to stud in America).

Sunbridge was by Bridge of Earn (two wins value £1,162 including the Newbury Cup), who was a better sire than racehorse. He had 20 seasons at stud during which he got the winners of some 270 races value about £80,000. His name was carried forward through his daughter Duccia di Buoninsegna (grandam of Donatello II), whilst another daughter Bridgemount, after winning £2,520, produced Cresta Run (£14,540, including the 1,000 Guineas) and Foxbridge (£2,250 and champion sire in New Zealand for 11 successive seasons).

SANTA BRIGIDA (1898)

Bridge of Earn was by Cyllene out of Santa Brigida by St. Simon. Santa Brigida won £4,724 and was also the dam of Bridge of Canny (£14,499, sire of the Cambridgeshire winner Cantilever, and later a great success in Argentina); Bridge of Sighs (dam of Light Brigade); Spean Bridge (dam of Knockando, second in 2,000 Guineas and did well as a sire in Venezuela, and Legionnaire, won £3,725 and a good sire in Australia); Bridge of Allan (dam of Mid-day Sun); and Brig of Ayr (grandam of 1,000 Guineas winner Brown Betty). Santa Brigida is also the ancestress of Tide-way (1,000 Guineas); Heliopolis (£14,792, twice leading sire in America); Shannon II (won £18,987 in the Antipodes and 211,611 dollars in America), Sun Stream (1,000 Guineas and Oaks); and Gulf Stream (Eclipse Stakes and £10,537, and successful sire in Argentina).

Light Brigade (1910) won £11,444 and went to stud at Newmarket in the early days of the First World War at a fee of only 18 guineas. After two seasons, when his achievements were confined to two unimportant winners, he was sold to America, where he proved a most successful sire before his death in 1933.

Windsor Lad's great grandam Sunshot bred nine winners who won 28 races between them but all of moderate value. She was by the great New Zealander Carbine.

1935

1 H.H. Aga Khan's b c Bahram (Blandford – Friar's Daughter).
2 Sir Abe Bailey's bl c Robin Goodfellow (Son and Heir – Eppie Adair).
3 Lord Astor's b or br c Field Trial (Felstead – Popingaol).
   16 ran. Time: 2 min 36 sec

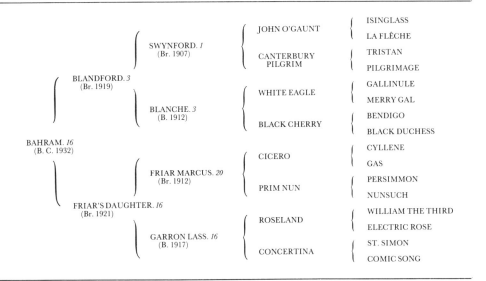

		JOHN O'GAUNT	ISINGLASS
	SWYNFORD. *1*		LA FLÊCHE
	(Br. 1907)	CANTERBURY PILGRIM	TRISTAN
BLANDFORD. *3*			PILGRIMAGE
(Br. 1919)		WHITE EAGLE	GALLINULE
	BLANCHE. *3*		MERRY GAL
	(B. 1912)	BLACK CHERRY	BENDIGO
BAHRAM. *16*			BLACK DUCHESS
(B. C. 1932)		CICERO	CYLLENE
	FRIAR MARCUS. *20*		GAS
	(Br. 1912)	PRIM NUN	PERSIMMON
FRIAR'S DAUGHTER. *16*			NUNSUCH
(Br. 1921)		ROSELAND	WILLIAM THE THIRD
	GARRON LASS. *16*		ELECTRIC ROSE
	(B. 1917)	CONCERTINA	ST. SIMON
			COMIC SONG

A very good-looking horse was Bahram—possibly a shade on the leg, like so many good racers—and his performances were impeccable. He was far and away the best of his age at two and three years old, winning in all nine races value £43,086. His success included the 2,000 Guineas, the Derby and the St. Leger, and he never knew defeat. He left the racecourse an absolutely sound horse and started at a stud as a four-year-old at a fee of 500 guineas a mare, and as the fourth son of Blandford to win the Derby in seven years.

BAHRAM (1932) AT STUD

I remember seeing him at the Egerton Stud, Newmarket, when he looked the "lord of all he surveyed"—and with the most charming and equable temperament. His time at Newmarket was not long as, after five seasons there he was sold to America for £40,000. For some reason breeders in the States did not take kindly to him and as a 14-year-old he went to Argentina where he died early in 1956. His stud career is most interesting—and alas! in some respects disappointing. In England in 1940 with only two crops of foals of racing age he was second on the list of sires of winners. Although never securing the championship, he occupied an honourable position for the next few years until the effects of his exportation were manifest and he dropped out of the list through lack of runners.

During his period in England he sired the 2,000 Guineas winner Big Game, champion sire in 1948 and an important sire of fillies; Persian Gulf, who was near the top of the sires' tree on several occasions; and the St. Leger winner but rather disappointing sire of winners Turkhan, who was sold to France in 1952. Turkhan mares, however, produced plenty of small winners in Britain, and one important one in Vienna, sire of Vaguely Noble.

BAHRAM MARES

Mares by Bahram produced Elpenor (Ascot Gold Cup); Migoli (£22,950 in England and 5,209,500 francs in France): Noor (£4,704 in England and 365,940 dollars in America), and others of high class. Mah Iran, the dam of Migoli, was also grandam of the brilliant filly Petite Etoile, winner of the 1,000 Guineas, Oaks and Champion Stakes. Yet during his stay in America, which was of equal length to his stud career in England, Bahram got nothing of particular note, and on transfer to Argentina fared no better.

I think we may sum up his stud career by saying that he was a great success at home but a failure overseas. The reverse is generally the case. Many a horse who has made no headway at stud at home has been a success on export. I attribute this to the fact that normally an English or Irish horse of suitable credentials on arrival in a new country gets plenty of their good mares, whereas at home he may have suffered from the lack of mates of class. Taking the broad view, I cannot help feeling that Bahram would have contributed a great deal more to bloodstock the world over if he had never left his native shores.

FRIAR'S DAUGHTER (1921)

I have already reviewed Blandford's history (see page 212). Bahram's dam Friar's Daughter was one of the least expensive of the Aga Khan's many purchases yet she turned out one of the most valuable. She was bred by Colonel F. Lort Phillips who sent her as a foal to the 1921 Newmarket December Sales. She then made 120 guineas to Mrs. E. M. Plummer who re-sold her as a yearling to the Aga Khan for 250 guineas. For him she won a two-year-old race value £168 and was sent to stud the following year. Friar's Daughter was strongly inbred, as her sire's dam was by a son of St. Simon, her own dam was by a grandson of St. Simon whilst her grandam was by St. Simon. In terms of the Vuillier Dosage System (see page 34 onwards) she carried 1,024 points of St. Simon which is nearly 150 per cent in excess to the standard dosage of 420.

DASTUR (1929)

At stud the Aga Khan's 250-guinea purchase produced eight winners of 26½ races value slightly over £57,600. The next in importance to Bahram was Dastur (by Solario), who won 6½ races worth £11,626, and was second in the 2,000 Guineas, Derby and St. Leger. He spent 19 years at stud with rather disappointing results, so far as siring winners is concerned, bearing in mind the excellent opportunities he had. In all his stock won some 250 races value about £85,000, but he made a better sire of broodmares, for Diableretta (£14,492), Darius (2,000 Guineas and £38,104) and The Cobbler (£17,011) were out of Dastur mares. His best racer was probably the wartime Gold Cup winner and Derby second Umiddad.

Friar's Daughter also produced Fille de Salut who failed to win, but is the grandam of Sunny Boy III (France) who made an astonishing rise to fame as a stallion during the 1954 racing season, when he sired the English Oaks winner Sun Cap and the French St. Leger and Prix de l'Arc de Triomphe winner Sica Boy to head the list of sires of winners in France with only his second crop.

Another of Friar's Daughter's non-winners, Fille d'Amour, became the grandam of The Phoenix, who won the Irish 2,000 Guineas and Derby and was only beaten once during two seasons in training. Sadri, a brother to Dastur, was a successful racehorse in South Africa.

FRIAR MARCUS (1912)

Friar's Daughter's sire Friar Marcus was bred and owned throughout his life by King George V. After winning nine sprint races value £9,435 during the First World War, he retired to stud and sired the winners of approximately 350 races value £140,000 during the 18 years he held court. The best of these was the 1,000 Guineas winner Brown Betty (six wins value £11,637), who went to America. Friar Marcus's son Beresford won £5,773 and sired a number of useful winners in Ireland, including the fast Portlaw (£10,855), who did quite well at stud and got Portobello (14 wins value £7,216). Other good sons of Beresford were Disarmament (£7,688 and later a leading sire in Chile) and Cat o' Nine Tails (£6,331 and sent to India).

Friar Marcus also got the very fast Lemnarchus (1928), who won nine races value £11,249 but only had a very short stud career on account of the war.

PERSIMMON (1893)

Friar Marcus enjoyed a high reputation as a sire of broodmares whilst his son (Beresford) and grandson (Portlaw) also show merit in this capacity. He was by Cicero (see page 168) out of Prim Nun who failed to score and produced no other winners. She was by Persimmon, who won seven races value £34,706 including the Derby, St. Leger, Eclipse and Ascot Gold Cup. In common with Prince Palatine, Golden Myth, Buchan (disqualified for Gold Cup) and others he achieved the double of the Gold Cup (2½ miles) and the Eclipse (1¼ miles) within a few weeks. It is a double which demands an exceptionally good and adaptable horse. Besides being an absolutely first-class racer Persimmon was an outstanding sire. He was head of the list of sires of winners four times (1902, 1906, 1908 and 1912) and head of the list of maternal grandsires thrice (1914, 1915 and 1919), yet he had a comparatively short stud career, dying as the result of a fractured pelvis when he was just 15 years old and at the height of his powers. His Classic winners were:

2,000 Guineas	Sceptre
1,000 Guineas	Sceptre
Oaks	Sceptre, Keystone II, Perola
St. Leger	Sceptre, Your Majesty, Prince Palatine

He got also the Ascot Gold Cup winners Prince Palatine (twice) and Zinfandel.

PRINCE PALATINE (1908)

His sire St. Simon, with the advantage of over double the period at stud, got the winners of 16 Classic races and three Gold Cups. St. Simon was leading sire of winners nine times and of winning broodmares on six occasions. Persimmon's best racing son Prince Palatine was a disappointing stallion who spent his stud career between England, France and U.S.A. before being unfortunately burnt to death in his stable in America at the age of 16. He however got in England Rose Prince who, as we shall see later (see page 325), carried on the sire line through Prince Rose and Prince Chevalier to the 1951 Derby winner Arctic Prince. In U.S.A. Prince Palatine's son Prince Pal (who was conceived in England and foaled in U.S.A.) got the good racer but indifferent sire Mate who won 20 races value 301,810 dollars.

Persimmon was a full brother to the Triple Crown winner Diamond Jubilee.

Bahram's grandam Garron Lass never won and died as a five-year-old, leaving behind no other winners except Friar's Daughter. She was by Roseland out of Concertina who produced nine winners including that great broodmare Plucky Liège, who won £1,811 and foaled 12 winners including the 1938 Derby winner Bois Roussel. Roseland only started four times, winning three races value £2,145. He spent many years at stud in England but his stock in all won barely £13,000. His fourth dam Electric Light was also fourth dam of Solario, and his dam Electric Rose also produced Rosa Croft, the dam of unbeaten Tolgus who sired the Oaks winner Lovely Rosa, in turn grandam of Wilwyn.

CHAPTER 18

1936 – 1937
The Tetrarch and Solario

1936

1   H.H. Aga Khan's gr or ro c Mahmoud (Blenheim – Mah Mahal).
2   H.H. Aga Khan's b c Taj Akbar (Fairway – Taj Shirin).
3   J. Shand's br c Thankerton (Manna – Verdict).
     22 ran. Time: 2 min 33.8 sec

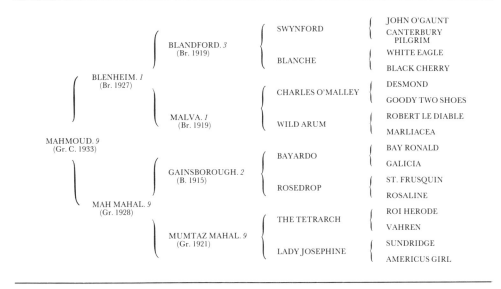

ONE OF BLENHEIM's second crop, Mahmoud was the first produce of his dam. He was bred by the Aga Khan and sent to the Deauville yearling sales but failed to make his reserve and went into training at Newmarket, where he was the second-best two-year-old of his year. His three wins as a juvenile earned £5,092 and placed him second to Bala Hissar in the Free Handicap. The Derby (£9,934)

was his only other victory, though he was beaten a short head in the 2,000 Guineas and finished third in the St. Leger.

Because of doubts about his staying the Derby distance, Mahmoud was less well fancied than the Aga Khan's Taj Akbar, but on very firm ground he won the race from his stablemate in record time. Thus the Aga Khan became the first since Colonel Peel in 1844 to own first and second in the Derby, and Mahmoud established a time record of 2 min 33.8 sec which still stands. How much Mahmoud owes his position to the vagaries of hand timing it is impossible to say, but since electrical timing was introduced to Epsom in 1964, the best time for the Derby is 2 min 34.68 sec clocked by Nijinsky in 1970, and the next best 2 min 35.04 sec by Snow Knight in 1974.

It is illuminating to notice how that very astute judge the Aga Khan valued his three Derby winners. Blenheim's stud fee was 400 guineas, Bahram's 500 guineas and Mahmoud's 300 guineas; they were later sold for £45,000, £40,000 and £20,000 respectively. The sale price of the latter two was doubtlessly enormously influenced by the fact that they were sold in the early days of the war, when the general outlook was at its blackest. I think this shows their owner considered Bahram the best of the three, yet the other two, possibly through better opportunities, have had a far greater beneficial effect on American and world bloodstock.

During his four seasons at stud at Newmarket Mahmoud's results were satisfactory but not startling. Mated to his owner's Qurrat-al-ain (£11,092), he sired the Irish 1,000 Guineas and Oaks winner Majideh, later the dam of Masaka (£21,997, including the English and Irish Oaks, Irish 1,000 Guineas) and other good winners. On transfer to America Mahmoud soon made a name for himself, reaching the top of their list of sires once, being second once and third once. His best American winners were Oil Capital (19 wins and 580,756 dollars), The Axe (two wins in England, and 13 in America for 393,391 dollars), Vulcan's Forge (nine wins and 324,240 dollars), First Flight (11 wins and 197,965 dollars), Mount Marcy (17 wins and 152,110 dollars) and several more who won over 100,000 dollars in stakes, including Cohoes (13 wins and 210,850 dollars) who sired the Belmont Stakes winner Quadrangle.

Before his death in his 30th year Mahmoud was also responsible for several fillies who have proved notable broodmares. They include: Boudoir (1938, second in the Irish 1,000 Guineas, and dam of Santa Anita Derby winner Your Host, sire of 39-race winner and five times American Horse of the Year, Kelso); Moonstone (1940, grandam of Irish 1,000 Guineas winner Even Star); Pontoon (1940, half-sister to Derby winner Pont l'Eveque; third dam of the Grand Prix de Saint-Cloud winner Sea Hawk, sire of the St. Leger winner Bruni, Italian 1,000 Guineas winner La Zanzara and Grand Prix de Paris winner Matahawk); Ghazni (1942, dam of Dunce, nine wins and 351,545 dollars including the American Derby); Mahmoudess (1942, dam of Promised Land, 21 wins and 541,707 dollars); Grey Flight (1945, grandam of Bold Lad (U.S.A.), see page 250); Koubis (1946, dam of Kentucky Derby winner Determine, winner of 18 races and 573,360 dollars and sire of Kentucky Derby winner Decidedly); Almahmoud (1947, grandam of Tosmah, who won 23 races and 612,588 dollars, and Northern Dancer, who will be discussed in relation to the Derby winner Nijinsky); and

Polamia (1955, dam of leading French two-year-old of 1964 Grey Dawn, and French 1,000 Guineas winner Right Away).

I referred earlier to Mahmoud's sire Blenheim. Mahmoud's dam Mah Mahal was not much of a racer, running mostly in very moderate five-furlong handicaps. She won once and dead-heated for first place once, earning £385 in stake money. After foaling Mahmoud, she produced Pherozshah (£1,360, did quite well as a sire in England and New Zealand); Khan Bahadur (£1,907, sire of a number of winners of limited class); Mah Iran (five wartime wins worth £1,944; dam of Arc de Triomphe winner Migoli, the sire of Belmont Stakes winner Gallant Man; grandam of Petite Etoile); and Mahee (£280). Gallant Man, who won 14 races for 510,355 dollars and has been a successful sire in America, had an interesting concentration of the Aga Khan's breeding. His sire Migoli was by Bois Roussel out of Mah Iran, by Bahram, whilst his dam Majideh was by Mah Iran's half-brother Mahmoud, by Blenheim.

MUMTAZ MAHAL (1921)

Mah Mahal was by Gainsborough out of the flying Mumtaz Mahal, a great filly who won seven races value £13,933 and although she could not truly stay more than six furlongs was second in the 1,000 Guineas. She was the best two-year-old of her year, being beaten only once, when giving the winner 7lb. I have sometimes heard it said she was the fastest filly ever to see a racecourse. It is, of course, impossible to substantiate such a claim but there is no doubt she was among the fastest.

In addition to Mah Mahal she foaled Badruddin (£4,561; third in the 2,000 Guineas; at stud in France for a short time and later in South America); Mirza II (£7,345, first at stud in France, taken by the Germans, recovered after the war and later at stud in England); Furrokh Siyar (won two small races in France and sired a number of winners in Ireland); Nizami (a non-winner but a good sire in New Zealand); Rustom Mahal (non-winner; dam of Abernant, who won £26,394 and did well at stud); and Mumtaz Begum, whose widescale influence, especially as the dam of Nasrullah, will be discussed when we come to the 1954 Derby winner Never Say Die.

THE TETRARCH (1911)

Mumtaz Mahal was by that wonderful horse The Tetrarch, who was bred in Ireland by Mr. E. Kennedy and sold as a yearling for 1,300 guineas—a big price for those days considering his background—to Major McCalmont. He won seven races value £11,336, never knew defeat but ran only as a two-year-old. He was winter favourite for the Derby and no attempt was made to run him for lesser spoils. In the middle of May, when taking part in an ordinary exercise gallop, he hit the joint of his off foreleg—an accident with minor effect which had happened before—and he could not be further trained.

He went to stud as a four-year-old at a fee of 300 guineas. His success was like his racing career, phenomenal but brief. His first two-year-olds ran in 1918 and their ability placed The Tetrarch ninth on the list of sires of winners, whilst the

Photo W. W. Rouch & Co Ltd

ST SIMON (1881), unbeaten on the racecourse and a sire of unsurpassed merit; and, below, CYLLENE (1895), so small as a yearling that he was never entered for the classics, but winner of the Ascot Gold Cup and sire of four Derby winners.

Photo W. W. Rouch & Co Ltd

SWYNFORD (1907), a champion sire who was lucky to go to stud since he broke a fetlock bone as a four-year-old; and, below, SON-IN-LAW (1911), not up to classic standard as a racehorse but an important influence for stamina at stud.

PHALARIS (1913), astonishingly successful during his 12 seasons at stud; and, below, CHAUCER (1900), many of whose daughters did particularly well when mated with Phalaris, so that the merit of one cannot be discussed without general reference to the other.

PHAROS (1920), whose maximum distance as a racehorse was one and a quarter miles; and, below, his brother FAIRWAY (1925), who stayed twice that distance. Though both bay, they bore little physical resemblance but as stallions sired classic winners of the highest standard.

Photo W. W. Rouch & Co Ltd

HURRY ON (1913), unbeaten on the racecourse and a champion sire who stood 17 hands; and, below, HYPERION (1930), a champion racehorse who was six times leading sire but stood only 15.1½ hands, with a girth of 68 inches, compared with Hurry On's 82½ inches.

Photo Rex Coleman

GAINSBOROUGH (1915), a wartime Triple Crown winner who sired Hyperion; and, below, AUREOLE (1950), a son of Hyperion and one of the most successful post-war sires in Britain, though not the best horse of his year.

Photo W. W. Rouch & Co Ltd

BLANDFORD (1919) survived pneumonia as a foal and had little racing because of leg trouble but sired four Derby winners, including, below, BLENHEIM (1927), an enormous influence in Europe as the sire of Mahmoud and Mumtaz Begum, and in America, where he was champion sire.

Photo W. W. Rouch & Co Ltd

Photo J. C. Meadors

MAHMOUD (1933), the record-breaking Derby winner and great sire in America whose make and shape can be compared with, below, FEDAAN, an Arab stallion.

Photo W. W. Rouch & Co Ltd

Photo W. W. Rouch & Co Ltd

NEARCO (1935), Italian-bred, unbeaten on the racecourse and one of the greatest names in stallion history, especially through his son, below, NASRULLAH (1940), the first horse to become champion sire in both Britain and the United States.

Photo W. W. Rouch & Co Ltd

BOLD RULER (1954), a son of Nasrullah and winner of 23 races before a brilliant stud career; and, below, NORTHERN DANCER (1961), a grandson of Nearco and one of the best horses bred in Canada, who made an immediate international impact as a stallion.

Photo W. W. Rouch & Co Ltd

THE TETRARCH (1911), whose careers as racehorse and stallion were both phenomenal but brief; and, below, his daughter MUMTAZ MAHAL (1921), a flying filly whose successful offspring included Mumtaz Begum, dam of Nasrullah, and Mah Mahal, dam of Mahmoud.

Photo European Racehorse

DJEBEL (1937), winner of the English and French 2,000 Guineas and four times leading sire in France; and, below, EMPERY (1973), the French-trained Derby winner whose British sire and South American dam illustrate the international aspect of modern racing and breeding.

Photo W. W. Rouch & Co Ltd

RIBOT (1952), a great Italian racer; and, below, VAGUELY NOBLE (1965), whose record price at auction was justified. Both were exported to America and triumphed spectacularly despite their talents apparently being at variance with the more precocious needs of the racing set-up there.

Photo W. W. Rouch & Co Ltd

ALLEZ FRANCE (1970), a brilliant American-bred filly who raced in France; and, below, her contemporary and great adversary DAHLIA (1970), twice winner of the King George VI and Queen Elizabeth Stakes. They met six times and Allez France won on each occasion.

BAHRAM (1932), the unbeaten Triple Crown winner who was at stud in England, America and Argentina; and, below, NIJINSKY (1967), the first Canadian-bred Derby winner and the first since Bahram to complete the treble of 2,000 Guineas, Derby and St Leger.

SEA-BIRD II (1962) won the Derby without coming off the bit; and, below, MILL REEF (1968), who had his racing career ended by a fractured foreleg. They are generally regarded as the best Derby winners in the period since 1959.

following year he was champion. His fertility during his stud career was always the cause of anxiety. He started fairly well and in his first three seasons sired 50 foals, but in 1919 32 of his 42 mares were barren. Things then improved slightly but in 1925 he only got one foal and then became completely sterile. In the ten seasons he was at stud he sired only 130 foals, but 80 of them were winners. He was kept as a pensioner at his owner's stud until he died at the age of 24.

Despite his lack of foals The Tetrarch sired four Classic winners, three in the St. Leger—Polemarch (went to Argentina and did well as a sire); Caligula (also very infertile, went to Germany) and Salmon-Trout (a disappointing sire who went to South Africa in his old age). Before export the last-named got King Salmon (four wins value £13,731 including the Eclipse), who after some seasons at stud in England was sent to Brazil where he soon became champion sire. King Salmon left behind him in England the 1943 St. Leger and 1,000 Guineas winner Herringbone, and Kingstone, who won nine races worth £6,391 in the Royal colours before going to stud in 1947.

The Tetrarch's other Classic winner, the 2,000 Guineas victor Tetratema, was champion sire in 1929 and second on the list three times. He got the 2,000 Guineas winner but moderate sire Mr. Jinks; Royal Minstrel, winner of the Eclipse Stakes and £21,549, and sire in America of First Fiddle (398,610 dollars), and the dams of Devil Diver (261,604 dollars), Fabius (Preakness Stakes) and others of note; Tiffin, unbeaten winner of eight races value £16,416, and dam of Merenda (£3,120) and died a fortnight after foaling this filly; Four Course, winner of the 1,000 Guineas and £14,074; Myrobella, winner of £16,143, dam of the 2,000 Guineas winner and champion sire Big Game, and grandam of the St. Leger winner Chamossaire; Thyestes, unbeaten winner of two races value £7,824 but an unsuccessful sire; Theft, winner of seven races for £10,628, sent to Japan; Foray II, whose eight wins totalled £9,674 before he did well as a sire in America; and others of high class.

BACTERIOPHAGE (1929)

A non-winning son of Tetratema deserving of mention is Bactériophage, who was foaled in France, ran once only and spent but two seasons at stud before dying. Yet despite this short racing and stud career he sired some very good winners, notably Teleferique (1934), the best two-year-old in France. The latter was taken by the Germans in 1940 and recovered in 1946. On return he got Alizier (third in the Ascot Gold Cup, stud in France), Cobalt (French 2,000 Guineas) and others of merit.

The significance of Bactériophage lies in the fact that the survival of the Castrel branch of the Herod male line in France is entirely dependent on his sons and grandsons. It is a curious turn of fate that 70 years ago Roi Herode was imported to Ireland from France to re-establish this stirp, which had long been dead in Britain, and then France became dependent on the issue of Bactériophage, who was conceived in Ireland of purely Irish parentage.

Tetratema's brother The Satrap was the best two-year-old of his year in England. He was then sent to America, where he spent some years at stud without special success. He returned to England in 1938 but did not get much patronage.

In fact I am told that during the war he was hawked round for sale at £50. However, he survived that indignity and before he died at the age of 27 in 1951 he sired Auralia (ten races value £12,118, including the Doncaster Cup and Ascot Gold Vase).

Auralia went to stud in England in 1950 but after the 1954 breeding season went to America, where breeders at one time showed a partiality for this line. The Tetrarch's son Stefan the Great spent some years on the other side of the Atlantic. He won his only two races as a two-year-old, worth £3,585. The following year he was greatly fancied in the 2,000 Guineas but hit a joint during the race and could not be further trained. He then went to stud at Newmarket, but in 1923, before any of his stock had run at home, he was sold to America. He left behind him such important animals as Fancy Free (dam of the Derby winner Blue Peter), Portree (dam of Portlaw) and unbeaten Tolgus (5½ wins value £5,511). Tolgus went to stud in Ireland and sired in addition to many small winners the Oaks victress Lovely Rose (grandam of Wilwyn, winner of the Washington International Stakes as well as 21 other races). Stefan the Great sired many winners in America but nothing of exceptional class and failed to enhance his reputation on his return to England in 1929.

Few English or Irish horses who after being at stud overseas, have done well on their return. There is an undercurrent of feeling that if they were in all respects satisfactory overseas, they would not usually have been sold for return here, and in consequence do not get sufficient mates of quality to re-establish themselves.

OTHERS BY THE TETRARCH

Others of note by The Tetrarch were Sarchedon (a very temperamental brother to Stefan the Great, won £5,400, went to Australia); La Dauphine (dam of Anita Peabody, one of the greatest race fillies of the century in America and winner of all but one of her 18 races and 113,105 dollars at a time when stakes were but a fraction of their present scale); Tetrameter (£3,968, sired a number of winners of moderate class including the very hardy Six Wheeler, who won 11 races value £7,023; Viceroy (£2,327, could have won more had he tried, sold to Hungary after a short and useless stud career in England); Paola (£7,171, no use as a broodmare); Tetrabazzia (£4,835 and dam of the St. Leger winner Singapore); Royal Alarm (£4,039, sent to Chile); Taj Mahal (dam of the 1,000 Guineas winner Taj Mah); and Moti Mahal (£11,307, dam and grandam of good winners).

REVIVAL OF THE HEROD SIRE LINE

The Tetrarch's breeder Mr. Kennedy was imbued with the idea of reviving the Herod sire line in the British Isles in the early days of the 20th century. It should not be thought that it was entirely dead here. It was represented by Book (who later made a name for himself as a sire of 'chasers), Le Souvenir and other French importations but none gave promise of siring stock of sufficient class to re-establish and maintain the line. Mr. Kennedy arranged with Mr. W. Cook and Mr. L. Robertson to send from Australia in 1904 The Victory who was their joint property to stand at Mr. Kennedy's stud in Co. Kildare.

The Victory had proved himself a top-class racer in Australia. He was a tail male descendant of that wonderful horse Fisherman who won no less than 70 races from 120 starts including the Ascot Gold Cup twice before export to Australia. The Victory and Roi Herode, who later succeeded in establishing the line, came from widely different branches of it, for the connecting link between the two was Castrel and he was seven generations back in each case. After spending four seasons at stud in Ireland without getting anything worthwhile The Victory died. Mr. Kennedy then bought Roi Herode (1904), who was by Le Samaritain out of the French Oaks winner Roxelane (1894), by the successful French sire War Dance (1887) out of Rose of York.

ROI HERODE (1904)

Rose of York was a half-sister to the Derby winner and very influential sire Bend Or (1877). It would be wrong to describe Roi Herode as a first-class racer. He always ran in good company but won very little. He put up a good show as a three-year-old when beaten a neck in the Prix du President de la Republique, and again as a five-year-old when second in the Doncaster Cup. It was after this race that Mr. Kennedy bought him for £2,000. He had previously won the Grand Prix de Vichy and he can roughly be described as a good staying handicapper—a type of horse who does not usually make a great sire.

After purchase his new owner kept him in training and during his preparation for the 1910 Chester Cup Roi Herode broke down and was promptly sent to Mr. Kennedy's stud. At that late hour his chances of getting many mares of decent class for that season were practically nil, but it so happened that Mr. Kennedy owned a mare called Vahren who was due to foal late to John O'Gaunt, so he mated her to Roi Herode who, one might say, was scarcely out of training. On April 22, 1911, the union produced a colt foal who was later named The Tetrarch. At birth he was a chesnut with black patches which later turned to the familiar grey with white patches—an unusual livery which caused the public to nickname him "The Rocking Horse".

VAHREN (1897)

Vahren was sold as a yearling at Doncaster for 60 guineas. She was in training for three years, winning three small races over distances up to 11 furlongs before Mr. Kennedy bought her as a broodmare for £200. She lived until she was 24 years old, but she was a very shy breeder, producing only five or six foals during her lengthy time at stud. Her only other winners besides The Tetrarch were Coupe d'Ore, who won the Russian Oaks, a race of some standing in those days, and Nicola, a good two-year-old who later produced the useful racer and sire Milesius (1918), also by Roi Herode. Vahren's dam Castania never ran, whilst her grandam Rose Garden was a half-sister to the 1,000 Guineas winner Briar Root (1885).

LE SAMARITAIN (1895)

Roi Herode, who started at a fee of 18 guineas per mare had a long innings at stud before dying in his 28th year, during which he sired the winners of 328 races value £119,506, but none except The Tetrarch of the highest class, and he left no other son who showed any likelihood of carrying on the line. His sire Le Samaritain (1895) won the French St. Leger but was not a great stallion although he got a good colt called Isard II (1910). The last-named sired the Grand Prix de Paris winner and Ascot Gold Cup second Filibert de Savoie and the French Derby winner Belfonds. After some years at stud in France with poor results, Filibert de Savoie went to Italy, but Belfonds did better and his name was found among the leading sires of winners in France for some years.

Le Samaritain was a beautifully bred horse. His sire Le Sancy (1884) met with such success as a stallion that even in those far-off days he commanded a stud fee of 500 guineas—a charge only equalled by St. Simon. Le Sancy had a fairly short stud career as he died in 1900. Le Samaritain's dam Clementina (1880) was a sister in blood to Carbine's dam Mersey, their dam Clemence (1865) being a half-sister to the 1,000 Guineas and St. Leger winner Imperieuse.

All honour to Mr. Kennedy for the unquestionable good he did in producing one of the greatest and most influential horses of modern times, but it must not be overlooked that exceptional good luck played its part. Whatever was at the back of Mr. Kennedy's mind, he could not have foreseen that Roi Herode would break down in his Chester Cup preparation, and so became available for mating with Vahren that year. Subsequent unions of the two produced no comparable results.

Like many exceptional horses The Tetrarch was inbred at his fourth remove. In fact, he had only 13 instead of 16 great-great-grandparents, the names of Doncaster, Speculum and Rouge Rose all appearing twice at that generation in his pedigree. During his brief active stud career The Tetrarch had no easy passage to fame, for as luck would have it, there were at least four other sires of exceptional merit in competition with him, namely Phalaris (1913), Gainsborough (1915), Hurry On (1913) and Son-in-Law (1911).

To return to Mahmoud's pedigree, his great-grandam Lady Josephine won four races value £3,636 before producing four winners. Two were of little account, but Mumtaz Mahal and Lady Juror more than made up for that deficiency. Lady Juror won £8,057 and was the dam of Fair Trial (£5,100 and champion sire); Jurisdiction (£4,632, third in the 1,000 Guineas); and other good winners. Lady Juror is also the ancestress of the Oaks winner Commotion and four above-average stallions—Tudor Minstrel (2,000 Guineas winner and sire of the Kentucky Derby winner but virtually sterile Tomy Lee, French 1,000 Guineas winner Toro, Eclipse Stakes winner King of the Tudors, America's best three-year-old filly of 1965 What a Pleasure, and two very fast horses in Sing Sing and Tudor Melody); unbeaten Combat (nine wins for £7,725); Faux Tirage (£8,744 and a leading sire in New Zealand), and Aristophanes (sold for 22,000 guineas in 1953 and leading sire in Argentina).

Lady Josephine was by Sundridge out of Americus Girl, who won over £8,000 and bred five other winners but none of equal merit to Lady Josephine. Americus, sire of Americus Girl, was bred in America, where he was named Rey de Carreras. He was brought to England by Richard Croker, and after winning a number of sprints went to stud. Except for Americus Girl, whose influence has been noted in Chapter 9, he made no show as a stallion.

1937

1 Mrs. G. B. Miller's b c Mid-day Sun (Solario – Bridge of Allan).
2 Mrs. F. Nagle's b c Sandsprite (Sandwich – Wood Nymph).
3 H.H. Aga Khan's b c Le Grand Duc (Blenheim – La Douairière).
21 ran. Time: 2 min 37.6 sec

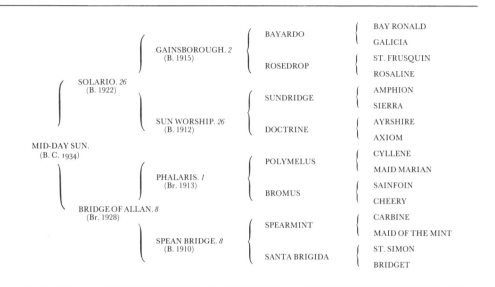

The three-year-olds of 1937 were not a particularly good lot, but the first Epsom Derby winner to carry a woman's colours, Mid-day Sun, showed himself to be the best. He did not run after the St. Leger, was rested as a four-year-old and started at stud at Newmarket in 1939 at a fee of 300 guineas. Circumstances were against him. The war broke out directly after his first season at stud and his owner was not a breeder with a large band of mares to tide him over the difficult period. In 1941 he got but 19 mares, and by 1948 his mates numbered only seven. Although patronage improved, he was getting on in years and had in fact missed the boat. He was sent to New Zealand after the 1950 covering season and died in 1954. He left behind him in England the winners of over £50,000 in stakes, which was not too bad considering everything, but none of his stock was of special merit. If any call for comment perhaps they are De Nittis (1940) who after winning some useful races in Italy occupied a prominent position in their list of sires of winners before transfer to France in 1951, and Sterope (1945), who won the Cambridgeshire twice and £14,844 in stakes.

SOLARIO (1922)

Mid-day Sun's sire Solario won £20,935 including the St. Leger and Ascot Gold Cup. He proved himself an exceptionally good racer before retiring to stud in 1927. He was an immediate success and with only two crops of foals of racing age,

1931 found his name third from the head of the list of sires of winners. The following year his owner died and Solario appeared in the sale ring at the Newmarket December Sales, where he was bought by an English syndicate for 47,000 guineas—an enormous sum considering the conditions. The world—and especially England and America—was in the throes of a considerable financial depression which caused bloodstock prices to drop to a level comparable with those prevailing during the first part of the Second World War. The syndicate were undismayed and put Solario back to stud at Newmarket at a fee of 500 guineas a mare. There he remained until he died in 1945 at the age of 23, having sired the winners of some £270,000 in stakes.

His get, in addition to Mid-day Sun, included Straight Deal (Derby); Exhibitionnist (Oaks and 1,000 Guineas, dam of good winners); Dastur (£11,626 and second in the 2,000 Guineas, Derby and St. Leger, see page 237); Tai Yang (unbeaten winner of £4,611, sire of winners of over 100 races); Tintoretto (£1,333 and a good sire in Argentina); Sind (£3,666, sire of many winners in France and Argentina); Andrea (£9,802, successful sire in Australia); Hanging Fire (dam of the St. Leger winner Ridge Wood); Orpen (£8,754, placed in three Classics and the Ascot Gold Cup, a stud failure); Solfo (£15,466); Traffic Light (£7,036, grandam of the Oaks winner Ambiguity); Starfaralla (dam of the St. Leger winner Tehran, sire of Tulyar); Solar Flower (dam of Peter Flower and Solar Slipper, and grandam of the Derby winner Arctic Prince); and Fille de Soleil (dam of Sunny Boy III).

SUN WORSHIP (1912)

Solario headed the list of sires of winners in 1937 and was second in 1936 and 1943, whilst in 1943 and 1949 his name stood at the top of the corresponding compilation of maternal grandsires. I have referred to his sire Gainsborough on page 188. His dam Sun Worship also produced Broken Faith (11½ wins value £4,811, a gelding); Voleuse (three wins value £1,497, dam of good winners including Theft £10,628 and grandam of good winners from the Aga Khan's stud); Imagery (£346 and dam of Phideas), and Bourbon (a non-winner who was tried as a stallion in Ireland but was a hopeless proposition).

Phideas (1934) who won £5,273 including the Irish 2,000 Guineas and Derby went to stud in 1939, sired the winners of about 185 races value some £85,000, including the Oaks winner Frieze (1949), before he died during the 1954–5 winter. Sun Worship was a shy breeder and when she was 18 years old her then owner—she had changed hands several times previously—sent her to the Newmarket December Sales, where she made 45 guineas to an Irish interest. The new owner re-sold her a month later in Dublin for 13 guineas and a few years later at the age of 22 she produced her final contribution to the bloodstock world—a bay filly by the Irish country sire Tour de Force.

Sun Worship was by Sundridge out of Doctrine, who won £5,037 including the Coronation Stakes. She produced very few foals but amongst them was Documentation (dam of Ornamentation £4,044). Doctrine was by the 2,000 Guineas and Derby winner Ayrshire, who sired any number of good horses without ever quite reaching the top of the tree as a stallion, out of Axiom.

Axiom was sister to Petrel (ancestress of the moderate sire Roseland, New Stakes winner Sir Archibald and unbeaten Tolgus), and to Dynamo (1893), who won £3,481 but made no show as a stallion. She was also half-sister to No Trumps (grandam of the great New Zealand gelding Gloaming, who raced until he was nine years old and won £43,100, without ever running in a handicap); Bill of Portland (a roarer who sired a good few winners in England and proved a useful stallion in Australia); and Lovely (12 wins value £5,610).

Mid-day Sun's dam Bridge of Allan won a small race and bred one other winner of no importance. She was by Phalaris out of Spean Bridge, who won three races value £337 and produced four other winners including Legionnaire, who won £3,725 at home and proved a useful sire in Australia. Spean Bridge was also the dam of the 2,000 Guineas second Knockando who only appeared on a racecourse twice. He spent five seasons at stud at home without achieving anything of note but on export to Venezuela did very well as a stallion. Spean Bridge was a half-sister to Bridge of Earn, being out of Santa Brigida, who bred eight winners of 31½ races and whose family was discussed in relation to Windsor Lad. The family nomenclature is somewhat confused by the fact that Spean Bridge had a half-brother as well as a daughter called Bridge of Allan. The half-brother was foaled in 1908 and although not much of a racehorse sired a few winners during a very short stud career.

DUPLICATED NAMES

The keepers of the Stud Books in the various countries are to be congratulated on the infrequency of duplicated names, but with the vast numbers of horses they have to review it is not surprising that confusing cases do sometimes occur. I have previously mentioned the two Mowerinas—one dam of the Triple Crown winner West Australian, the other a great-granddaughter of the first Mowerina and an equally famous broodmare who produced the Derby and St. Leger winner Donovan and the 1,000 Guineas winner Semolina.

There were two Blenheims at one time at stud in America, both by Blandford. The famous one was foaled in 1927 out of Malva by Charles O'Malley, and the other, winner of 45,450 dollars, appeared in 1928 out of Flying Squadron by Light Brigade. Blandford had two male grandsons at stud named Whirlaway. The best-known was born in America in 1938 and was by Blenheim out of Dustwhirl by Sweep. Winner of the American Triple Crown, he was a far better racer than a sire and died a failure in France at the age of 15. The other Whirlaway was foaled in England in 1940 by Bahram out of Jury by Hurry On. He won a small race before export to Australia, where he turned out to be a far better stallion than a racer.

Teheran (1921), dam and grandam of winners and half-sister to The Tetrarch, and Tehran (1941), the St. Leger winner and sire of Tulyar, were both one-time inmates of the Aga Khan's stud and bound to cause confusion; and breeders examining the useful American stallion Ambiorix—by Tourbillon out of Lavendula, leading American sire in 1961 after heading the French Free Handicap in 1948—should beware not to mix their facts with those of Ambiorix (1940), who was by Tourbillon out of Alfane and could be found in English amateur Flat races after the end of the Second World War.

There are two Flamingos sometimes encountered in pedigrees. One by Flamboyant won the English 2,000 Guineas in 1928, the other by Figaro was the sire of many winners in France. There were also two Caligulas—one by The Tetrarch won the St. Leger, the other by Town Guard was a useful winner and sire on the other side of the Channel—and so on.

Although it may sound elementary, these cases show that when we are looking up the records of a horse with which we are unfamiliar, we must exercise care to ensure we are investigating the right one, and especially so if foreign horses are involved. The spread of international interchange has increased the risk of confusion, though the introduction in the late 1960s of international suffixes to the names of horses racing outside the country in which they were born in theory solves the problem. The value of this idea is best illustrated by reference to Bold Lad (U.S.A.) and Bold Lad (Ire), both sons of the great American stallion Bold Ruler, and both born in 1964.

Until the addition of suffixes, there was nothing to distinguish the two Bold Lads by name only. Now, however, it is possible to differentiate between Bold Lad (U.S.A.)—1964, by Bold Ruler out of Misty Morn, by Princequillo; champion American two-year-old; retired to stud in America in 1967, transferred to France in 1973; sire of French 1,000 Guineas winner Bold Fascinator and Cheveley Park Stakes winner Gentle Thoughts—and Bold Lad (Ire)—1964, by Bold Ruler out of Barn Pride, by Democratic; unbeaten two-year-old trained in Ireland; retired to stud in Ireland in 1968; sire of 1,000 Guineas winner Waterloo in his first crop. Providing the suffixes remain part of the horse's name, duplication and confusion can be avoided, but it is up to keepers of the Stud Books and students of breeding to maintain their vigilance in these matters.

CHAPTER 19

1938 – 1939
Revival of the St. Simon Line

1938

1   P. Beatty's br c Bois Roussel (Vatout – Plucky Liège).
2   J. V. Rank's b c Scottish Union (Cameronian – Trustful).
3   H. E. Morriss's b c Pasch (Blandford – Pasca).
       22 ran. Time: 2 min 39.2 sec

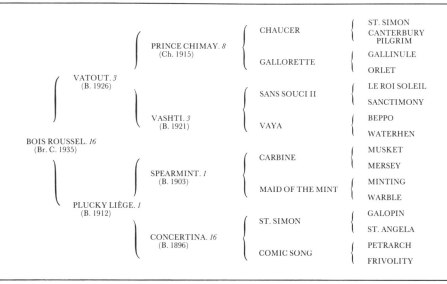

A SPECIAL INTEREST in the 1938 Derby lies in the fact that the winner revived the St. Simon sire line, which was once so powerful. St. Simon line horses had won in 1913 and 1914, but their blood was not available to British breeders. The 1913 victor Aboyeur went to Russia, whilst the 1914 winner Durbar II, who retired to France, was not eligible for entry in the Stud Book according to the rules then in force.

Bois Roussel, the 1938 winner, was bred in France. He did not run as a two-year-old but the following year won the Prix Juigne at Longchamp from 17 others and worth about £1,000. He was then bought for £8,000 by Peter Beatty, who brought him to England and put him in Fred Darling's charge. He never ran before or after the Derby in England, and his only other race was when third to unbeaten Nearco and Canot in the Grand Prix de Paris.

Bois Roussel was rested as a four-year-old and went to stud in 1940 at a fee of 300 guineas. His stud career was satisfactory—one reference credits him as champion sire in 1949—with his colts shining above his fillies. His best get were Tehran (£7,258, including the St. Leger, placed in the Derby, 2,000 Guineas and Gold Cup, sire of Tulyar); Ridge Wood (seven wins for £21,658, including the St. Leger, but a stud failure); Migoli (Arc de Triomphe, Eclipse Stakes, second in the Derby, £22,950 in England and 5,209,500 francs in France; a successful sire); French Beige (Doncaster Cup, third in the St. Leger); Swallow Tail (six wins for £12,440, third in the Derby and exported to Brazil); Fraise du Bois II (Irish Derby, second in the St. Leger, died as a four-year-old); Hindostan (Irish Derby, four years at stud in Ireland before being sent to Japan, where he became a leading sire of winners and successful broodmares); and Delville Wood (leading sire in Australia).

Though Bois Roussel's fillies did not distinguish themselves, several made successful broodmares, the best-known being Castagnola (dam of the wayward Zucchero, who won ten races); Rustic Bridge (dam of the St. Leger winner Cantelo); Edie Kelly (dam of the Derby winner St. Paddy) and Star of Iran (dam of the Oaks winner and outstanding filly Petite Etoile).

Bois Roussel was destroyed in October 1955 because of severe laminitis.

Bois Roussel's sire Vatout won five races as a three-year-old, including the French 2,000 Guineas, and was second in the Prix du President de la Republique and in the English Cambridgeshire, where he was beaten a neck by Double Life (later the dam of Precipitation, Persian Gulf etc.), to whom he gave 3lb. He went to stud in his native France in 1931 as a five-year-old but died six years later. During the comparatively short time he held court, he had most satisfactory results.

Amongst his first crop was William of Valence, who won 104,275 francs in France and £5,069 in England, including the City and Suburban carrying 9st 6lb—until 1967 the highest weight successfully carried since the race began in 1851. In Vatout's second crop came Vatellor, sire of the Derby winners Pearl Diver and My Love, to whom I will refer later. After an interval of one year came Bois Roussel, then after two more crops Vatout died.

His daughters produced such as Amour Drake (3,783,377 francs, including the French 2,000 Guineas, plus £2,325 in England, and second in the English Derby) and Mistral (French 2,000 Guineas and sire of a number of useful winners in France).

PRINCE CHIMAY (1915)

Vatout's sire Prince Chimay was bred in England, where he won four races value £6,673, including the Jockey Club Stakes, beating the Triple Crown winner

Gainsborough. He went to stud in France but except for Vatout made no mark as a stallion. He comes from a tail female line which has a record of producing some very useful racers but shocking bad sires. His dam Gallorette (1907) was a small winner herself but foaled no other winners. She was a half-sister to Amphlett, who won £2,132 and was the grandam of Invershin (1922), whose nine wins, including the Ascot Gold Cup twice, netted £11,974, and great-grandam of Delius (1923), who gained seven wins for £5,891—both of whom were useless at stud. Gallorette also was half-sister to Orphrey, dam of Orpiment (1907), who won nine races but got very few winners. Their dam Orlet was a half-sister to the Prince of Wales Stakes (Goodwood) winner Tarporley (1892), who also failed to come up to expectations as a sire.

Another stud failure from this stirp was Deuce of Clubs (1883), whose name, however, appears in the pedigree of Snoot, dam of the almost impotent St. Leger winner Caligula, the Aga Khan's famous broodmare Eagle Snipe, the Irish Oaks winner Snow Maiden (dam of useful winners) and Lady Comfey (dam of American Flag, a big winner in America, including the Belmont Stakes, and a useful sire).

Prince Chimay's tail female line cannot boast a single successful sire in all its various branches until we go back to Tipple Cyder, foaled in 1788, and then the connection is remote for Tipple Cyder's half-sister—a mare by Spadille—was but the great-great-grandam of Newminster (1848) and of Honeysuckle (dam of Woodbine).

SANS SOUCI II (1904)

Vatout's dam Vashti produced nothing else of note. She was by the 1907 Grand Prix de Paris winner and great sire Sans Souci II, descendant of the Derby winner and seven times champion sire Hermit, whose son Tristan sired Canterbury Pilgrim, the dam of Chaucer and Swynford.

Amongst other excellent get Sans Souci II sired La Farina, who was about the best racer of his generation in France. He did not run for either the English or French Derby, but in a terrific race for the Grand Prix de Paris was beaten only a neck by that great performer Sardanapale, whom he had previously beaten twice. At stud La Farina produced first-class results and amongst his successful sons was Bubbles, who did well as a stallion up to the war, when the Germans took him at the age of 15. He was recovered when he was 21 and four years later his son Guersant (French 2,000 Guineas, and successful sire) was born, whilst Guersant's unbeaten brother Ocarina appeared two years earlier.

Vatout's grandam Vaya (1909) bred seven winners, the best of whom was Vatel (dead heated with the French Derby winner and much superior racer Ksar in the Prix Edgard Gillois). She also produced Variété, dam of the French Derby and St. Leger winner Verso II, who sired the English Derby winner Lavandin.

Vaya was by Beppo (1903), who won about £14,000, including the Jockey Club Stakes, and was third in the St. Leger. Beppo was rather a disappointing stallion and his best get, until his old age, was the Oaks winner My Dear, who was also second in the St. Leger and 1,000 Guineas. When he was 18 years old and apparently on the wane, he suddenly sired an extremely good colt in Picaroon, who was undefeated as a two-year-old, won five of his seven races as a three-year-old and then died, having netted £11,034 in stakes. Beppo was a half-brother to Contessina, dam of the good American racer and sire Reigh Count, who came to England to win the Coronation Cup at Epsom. Their grandam Florence (1880) won the Cambridgeshire and was a sister to Gravity, who produced the Ascot Gold Cup winner William the Third.

Vatout's great-grandam Waterhen (1894) won £4,488 and was the ancestress of Flowership (French 1,000 Guineas) and others of merit. She was half-sister to Queen's Bower, who produced the 1912 French Oaks winner Qu'elle est Belle II.

PLUCKY LIEGE (1912)

To turn to the distaff side of Bois Roussel's pedigree, his dam Plucky Liège was one of the greatest broodmares of modern times whose influence has been felt on bloodstock throughout the world. She was bred in England in 1912 by Lord Michelham, for whom she won four two-year-old races value £1,811. She first ran under the designation the Concertina filly, a few months later was given the name Lucky Liège, and then became Plucky Liège on account of her owner's admiration for the defence of Liège during the early days of the First World War.

After her racing career Plucky Liège went to France. The first of her 12 produce of particular note was Sir Gallahad III, who was born when she was eight years old. He won the French 2,000 Guineas and other good races including the Lincolnshire Handicap, and was placed in the French Derby and St. Leger. He spent a couple of seasons at stud in France, where he sired Galaday II (dam of the English 1,000 Guineas and Oaks winner Galatea II and great-grandam of the Derby winner Never Say Die), and was then sent to America, where he proved to be an exceptional stallion, heading the list of sires of winners four times and the corresponding compilation of maternal grandsires on 12 occasions. He was by Teddy, who was by the French Derby and Grand Prix de Paris winner Ajax out of the English mare Rondeau, by Bay Ronald.

Rondeau started life as a two-year-old selling plater but later showed greatly improved capacity, earning in all £4,517. After retiring to the paddocks she was sold for 4,000 guineas to France, when in foal to the St. Leger winner Rock Sand. This foal died soon after birth, the next one was useless, then came three in succession who were all born dead. Her next effort was a filly who never raced, followed in 1913 by Teddy. When Teddy was due to go into training, the First World War had started, whilst in 1915 there was no racing at all in France, and for the remaining years it was on only a very modest scale.

TEDDY (1913)

Teddy's breeder, M. Edmond Blanc was by this time getting a little tired of the ill luck that had dogged Rondeau's stud career and sold Teddy as an untried two-year-old for slightly over £200. The following year Teddy won a kind of substitute French Derby and then went to Spain, where he won the San Sebastian St. Leger and the Grand Prix de San Sebastian—races of some standing in those days. He went to stud in France and quickly made a great name for himself, siring in addition to Sir Gallahad III such as Asterus (French 2,000 Guineas, etc., and an outstanding stallion); Ortello (a great sire in Italy whose blood was considered so valuable that a special decree of the Fascist Government prohibited his export, but he went to America after their downfall); Aethelstan (nine races in France, and sire of good winners there before going to America); Shred (useful winner and sire in France); Brumeux (Jockey Club Cup etc., and sire of Borealis, who won the Coronation Cup and was second in the St. Leger); and Rose of England (Oaks winner and dam of Chulmleigh). When Teddy was 18, he was sold to America, where he made his influence felt before dying in his 24th year.

Before leaving France he sired Bull Dog and Quatre Bras II to further matings with Plucky Liège, and like Sir Gallahad III, both went to America. Bull Dog's results at stud in America were comparable to his elder brother's. His son Bull Lea, after winning ten races value 94,825 dollars, proved the most successful American stallion of all time until Bold Ruler came on the scene. Bull Lea, champion sire five times and leading broodmare sire four times, was represented by 24 crops which won 13,589,021 dollars, and included two Kentucky Derby winners Citation (Triple Crown and 1,085,760 dollars) and Iron Liege, as well as Bewitch (champion filly of 1947, 20 wins for 462,605 dollars), Armed (817,475 dollars) and Waltown (Horse of the Year in 1949, winner of 415,676 dollars).

Quatre Bras II won the Prix Yacowlef in France plus ten events in America, but his achievements at stud are on an altogether lower plane than those of either Sir Gallahad III or Bull Dog.

Additionally Plucky Liège produced Bel Aethel, whom I have already mentioned, and Admiral Drake (by Craig an Eran), who won five races value 1,413,334 francs including the Grand Prix de Paris. Admiral Drake spent some 20 years at stud in France with satisfactory but not phenomenal results, and I will detail some of his good winners in relation to his Derby victor Phil Drake.

Plucky Liège also produced the winners Marguerite de Valois (four wins and dam of winners in France and America); Chivalry (two wins); Elsa de Brabant; Noble Lady (two wins); Noor Jahan (three wins in France, dam and grandam of good winners); and also Diane de Poitiers, her only non-winning foal but dam of winners in America. Apart from the fact that the production of all these winners— including an English Derby winner at the age of 23 and three Classic winners in all—is a remarkable performance, the outstanding point in her record is that all her sons were good stallions. There cannot be a major bloodstock breeding country in the world that has not had one or more horses holding court by Sir Gallahad III, Bois Roussel, Bull Dog or Admiral Drake.

Plucky Liège's dam Concertina produced nine winners but none of top class, and also the non-winning Garron Lass, ancestress of Bahram and Dastur. Concertina's grandam Frivolity was a sister to Windermere (the dam of Kendal, sire of the Triple Crown winner Galtee More), and sister in blood to Lily Agnes (dam of the Triple Crown winner Ormonde, 1,000 Guineas winner Farewell, and grandam of Sceptre). In order to avoid confusion I had better add that this Concertina (bay, 1896, by St. Simon out of Comic Song) is not the same as Lord

Astor's brown filly of the same name who was foaled in 1920 by Sir Eager or Florentino out of Conjure, and won £5,473.

Once again we have a Derby winner who is inbred at the third and fourth remove. In Bois Roussel's case his grandam was by St. Simon and his paternal great-grandsire by St. Simon.

SOME SUCCESSFUL OLD PARENTS

Plucky Liège has been erroneously credited with being the oldest mare to have bred a Derby winner, being 23 when Bois Roussel was foaled. Plucky Liège did many successful things, but this distinction is, in fact, held by Horatia, who was 25 when she produced the 1806 winner Paris. However, Plucky Liège comfortably holds the record for this century, since the oldest subsequently successful mares have been Lavande and Pamplona, who produced Lavandin and Empery respectively at the age of 17. On the male side, the Derby winner The Little Wonder (1837) was 27 years younger than his sire Muley, whilst this century Admiral Drake takes the honours for being 21 years old when Phil Drake was foaled. If it is of any significance, the average age of the Derby-winning sires since 1938 is 10, and of the dams 9.5.

1939

1   Lord Rosebery's ch c Blue Peter (Fairway – Fancy Free).
2   E. Esmond's ch c Fox Cub (Foxhunter – Dorina).
3   Lord Derby's b c Heliopolis (Hyperion – Drift).
        27 ran. Time: 2 min 36.8 sec

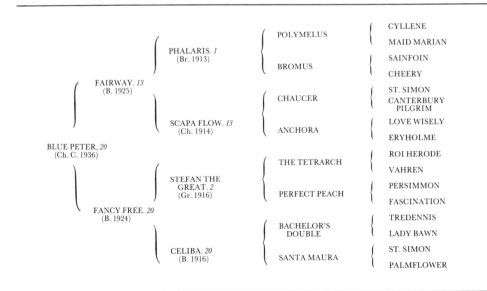

					CYLLENE
			POLYMELUS		
		PHALARIS. *1*			MAID MARIAN
		(Br. 1913)			SAINFOIN
			BROMUS		
	FAIRWAY. *13*				CHEERY
	(B. 1925)				ST. SIMON
			CHAUCER		CANTERBURY PILGRIM
		SCAPA FLOW. *13*			LOVE WISELY
		(Ch. 1914)	ANCHORA		
BLUE PETER, *20*					ERYHOLME
(Ch. C. 1936)					ROI HERODE
			THE TETRARCH		
		STEFAN THE			VAHREN
		GREAT. *2*			PERSIMMON
		(Gr. 1916)	PERFECT PEACH		
	FANCY FREE. *20*				FASCINATION
	(B. 1924)				TREDENNIS
			BACHELOR'S DOUBLE		LADY BAWN
		CELIBA. *20*			ST. SIMON
		(B. 1916)	SANTA MAURA		
					PALMFLOWER

As a two-year-old Blue Peter ran only twice—once unplaced and once second, beaten a length and a half in the Middle Park Stakes—but in the Free Handicap he was rated second-best of his age. The following season he won all four races including the 2,000 Guineas, Derby and Eclipse. Before he could show his ability in the St. Leger, war broke out and the race was not run. He was then taken out of training and went to stud as a four-year-old at a fee of 400 guineas.

The manner in which he won his four races as a three-year-old rightly earned for him the reputation of being an outstandingly good performer, and judged against that he must be regarded as a slightly disappointing stallion. Though he sired plenty of winners, he got only one in a Classic, the wartime Derby victor Ocean Swell, who went on to win the Ascot Gold Cup but failed to make the grade as a stallion. When Blue Peter died in 1957 his stock had won 436 races worth over £265,000, and also included the Italian-bred Ascot Gold Cup winner Botticelli, unbeaten Blue Train, the Hardwicke Stakes winner Peter Flower, and Bowsprit, a modest racehorse who proved a useful sire of 'chasers.

Blue Train is of interest because he was a difficult horse to train on account of a ringbone, and he ran only three times, winning £3,736 for his three victories. Bred and owned by the National Stud, he was out of Sun Chariot (1,000 Guineas, Oaks and St. Leger) and on retirement was sold privately to an Irish stud. This transaction caused considerable discussion, as it was said a horse of his standing ought to have been offered for sale to English breeders either by tender or by auction. He was, after all, the taxpayers' property. The National Stud authorities defended their action on the grounds that he was not a sound horse and by implication was thus best out of the way. Personally I do not consider unsoundness is necessarily a bar to a horse having a great beneficial effect on his breed. Sundridge, Gallinule and Hermit are only three of many unsound sires who, without doubt, did a great deal of good. Each case must be judged on its merits and consideration given to racing merit, pedigree, general confirmation, temperament etc., factors which may far outweigh any technical unsoundness.

However, Blue Train's departure hardly proved an insurmountable loss. He started his career normally, siring stock which gave promise that he was going to turn out a first-class stallion, but in his third season, alas, he had only eight foals, 29 mares being barren and five sent abroad or with no returns. All efforts to restore his fertility failed and he was put back into training, but he suffered a most unusual accident to a hock which rendered him incapable of withstanding further exertion. His significant influence nowadays is as the sire of Queen of Speed (dam of the 2,000 Guineas winner Kashmir II) and Blue Mark (dam of the Cheveley Park Stakes winner and dam of winners Lalibela).

Blue Peter's failure to produce some of his own calibre has been balanced by the success of his daughters as broodmares, which enabled him to head this particular list in 1954. The most important mares by him were: Glen Line (who whipped round at the start of her only race but at stud produced the Eclipse Stakes winner King of the Tudors and the 2,000 Guineas winner Our Babu); Cottesmore (dam of the useful sprinter Kelly); Snow Bunting (whose sons Sandiacre and Come to Daddy both won the Cesarewitch); All Aboard (who produced four good-class handicappers for Lord Rosebery in Captain Kidd, Copenhagen, Smugglers Joy and The Bo'sun); Eyewash (dam of 12 winners including the Park Hill Stakes winner Collyria; Varinia, third in the Oaks; Aranda, grandam of the Italian Derby winner Ardale; and the non-winning Visor, dam of the Ascot Gold Cup winner Raise You Ten); Blue Prelude (winner of the Lancashire Oaks and dam of Champion Hurdle winner Magic Court, and grandam of triple Classic-winning filly Altesse Royale, and Imperial Prince, second in the English and Irish Derbys); and Blue Line (dam of the useful stayers Allenheads, Hornblower and Alignment).

FAIRWAY (1925)

I have already referred to the trials and tribulations suffered by Blue Peter's sire Fairway in the Derby. I will now add that he was bracketed with two others at the top of the Free Handicap of his year, was beaten only once the following year and once more as a four-year-old. His victories included the St. Leger, Eclipse Stakes and Jockey Club Cup. He won 13 races between the age of two and four from 16 starts, earning £42,722. A tendon gave out as he was being prepared for the Ascot Gold Cup as a five-year-old and he started at stud the following season.

Success came early, as amongst his second crop of foals were the 2,000 Guineas winner Pay Up (who later proved a moderate sire), the 1,000 Guineas winner Tide-way (dam of Gulf Stream), and Taj Akbar, who won £6,681 but proved an indifferent sire first in England, then in France, and was finally banished to Hungary.

As a result Fairway headed the list of sires of winners in 1936—a position he also held in 1939, 1943 and 1944. In all his stock won nearly £315,000 in stakes, and included the winners of seven Classic races—Pay Up, Blue Peter, Kingsway and the filly Garden Path in the 2,000 Guineas; Tide-way in the 1,000 Guineas; and Blue Peter and Watling Street in the Derby.

His sons Fair Copy (sire of the Cambridgeshire winner Sayani while based in France), Channel Swell (in Australia) and Grassgreen and Fair Test (in South Africa) did well at stud abroad, and Ribbon, after being the best two-year-old filly of her year, ran second in the 1,000 Guineas, Oaks and St. Leger; but it is to Fair Trial and Honeyway that Fairway owes his position as one of the most important modern influences.

FAIR TRIAL (1932)

Fair Trial, though third in the Eclipse Stakes, barely stayed ten furlongs and having won £5,100, at stud proved an important source for speed. He quickly emulated Fairway as leading sire, in 1950, when Palestine won the 2,000 Guineas to complete a treble in the race for Fair Trial, after Lambert Simnel and Court Martial. Festoon won the 1,000 Guineas whilst Petition won the Eclipse Stakes, and it is in this distance range that Fair Trial has done most to promote the Fairway line.

Court Martial, a horse full of quality, was twice leading sire and bred such winners as Major Portion (the best miler of his year), Rosalba (second in the 1,000 Guineas) and Timandra (French Oaks). Major Portion has been Court Martial's best son at stud, producing the smart sprinters Majority Blue and Potier, as well as the good miler and successful stallion Double-U-Jay and the City and Suburban winner Minor Portion; but Counsel, Epaulette and Ratification are others of his produce who have made some mark at stud.

Petition, who sired the marvellous filly Petite Etoile, and Palestine, whose son Pall Mall emulated his sire by winning the 2,000 Guineas, strengthened the emphasis on speed in this line, but more recent events have suggested that stamina is making a welcome resurgence.

Leading sire in 1959, Petition owed his position largely to Petite Etoile, an impressive looker who was undefeated as a three-year-old in the Free Handicap (under top weight), 1,000 Guineas, Epsom Oaks, Sussex Stakes, Yorkshire Oaks and Champion Stakes, to earn the distinction of having won more prize money than any other filly before her. The following year she confirmed her ability over a mile and a half by winning the Coronation Cup and being narrowly—some would say unluckily—beaten in the King George VI and Queen Elizabeth Stakes. She stayed in training as a five-year-old and took her earnings to £68,000—bettered only by three colts at that time—by again winning the Coronation Cup. Her failure as a matron has been a bitter disappointment—not least presumably to her connections, who were reported to have turned down an American offer of £320,000 to keep her at stud.

As a sire of sires Petition has been responsible for speed through the useful sprinter Runnymede, but his line through March Past and Queen's Hussar has recently produced the smart middle-distance horses Brigadier Gerard (winner of 17 races from 18 outings from five furlongs to a mile and a half, including the 2,000 Guineas, Eclipse Stakes, King George VI and Queen Elizabeth Stakes and Champion Stakes) and Highclere (1,000 Guineas and French Oaks). His son White Fire III (a three-parts brother to Petite Etoile who stood in England for three seasons at 250 guineas before export to Japan) sired the Eclipse Stakes winner Coup de Feu; and another son Petingo (winner of the Middle Park Stakes and Sussex Stakes, and second in the 2,000 Guineas) was responsible for English Prince (Irish Sweeps Derby) and Miss Petard (Ribblesdale Stakes) before his premature death in 1976.

Palestine, a precocious two-year-old, was less likely than Petition to get middle-distance stock, but his son Pall Mall, who did not get beyond a mile, got the Champion Stakes winner Reform, who in turn sired two middle-distance Classic winners in Roi Lear (French Derby) and Polygamy (Oaks).

HONEYWAY (1941)

Though Fair Trial can claim the major responsibility for enhancing the reputation of Fairway, the success of Honeyway should not be ignored, especially since he bridged a 16-year gap on the list of leading sires of winners for the Fairway line. Honeyway caused his first stir when after three seasons in which he proved himself a tip-top sprinter, he won the Champion Stakes from a strong field of middle-distance horses. It had been assumed that since he was originally trained as a sprinter, he would never be anything else. Yet he had a stamina probability figure of 11 furlongs—his dam Honey Buzzard (by Papyrus) was beaten a neck in the one-mile Coronation Stakes when conceding 6lb to the winner, and appeared to stay a mile and a half—and ten furlongs proved no problem at Newmarket, where as was custom, even in sprints, he was ridden from behind.

Retired to stud at the end of 1946, having won 13 races, Honeyway ran into fertility problems and was returned to training, winning three of his four outings in 1947 before going lame. He returned to stud in 1949, having been successfully treated, and went about his work in satisfactory fashion, showing that early sterility need not be permanent in a stallion, and that a period covering mares does not necessarily take the edge off a horse's enthusiasm for racing.

At stud Honeyway's best get were the 1,000 Guineas winners Honeylight (England) and Dictaway (France), Typhoon (leading two-year-old colt of 1966) and Sunny Way (a useful stayer whose effectiveness was perhaps not realized fully because of his job as lead horse to St. Paddy, and subsequently the sire of Eclipse Stakes winner Scottish Rifle). As a sire of broodmares Honeyway is notable for Sweet Angel (dam of the Dante Stakes winner Sweet Moss, Royal Lodge Stakes winner Soft Angels, and Sucaryl, second in the Irish Sweeps Derby); Brabantia (dam of the sprinter and sire Polyfoto); Vauchellor (dam of the dual Cambridgeshire winner Prince de Galles) and Too Much Honey (dam of the smart sprinter and successful sire Sweet Revenge).

However, Honeyway's most significant contribution to the story of Fairway came late in life through Great Nephew, who was 22 years younger than his sire. Great Nephew, beaten a short head in the 2,000 Guineas, won five races—four of them in France—and is the sire of Derby winner Grundy, whose success put Great Nephew at the head of the list of sires of winners in 1975. Thus Great Nephew became the first from the Fairway line to be champion sire since Petition in 1959.

Fairway's mares placed him at the top of the tree in that department, and they included the dams of Gulf Stream (£10,537, second in the Derby), Sandjar (French Derby), Hyperbole (£10,799), Trimbush (£9,203) and Chivalry (£9,637).

Fairway died at the age of 23, and his success in the 2,000 Guineas is worth recording. He himself was responsible for four winners, one second and two thirds between 1936 and 1943; the winners in 1941, 1945 and 1950 were by his son Fair Trial, in 1951 by Fair Trial's son Ballyogan, in 1958 by Palestine, in 1962 by Court Martial's son Counsel, and in 1971 through Petition's son March Past and Queen's Hussar.

THE BROTHERS FAIRWAY AND PHAROS COMPARED

Fairway was a brother to Pharos (see page 220) and comparison between the two affords a useful illustration of the transmission of hereditable factors. I have outlined in Chapter 1 the general principles of heredity which may be summarized as follows. When we arrange a mating we are taking a dip into the well of the ancestral factors of both parents. We know the ingredients in the wells, but we cannot accurately foretell which of these will be drawn out.

In the case of Fairway and Pharos, the dip naturally was into the same parental wells. Although they were both bay in colour, they bore little physical resemblance, and Fairway was incomparably the better racer. Pharos's maximum distance was about a mile and a quarter, whilst Fairway could stay double that distance. So roughly speaking, we can say that the genes which formed their somatic cells, or external make-up, produced horses of considerable variation. When they went to stud, the results produced by the two brothers were remarkably level. Both were first-class stallions who sired Classic winners of the highest standard. The average winning distance of Pharos stock, three years old and over, was 8.53 furlongs, whilst the corresponding figure for Fairway was 8.94. This is noteworthy bearing in mind the wide discrepancy of the stamina of the pair on the racecourse. It shows that although the genes which formed their respective somatic cells were different, to a considerable degree those which entered their germ plasm were far closer allied.

In course of time both have succeeded in carrying on their sire lines, to the extent that of the 32 leading sires from 1945 to 1976 exactly half are attributed to these two stallions, a remarkable concentration in view of the numbers of stallion sources available. To emphasize the extent of their influence, the 182 sires for whom long pedigrees were detailed in the 1976 Register of Thoroughbred Stallions include 73 who trace in male line to either Pharos or Fairway.

However, there is no doubt that Pharos, an inferior racer to Fairway, has had greater influence than his brother at stud, most notably through Nearco but also with Pharis II. The achievements of Nearco will be chronicled later, and those of Pharis II were dealt with in Chapter 16, but it is worth noting that together they provide 51 of the 73 stallions mentioned above in relation to the Register of Stallions, and ten of the 32 leading sires of winners between 1945 and 1976. It is a record without parallel in modern times.

FANCY FREE (1924)

To return to Blue Peter's ancestry, his dam Fancy Free was a hardy filly who ran 24 times between the ages of two and four, winning four times for £2,477. At stud she produced eight winners, including in addition to Blue Peter, his brother Full Sail (£7,037 and a leading sire in Argentina); Tartan (£4,405, a good sire of 'chasers who got the 1954 Grand National winner Royal Tan and the 1961 Cheltenham Gold Cup winner Saffron Tartan); and Springtime (dam of 12 winners and fourth dam of Blakeney), as well as the non-winner Flapper (dam of six winners of over £15,000).

Fancy Free was one of Stefan the Great's stock sired before his visit to America. In addition to being a very fast racer, Stefan the Great was a very well-bred horse, as not only was he by The Tetrarch and his dam Perfect Peach by Persimmon, but his grandam Fascination (1896) won the Coronation Stakes and four other good races besides being second in the 1,000 Guineas. Fascination was a half-sister to Captivation, and their dam Charm (1888) was a sister to the 1,000 Guineas and Oaks winner Amiable, whilst Tact (1882), dam of Amiable and Charm, was a sister to Gravity (1884), who produced William the Third (1898). A further sister to Tact and Gravity was the Cambridgeshire winner Florence (1880), whose daughter Pitti produced Beppo.

Fancy Free's dam Celiba bred six winners, though none as good as Fancy Free, and also the non-winner Micmac, who was third in the Oaks. Celiba was by Bachelor's Double out of Santa Maura, a good winner, second in the 1,000 Guineas and dam of six winners, including Leucadia (also a good winner and second in the 1,000 Guineas) and Cyllene More (£2,544 but a poor sire who finished his days in Australia). Santa Maura was a sister to St. Florian, whose only claim to fame is that he sired the Derby winner Ard Patrick, and to Siphonia (1888), who produced Symington, sire of Tetratema's dam. Santa Maura was also a half-sister to the 1899 Oaks winner Musa, who produced Mirska, winner of the corresponding Classic in 1912.

Blue Peter is inbred to St. Simon at his fourth remove and carries auxiliary strains of that great horse at more remote incidences.

CHAPTER 20

1940 – 1942
The Son-in-Law Tribe

1940

1 F. Darling's b c Pont l'Eveque (Barneveldt – Ponteba).
2 H.H. Aga Khan's b c Turkhan (Bahram – Theresina).
3 Lord Derby's br c Lighthouse II (Pharos – Pyramid).
   16 ran. Time: 2 min 30.8 sec   Run at Newmarket.

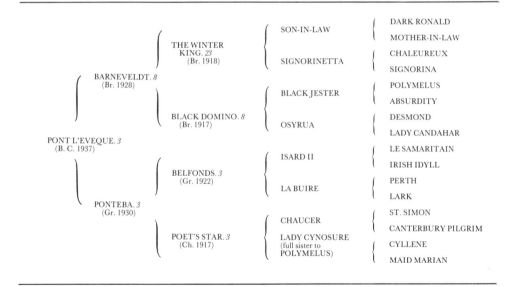

ONCE AGAIN THE country was at war and facing dire adversities, but the Derby had always been run on its traditional date and, come what may, to hell with anyone who deigned to interfere with it! The only difference was that the Derby from 1940 to 1945 inclusive was run at Newmarket, as the Epsom course was requisitioned by the Government.

Pont l'Eveque's task was probably simplified by the fact that M. Boussac's Djebel, who won both the French and English 2,000 Guineas, could not be transported from France owing to the war conditions. Pont l'Eveque, who was not foaled until May 25, ran twice unplaced as a two-year-old, and the following season won three of his five starts, earning £6,389. He did not run for either the 2,000 Guineas or St. Leger and went to stud at Newmarket as a four-year-old, but after two seasons was sold to Argentina. There he was not a conspicuous success, though he sired Antinea, dam of Atlas, whose son Forli is currently proving a worldwide success.

THE WINTER KING (1918)

Pont l'Eveque's sire Barneveldt finished third to Tourbillon in the French Derby but got his revenge in two of their three subsequent meetings, including in the Grand Prix de Paris. Barneveldt went to stud in his native France as a five-year-old and died at the age of 14, but except for Pont l'Eveque did not do well as a stallion. Barneveldt's sire The Winter King, whose dam Signorinetta won the 1908 Derby and Oaks, was bred and raced by Lord Rosebery. He was far removed from Classic class but scored four times on the Flat and twice over hurdles. He went to his breeder's stud in 1923, as a five-year-old, at a fee of 19 guineas, but went to the Newmarket December Sales six years later when he made 200 guineas. He then stood in Yorkshire for a couple of seasons and later moved to France. A short stay there was followed by his sale for £2,500 and transfer to Italy, where his dam's half-brother Signorino (1902), who was second in the English 2,000 Guineas and third in the Derby, had been a leading sire for some years. There The Winter King died, and although his many perambulations cannot have helped his stud career, he must be regarded as a failure as a stallion, except for his high-class son Barneveldt, and except that he was leading sire of jumpers in Italy in 1939.

The Winter King's sire Son-in-Law established a patriarchy of high-class stayers who have had an enormous influence on bloodstock the world over. Foaled in 1911, he won eight races value £5,546 from a mile to 2 miles 5 furlongs, including the Cesarewitch with 8st 4lb, the Jockey Club Cup twice and the Goodwood Cup. He scored five times as a three-year-old, twice at four and once at five. He did not run for the Classics—and was in fact considerably below that class.

He went to stud in 1917 at a fee of £98 a mare, so The Winter King was one of his first crop of foals. By the time these were three-year-olds, it became apparent that he was likely to be a stud success, for 1921 found his name 13th on the list of sires of winners. He lived until he was just over 30 years of age and during his long stud career sired 223 individual winners of some 650 races value nearly £400,000. He was champion sire in 1924 and 1930. His only Classic success was the Oaks winner Straitlace, and it is a remarkable fact that Pont l'Eveque is the only Derby winner to date in direct male line from Son-in-Law.

Straitlace, who helped Son-in-Law to become leading sire in 1924, set a record of 17,000 guineas for a broodmare at public auction when sold to French breeder

Edward Esmond in 1925, and bred Necklace II (Prix Robert Papin and Prix Morny); Staylace (dam of Foxglove who went to Argentina and sired Veneta, grandam of Forli); Lovelace (second in the French Derby); and Silverlace II (second in the Oaks and ancestress of Oaks winner Sleeping Partner and Oncidium, leading sire in New Zealand).

Among Son-in-Law's other influential daughters were two sisters to Foxlaw— Foxberry, who bred the Sydney Cup winner Proctor, and Aloe, who threw the famous half-sisters Feola and Sweet Aloe—and Trilogy (dam of the Oaks winner Light Brocade), and most significantly Lady Josephine (dam of eight winners including Fair Trial).

However, it was Son-in-Law's colts who made his name, especially by their domination of the major staying races. His male-line successes in events of this nature and class can be gauged from the following list of big-race winners between 1921 and 1956:

ASCOT GOLD CUP

1927 Foxlaw (by Son-in-Law)
1930 Bosworth (by Son-in-Law)
1931 Trimdon (by Son-in-Law)
1932 Trimdon
1933 Foxhunter (by Foxlaw)
1935 Tiberius (by Foxlaw)
1953 Souepi (by Epigram, by Son-in-Law)
1956 Macip (by Marsyas II, by Trimdon)

GOODWOOD CUP

1921 Bucks (by Son-in-Law)
1935 Tiberius
1936 Cecil (by Foxlaw)
1937 Fearless Fox (by Foxlaw)
1938 Epigram (by Son-in-Law)
1946 Marsyas II (by Trimdon)
1953 Souepi

FRENCH GOLD CUP

1939 Foxlight (by Foxhunter)
1944 Marsyas II
1945 Marsyas II
1946 Marsyas II
1947 Marsyas II

CESAREWITCH

1934 Enfield (by Winalot, by Son-in-Law)
1935 Near Relation (by Son-in-Law)
1936 Fet (by Son-in-Law)
1940 Hunter's Moon IV (by Foxhunter)
1954 French Design (by Coup de Lyon, by Winalot, by Son-in-Law)

QUEEN ALEXANDRA STAKES

1923 Bucks
1935 Enfield
1936 Cecil
1937 Valerian (by Son-in-Law)
1938 Epigram
1946 Marsyas II

ASCOT STAKES

1936 Bouldnor (by Son-in-Law)
1937 Valerian
1938 Frawn (by Foxlaw)
1939 Frawn
1946 Reynard Volant (by Foxhunter)
1947 Reynard Volant.

FOXLAW (1922)

Foxlaw was not the most fortunate of animals. He fractured a pelvis as a two-year-old and was held up for seven months; he broke down in winning the Ascot Gold Cup; and finally he broke a leg and died at the age of 13 in 1935. He did well during his comparatively short life at stud.

Trimdon, whose dam also produced Foxhunter, got nothing of special note except Marsyas II, who in addition to his English victories won the French Gold Cup (Prix du Cadran) a record four times; Trimbush, winner of 14 races including the Doncaster Cup; and Triumvir, winner of the Queen's Prize at Kempton and on export to New Zealand maternal grandsire of the local St. Leger winner Cracksman.

Foxhunter, who was at stud in France from 1933 to 1938, in England from 1939 to 1945 and in Argentina afterwards, was a success in all his locations. In England he left behind two useful fillies in Foxtrot, winner of the Ebor Handicap and grandam of the Ascot Gold Cup winner Random Shot, and Young Entry, winner of the Lancashire Oaks and dam of Atlas (Doncaster Cup) and Apprentice (Yorkshire and Goodwood Cups). In France Foxhunter was the maternal sire of Fast Fox (Doncaster Cup and sire of Ascot Gold Cup and French St. Leger winner Wallaby II).

Tiberius, who retired as a replacement for his deceased sire at stud, made no important mark as a sire of winners, except under National Hunt Rules, though his daughter Catherine bred Djebe, sire of Midget II (winner of the Cheveley Park Stakes and dam of the similar winner Mige).

Epigram's sole contribution—outside the Irish winners Shagreen (Grand National) and Esprit de France (St. Leger)—was Souepi, who was sold out of training at the 1954 Newmarket December Sales for 2,100 guineas for export to Chile. He cannot be regarded as a well-bred horse, but at the same time, during a record-breaking sale, to see a six-year-old who had won £23,046, including the Ascot Gold Cup, Gold Vase, Goodwood Cup and Doncaster Cup (in a dead heat), sold for such a small sum provides food for thought. To the best of my belief no young Ascot Gold Cup winner has ever previously been sold at anywhere near approaching such a low figure.

Cecil, who came from the same male and female lines as The Winter King, died after one season at stud but got Cecily, great-grandam of the Arc de Triomphe winner Rheingold. Fearless Fox made little impact, though he became the sire of Namesake, grandam of the 1966 French Derby winner Nelcius; and Fet and Near Relation were geldings. Hunter's Moon IV went to America where he sired Moonrush (about 430,000 dollars) and other useful stock. Marsyas II retired to his native France but did less well than other Boussac stallions, his best get being Marsyad (Dewhurst Stakes but broke a leg in the Derby), Childe Harold (Yorkshire Cup), Kirkes (second in the Grand Prix de Paris) and Macip (French St. Leger and Ascot Gold Cup).

Valerian spent a few seasons in England with poor results before being sent to Brazil; but Enfield who went to Australia, got two winners of the Melbourne Cup—Sirius and Rimfire—and other good horses during a comparatively short time at stud.

Winalot was not a typical Son-in-Law as he was a middle-distance horse who won seven races value £8,964 and sired the winners of about 350 races value nearly £130,000, which are the highest relative figures of any Son-in-Law stallion at home. Coup de Lyon, who never won beyond a mile and a quarter, secured six races value £5,177 and retired to stud in Ireland at the modest fee of £25 a mare, but lived to get the winners of close on £100,000 in stakes, including the Irish Classic winners Etoile de Lyons, Spam and Mighty Ocean.

Young Lover, who was by Son-in-Law out of a half-sister to Trimdon and Foxhunter, was also untypical of his sire's stock, since he came to hand early enough to win the Gimcrack Stakes as a two-year-old. He was no good at stud in either England or Ireland.

Other Son-in-Law offspring who belied the family's reputation for Cup horses and slow plodders included Maureen (Queen Mary Stakes), Pandarus (Woking-ham Stakes) and Knight of the Garter (Coventry Stakes). Knight of the Garter had a most versatile stud career, during which he got the winners of 350 races on the Flat and 470 under National Hunt rules, and his produce included Knight's Caprice (Stewards Cup), Bright Cherry (dam of the great 'chaser Arkle), Lady Walewska (dam of the Grand National winner Team Spirit) and My Aid (grandam of Le Levenstell, whose influence on the Son-in-Law story will be discussed later).

SON-IN-LAWS IN FOREIGN LANDS

Though Son-in-Law twice topped the list of leading sires in Britain, none of his male-line issue has followed suit, but one to do so abroad was Beau Pere, whose dam was the 1,000 Guineas winner and Oaks second Cinna. He was a racer of most limited ability who scored in three races value £974 and then went to stud at Newmarket at a fee of nine guineas. Even on those terms he received the poorest patronage and was sent to the December Sales of 1933, changing hands for 100 guineas. He was then exported to New Zealand, where he quickly made a name for himself, later being sold to Australia and finally to America for 100,000 dollars. He was champion sire twice in New Zealand and in Australia three times, whilst in America his daughter Iron Reward bred the great Swaps, and Flower Bed, the grandam of successful stallion Graustark.

Foxlaw's son Foxbridge did excellently in New Zealand, where he was leading sire 11 times and leading sire of winning broodmares ten times, and sired the winners of 23 Classic races. He won only one race value £2,520 at home but was a half-brother to the 1,000 Guineas winner Cresta Run, tracing to Adula, a sister to Pretty Polly.

The English Derby second Robin Goodfellow also did well in New Zealand. He was by Son and Heir, who never won but was third in the St. Leger and died after four seasons at stud in Newmarket. The latter was by Son-in-Law out of Cinderella, a sister in blood to Myrobella, the dam of Big Game.

In Argentina, Foxhunter, Rustom Pasha (£14,921, by Son-in-Law) and Fox Cub (won £950, second in the Derby, by Foxhunter) have achieved fame, while in Chile, The Font (ten races value £3,514, by Son and Heir) was among the leading sires of winners for some years. In France the line was represented by Foxhunter, Marsyas II, Barneveldt, Comedy King (£7,442, by Son-in-Law and a half-brother to the near-sterile Derby winner Call Boy), and Plassy (£9,688).

DARK RONALD (1905)

Son-in-Law's sire Dark Ronald, who was by Bayardo's sire Bay Ronald and cost 1,300 guineas as a yearling, won £8,238 in stakes. He broke down in preparation for the Derby and it was necessary to fire the tendons on both forelegs. His owner Sir Abe Bailey could not get away from South Africa at the time, and had his horse sent out to convalesce, whereupon he returned to race as a four-year-old, and landed a huge gamble in the Royal Hunt Cup and won the Princess of Wales's Stakes at Newmarket. In his last race he was beaten less than two lengths into third place in the Doncaster Cup. He retired to stud in England at a fee of £98 but in June 1913 was sold for £25,000 to Germany, where he was champion sire five times.

DARK LEGEND (1914)

Before his export, in addition to Son-in-Law, Dark Ronald sired Dark Legend, to whom brief reference was made in Chapter 13. Dark Legend, who came from Dark Ronald's final crop in England, proved an excellent stallion, his get including: Duplex (French Derby); Fairy Legend (French 1,000 Guineas and Oaks); Mary Legend (French Oaks); Dark Lantern II (French 2,000 Guineas); Dark Japan (Goodwood Cup, etc., and sire of many winners in France); Galatea II (English 1,000 Guineas and Oaks); Easton (£3,369 and 600,000 francs in Belgium, second in the 2,000 Guineas and Derby, a poor sire who eventually went to America); Legend of France (15,800 francs in France and £2,825 in England, sent to Brazil after ten seasons at stud in Ireland); and Rosy Legend (dam of Dante and Sayajirao).

OTHERS BY DARK RONALD

The brothers Ambassador IV (£2,779 including the July Stakes) and Brown Prince II (£2,342 including the Cambridgeshire and Jockey Club Cup on successive days), who both did quite well as sires in America, were also got by Dark Ronald during his few seasons in England, as was Magpie (a minor winner in England and second in the 2,000 Guineas, and a good winner, including the Caulfield Cup, in Australia, where he became a leading sire, heading their list in 1929). Dark Ronald's sole Classic winner was Vaucleuse (1,000 Guineas and dam of Doncaster Cup winner Bongrace) but Magpie's full sister Popingaol bred 11 winners, including Pogrom (Oaks) and Book Law (St. Leger) and was the ancestress of many high-class winners bred at Lord Astor's Cliveden Stud.

In Germany Dark Ronald sired a number of their Classic winners, including Prunus (2,000 Guineas and St. Leger), who sired Oleander, one of the best horses ever bred in Germany. Lord Astor sent Plymstock (dam of the Oaks winner Pennycomequick) to Oleander when she was 21 years old and the result was Pink Flower, who belied an early sale for 18 guineas by winning a number of races and losing the 2,000 Guineas by the shortest of short heads. Pink Flower went to stud at Newmarket in 1945, his best get being Wilwyn (20 wins in England value £12,940 and the Washington International Stakes). At stud Wilwyn made no major contribution, and it has been left to others to keep the Son-in-Law line alive.

In Germany, as well as through Prunus, it has been boosted by Wallenstein and the German Derby winner Harold, from whom is descended the 1958 German Derby winner Birkhahn, who once stood at the German National Stud where Dark Ronald was based. Leading sire four times before he died at the age of 20, Birkhahn sired three fillies to win German Classics—Indra, Bravour and Literat—as well as Priamos, a good winner in France who emulated his successful sire line by producing the 1976 German Derby winner Stuyvesant. The future of the Son-in-Law line, in Germany at least, seems assured, and Birkhahn has further distinguished himself as a sire of broodmares through the German Derby winners Konigsee, Lauscher and Alpenkonig, all being out of mares by Birkhahn.

THURIO (1875)

Dark Ronald's dam Darkie was a selling plater who could not win even in that class, but as well as Dark Ronald she bred six winners including Desiree (Gimcrack Stakes). Darkie was by Thurio, who won 13 races out of 29 starts, during four years campaigning, including the Grand Prix de Paris. He was English bred (by the Derby winner Cremorne, 1869, out of Verona, 1854, a mare closely related to the Derby and Ascot Gold Cup winner Teddington), but spent his stud career between France and England. He was a disappointing sire of winners, but left an indelible mark on bloodstock history as he got in addition to Darkie, Hurry On's grandam Star, also a modest performer.

MOTHER-IN-LAW (1906)

Son-in-Law's dam Mother-in-Law won five of her nine races as a two-year-old for £2,025, but failed in all eight outings the following year and was sold to Sir Abe Bailey, for whom she produced four other winners of lowly class—Monaghan (£197), Queen Wasp (£846), My Stars (£430 on the Flat and £137 under National Hunt Rules) and After Dark (a winner in South Africa, returned to England in 1922, and ancestress of the tip-top sprinter Be Friendly). My Stars, by Sunstar, was an entire horse who came down to the lowest class of selling hurdle races, where he managed to win an event value £45, in which he carried bottom weight bar one and scraped home by a short head. At the subsequent auction he was bought in for 75 guineas. In due course he was tried at stud but was a hopeless proposition.

Mother-in-Law's dam Be Cannie (1891) won seven races as a two-year-old but then trained off and was later sold with a colt foal for a total of 30 guineas. Be Cannie's dam Reticence (1874) was a half-sister to the Derby winner and wonderful sire Hermit (1864), champion sire for seven consecutive seasons. Mother-in-Law was put to Dark Ronald as a maiden and was the first mare he covered—so Son-in-Law was the product of two virgin parents.

BOSWORTH (1926) and THE FUTURE FOR SON-IN-LAW

Son-in-Law was yet another highly influential horse inbred to a common ancestor at the third and fourth remove, in his case to the Derby and St. Leger winner and four times champion sire Blair Athol (1861).

The very domination of Son-in-Law in the staying races cited earlier may well have acted against him and his sons, for breeders, fearing they were one-paced plodders capable only of similar offspring, did not afford his line the opportunities they might have given to speedier sires. As a result his sire line went into decline.

There was never a danger of Son-in-Law's name disappearing completely, since Fair Trial, whose success was discussed in Chapter 19, was out of the Son-in-Law mare Lady Juror. And in more recent years the success of Le Levenstell highlighted the name of Son-in-Law. Le Levenstell, who began his stud career at a fee of less than £100 and was later priced at £6,000, was a better sire than racer, and his get ranged the distance scale, from My Swallow, leading two-year-old of 1970, to Levmoss, who completed the Ascot Gold Cup and Arc de Triomphe double. Le Levenstell's pedigree contained three crosses of Son-in-Law, by his sons Rustom Pasha and Knight of the Garter in the third generation and by his daughter Lady Juror in the fourth.

Le Levenstell died of cancer in 1974 at the age of 17, and neither of his best sons looks like achieving his status at stud. My Swallow is siring stock who take more time to mature than he did, and Levmoss died of a haemorrhage in 1977 at the early age of 12. However, all is not lost for Son-in-Law.

The resurgence of Son-in-Law's sire line can be attributed to a slightly unexpected source—his son Bosworth, a half-brother to Selene (dam of Hyperion) and Tranquil (1,000 Guineas and St. Leger). Bosworth spent about 13 seasons at Lord Derby's stud, but was overshadowed by Fairway and Hyperion, and towards the end of his days sank to the task of being a glorified teaser for his illustrious companions. Bosworth did sire two Classic winners—Boswell (St. Leger, and a poor sire in America and Canada) and Superbe (Irish Oaks)—as well as Filator (Cesarewitch) and Lavinia (great-grandam of the St. Leger winner Intermezzo), but he was not a great success. However, through his son Plassy he has maintained the Son-in-Law line.

PLASSY (1953)

Plassy was out of Pladda, a daughter of Scapa Flow's half-sister Rothesay Bay, and gained his finest hour when he won the Coronation Cup from Cecil (by Foxlaw) and Robin Goodfellow (by Son and Heir). Plassy retired to stud in France, and despite a certain lack of patronage got Nepenthe (French Gold Cup and second in the Grand Prix, French St. Leger and Arc de Triomphe) in his first crop. After four seasons at stud in France, where he got Ysard (French Gold Cup) and Pantomime (dam of the French Derby winner Amber), Nepenthe moved to England, where his only product of note was the filly Ash Plant, dam of the Belmont Stakes winner and successful sire Celtic Ash.

In his second crop Plassy got Arcot, a useful racer whom injury kept out of the middle-distance French Classics. Retired to stud alongside Plassy both Arcot and his sire were killed by a shell attack during the Second World War invasion of Normandy. The furtherance of the Son-in-Law line was thus left to Plassy's son Vandale (1943), who won 2,284,375 francs from five wins. Vandale, leading sire in France in 1959, got Douve (French Oaks); Fric (Coronation Cup, second in the Arc de Triomphe, and sire of French Gold Cup winner Fantomas); Tapioca (winner of 12 races, and sire of Grand Prix de Saint-Cloud winner Taneb); Taine (dual winner of the French Gold Cup); Braccio da Monte (Italian Derby, and sire of Classic winners in his native country); and Accrale (Gran Premio d'Italia, and sire of Italian Derby winner Ardale).

Pride of place among Vandale's stock, and the main branch on which Son-in-Law's line lives on, must go to Herbager (1956), winner of the French Derby and Grand Prix de Saint-Cloud, and third despite breaking down in the Arc de Triomphe. After four seasons at stud in France, Herbager was sold for a reputed £265,000 to America, where he died in 1976.

Herbager's most important get are: Grey Dawn (leading two-year-old in France, a comparative flop when raced in America at four years but now emerging as a sire of merit); Lionel (twice winner of the Grand Prix de Saint-Cloud); Sea Hawk II (Grand Prix de Saint-Cloud, retired to stud in Ireland and before export to Japan the sire of Ascot Gold Cup winner Erimo Hawk, St. Leger winner Bruni, Grand Prix de Paris winner Matahawk, and Italian 1,000 Guineas winner La Zanzara); and Appiani II (Italian Derby, at stud in Ireland and France, and sire of Arc de Triomphe and Eclipse Stakes winner Star Appeal).

The Son-in-Law line thus is still evident, even though it has required boosts from France and Italy to rejuvenate an essentially British line. The chances are that as a result of its modern members, several of whom are showing signs of being important sires, it will be around for some years yet.

BLACK DOMINO (1917)

I have strayed long enough from Pont l'Eveque's pedigree, but I make no apology for so doing, as Son-in-Law's name is familiar in every part of the world where bloodstock is bred or raced.

Barneveldt's dam Black Domino (1917) was bred in England and sent to France, where she produced three other winners, including a useful racer called Bartolo (230,000 francs). She was by the St. Leger winner but indifferent sire Black Jester (1911). Black Domino's dam Osyrua bred six winners including Dominion, who won £4,573 and carried his owner's first colours in the Derby won by his stablemate Grand Parade. Dominion, who was also third in the 2,000 Guineas, was sent to Spain and later to France but did not do well as a sire. The female line is not a distinguished one, but Black Domino has reappeared in recent years as the fifth dam of 1976 French Derby winner Youth.

BELFONDS (1922)

Pont l'Eveque's dam Ponteba was bred in France and came to England as a six-year-old carrying her Derby-winning son. She won one race in her native land and also produced Pontoon (£1,406 and dam of winners). She was by the French Derby winner Belfonds (1922), who did well at stud, siring Commanderie (a fine race filly who won the Grand Prix de Paris, French Oaks, etc.); Ligne de Fond (French 1,000 Guineas); Vendage (French Oaks); Peniche (French Oaks); Lysistrata (French Oaks) and others of class. Belfonds was sent to America in his old age. He was by Isard II, a good winner in a year when the French had a particularly good crop of horses racing. He was by Le Samaritain, who got The Tetrarch's sire Roi Herode.

Pont l'Eveque's grandam Poet's Star was bred in England, being first covered as a two-year-old (by Bachelor's Double) and sent to France at that age. There she produced nine winners, including Pervencheres (81,950 francs and second in the French 1,000 Guineas) and Priok (a gelding who won the Royal Hunt Cup and six races in France). Pervencheres is the tail female ancestress of Wild Risk (1940), by Rialto by Rabelais by St. Simon.

WILD RISK (1940)

Wild Risk won 4,859,395 francs. He was a useful handicapper on the Flat but a large portion of his earnings were due to his great ability as a hurdler. It is unusual for an ex-hurdler to succeed as a stallion, but he went to stud in France in 1946 and did exceedingly well, his most important get being Worden II; Vimy (King George VI and Queen Elizabeth Stakes, second in the French Derby); and Le Fabuleux (French Derby, spent six seasons at stud in France before export to America).

Vimy, bought by the Irish National Stud for £105,000, got the Irish St. Leger winner Vimadee, and the Eclipse Stakes winner Khalkis, as well as the dams of Busted, High Top and Linden Tree.

Worden II, who died in 1969, was a good winner in France, Italy and America, and has become an influential sire of winners and more recently of successful broodmares. His most important offspring were Armistice (Worden II's only male Classic winner who took the Grand Prix de Paris); Bon Mot III (Arc de Triomphe); Karabas (Washington International Stakes); Barquette (French Oaks); and Angers (Grand Criterium, broke a leg when favourite in the Epsom Derby).

Worden II's emergence as one of the most important broodmare sires of the moment was built on the success of Behistoun (Washington International Stakes, third in the French Derby); Dhaudevi (Grand Prix de Paris and French St. Leger); Sancy (second in the French Derby), and Rarity (second in the Champion Stakes), all being out of mares by Worden II, who went on to reach his peak in 1975. That year, Grundy (out of Word from Lundy, by Worden II) won the Derby; Juliette Marny (out of Set Free, by Worden II) the Oaks; Val de l'Orne (out of Aglae, by Armistice, by Worden II) the French Derby; and Wollow (out of Wichuraiana, by Worden II) the Dewhurst Stakes. Wollow went on the following year to win the 2,000 Guineas, Sussex Stakes and Benson & Hedges Gold Cup outright, and was awarded the Eclipse Stakes on the disqualification of Trepan.

1941

1   The Hon. Mrs. R. Macdonald-Buchanan's br c Owen Tudor (Hyperion – Mary Tudor II).
2   H.H. Senior Maharani Sahib of Kolhapur's gr c Morogoro (Felicitation – Moti Begum).
3   Sir William Jury's b c Firoze Din (Fairway – La Voulzie).
     20 ran. Time: 2 min 32 sec   Run at Newmarket.

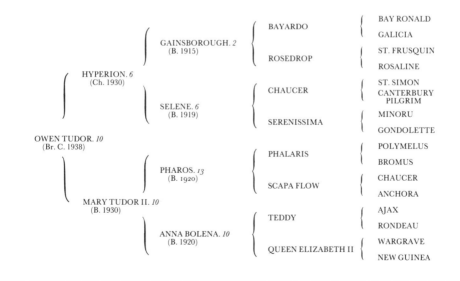

Owen Tudor won once as a two-year-old, three times as a three-year-old and twice as a four-year-old, including the substitute Gold Cup. He was unplaced in the 2,000 Guineas and St. Leger, earning in all £7,670 on the course. He went to stud as a five-year-old and his significance as the only Derby-winning son of Hyperion was discussed in Chapter 17.

MARY TUDOR II (1930)

Owen Tudor was out of the French-bred Mary Tudor II, who won the 1,000 Guineas of her native land and five other races, value 503,385 francs, besides being second in the French Oaks before importation. At stud she had five other winners to her credit and was destroyed at the age of 24 having produced her last foal three years previously. These winners include Solar Princess (£498 and third in the Oaks); Edward Tudor (£6,155); Tudor Maid (unbeaten as a two-year-old, winner at three and dam of Royal Forest, a leading sire in Brazil); and 'chaser sire King Hal. The last-named was of a quite different calibre to the others as during four seasons in training he managed to win one race value £83 in Ireland.

An interesting point about King Hal's stud career is the large number of mares he covered. In 1950 he had 60 mares, and in 1951 62.

The 'chaser sires Cottage (1918) and Steel-point (1923) also covered between them 50 and 60 thoroughbreds in some seasons. In 1938 Cottage, then 20 years old, covered 55 mares, and got the Grand National winner Sheila's Cottage. However, none of these comparative old-timers could hold a candle to Menelek (1957, by Tulyar), who between 1974 and 1976 gradually increased his annual activity from 118 mares to 119 and finally to a remarkable 135. It is unusual for predominantly Flat-race sires to cover 64 mares, as Busted did in 1976, or 67, as Green God and Jukebox did in 1975, but even Menelek's more lowly status as a 'chaser sire—and a successful one at that—cannot discredit his obvious vitality and stamina.

Mary Tudor II was by Pharos out of Anna Bolena, who won three races value 214,775 francs and produced five winners including Jumbo (452,160 francs). She was by the great French sire Teddy out of Queen Elizabeth II (1908).

Queen Elizabeth II was bred in England and sent to France as a five-year-old. She was by the Cesarewitch winner Wargrave (1898), a very indifferent sire of winners but a brother in blood to the 1906 Derby and Grand Prix de Paris victor Spearmint.

Owen Tudor is inbred to Chaucer at the third and fourth remove with more St. Simon blood farther back in his pedigree.

1942

1   Lord Derby's b c Watling Street (Fairway – Ranaï).
2   Lord Rosebery's ch c Hyperides (Hyperion – Priscilla).
3   A. E. Allnatt's b c Ujiji (Umidwar – Theresina).
      13 ran. Time: 2 min 29.6 sec   Run at Newmarket.

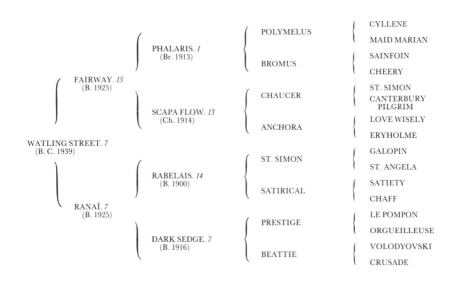

As a two-year-old Watling Street won two races, two the following season and was second in both the 2,000 Guineas and the St. Leger before retiring to stud as a four-year-old. He was not the best of his generation. Pride of place must be given to that excellent filly Sun Chariot who won the 1,000 Guineas, the Oaks and the St. Leger but did not contest the 2,000 Guineas or the Derby. The filly beat the colt every time they met. Considering the opportunities he had in Lord Derby's stud Watling Street was a disappointing stallion. After the 1952 breeding season he was sent to America and died the following year.

RANAI (1925)

He was by Fairway out of Ranaï, who was bred in France and bought by Lord Derby as a yearling at the Deauville sales. She won two races value 46,550 francs. At stud she produced nine other winners in addition to Watling Street, including Garden Path (£4,755 and the 2,000 Guineas, dam of five winners), and Correa (three races value £1,459). Correa, by the Derby and St. Leger winner Coronach, was the dam of the Ascot Stakes etc., victor Corydalis (a gelding), Gaekwar's Pride (nine wins value £2,688, to stud in Australia), and Ruthless (a non-winner by Hyperion but champion sire in New Zealand).

SARDANAPALE (1911)

Ranaï was by the English-bred but French-located sire Rabelais (see page 181) out of Dark Sedge, who did not win but produced six moderate winners. Dark Sedge was by the French racer and successful sire Prestige, who did not run in the Classics but won all his 16 races with ease. At stud he headed the French list of sires of winners, his most important son being the Grand Prix de Paris and French Derby victor Sardanapale, himself champion sire in France and getting amongst others Fiterari (1924), Apelle (1923) and Zariba (1919).

Fiterari won the Grand Prix de Paris, French 2,000 Guineas, French St. Leger etc., and did well at stud without quite reaching the top of the tree.

Zariba won 13 races value 458,975 francs and was second in the French Oaks. Retiring to the paddocks, she became one of the foundation mares of M. Marcel Boussac's famous stud. Her produce included that great filly Corrida (see page 206); Goya II (eight wins in France, England and Germany, second in the English 2,000 Guineas, champion sire in France, and sire of such as English Oaks winner Asmena, French Derby winners Sandjar and Good Luck, and French 1,000 Guineas and Oaks winner Corteira); Goyescas (£6,570 in England, won Champion Stakes, second in English 2,000 Guineas and Eclipse Stakes (twice), won 350,000 francs in France, broke a pastern when racing and was destroyed); and Abjer (£3,427, died after three seasons at stud in France) Zariba is the tail female ancestress of such as Coaraze (French Derby), Galcador (English Derby), Galgala (French 1,000 Guineas) and Crepallana (French Oaks).

APELLE (1923)

Apelle was bred in Italy where he won their Derby and 860,000 lire, in addition to £5,985 in England and 220,750 francs in France. He was then bought as a stallion for England, starting his stud career as a six-year-old at a fee of £49 per mare. On these terms he was quickly booked up and after a couple of seasons his fee was raised to £250. Amongst his first crop he sired the Grand Prix de Paris winner Cappiello (foaled in France) and the English 1,000 Guineas second and Oaks third Fur Tor.

The victories of Cappiello and one or two others in France placed Apelle's name at the head of the list of sires of winners in France for 1933. After his initial successes he did not maintain the same high standard and Anglo-Irish breeders soon lost confidence in him with the result that he was sold back to Italy in 1937. During his sojourn in England he left some fillies who turned out excellent broodmares including Good Deal (£4,194 and dam of the Derby winner Straight Deal); Lapel (£1,776, including the Irish 1,000 Guineas, and dam of Durante, a remarkably consistent and durable gelding who in seven seasons had 51 races, won 14 and was placed on 23 other occasions; Val d'Assa, £6,199 including the Royal Hunt Cup; and Cassock, Irish St. Leger); and Tofanella.

Tofanella was bred in England and sold as a yearling at the December Sales for 140 guineas to that wonderful judge of bloodstock the late Senator Tesio. For him she won three races in Italy value 43,100 lire and one in Germany worth 14,000 marks. She was not a prolific broodmare, being frequently barren, but made up for this by the quality of the comparatively few foals she did produce, the most famous being Tenerani, who won 12,345,000 lire, including the Italian Derby and St. Leger, plus £9,381 in England from the Queen Elizabeth Stakes and Goodwood Cup. He went to stud in Italy in 1949 and moved to England two years later as the property of the English National Stud.

Whilst in Italy Tenerani got the unbeaten Italian-bred but English-born Ribot (1952), whose 16 victories included the Arc de Triomphe (twice) and King George VI and Queen Elizabeth Stakes. His career will be discussed in relation to the 1973 Derby winner Morston, who was by Ribot's son Ragusa.

Tenerani's sire Bellini won the Italian Derby and St. Leger and good races in Germany. He went to stud in Italy, but the year Tenerani was foaled was taken by the Germans and eventually fell, unidentified, into Russian hands.

Bellini (1937) was by unbeaten Cavaliere d'Arpino (1926), who won six races worth 604,000 lire, and was the horse Tesio was inclined to rate the best he ever bred. Cavaliere d'Arpino was by the French-bred but Italian-located Havresac II, by Rabelais by St. Simon.

Tofanella's daughters have continued to serve the Dormello Stud well, notably through Tokamura, who won nine races including the Italian 1,000 Guineas and St. Leger. Tokamura's 13 winning produce from 15 foals included Toulouse Lautrec (seven races including the Gran Premio d'Italia, and successful sire in Italy and America); Tommaso Guidi (20 races including the Italian St. Leger, and sire of winners in Italy and America); and Theodorica (six races including the Italian 1,000 Guineas and Oaks). Another of Tofanella's daughters, Trevisana, was unbeaten in five races as a two-year-old, won the Italian St. Leger and produced three similar winners in Tiepolo, Tiziano and Tavernier, as well as the Champion Stakes and dual Italian Classic winner Tadolina. The record clearly brings out the nature of the bargain Senatore Tesio struck at Doncaster!

COLONIST II (1946)

Returning to Watling Street's pedigree, his great-grandam Beattie won in England and was sent to France, where she produced L'Avalanche, who won 90,000 francs on the course and at stud produced Rienzo. The latter won a number of races in Egypt and then held court in France, where he sired the gallant Colonist II, who won 13 races value £11,938 in Sir Winston Churchill's colours. It is most unusual for a horse who raced in Egypt to sire anything approaching the merit of Colonist II. The latter was of no great shakes at stud and descended to a fee of 25 guineas, with the Cesarewitch winner Mintmaster his most notable success on the Flat, and the grey 'chaser Stalbridge Colonist, winner of the Hennessy Gold Cup, his best under National Hunt Rules.

Beattie was by the Derby winner Volodyovski who was by Florizel II, whilst she traces in tail female to La Belle Helene. The last-named was also the third dam of Florizel II, Persimmon and Diamond Jubilee, her inbreeding being at the third and fourth remove. Watling Street is himself inbred at the third and fourth remove to St. Simon, with auxiliary lines of that horse farther back in his ancestral tree.

CHAPTER 21

1943 – 45
Ormonde and Nearco

1943

1 Miss D. Paget's b c Straight Deal (Solario – Good Deal).
2 H.H. Aga Khan's b c Umiddad (Dastur – Udaipur).
3 H.H. Aga Khan's b c Nasrullah (Nearco – Mumtaz Begum)
    23 ran. Time: 2 min 30.4 sec   Run at Newmarket.

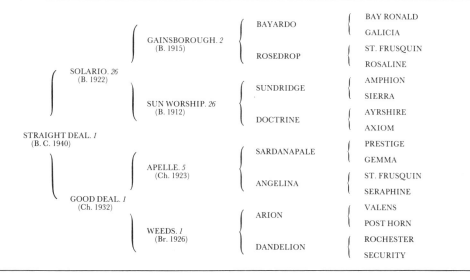

AS A TWO-YEAR-OLD Straight Deal won two races and was assessed at 9st 1lb in the Free Handicap, having above him Lady Sybil (who did not win after her first season) at 9st 7lb, Nasrullah (far more famous as a stallion than as a racer) at 9st 6lb, Umiddad (later a substitute Gold Cup winner) at 9st 5lb, and Ribbon (a good winner and narrowly beaten for the 1,000 Guineas, Oaks and St. Leger) at 9st 3lb.

As a three-year-old Straight Deal won two races in addition to the Derby—in which he provided his owner Miss Dorothy Paget with her only Classic success—and was third in the St. Leger, but it must be admitted that the colts of his generation were somewhat below the normal standard. He went to stud in England as a four-year-old, at a fee of £248, and moved to Ireland in 1953. He died in 1968.

The major feature of his stud career is that he was a better sire of fillies than colts, and more particularly of broodmares than winners. His only high-class sons were the Irish-trained Sicilian Prince (the first foreign-based horse to win the French St. Leger) and the Doncaster Cup and Queen Alexandra Stakes winner Aldborough, whereas he was responsible for three top-rate fillies—Ark Royal, winner of £12,400 in stakes as a three-year-old, and Kebab, who finished second in the Oaks to Meld and Ambiguity respectively, and Above Board, winner of the Yorkshire Oaks and Cesarewitch.

Both Ark Royal and Above Board went on to distinguish themselves as broodmares—Ark Royal as the dam of Ocean (Coronation Stakes) and Hermes (Great Voltigeur Stakes), and Above Board as the dam of Doutelle (Ormonde Stakes) and Above Suspicion (St. James's Palace Stakes and Gordon Stakes). Doutelle's death after only four seasons at stud was a serious blow, for he had made an impressive start by siring Fighting Ship (Greenham Stakes, third in St. Leger); Canisbay (Eclipse Stakes); Pretendre (Dewhurst Stakes, Timeform Gold Cup and beaten a neck in the Derby); and Dites (Cambridgeshire). The sense of loss was further highlighted in Doutelle's role as a sire of successful broodmares, among his daughters being Amicable (a useful racemare who bred the Lancashire Oaks winner Amphora); Erisca (dam of the smart two-year-old and miler Murrayfield); Hiding Place (another useful winner whose first five foals, all winners, included the Royal Hunt Cup victor Camouflage and the smart stayer Smuggler); and Ship Yard (whose son Bustino won five races in England for £116,446, including the St. Leger and Coronation Cup, and was beaten half a length by Grundy in an epic race for the King George VI and Queen Elizabeth Stakes).

Straight Deal's other important broodmare successes included Double Deal (dam of Silly Season, who won seven races worth £61,948, including the Champion Stakes, and finished second in the 2,000 Guineas); Desert Girl (dam of the Gordon Stakes winner Tiger, the Richmond Stakes and St. James' Stakes winner Berber, and Desert Beauty, one of the best three-year-old fillies of her year at a mile and a quarter); Capital Issue (dam of the Irish 2,000 Guineas winner and Epsom Derby third Kythnos); Shrubswood (whose son Dart Board finished third in both the Epsom Derby and Irish Derby); and Wayfarer (chiefly noted for Royal Avenue, who won the John Porter Stakes and Grand Prix de Printemps at the age of six).

GOOD DEAL (1932) AND HER FAMILY INFERTILITY

Straight Deal was by Solario (see page 247) out of Good Deal, who won seven races value £4,194 but produced only four foals, none of whom won, except Straight Deal. As we will see her infertility is a family trait. She was by Apelle (see page 276) out of Weeds, who won eight races value £2,852 and also had only four foals, of which Good Deal and Dent de Lion (£1,345 and produced only two foals to live) were winners. Dent de Lion had a daughter called Roman Light and she in turn produced only two live foals. Weeds was by Arion out of Dandelion, who did not win and had only the one foal. So the four generations of broodmares Roman Light, Dent de Lion, Weeds and Dandelion produced nine foals between them— or the number which one good fertile mare frequently breeds in her lifetime.

HEREDITARY INFERTILITY

The line traces in tail female to Sonsie Queen (1875) by Musket which produced a number of first-rate performers in addition to Straight Deal, such as Cameronian (2,000 Guineas and Derby), Cherimoya (Oaks), Jacopa del Sellaio (Italian Derby, 1,000 Guineas, 2,000 Guineas, St. Leger and Oaks), Lord Burgoyne (French 2,000 Guineas), Hannibal (German Derby and a great sire there), Haut Racine (Belgian Derby), Galvani by Laveno (Middle Park Plate Newmarket and £11,686), Couvert (Royal Hunt Cup), etc. I have come across other mares of this stirp with very poor foaling records and I think that the operation of hereditary infertility in it is well marked. The fact that it has produced many good winners in spite of this defect—and consequent numerical inferiority—shows that in other respects it is an excellent one. In fact what it produces is good, but it does not produce in abundance.

The whole question of hereditary infertility is one of the many things concerning which I am confident an exhaustive research into the history of the Stud Book would produce interesting and useful results. So far as I am aware the defect is not transmitted from daughter to son. I feel sure, however, that the transmission of infertility from mother to daughter, as in the Roman Light to Dandelion case, is often beyond dispute. Then again is there a tendency for infertile sires to get infertile sires? Most of The Tetrarch's sons such as Tetratema, Salmon-Trout, Ethnarch, Blue Ensign, Arch-Gift were not amiss in this respect although Caligula was. But one swallow does not make a summer. Caligula's son Prestissimo showed normal fertility. On the other hand the full brothers Clairvaux (1880) (never beaten) and Grand Prior (1888), by Hermit out of Devotion, both left very few foals, whilst the Kentucky Derby winners Twenty Grand and Assault, who were respectively son and grandson of St. Germans, were both sterile.

The whole matter requires most exhaustive investigation and the results could not be other than of great value to breeders. If it could be established that the defect is handed down in one direction or the other the wise breeder would probably eliminate contaminated strains from his stud. But if it were shown it is not hereditary, breeders could use in their studs the stock of infertile parents with greater confidence.

ARION (1915)

Weed's sire Arion won six races value £5,928 and during his 12 seasons at stud got the winners of some 50 races value £10,000. He can therefore be regarded as a stud failure, which is hardly surprising in view of his family history. The weakness of his pedigree lies in his tail female line, whilst his paternal connections were disappointing stallions. His dam Post Horn who won a small race, was a member of No. 29 Family according to the Bruce Lowe classification. The various branches of this stirp have, except for Arion, hardly produced a single horse of standing for well over 100 years. In the history of the Turf their combined efforts have resulted in only three Classic winners: Ashton (1809 St. Leger), Landscape (1816 Oaks) and Rowton (1829 St. Leger).

By the time Arion went to stud in 1921 the good-class winning strain in the family was so remote as to leave a void for many generations. Post Horn's only other winner was Serenade, who scored in two small selling plates, although she was the grandam of Vieux-Turnhout, who won 12 races value about £2,830 in Belgium. Post Horn's dam Belgravia won one race and bred seven winners— mostly of the one race apiece selling-plate class—although her son Father Vaughan won the Johannesburg Summer Handicap of £1,500. Belgravia was the only winner produced by her dam Gammer, who failed to score. So Arion, with the handicap of so many moderate animals in his family tree, could hardly be expected to make good in competition with sires whose pedigrees teemed with high-class winners.

VALENS (1906)

Arion's sire Valens (1906) was second in the St. Leger and won £4,805 in stakes but was not a great stallion. In his 15 years at stud he got the winners of some £70,000 in stakes at home and a good horse called Violoncello who won the Caulfield Cup and over £10,000 in Australia. Valens was by Bend Or's son Laveno (1892) (Jockey Club Stakes, etc. A prolific stallion but not a top-class one) out of Valenza who won nine races value £3,440. Valenza's dam Bellinzona (1889) was a half-sister to the St. Leger winner Kilwarlin (1884) and to the hardy Bendigo (1880) who withstood six years racing and won the Cambridgeshire, Kempton Park Jubilee, Hardwicke Stakes, Eclipse and other good races. Both Kilwarlin and Bendigo were better racehorses than stallions.

ROCHESTER (1907)

Dandelion's sire Rochester won the Newbury Spring Cup and other races. He spent over 20 years at stud and got the winners of some 250 races, value about £75,000. His stock were lacking in class—the average value of the races they won being £300—but they were mostly good, honest, hardy horses. He was sired by the Triple Crown winner Rock Sand (see page 166) before export. Dandelion's dam Security won six races value £2,295 and bred four winners of small account, but is the ancestress of Darius, who won £38,105 from wins that included the 2,000 Guineas and Eclipse Stakes, and sired Classic winners Pia (Oaks), Pola Bella (French 1,000 Guineas) and Varano (Italian Derby), as well as Derring-Do, a leading miler who won six races for £20,316 and sired Classic winners High Top (2,000 Guineas) and Peleid (St. Leger), along with the outstanding sprinter Huntercombe. Darius was also responsible for Dart Board, third in the Derby and beginning to make a name for himself as a stallion in Argentina.

Security was by St. Simon's son of his old age Simon Square, who won ten races as a two-year-old and lived to around 25. His stock won in the neighbourhood of

£56,000 but he left a more important imprint on the Stud Book through his daughter Simon's Shoes, whose significance was discussed in Chapter 9.

Simon Square's son Simon Pure (£5,022) was the sire of Simonella, whose daughter Allure (£2,732, third in the 1,000 Guineas) produced Infatuation, who won £9,954 before retiring in 1955 for a stud career which was distinguished only by Showdown (a high-ranking two-year-old who on export became leading sire in Australia) and Tintagel II (winner of the Ebor Handicap).

Simon Square and Rock Sand's dam Roquebrune were brother and sister in blood, their mutual sire being St. Simon, while St. Marguerite (1879) was the dam of Roquebrune and grandam of Simon Square. So Dandelion was inbred to St. Simon at her third and fourth remove and to St. Marguerite at the fourth generation, whilst Straight Deal was inbred at his fourth remove to St. Frusquin (2,000 Guineas and £33,960, and champion sire of his time).

In order to avoid confusion I had better point out that there were two different mares named Simonella connected with Simon Pure. There was his daughter, the grandam of Infatuation. This Simonella (1927) is a tail female descendant of Ornament (the dam of Sceptre). Then there was Simonella (1897) by St. Simon who was the grandam of Simon Pure (1919) and also the grandam of the Grand Prix de Paris and French St. Leger winner Fiterari (1924). This bearer of the name traces to Lady Mary (1838), a half-sister to the 1,000 Guineas and Oaks winner Mendicant (1843). As Simon Pure was by a son of St. Simon he was thus incestuously bred to that horse, his dam being a half-niece to his sire—if I may use the terms.

1944

1 Lord Rosebery's b c Ocean Swell (Blue Peter – Jiffy).
2 H.H. Aga Khan's b c Tehran (Bois Roussel – Stafaralia).
3 W. Hutchinson's br c Happy Landing (Windsor Lad – Happy Morn).
　　20 ran. Time: 2 min 31 sec　Run at Newmarket.

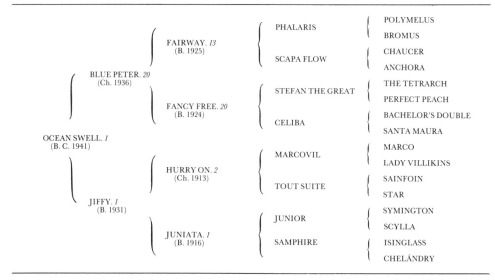

		PHALARIS	POLYMELUS
	FAIRWAY. *13*		BROMUS
	(B. 1925)	SCAPA FLOW	CHAUCER
BLUE PETER. *20*			ANCHORA
(Ch. 1936)		STEFAN THE GREAT	THE TETRARCH
	FANCY FREE. *20*		PERFECT PEACH
	(B. 1924)	CELIBA	BACHELOR'S DOUBLE
OCEAN SWELL. *1*			SANTA MAURA
(B. C. 1941)		MARCOVIL	MARCO
	HURRY ON. *2*		LADY VILLIKINS
	(Ch. 1913)	TOUT SUITE	SAINFOIN
JIFFY. *1*			STAR
(B. 1931)		JUNIOR	SYMINGTON
	JUNIATA. *1*		SCYLLA
	(B. 1916)	SAMPHIRE	ISINGLASS
			CHELÁNDRY

Ocean Swell won only one race as a two-year-old, earning 8st in the Free Handicap, 21lb below top weight. The following season he won the Derby (by a neck), Jockey Club Cup (2¼ miles) and Column Stakes (1¼ miles). He was third in the St. Leger and unplaced in the 2,000 Guineas. As a four-year-old he won the Ascot Gold Cup (2½ miles) and another race, besides being second in his other three starts. In all he won £9,095. His Derby success was scored over the Newmarket course but by the time of the Gold Cup the war was over and the race was back at its traditional Ascot setting.

He went to stud as a five-year-old in 1946 at a fee of 300 guineas, and was destroyed after an accident in 1954. Considering his pedigree and performance, he must be regarded as a failure, his only important winners in England being Fastnet Rock (Victoria Cup and Rosebery Stakes) and Sea Parrot (Yorkshire Oaks; dam of the Craven Stakes winner Shearwater and the useful middle-distance three-year-old filly Green Opal). Ocean Swell also sired St. Vincent, who showed decent two-year-old form in Britain before being sent to America, and his other important broodmare was Tahiti, dam of two Irish St. Leger winners in Barclay and Christmas Island.

Ocean Swell was from the first crop of Blue Peter (see page 257) out of Jiffy by Hurry On. Jiffy did not win but was placed several times. She distinguished herself at stud by producing nine winners, including Ocean Swell; Iona (second in Oaks and third in 1,000 Guineas); Parhelion (won £3,682 before export to Australia); and Staffa (won £2,729, and dam of 50-1 Irish Champion Stakes winner Sail Cheoil).

JUNIOR (1909)

Jiffy's dam Juniata could manage to win only a small race value £176 and produced two winners of limited ability. She was by Junior, who won ten races value £7,835 including the Atlantic Stakes at Liverpool and the Ebor, besides being third in the Ascot Gold Cup. He had a long innings at stud (21 seasons), mostly in Ireland, during which he sired the winners of 258 races value £63,123 in all, but none call for special mention. He was a well-bred horse being by Symington (1893) out of Scylla.

SCYLLA (1903)

Scylla was a full sister to the 1,000 Guineas winner Electra (1906), who produced Orpheus (£11,972; an evil-tempered horse and a bad sire) and Salamandra (won three races; second in the Oaks and third in the 1,000 Guineas; dam of the St. Leger winner Salmon-Trout). Their dam Sirenia (1895) won nine good-class races and was the ancestress of Snow Marten (Oaks) and Comrade (Grand Prix de Paris). She was a full sister to the Cesarewitch winner Hammer-kop (dam of the Derby winner Spion Kop) and a sister in blood to Llangibby (£13,907, but a disappointing stallion).

SYMINGTON (1893)

Junior's sire Symington won the Duke of York Stakes and Exeter Stakes. He went to stud at Newmarket but meeting with poor patronage there was sold to

Ireland to Mr. E. Kennedy, breeder of The Tetrarch. Amongst Symington's first crop of Irish-bred foals was Scotch Gift, who won £1,033 and earned enduring fame as the dam of the 2,000 Guineas winner and champion sire Tetratema, besides The Satrap, Arch-Gift etc.—all by The Tetrarch and all bred by Major D. McCalmont.

Symington, influential sire that he was, never sired a Classic winner, nor worked his way up to the top of the tree either in the list of sires of winners or in the parallel computation of maternal grandsires. His results can be considered as very good when it is borne in mind that he himself was far from a top-class racer and that he had, shall I say, a false start at stud. He was a 13-year-old when he made his first season in Ireland and was not a particularly long-lived horse, so nearly half his stud life was a write-off.

He was by the 2,000 Guineas and Derby winner Ayrshire, whose sire Hampton started his racing career as a selling plater but later so improved that he won the Goodwood Cup, Doncaster Cup, Great Metropolitan and other high-class events. In due course he became champion sire. His line was not carried on by any of his three Derby winners Merry Hampton, Ayrshire or Ladas, but by his handicapper son Bay Ronald.

Symington's dam Siphonia was a sister to St. Florian, who sired the Derby winner Ard Patrick (see page 165), and a half-sister to the Oaks winner Musa (dam of Mirska, who also won the Oaks). St. Florian himself was strongly inbred to Galopin, who got both his dam's sire and his sire's dam. He thus had only seven great-grandparents instead of eight.

ISINGLASS (1890)

Ocean Swell's third dam Samphire did not win but produced six winners. The only one of note was Wrack, who won ten Flat races and six hurdle races value £6,869. He went to America where he proved a highly successful stallion, amongst his daughters being Flambino (dam of the Ascot Gold Cup winner Flares, and also Omaha who was second in the corresponding race behind Quashed). Samphire was by the Triple Crown and Ascot Gold Cup winner Isinglass, whose total earnings on the course amounted to £57,455, garnered between 1892 and 1895 and an all-time British record until Tulyar came along 60 years later.

Isinglass was by Isonomy, a small horse—barely 15.2 hands—who won the Ascot Gold Cup twice, Goodwood Cup, Doncaster Cup and Cambridgeshire. He did not run in the Classics as his owner was partial to a tilt at the bookmakers and preferred to keep him a dark horse until the Cambridgeshire. He is reputed to have won £40,000 when his plan came to fruition. Isonomy did well at stud and it is a curious fact that although he sired two Triple Crown winners—Isinglass and Common—he was never champion sire.

Deadlock (1878), dam of Isinglass, had an odd history. She never ran and came into the hands of Mr. Gervas Eyre, who sold a hunter to Mr. Tatton Willoughby and Deadlock was thrown in as part of the bargain.

For her new owner she produced a couple of foals and in between times was used as a hack. She was then sold to the astute Captain Machell with a foal at foot, for

£65 the pair. Capt. Machell sold her to Colonel H. McCalmont, for whom she produced Isinglass. Her other foals were not much good, although Gervas by Trappist won £2,930. Her tail female line was very weak, a circumstance which was reflected in Isinglass's rather disappointing stud results. It is true that he got Cherry Lass (1,000 Guineas and Oaks); Glass Doll (Oaks); John O'Gaunt (second in the 2,000 Guineas and Derby, sire of Swynford); Louvois (2,000 Guineas); and Star Shoot (a great sire in America). In all Isinglass had winners of 320 races value £166,892, but as he was mated to the best mares in England and Ireland more was expected of him. He died a few days short of his 23rd year.

CHELANDRY (1894)

Samphire's dam Chelándry won the 1,000 Guineas and £13,183 before retiring to the paddocks, where she produced six winners who netted £42,675 in stakes between them, including Neil Gow (2,000 Guineas and £25,771; died after eight seasons at stud). Chelándry is one of the great tap roots of modern bloodstock the world over, and her record of success can be found in Chapter 9.

Chelándry, a half-sister to the 2,000 Guineas and Derby winner Ladas, was by Goldfinch, who was out of the same dam as the Triple Crown winner Common (1888) and the wayward filly but St. Leger winner Throstle (1891). Goldfinch won two races value £2,464 and spent three seasons at stud in England, getting in all only 27 live foals, before being sold to America.

ORMONDE (1883)

Goldfinch was unsound in his wind, as was his sire Ormonde, who began well at the first Duke of Westminster's stud in 1888 by siring Orme (tail male progenitor of Teddy, Asterus, Gold Bridge, Panorama, etc.). Ormonde was then moved to Newmarket, became ill and left only one foal the following season. At the end of that year, as a six-year-old, he was sold to Argentina for £12,000, but did no good there. His Argentinian owner sold Ormonde as a nine-year-old to America for approximately £31,250. There he remained until he was destroyed at the age of 21. During the 11 seasons he was at stud in America he left only 16 foals, so he was an expensive failure, though his American get included Ormondale (1903), who was quite a useful racer and sire. Ormonde's full sister Ornament produced Sceptre.

VAMPIRE (1889)

The first Duke of Westminster had a wonderful stud and held the distinction of both breeding and owning two Triple Crown winners, Ormonde and Flying Fox, an achievement unequalled. Flying Fox was by Orme out of Vampire by Galopin. This mating was arranged simply because the evil-tempered Vampire had foaled to one of the Duke's other stallions, and she stayed on the premises. The mating was repeated and produced Vamose (£5,604 and at stud in France with limited success); Flying Lemur (£1,325, a stud failure); Pipistrello (a non-winner and useless as a stallion); and Vane (dam of the Royal Hunt Cup and Ebor Handicap winner Weathervane, who went to New Zealand but did not do well as a sire).

The proof of the pudding is in the eating, but on the face of it the mating of Vampire to Orme appeared extremely unconventional, as she was by the highly-strung Galopin whilst his dam was by the same horse. Furthermore Orme's dam Angelica was a sister to St. Simon, who was not above showing his temper on occasion, and as I have related Vampire herself was notoriously wicked. It therefore produced a very close inbreeding to evil-dispositioned stock.

1945

1 Sir E. Ohlson's br c Dante (Nearco – Rosy Legend).
2 Lord Rosebery's b c Midas (Hyperion – Coin of the Realm).
3 Lord Astor's ch c Court Martial (Fair Trial – Instantaneous).
    27 ran. Time: 2 min 26.6 sec   Run at Newmarket.

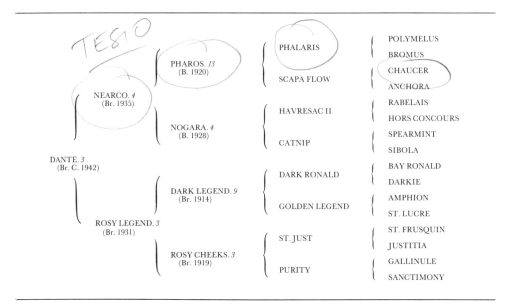

Dante was an exceptionally good-looking horse and proved an equally good racer, in spite of defective vision which ended in blindness in the early days of his stud career, though when he was syndicated as a seven-year-old at £2,500 per share this did not prevent nearly 100 applications being received. As a two-year-old he won the first race at Stockton on the opening day of the English Flat season, an hour after the season had been opened at Windsor.

During his first season he won all six races, and headed the Free Handicap. As a three-year-old he ran three times, winning twice and being second in the 2,000 Guineas, beaten a neck by Court Martial. The Derby proved to be his last race, and he retired as the first Yorkshire-trained Derby winner since 1869 and with a record of eight victories worth £11,990. He went to stud as a four-year-old and lost no time in making his mark, as among his first crop was Diableretta, the best two-year-old filly of her generation, winning eight races to the tune of £14,491, and subsequently dam of the French 1,000 Guineas winner Ginetta. Next season came Chinese Cracker (five wins worth £7,856, second in the Oaks), the useful Italian stayer Toulouse Lautrec (third in the Italian Derby and winner of seven races), and the hardy Durante (14 wins worth over £14,000), followed in 1951 by the 2,000 Guineas and Eclipse Stakes winner Darius. Successful rather than out-standing, Dante died suddenly of heart failure after the 1956 breeding season, at which point he had sired the winners of 256 races worth almost £200,000. The following year his daughter Carrozza won the Oaks, and later she produced Carromata, dam of the Grand Prix de Paris winner Matahawk.

NEARCO (1935)

When Dante provided his sire Nearco with his first Classic success in 1945, no-one could have predicted the impact that the latter would have on the destiny of subsequent thoroughbred breeding, but Nearco now ranks with the greatest names in stallion history. Putting his achievements into a sentence is difficult; but the breadth of his influence can be gauged from the fact that nine Epsom Derby winners, eight Kentucky Derby winners and five Arc de Triomphe winners came from his sire line up to 1976, and the depth can be judged from the following table of English Classic winners emanating from the same source:

2,000 GUINEAS: Nimbus, Nearula, Darius, Gilles de Retz, Royal Palace, Sir Ivor, Nijinsky, High Top, Nonoalco, Wollow (Total 10).
1,000 GUINEAS: Musidora, Belle of All, Happy Laughter, Never Too Late II, Night Off, Fleet, Humble Duty, Waterloo, Nocturnal Spree, Flying Water (Total 10).
DERBY: Dante, Nimbus, Never Say Die, Larkspur, Royal Palace, Sir Ivor, Nijinsky, Mill Reef, Roberto (Total 9).
OAKS: Masaka, Musidora, Neasham Belle, Carrozza, Never Too Late II, Noblesse, Pia (Total 7).
ST. LEGER: Sayajirao, Never Say Die, Ballymoss, Indiana, Nijinsky, Peleid (Total 6).

In 32 seasons from 1945 to 1976 the Nearco male line provided the winners of 42 English Classics, or almost 25 per cent of the total. Other statistics are equally revealing: Nearco was leading sire of winners twice, in 1947 and 1948, but was in the top eight for 15 consecutive years, from 1942 to 1956; and between 1951 and 1961 he was never lower than third in the list of sires of winning broodmares. At the time of his death in June 1957 he was responsible for the winners of 571½ races worth £427,662, and 80 of his sons were standing at studs all over the world.

Nearco, bred in Italy by Senator Federico Tesio, won 14 races from five furlongs to 1¾ miles and never knew defeat. His victories included the Italian Derby by a distance, and on his only appearance outside Italy the Grand Prix de Paris, where he beat the English Derby winner Bois Roussel, the French Derby winner Cillas, the French 1,000 Guineas and Oaks winner Feerie, and others of note. Nearco was then bought by the bookmaker and breeder Mr. Martin Benson for £60,000, an inspired piece of business to which British breeding remains indebted. Bookmakers have not always been viewed with the highest regard among the ranks of racing professionals—but Mr. Benson deserved to be judged as a most honourable exception!

Tesio's private stud book described Nearco thus: "Beautifully balanced, of perfect size and great quality. Won all 14 races as he was asked. Not a true stayer though he won up to 3,000 metres. He won these longer races by his superb class and brilliant speed." Such a horse would be expected to sire some above-average stock given normal fertility, but Nearco did more than that, though future generations may look up the records, see that he was champion sire only twice—to, say, Hyperion's six times—and deduce he was no more than a good-class sire. The overall figures show otherwise, and no other stallion can boast Nearco's international success.

Nearco, who began at stud at a fee of 400 guineas, spent the whole of his career at Newmarket. His Classic winners in England were Dante, Nimbus, Sayajirao, Masaka and Neasham Belle. His other important money-spinners were: Hafiz II (Champion Stakes); Infatuation; By Thunder! (£9,197 including the Ebor Handicap); Krakatao (£7,697) and Narrator (£6,451); while Nearctic, voted Canadian Horse of the Year at the age of four, won 21 races worth 152,384 dollars.

Strangely, three of Nearco's sons who did most to further his prowess cannot be included on a list of big-race winners—Nasrullah and Royal Charger, who blazed the Nearco trail in America, and Mossborough. All three became closely associated with the sire lines of Derby winners—Nasrullah with Never Say Die; Royal Charger with Sir Ivor, and Mossborough with Royal Palace—and will be discussed later.

Derby winners Tulyar and Arctic Prince were out of mares by Nearco, as were Rose Royale (1,000 Guineas); Saint Crespin III (Arc de Triomphe, Eclipse Stakes); Aggressor (King George VI and Queen Elizabeth Stakes); Miralgo (Hardwicke Stakes, Timeform Gold Cup); Sheshoon (Ascot Gold Cup); Ambergris (Irish Oaks); Track Spare (Middle Park Stakes); Tamerlane (St. James's Palace Stakes) and Vaguely Noble (Arc de Triomphe, and one of the most successful modern sires).

Together this makes a catalogue of phenomenal all-round success in the highest grade, and adds enormous lustre to the name of Nearco's dam, the Italian-bred Nogara, who won her local 1,000 and 2,000 Guineas. In addition to Nearco, Nogara bred five winning colts (as detailed on page 101) and the successful filly Nervesa (see page 123), making her a mare with tremendous influence in all parts of the world. In fact, almost every country where thoroughbred breeding is conducted has at stud one or more sires by Nogara's sons or from their direct male descent.

CATNIP (1910)

Nearco's grandam Catnip was by the Derby winner Spearmint. She was bred in Ireland and was a weedy sort of filly who won a £100 two-year-old race at Newcastle. She was then sold, covered by Pretty Polly's brother Cock-a-Hoop, at the 1915 Newmarket December Sales for 75 guineas to Senator Tesio, and in Italy, in addition to Nogara, she produced Nesiotes (15 wins value 615,300 lire and a successful sire in Italy); Nomellina (won eight races worth 94,075 lire in Italy, but failed as a broodmare when brought to England); and Nera do Bicci (nine wins value 212,500 lire in Italy and ancestress of many good horses).

Catnip's descendants not through Nogara include the Italian Oaks winner Neroccia and their 1,000 Guineas victress Nanncia. Catnip's dam Sibola was bred in America and came to England to win the 1,000 Guineas and other races value £5,864. She was also second in the Oaks. She went to stud in Ireland, where she produced several winners of small account, and Baltinglass (£2,154 and second in the St. Leger; a stud failure at home and exported).

SIBOLA (1896)

Sibola was by the Cambridgeshire winner The Sailor Prince (1880) who was sent to America, out of the American-bred Saluda (1883). Saluda was by Mortemer (1865) who was bred in France, came to England to win the Ascot Gold Cup and stood at stud in his native land and in America.

Nearco's pedigree which shows an inbreeding to St. Simon at the fourth remove, affords a useful illustration of the cosmopolitan aspect of modern bloodstock breeding. I will therefore recapitulate this briefly.

Nearco: Bred in Italy. At stud in England.

FIRST GENERATION:
Pharos: Bred in England. At stud in England and France but mostly in France. **Nogara:** Bred in Italy. At stud in Italy until her old age when she went to Ireland.

SECOND GENERATION
Phalaris: Bred and at stud in England. **Scapa Flow:** Bred and at stud in England. **Havresac II:** Bred in France. At stud in Italy. **Catnip:** Bred in Ireland. At stud in Italy.

THIRD GENERATION
Polymelus: Bred and at stud in England. **Bromus:** Bred and at stud in England. **Chaucer:** Bred and at stud in England. **Anchora:** Bred in Ireland. At stud in Ireland and England. **Rabelais:** Bred in England. At stud in France. **Hors Concours:** Bred and at stud in France. **Spearmint:** Bred in England. At stud in Ireland. **Sibola:** Bred in America. At stud in Ireland.

FOURTH GENERATION:
Cyllene: Bred in England. At stud in England and Argentina. **Maid Marian:** Bred and at stud in England. **Sainfoin:** Bred and at stud in England. **Cherry:** Bred and at stud in England. Went to Germany in her old age. **St. Simon:** Bred and at stud in England. **Canterbury Pilgrim:** Bred and at stud in England. **Love Wisely:** Bred and at stud in England. **Eryholme:** Bred and at stud in Ireland. **St. Simon:** Bred and at stud in England. **Satirical:** Bred and at stud in England. **Ajax:** Bred and at stud in France. **Simona:** Bred and at stud in England. **Carbine:** Bred in New Zealand. At stud in Australia but mostly in England. **Maid of the Mint:** Bred and at stud in England. **The Sailor Prince:** Bred in England. At stud in America. **Saluda:** Bred and at stud in America.

So the Italian-bred Nearco has only one Italian-bred progenitor, 20 who came from England, three from Ireland, three from France, two from America and one from New Zealand among his 30 nearest ancestors.

ANGLO-IRISH BLOOD IN FOREIGN HORSES

When foreign-bred or trained horses come to win some of the important English races, some people get the impression they are entirely foreign products, and that the nationality of their forebears is as deeply rooted in their native land as, for instance, the ancestors of the average Englishman or of the average Eskimo. This is hardly ever the case. They are cosmopolitan and an examination of their pedigrees on many occasions shows a predominance of Anglo-Irish blood.

For well over 100 years foreign horses have won our big events time and time again, yet judging from the demand for our good-class horses for export, it is patent that foreign breeders have no misgivings as to where to come to replenish their stocks. Obviously it would do the long-term future of British bloodstock no good if all our best horses were to go overseas to the highest bidder, and careful watch must be taken to ensure that we are not depleted of our best mares and potential stallions, with appropriate steps taken by breeding authorities when such an instance may arise. But positive restrictions on the export of horses would be most difficult to observe. No-one can be certain about the outcome of a stallion or a mare going to stud; one can only estimate the probabilities and work accordingly, though even then opinions may differ from one breeder to another. At the same time, British breeders should not develop a totally insular attitude. The example of Nearco on import to this country is still a potent reminder of what might be achieved by observing events overseas.

ROSY LEGEND (1931)

To return to Dante: his dam Rosy Legend was bred in France, where she won four races value 32,010 francs, and came to England as a five-year-old. Five years later Sir Eric Ohlson bought her, carrying Dante, for 3,500 guineas. As well as Dante, she bred his brother Sayajirao, who three months after Dante's success in the Derby was sold for the then record yearling price of 28,000 guineas, and five other winners of considerably lesser ability. Sayajirao won £19,343 including the St. Leger, but did not have the speed of his brother, and finished third in both the 2,000 Guineas and Derby. He retired to stud in Ireland and though siring the Irish 1,000 Guineas winner Dark Issue, proved a useful source of stamina, his best winners being Gladness (Ascot Gold Cup), Indiana (St. Leger) and Lynchris (Irish Oaks and St. Leger).

Rosy Legend's sire Dark Legend was bred in England and I have outlined on page 268 his racing and stud career, but I will now add a note about his tail female ancestors. His dam Golden Legend (1907) by Amphion also produced Golden Orb (£2,223, a very moderate stallion except for the fact he got the dam of Bellacose, the best sprinter of his time and sire of nearly £50,000 in stakes).

Amphion won 14 races during his four seasons in training and made an indelible mark on the Stud Book by siring Sundridge (see page 177) and other useful horses. He was never near the top of the tree as a sire of winners—in his best season (1899) he was 11th on the list. Golden Legend's dam St. Lucre (1901) was by St. Serf (1887) (£5,814, sire of the St. Leger winner Challacombe and the 1,000 Guineas winner Thais).

FAIRY GOLD (1896)

St. Lucre was a mare who exercised an enormous influence on the bloodstock of the world as she produced Zariba (1919) the dam of Corrida, Goyescas, Goya II, Abjer, etc., and ancestress of Galcador, Galgala, Coaraze, etc., who I have detailed on page 276. St. Lucre's dam Fairy Gold (1896) produced the great American racehorse and sire Fair Play (1905) and also Friar Rock (1913), the best racer of his time in America. After his racing days Friar Rock was sold for 50,000 dollars and proved an excellent sire in his native America. Fairy Gold was also the ancestress of Astérus (see page 328), Fastnet (see page 220), Bateau and others of merit.

BATEAU (1925)

Bateau was by Fair Play's son Man O' War, whilst her grandam was out of Fairy Gold, so she carried a heavy dosage of the blood of that great broodmare. She was one of the best racing fillies bred in America in this century, winning 11 good races and being placed in 14 others. Unhappily she proved to be a non-breeder and after many attempts to get her in foal she was used as a hack. Amongst others Man o' War (1917) got War Admiral, the sire of Never Say Die's dam. I will refer to this line of blood later (see page 353).

ROSY CHEEKS (1919)

Dante's grandam Rosy Cheeks won four races value 24,935 francs and produced four winners, including Papillon Rose, who scored nine times to the tune of 282,470 francs, and Double Rose II (1934) by Pharos who was tiny and ran in pony races but was the dam of six winners under Jockey Club Rules including Daily Mail (£2,877 and exported to one of the smaller bloodstock breeding countries). Rosy Cheeks was by St. Frusquin's son St. Just (1907), who won a whole lot of races but was not quite a top-class horse. He stood at his owner Baron de Rothschild's famous Haras de Meautry and made more of a mark as a sire of broodmares than of winners.

Dante's grandam Purity, who was a winner in France, bred five winners. She was a half-sister to the Grand Prix de Paris winner Sans Souci II (1904), who was a highly successful stallion. Their third dam Gardenia (1871) was a half-sister to the St. Leger winner Rayon d'Or (1876) (won £23,754 in England and France, sold to America); the 2,000 Guineas winner Chamant (1874); the 1,000 Guineas and Oaks winner Camelia (1873); and Wellingtonia (1869), sire of that wonderful mare Plaisanterie (1882), who won 16 races value £11,295 including the Cesarewitch and Cambridgeshire in the same year. Plaisanterie later became the dam of the Cesarewitch winner Childwick, who did not make much headway as a stallion in England, but later went to France where he sired their Derby winner Negofol and others of quality. Plaisanterie was also grandam of the St. Leger winner Tracery.

CHAPTER 22

1946 – 1947
Post-War Conditions

1946

1 J. E. Ferguson's gr c Airborne (Precipitation – Bouquet).
2 Lord Derby's b c Gulf Stream (Hyperion – Tide-way).
3 T. Lilley's ch c Radiotherapy (Hyperion – Belleva).
    17 ran. Time: 2 min 44.6 sec

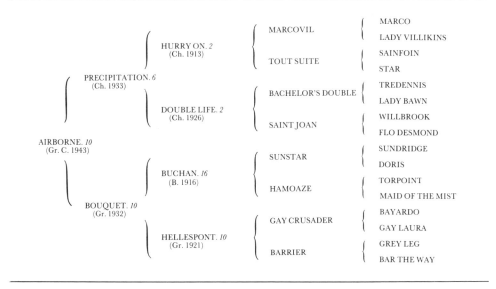

ONCE AGAIN THE Derby was back in its traditional Epsom setting. Racing was booming to a degree only equalled perhaps in the days immediately following the First World War. The world and his wife were in a spending mood. All kinds of restrictions, of a severity never previously experienced, were the order of the day and taxation had reached an unsurpassed level. But probably owing to many years of regimentation these curbs on the enjoyment of life did not seem so irksome as similar, though not so drastic conditions following the First World War, when their impact fell on a people who had been accustomed to a large measure of personal freedom.

Between 1939 and 1945 a great number of broodmares had disappeared from the Stud Book and breeding operations had been conducted on a much reduced scale. In 1945 there were 779 races run under Jockey Club Rules whilst from 1940 to 1944 the average annual number was 523. In 1946 the figure increased to 1,795, followed by 2,096 in 1947 and 2,203 in 1948. The natural result of this was that immediately after the war there were insufficient horses to meet requirements. The top grade were not affected as their production had been only slightly interrupted in the lean times, but the second- and third-class horses were fetching a great deal above their true value.

In this category races were comparatively easy to win—which is a point worth bearing in mind when assessing the ability of horses who raced at this period. I do not wish to suggest that all winners then were below par—far from it—but there were many horses who earned brackets in those two or three years who would not have done so, except possibly sometimes in lower-class races, under normal conditions. The position caused an enormous increase in the numbers of horses and mares put to stud to cope with this demand—many of them being of the most doubtful credentials. It seemed easy money for the speculator. Demand was greatly in excess of supply and demand appeared to be not very particular provided the animal for sale was sound enough to race.

This state of affairs did not last long and righted itself largely through financial and other controls which severely curtailed—in fact almost eliminated—the export trade. With this drain stemmed, before very long, instead of there being a shortage of horses available to fill races there was a surplus. The normal effects of competition had their results. Races were easily filled with the best horses of their class and the others were not needed. For the indifferent animal the bottom of the market fell out and the speculators who had started breeding immediately after the war with low-class stock soon gave up. Thus the normal state of affairs was restored.

AIRBORNE (1943)

Airborne, bought for 3,300 guineas as a yearling, ran four times as a two-year-old without winning. The following season he won five of his seven starts, including the Derby (at 50-1) and the St. Leger, earning £20,345 in stakes. He did not run for the 2,000 Guineas, nor could he be trained as a four-year-old, and he went to stud as a five-year-old in 1948 at a fee of 400 guineas. He can be deemed a failure as a stallion, with the narrowly-beaten Oaks second Silken Glider (dam of the useful stayer Alciglide) perhaps his best produce.

PRECIPITATION (1933)

Airborne's sire Precipitation, who died in March 1957, won seven races value £18,419 including the Ascot Gold Cup. He did not race as a two-year-old nor did he appear in the Classics. He went to stud as a five-year-old, and the chief results of his efforts were detailed on page 197.

The inclusion of Supreme Court among his winners—which also included Amber Flash, subsequently the dam of Oaks winner Ambiguity—cannot be wholly substantiated, since Forecourt, the dam of Supreme Court, was covered by Persian Gulf and Precipitation in the same breeding season. Although there can be no certainty as to the male parent in such cases, the date this mare was served by the respective horses, brought into line with the date of Supreme Court's birth, makes it extremely likely that the effective service was by Precipitation.

DUAL MATINGS

Taking it fine and large when a mare is covered by more than one stallion in the same year, any foal she produces next breeding season is usually the result of the last mating. But this is by no means always so. Occasionally a mare who is in foal will come into use and accept service time and time again. If the subsequent services are by a different stallion, who is the father of the foal? The relative dates are not necessarily conclusive evidence. The normal period of gestation is 11 months but it is not uncommon for this time to be prolonged by a month, and in some cases it is considerably longer.

There is the example of the nicely-named Post-Tempore (1846), who was carried by his dam for 13 months and ten days. So if a mare is covered by two different sires and apparently foals to the date of her last service, it is quite feasible that the foal is actually the result of an earlier mating and it has been carried beyond its normal time.

Usually when a mare is in foal and comes into use again, the subsequent periods of heat are confined to the same breeding season, but even this is not always so. For example, a mare called Agnostic (1916) by Senseless was covered on May 25 and May 29, 1934, by Vanoc. She was then sent home and was thought to be barren. She came into use in the middle of June the following year and was mated to Friendship on June 19, 1935. On July 3 of that year she dropped a perfectly normal colt foal. She was barren in 1936 to her union with Friendship. All this shows we cannot be dogmatic as to parentage in the case of dual matings.

DOUBLE LIFE (1926)

Precipitation's dam Double Life won six races value £5,647 including the Cambridgeshire. At stud, in addition to Precipitation, she produced Casanova (£2,555, sire of winners of about £50,000 in stakes during the ten seasons he stood at Newmarket, then exported to Norway); Persian Gulf (£2,123, including the Coronation Cup, went to stud in 1946, and sire of the 1959 Derby winner Parthia); and Doubleton (a non-winner, dam of five winners including Double Eclipse, who won £4,557 and was third in the Derby before export to South Africa, and grandam of the 1,000 Guineas, Oaks and St. Leger winner Meld).

WILLBROOK (1911)

Double Life was by Bachelor's Double (see page 207) out of Saint Joan by Willbrook. Saint Joan won two races value £262 as a two-year-old. She was then sent to Germany but fortunately returned as, through Double Life and the latter's sons, she has exercised a considerable beneficial influence on the breed. She produced three other winners but none in the same class as Double Life. Willbrook, whose name survives only through his daughter Saint Joan, was a hardy sort who ran 24 times between the ages of two and five, winning on 11 occasions including the Doncaster Cup of 1914. He was also second in the Ascot Gold Cup, went to stud in Ireland in 1917 as a six-year-old and died the following year. He was by Bend Or's son Grebe, who stood in Ireland for over 20 years at a fee ranging from five guineas to 19 guineas a mare, which indicates the valuation breeders of those days placed on his services. Willbrook's great-grandam was a half-sister to the dam of the 2,000 Guineas winner Kirkconnel (1892).

BOUQUET (1932)

Airborne's dam Bouquet never ran but produced seven other winners of no special significance except Fragrant View (six wins value £1,811 and dam of winners including Summer Rain, who was by Precipitation, and won £6,766 but died as a four-year-old). Bouquet was by Buchan (see page 224) out of Hellespont by Gay Crusader (see page 186). The best Hellespont could do on the racecourse was to win a small maiden two-year-old race. At stud she produced no winners on home courses, but two capable of scoring in one little race each on the less competitive Danish tracks.

BARRIER (1910)

Hellespont's dam Barrier won four races value £836 and foaled five winners. These included the Irish country sire Barocco (five races value £1,294), Barrulet (five races value £2,449, dam and grandam of useful winners) and Indolence (the dam of that good Continental racer and sire Prince Rose (see page 323). Barrier, who reappeared more recently as the sixth dam of the Arc de Triomphe winner Rheingold, was by the aptly named Grey Leg (1891) for he was a grey and it is from him, via Barrier, Hellespont and Bouquet that Airborne derived his coat colour.

TRANSMISSION OF GREY COAT COLOUR

There is no known case when a thoroughbred grey horse has been foaled without one or both of its parents being grey, although the union of two greys by no means necessarily produces an offspring of the same hue. Greys are born bay, brown, black, chesnut or dun, and assume the grey livery usually within the first 18 months of their lives. Year by year they become lighter in colour until they are almost white.

From time to time animals have been returned in the Stud Book as greys without apparently having a grey parent. Careful investigation into these cases has always shown that this was an error. Either it is found that one of the parents was in fact a grey, but being born of another colour was returned to the Stud Book as of that colour, and his or her subsequent change of hue was never recorded; or, and less frequently, some horses and mares of various colours have a heavy sprinkling of white hairs in their coats, usually termed Birdcatcher ticks. Sometimes animals with this latter peculiarity are incorrectly recorded as greys, whereas they are no more greys than a bay, brown or chesnut who has a white leg or blaze. In the one case the white hairs are distributed all over the body and in the other they are concentrated in one patch.

Whilst I have noted that grey horses always have one or both parents grey, and are born of various colours, there have been a very few cases of white horses. This is a different matter. These animals are born white and may be albinos, or pure whites if the colour genes of both parents are absent or have been destroyed at the time of fertilization.

Albinos are extremely rare, and until 1963 I knew of only two cases when white foals, who were not albinos, had made their appearance. One was dropped at the late Lord Middleton's stud in 1914 from two bay parents, but I have no record of their names or breeding. The other was a filly born in Germany in 1925 and with the name Woher? She was by Pergolese out of Lonja, neither of whom was grey. Her appearance caused widespread interest and most searching inquiries were made into her case. All concerned agreed that she was not an albino, that she was born white and that there was no mistake in her parentage. Moreover, if she had been a grey, from a grey parent, she would have been born of another colour.

Then in 1963 there came, for no apparent reason, three white foals—two on a stud in Kentucky sired by Ky Colonel and one in France by Murghab. The Murghab colt, named Mont Blanc II, is easily the most interesting since he was raced in England as a three-year-old, winning twice and finishing third in the Brighton Derby Trial, and went to stud in France in 1969. Mont Blanc II sired several white offspring, including some winners, of whom the sprinter White Wonder made his way to England, and the inference is that he owed his colouring to a mutated gene. Incidentally, the link between the three white foals of 1963 is that none had a grey ancestor in the first three generations of its pedigree. In Mont Blanc II's case the nearest grey was The Tetrarch, who appeared twice in the third generation of Murghab.

GREY LEG (1891)

Grey Leg was a good racer who won £6,664 including the City and Suburban and the Portland Plate (Doncaster) and started his stud career as a Queen's Premium hunter sire. He was later promoted to bloodstock and although never a top-class stallion, worked his way up until his name appeared on several occasions among the leading 20 sires of winners. Similarly to Loved One (see page 202), he was one of the last surviving tail male descendants of the Highflyer branch of the Herod sire line. Grey Leg's dam Quetta was a sister in blood to Dongola, who produced Gondolette the ancestress of Hyperion, and traced back to Blink Bonny's dam Queen Mary.

1947

1   Baron G. de Waldner's b c Pearl Diver (Vatellor – Pearl Cap).
2   H.H. Aga Khan's gr c Migoli (Bois Roussel – Mah Iran).
3   Gaekwar of Baroda's br c Sayajirao (Nearco – Rosy Legend).
     15 ran. Time: 2 min 38.4 sec

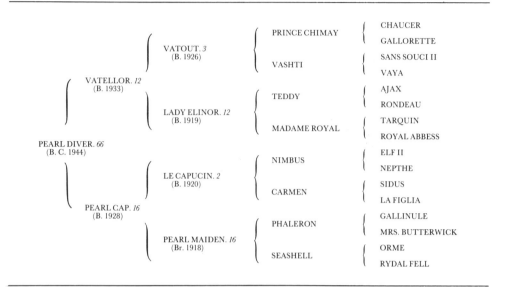

				CHAUCER
		PRINCE CHIMAY		GALLORETTE
	VATOUT. *3*			SANS SOUCI II
	(B. 1926)	VASHTI		VAYA
VATELLOR. *12*				AJAX
(B. 1933)		TEDDY		RONDEAU
	LADY ELINOR. *12*			TARQUIN
	(B. 1919)	MADAME ROYAL		ROYAL ABBESS
PEARL DIVER. *66*				ELF II
(B. C. 1944)		NIMBUS		NEPTHE
	LE CAPUCIN. *2*			SIDUS
	(B. 1920)	CARMEN		LA FIGLIA
	PEARL CAP. *16*			GALLINULE
	(B. 1928)	PHALERON		MRS. BUTTERWICK
	PEARL MAIDEN. *16*			ORME
	(Br. 1918)	SEASHELL		RYDAL FELL

As a two-year-old Pearl Diver won one race in France; as a three-year-old he won the Derby (run on a Saturday for the first time in peace time and in which Tudor Minstrel started 7-4 on but failed to stay) and one small race in France; and as a four-year-old he won the Prix d'Harcourt. His earnings on the course amounted to 1,257,425 francs in France and £9,601 in England. He was unplaced in the Grand Prix de Paris and the English St. Leger. Bought as a stallion for England, he started his new career as a five-year-old at a fee of 300 guineas but was a failure and in 1957 left for Japan, being one of the first European big-race winners to land in a country whose pre-occupation with British stallions was soon to grow.

VATELLOR (1933)

Bred in France, he is by Vatellor who was an exceptionally game, hardy horse, who won six races and was placed on 11 other occasions. Amongst his victories was the Prix du President de la Republique, whilst he was second in the French Derby and the French Gold Cup. His earnings on the course amounted to about 1½ million francs (pre-war value). He made his first season at stud in 1939 and quickly showed his merit by siring the French Oaks winner Pointe à Pitre amongst his second crop.

He later got such as My Love (English Derby, Grand Prix de Paris); Nikellora

(French 1,000 Guineas and Oaks, and Arc de Triomphe); Real (French 1,000 Guineas); Val Drake (Goodwood Cup); Fauborg II (five wins in France and Germany value about 3.7m francs, third in French 2,000 Guineas and English Derby); Felix II (Jockey Club Cup); Honorable II (Ascot Stakes); Vattel (Grand Prix de Paris); and Vamos (second in King George VI and Queen Elizabeth Stakes and sire of Grand Prix de Paris winner Vamour).

Vatellor's daughters showed their worth as broodmares and included the dams of Royal Drake (six wins, second in English Derby); L'Amiral (Prix Hocquart); Fine Top (consistent winner in France including Prix de la Fôret, Prix Boiard, etc.); Bozet (one of the best two-year-olds of his generation in France); Phil Drake (English Derby and Grand Prix de Paris); Supreme Sovereign (Lockinge Stakes); and Tello (French Gold Cup).

Taking it fine and large Vatellor proved a good-class sire who showed his *métier* in excess to his capabilities as a racer. He was by Vatout out of Lady Elinor, who won six races value 76,200 francs and also produced Lord Bob (seven wins and sire of a number of small winners in France) and Vanor (two wins). She was by Teddy out of Madame Royale, who did not win but foaled Dauphin (12 wins value 371,400 francs but did not do well as a stallion in France). Madame Royale was by the French-bred Tarquin (1901), a son of that remarkable horse The Bard, who stood barely 15 hands but succeeded in winning no less than 16 consecutive races as a two-year-old, at which age he was never beaten.

THE BARD (1883)

The following year The Bard ran second to unbeaten Ormonde in the Derby and won the Doncaster Cup, Goodwood Cup etc. His generation was an exceptionally good one as it also included the English-bred and owned Grand Prix de Paris winner Minting, who was generally considered to possess ability in excess to the average Derby winner.

Madame Royale's dam Royal Abbess was bred in England and neither won nor produced a winner, but she was a half-sister to the 1,000 Guineas winner Nun Nicer, who won £10,062 during her racing career but was a bad broodmare.

PEARL CAP (1928)

Pearl Diver's dam Pearl Cap was an exceptionally good racer, who won 11 of her 13 starts and was second in the others. Her successes included the French 1,000 Guineas and Oaks, and the Arc de Triomphe. In all she won 2,167,314 francs. At stud she at first looked like being a failure, as she produced nothing of note until she was 16 years old, when Pearl Diver appeared. Six years later she produced Seed Pearl, who was not of much account on the racecourse but became the dam of French Oaks winner Fine Pearl.

LE CAPUCIN (1920)

Pearl Cap's sire Le Capucin was a good winner at two, three, four and five years of age. His victories included the French Derby. He made an excellent start at stud, getting in his first season Le Flambeau (won over two million francs) and Pearl Cap the next year. These initial successes were not, however, sustained and although he sired plenty of useful winners, his subsequent record was not as good as appeared probable in the early days of his stud career. He was by the French Gold Cup winner Nimbus (1910), who went to Germany after a few seasons at stud in France. Nimbus was a half-brother to the Grand Prix de Paris winner Nuage (1907), who had preceded him to Germany. Nimbus was by Elf II, who came from France to win the Ascot Gold Cup of 1898 and whose English-bred dam Analogy (1874) was a full sister to that great filly Apology (1871). The last-named won the 1,000 Guineas, St. Leger, Ascot Gold Cup and Oaks.

THE DOLLAR SIRE LINE

Elf II sired some good winners including Sea Sick (1905), who dead-heated for the French Derby. Sea Sick was by Upas (1883) by Dollar (1860). Dollar's advanced age when siring this influential son should be noted. This sire line did well for close on 50 years but possibly due to the defection of Nimbus to Germany went into decline so far as high-grade sires are concerned. The existing and highly successful branch of the Dollar male line comes down as follows: Dollar (1860)—Androcles (1870)—Cambyse (1884)—Gardefeu (1895)—Chouberski (1902)—Brûleur (1910)—Ksar (1918)—Tourbillon (1928) and his various sons and grandsons who will be discussed in the next chapter.

PEARL MAIDEN (1918)

Pearl Diver's grandam Pearl Maiden was bred in England and never ran. She was sold to France for 1,000 guineas at Newmarket in 1925, 250 guineas more than she fetched the previous year and a high price considering she had an odd foot, that neither she, nor her dam, nor yet her grandam, were winners, and that she was by a most indifferent sire. Be that as it may, her subsequent record shows she was a gift, as she turned out to be one of the great broodmares of modern times.

Her produce included Muci (won six races in Italy and ancestress of Italian 1,000 Guineas winners Meda and Miranda, and Italian 2,000 Guineas winner Murghab); Pearlweed (French Derby); Pearl Cap; Bipearl (French 1,000 Guineas and other races, dam of the English Oaks second White Fox); Pearl Opal (won two races in France and dam of the Cesarewitch winner Hunter's Moon); Pearl Drop (grandam of the French 2,000 Guineas and St. Leger winner Tourment, and the Eclipse winner Flocon); Lost Pearl (won four small races in England); Pearl Ash (won three races value 150,800 francs); and Motilal (winner in France and Italy).

PHALERON (1906)

Pearl Maiden was by the beautifully-bred but wayward Phaleron, who won races worth £10,376 and was second in the 2,000 Guineas, and would probably have won more but for his unfortunate habit of putting his jockey on the floor. As a stallion he was a hopeless failure, like so many other sons of Gallinule.

MRS. BUTTERWICK (1890)

Phaleron's dam Mrs. Butterwick by St. Simon was a very small filly who won the Oaks and five other races before breeding six winners and becoming the ancestress of such as Singapore (St. Leger, sent to Brazil after 11 seasons at stud in England); Royal Alarm (£4,039, did well as a sire in Chile); Portlaw (£10,855 and a fairly good sire, particularly of broodmares); and Hyndford Bridge (second in 1,000 Guineas and dam of Falls of Clyde, who won £4,761).

KENDAL (1883)

Pearl Maiden's dam Seashell was by Orme out of Rydal Fell (a non-winner) by Ladas out of Rydal. The last-named produced several winners and also Rydal Mount, dam of the St. Leger winner Troutbeck, who was also by Ladas. Troutbeck went to America after nine seasons at stud in England but with most unsatisfactory results in both countries. Rydal was a full sister to Kendal, who won the July Stakes at Ascot and other races before becoming a most successful sire, first in England and then in Argentina, where he was sent at the age of 18. His best son was the Triple Crown winner Galtee More (1894), who went first to Russia and then to Germany. But the greatest worldwide influence Kendal exercised was through Tredennis and the latter's son Bachelor's Double, and particularly through the daughters of these two.

LADAS (1891)

Ladas was a half-brother to the 1,000 Guineas winner Chelándry, and to Gas (dam of the Derby winner Cicero). He won the 2,000 Guineas, Derby and other races worth some £18,000, besides being second in the St. Leger. He was a Classic winner above the normal standard, a very good-looking horse and a beautifully bred one, but alas his stud record was disappointing. His only Classic winners were Troutbeck and the 2,000 Guineas winner Gorgos (1903), who also met with limited success as a sire. Ladas lived until he was 23 years old, but his name appears only three times among the 20 leading sires of winners of his period.

PEARL MAIDEN'S REMARKABLE HISTORY

To end the discussion of Pearl Diver reference must be made to the remarkable early history of his grandam Pearl Maiden, whose dam Seashell was bred by the Duke of Westminster and sold to Mr. R. C. Thompson, for whom she bred three foals, all non-winners. In 1916 when she was an eight-year-old Mr. Thompson disposed of Seashell, when covered by Phaleron, to an innkeeper at Long Melford some 20 miles from Newmarket. She duly produced a filly foal afterwards named Little Winkle. That year—1917—she was again mated to Phaleron, and in 1918 produced Pearl Maiden. Seashell was then covered by Dairy Bridge, and in 1919 produced a chestnut colt by him. The colt was deformed and subsequently destroyed.

Mr. Harvey Leader, the Newmarket trainer, happened one day to be in the Long Melford district and, very likely because Phaleron stood at his stud, he went to have a look at Seashell. He found her, Little Winkle, Pearl Maiden and the deformed foal all stabled loose in a barn together, the youngsters apparently not even halter broken. Being partial to a horse deal he bought the lot for a small sum. Little Winkle was sent to Egypt as a five-year-old but does not appear to have won there. Having been bought in by Mr. Leader for 750 guineas in 1924, Pearl Maiden was sold for 1,000 guineas to France as a seven-year-old covered by Rocksavage, with the result she became a great broodmare. The Dairy Bridge foal was destroyed, and Seashell passed into the hands of Lady Wilton but had no further produce.

CHAPTER 23

1948 – 1949
Some Important French Sires

1948

1 H.H. Aga Khan's b c My Love (Vatellor – For My Love).
2 L. Volterra's b c Royal Drake (Admiral Drake – Hurry Lor).
3 H.H. Aga Khan's br c Noor (Nasrullah – Queen of Baghdad).
    32 ran. Time: 2 min 40 sec

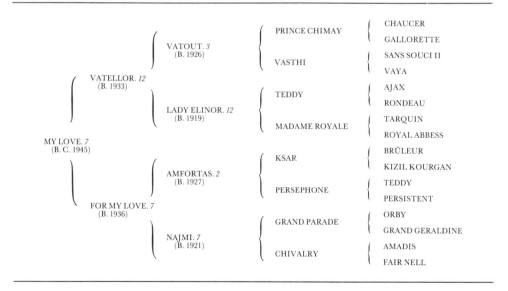

AS A TWO-YEAR-OLD My Love did not win. As a three-year-old he scored in the Prix Hocquart, English Derby (providing humiliation for Britain since there were four French-bred horses in the first six) and Grand Prix de Paris. He was unplaced in the St. Leger, which was the only other race in England besides the Derby which he contested. He did not run after his three-year-old days and started at stud in France the following year. After five seasons, he was sold to the Argentine Government who shipped him to Buenos Aires and put him up for auction when he fetched the equivalent to about £46,250 in June, 1954. He was no more of a success in Argentina than he had been in France. In the 20th century only three horses have achieved the double of the English Derby and Grand Prix de Paris, namely Spearmint (in 1906), My Love (1948) and Phil Drake (1955). Prior to that the Hungarian-bred Kisber (1876), Cremorne (1872) and the French-bred Gladiateur (1865) were the only others in Turf history to do the same. Nowadays the feat is rarely attempted, though Charlottesville, Sanctus, Reliance and Rheffic all completed the French Derby–Grand Prix de Paris double between 1960 and 1971.

My Love completed a Derby double for his sire Vatellor, following Pearl Diver the year before, the seventh time this century that a sire has been responsible for two Derby winners in succession—the others being Cyllene (1909–10), Polymelus (1915–16), Bayardo (1917–18), Hurry On (1926–27) and Blandford (1929–30 and 1934–35). However, Vatellor's feat is also the last to date, and illustrates the increased competition a stallion faces these days.

My Love's dam For My Love by Amfortas was a useful two-year-old winner in France, but at stud her only produce to score, besides My Love, was the one-race winner Mon Cheri.

AMFORTAS (1927)

Amfortas won four useful races in his native France and was second several times, including in the Ascot Gold Vase. He can roughly be described as a good staying handicapper but considerably below Classic standard. He sired a number of winners of limited ability but was far from a good stallion. Nor did his daughters, except for For My Love, make especially good broodmares. He was by Ksar, who was bought for a record 151,000 francs (£6,000) as a yearling and won the French Derby, St. Leger and Gold Cup, and the Arc de Triomphe twice. He proved himself a very good racer indeed before retiring to stud, where he did equally well.

KSAR (1918)

Ksar blotted his copy-book once when he ran badly in the Grand Prix de Paris won by the English-trained Lemonora. This mishap, rightly or wrongly, was attributed to an error of judgement on the part of his jockey, but there is little doubt that he was vastly superior to the English horse under normal conditions. The blot was particularly unfortunate as both Ksar's sire and dam were Grand Prix winners.

He himself soon showed that he was a stallion of great ability and will be remembered in Britain as the sire of Ut Majeur (Cesarewitch etc., sent to Hungary in 1936 after three or four seasons at stud in England); Le Ksar (2,000 Guineas, at stud in England from 1939 to 1943 and then exported to Argentina); Yenna (dam of the 2,000 Guineas winner Kingsway); Thor II (French Derby and Gold Cup, second in the Ascot Gold Cup); and Ukrania (French Oaks).

INFLUENCE OF TOURBILLON

Ksar's outstanding contribution to the build-up of modern blood lines was when he begat Tourbillon (1928), who won the French Derby and other races but was probably an inferior racer to Pearl Cap, whilst whether or not he was the equal of Barneveldt is a moot point.

It is as a sire not as a racehorse that Tourbillon's reputation rests. He was champion sire in France on four occasions before he died in 1954, having commenced stud duties in 1932. A picture of his influence as one of the main factors for the ascendancy of French-bred horses in the 1940s and 1950s, and even later, can be gained by reference to the following table of French Derby winners:

    1931 – Tourbillon
    1938 – Cillas—by Tourbillon
    1945 – Coaraze—by Tourbillon
    1947 – Sandjar—by Goya II by Tourbillon
    1952 – Auriban—out of Arriba, by Tourbillon
    1955 – Rapace—by Djefou by Djebel by Tourbillon
    1956 – Philius II—out of Theano, by Tourbillon
    1963 – Sanctus II—out of Sanelta, by Tourment by Tourbillon
    1966 – Nelcius—by Tenareze (by Goyama by Goya II by Tourbillon) out of Namagua (by
           Fontenay by Tornado by Tourbillon)
      *Also Tornado, Tourment and Ambiorix II (all by Tourbillon) were second.

Tourbillon's own French Derby winners differed widely in their future careers—Cillas dying after a few seasons at stud in Ireland during the war, and Coaraze (see page 206) siring La Mirambule (dam of Tambourine and Nasram) and Alba Nox (dam of the Grand Prix de Paris winner White Label II) before being sent to Brazil, where his unbeaten son Emerson distinguished himself to such purpose he was brought to Europe and sired the French Oaks winner Rescousse to signal his return to the French Classic roll. Coaraze died in 1970 at the age of 28, a year before he led Brazil's list of winning broodmares.

Tourbillon's French Derby runners-up were all successful at stud. Tourment, winner of the French 2,000 Guineas and St. Leger and beaten a head by Goya II's son Sandjar in the French Derby, sired Cerisoles (French Oaks); Seed Pearl (dam of the French Oaks winner Fine Pearl), and Sanelta (dam of the French Derby winner Sanctus II). Tornado sired Aquino II (Ascot Gold Cup, a bad-tempered horse who was sent to Poland); the French 2,000 Guineas winners Tyrone and Thymus; and Torbella (dam of the Sussex Stakes winner Carlemont). And Ambiorix II, as well as siring Fantan, dam of the very smart Ragusa, made his mark in America, where he was leading sire in 1961.

Outside the Derby, Tourbillon's other French Classic winners were Djebel (2,000 Guineas and English 2,000 Guineas, and Arc de Triomphe, whose stud career will be discussed in relation to the English Derby winner Galcador); Gaspillage (2,000 Guineas); and Esmerelda (1,000 Guineas, dam of the fine filly Coronation V, see page 23). He was also responsible for such high-class winners as Caracalla (unbeaten winner of the Ascot Gold Cup, Grand Prix de Paris etc., but a failure at stud); Turmoil (French Gold Cup, sire of the Grand Prix de Paris winner Popof from his first crop); Goya II (won races in England, France and

Germany, second in the English 2,000 Guineas, champion sire in France before export to America); and Cagire II (£13,204, sire of the Irish 2,000 Guineas winner El Toro, and Northumberland Plate winner Cagirama but little else of quality).

While the greater part of Tourbillon's worldwide influence can be attributed to Djebel, his other sons have ensured him lasting fame. Considering that for nearly the whole of his career Tourbillon was not considered a thoroughbred horse according to English classification, because of his so-called impure American ancestors, this is indeed remarkable.

Tourbillon's dam Durban was the best two-year-old of her generation in France, winning in all some 163,675 francs (pre-war value). At stud besides Tourbillon she produced Banstar, who won five races value 421,450 francs and proved a very useful sire in France, and Diademe, a good winner and second in the French 1,000 Guineas. Diademe, a sister to Tourbillon, is the grandam of Caravelle, who raced in France during the war and secured their 1,000 Guineas, Oaks and other good stakes to the value of 4.4m francs. Durban was by the French-bred English Derby winner Durbar II out of the French 1,000 Guineas winner Banshee.

BRULEUR (1910)

To return to Ksar's history, he spent some 13 years at stud in France and then went to America, where he died three years after arrival, too soon to have an effect on American breeding. Ksar was by the Grand Prix de Paris and French St. Leger winner Brûleur, a bad-tempered horse but who sired four French Derby winners—Ksar, Madrigal III (by no means the best of his generation and no good as a stallion), Hotweed (also won the Grand Prix de Paris, at stud first in France and then in England with moderate results except for the French Derby winner Pearlweed, who held court in England from 1937 to 1943 and then went to Argentina), and Pot au Feu (at stud for one year in England, two in France and afterwards in America, but a failure in all locations).

BIRIBI (1923)

Brûleur's stock had the reputation of being highly strung—a factor which is occasionally noticeable in his descendants and horses inbred to him, such as the Prix de l'Arc de Triomphe and French 1,000 Guineas winner Coronation V (1946); Goya II (1934); Aquino II (1946) and Tenareze (1953). Brûleur's sire Chouberski (1902) ran only once and won. He left a considerable imprint on the breed as in addition to Brûleur he sired La Bidouze, dam of Biribi, who won the French St. Leger and 1,388,775 francs. Biribi (by Rabelais) went to stud in France and headed their list of sires of winners, his get including Le Pacha (French Derby, Grand Prix de Paris, etc., and sire of many winners); Bipearl (French 1,000 Guineas); Perruche Bleue (French Oaks); Un Gaillard (4,453,285 francs, including the Prix du President de la Republique, and a leading sire in France); and Berikil (won 594,816 francs in France and good races in Germany as a prisoner of war; repatriated after the war and sired in France the excitable French Gold Cup winner Mat de Cocagne, who in turn sired the similar winner Tello).

CHOUBERSKI (1902)

Biribi, who had a deformed foot, went to stud in France in 1928, was sequestered by the Germans in 1940, recovered in 1946 and died that year. Chouberski was a brother in blood to Codoman (1897) who won the equivalent to about £10,000 in France and was narrowly beaten by Berrill in the 1900 Cambridgeshire. Chouberski and Codoman were scions of No. 28 Family according to the Bruce Lowe classification. The family has gradually been almost eliminated, in the process of competition, amongst good-class horses in Europe. Chouberski was by the 1898 French Derby winner and good sire Gardefeu.

Brûleur's dam Basse Terre (1899) by Omnium II produced a very good filly called Basse Pointe, who won 459,000 francs in stakes and became the dam of La Bahia (117,960 francs), Bou Jeloud (230,875 francs and a big winner in Rumania) and Bonny Boy II (Ascot Stakes, Ebor Handicap, and a moderate sire in France), whilst she is the ancestress of Taicoun (French St. Leger); Formasterus (a good winner in France and very successful sire in Brazil); Wild Mec (won about 6m francs in France and England, started well at stud in France but died after four seasons); Point à Pitre (French Oaks), and Balmoral (570,000 francs, and sire of winners in France).

BREEDERS' PREMIUMS

Incidentally, Bonny Boy II (1924) was by the Irish-bred Comrade (1917), who was sold by auction as a yearling for £25 and earned some £25,000 on the course, including the Grand Prix de Paris and the first running of the Arc de Triomphe, for his lucky owner. Comrade provides a striking example of the need for breeders' premiums!

They would serve a particularly useful purpose in the case of long-distance races. At present in the foreseeable future the trend of the yearling market provides small inducement for the breeder for sale to produce stayers. But the breeding of stayers plays now—and always has done—a vitally important part in the production of our best-class horses. Strains of staying blood are essential if our standard is to be maintained.

In these days the breeder for sale produces comfortably the highest proportion of our annual output, and if he could look forward to an increment in the shape of a breeder's premium on a successful product, it could but act as an incentive to him to breed a proportion of stayers. I think this is one of the reasons why so many good stayers have been bred in France.

If we are to maintain our worldwide bloodstock market, it is vital we should breed a proportion of stayers and I can think of no more practical inducement to breeders to do this than to reward them with premiums when they produce a successful animal.

A sale yearling is very generally a "perishable commodity". The vendor is more or less forced to take what he can get for it. It is not a practical proposition for most breeders for sale to take home, and race themselves, animals who do not make a fair price. It seems therefore hard that a breeder faced with these conditions should more or less have to give away a youngster who may well add lustre to the

Turf and be the means of introducing untold gold to the pockets of those not so situated. On the other hand there are many yearlings sold for very high prices who turn out to be useless. Taking it all in all, breeders' premiums would act as a form of payment for results to the breeder—which surely must be good.

KIZIL KOURGAN (1899)

Now to get on with Ksar's antecedents: his dam Kizil Kourgan was one of the best fillies to appear in France. Her victories included the French 1,000 Guineas and Oaks, the Grand Prix de Paris and other races. She proved herself superior to the colts of her generation. She went to the paddocks as a four-year-old and at first seemed to be only a moderate broodmare, as it was not until she was 19 years old that she threw Ksar. Her previous best produce was Kenilworth (1905), who won France's marathon race—the-then four-mile Prix Gladiateur—and did well as a stallion in Australia. But Kizil Kourgan also foaled a number of fillies who though not much good on the racecourse turned out first-class broodmares. Through these she became the ancestress of such as Kandy (English 1,000 Guineas); Kipling (Durban July Handicap and champion sire in South Africa); Kitty (Belgian Derby and St. Leger); Kitty Tchin (a good winner in France, second in their Oaks) and Kalgan (Belgian St. Leger). Kizil Kourgan was by the French Derby winner Omnium II (1892) out of the French Oaks winner Kasbah (1892).

Omnium II died after four seasons at stud. It will be noticed that he was the sire not only of Ksar's dam but also of Ksar's sire's dam, so that the horse was incestuously bred. His pedigree reads:

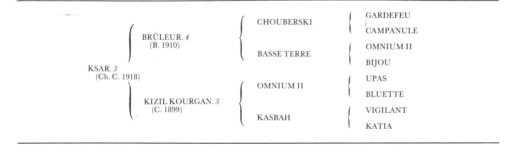

PERSEPHONE (1921)

Persephone by Teddy, the dam of Amfortas, won 34,500 francs and produced nothing else of note, but was a full sister to Fleche d'Or II, who won three races value 57,240 francs in France and foaled several useful winners in Ireland, including Golden Sovereign (£4,307 and sire of winners of about £125,000 in stakes in Australia), Nadushka (£2,386) and Zanzibar (£2,355). Persephone was sister in blood to Royal Mistress, whose son Atout Maître (1936) won the Ascot Gold Vase, Jockey Club Cup etc., before retiring to stud in England in 1941. His English-sired stock won about £40,000 in stakes and he was sent to Australia in 1952.

My Love's grandam Najmi was bred in Ireland and did not score but produced three winners in England before being sold for 165 guineas to My Love's breeder M. Leon Volterra, for whom she also bred D'Adjimir (440,485 francs). Najmi was by the Derby winner Grand Parade out of Chivalry (a non-winner who produced no winners of note) by the Doncaster Cup etc., victor Amadis (1906). Amadis had a long stud career in England, during which his fee ranged between £25 and £50, which indicates his class as a stallion. His best son was Chivalrous (1918), who won six races value £7,340 and after a few seasons at stud in England with poor results went to Australia. Amadis also sired Love Oil, dam of eight winners who collectively won £10,812 and ancestress of Medieval Knight (£8,014 and sent to Australia) and Zarathustra (Ascot Gold Cup, Irish Derby and St. Leger).

TANGIERS (1916)

Chivalry was a half-sister to Orange Girl, dam of Tangiers, who won five races value £8,258 including the Ascot Gold Cup on the disqualification of the greatly superior Buchan, who was first past the post. As I have already related, Tangiers was a very bad sire, as may be judged from the fact that when he was 16 years old and his ability as a stallion fully exposed, he fetched but 35 guineas at auction. Orange Girl was the ancestress of the Oaks winner Lovely Rosa (1933), and the hardy Wilwyn (1948).

My Love is inbred at the third and fourth remove to Teddy.

1949

1   Mrs. M. Glenister's b c Nimbus (Nearco – Kong).
2   L. Volterra's b c Amour Drake (Admiral Drake – Vers l'Aurore).
3   Lord Derby's b c Swallow Tail (Bois Roussel – Schiaparelli).
        32 ran. Time: 2 min 42 sec

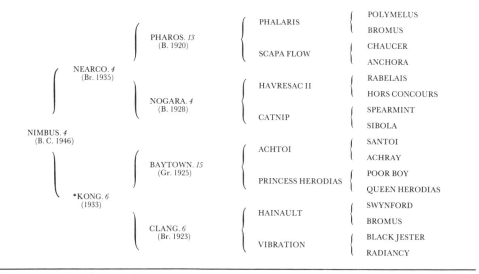

* First registered as a brown, later changed to grey.

Nimbus, who was sold by his breeder, the bookmaker William Hill, for 5,000 guineas as a yearling, won two of his five races as a two-year-old, and was placed in the remainder, earning him 8st 11lb in the Free Handicap, 8lb below the best of his age. The following season he was the unbeaten winner of four races, including the 2,000 Guineas and the Derby (in which for the first time the placings were determined by the photo-finish). He was not entered in the St. Leger but it was intended to run him in the Prix de l'Arc de Triomphe. Unfortunately during his preparation he hurt his leg and was taken out of training. He was then syndicated at a capitalized value of £140,000. His earnings on the racecourse added up to £33,075, so he paid a handsome dividend on his yearling purchase price.

He started at stud at Newmarket as a four-year-old but did not live up to expectations and in 1962 was exported to Japan, to where several of his near relations on his dam's side later found their way. His best offspring were Nucleus (second in the St. Leger but died as a four-year-old) and Nagami (winner of the Coronation Cup and races in France and Italy, and third in the 2,000 Guineas, Derby and St. Leger). Mares by Nimbus have been a better source of winners, most being noted for speed, such as the dams of Blast (good miler); Whistling Wind (unbeaten in three five-furlong races before splitting a pastern and being retired to the Irish National Stud); Daylight Robbery (July Cup), and Mountain Call (smart sprinter, sent to Australia in 1976), though Girandole showed less of the family trait by winning the Goodwood Cup.

Nimbus was by Nearco out of Kong, who started in 19 races between the ages of two and four, winning four value £2,878. She was first mated as a four-year-old when in training, and continued racing up to the end of July, scoring in the Wokingham Stakes at Ascot in mid-June.

KONG (1933) and HER COAT COLOUR

Kong, though a grey, was originally registered as a brown and was always so described during her racing career. The error was due to the fact that Kong was a brown as a foal and later changed her colour to that of her grey sire Baytown. The alteration was not notified to the Keepers of the Stud Book until some years later. As it happened this mare was much in the public eye both as a racer and as a broodmare, so the mistake was common knowledge. But supposing she had never run and fallen into the hands of a stud which either did not bother about, or were unaware of the original registration, she might easily have gone through her stud career as a brown and produced grey offspring by a non-grey sire. In due course her grey fillies might have thrown grey issue and that colour been accepted—and accepted correctly—in her line for generations. Then, may be, long afterwards someone would trace back the coat colour of his grey mare and discover that it originated, according to the records, from the mating of two non-greys.

Kong's comparatively hard racing career and the fact that she was raced when in foal did not jeopardize her stud career, as in addition to Nimbus she bred four other winners, including Grey Sovereign, who won eight races value £8,162, and Congo, the dam of Byland (11 wins value £5,838).

GREY SOVEREIGN (1948)

Grey Sovereign was a three-parts brother to Nimbus, being by Nearco's son Nasrullah out of Kong, but though his racing career did not match his illustrious Classic-winning relative—he was a horse of great speed who did not number big races among his eight victories—his influence at stud was far greater. Grey Sovereign, who had badly-shaped forelegs and was back at the knees, spoiled his racing career by behaving badly at the start on several occasions, but neither his physical nor mental faults held him back at stud, proving that sound confirmation and unfailing honesty are not always the prerequisites for breeding success.

By the time of his death in retirement in January 1976, at the age of 28, Grey Sovereign had proved himself the most influential son of Nasrullah in Britain, with several of his sons ensuring his lasting fame as a sire of sires. Grey Sovereign's chief attribute was as a source of speed and early maturity, as reflected by La Tendresse (1959) and Young Emperor (1963), who both headed the English and Irish Free Handicaps of their year. Others who demonstrated this trait were: Algaiola (Italian 1,000 Guineas); Silver King (eight wins worth £14,941 including the Wokingham Stakes, sent to America); Queensberry (Cheveley Park Stakes); Cynara (Queen Mary Stakes); Silver Tor (King's Stand Stakes, sent to South Africa); Fortino II (smart sprinter in France); Gustav (Middle Park Stakes, at stud in France, America and England); Sovereign Lord (Gimcrack Stakes, disappointing at stud and sent to Japan where he died shortly after); Janeat (Ayr Gold Cup); Matatina (smart sprinting filly); Zeddaan (brilliantly fast two-year-old, and winner of French 2,000 Guineas); Raffingora (23 wins for £23,746, including world record for electrically-timed five furlongs, exported to Japan after a few seasons at stud in England); Don II (French 2,000 Guineas, exported to Japan after standing in Ireland); Grizel (Queen Mary Stakes); and Sterling Bay (Swedish 2,000 Guineas and five other races, at stud in Ireland).

Grey Sovereign's influence on the French 2,000 Guineas for a six-year period between 1968 and 1973 is worth recording. His sons Zeddaan and Don II won in 1968 and 1969; Caro (by Fortino II) won in 1970 on a disqualification; and Zeddaan's son Kalamoun was successful in 1973.

These examples show that Grey Sovereign did get offspring who showed their prowess over further than he himself did, and others were Grey Monarch (13 wins in North America for 216,146 dollars, exported to Japan after failing at stud in Canada); Sovereign Path (eight wins, £10,751, including the Lockinge Stakes and Queen Elizabeth II Stakes); and Negus (Cambridgeshire). A mile was, however, just about the limit for the majority of Grey Sovereign's produce.

A number of Grey Sovereign's sons have had some success at stud abroad, and Fortino II (sire of Caro) and Zeddaan (sire of Kalamoun) produced at least one top-class son in Europe. But the lasting influence of Grey Sovereign rests on Sovereign Path (1956), who at the time of his death in December 1977 had bred the winners of some 750 races worth over £1m, at home and abroad. Bought for 700 guineas as a yearling and syndicated for £48,000 in 1960, Sovereign Path was leading first-season sire in 1964, when his earnings of £28,161 (from 11 races) enabled him to beat the record first-season figures set by St. Simon as long ago as

1889—a tribute to St. Simon as much as anything else in view of the falling value of the pound. Sovereign Path's chief money-spinner was the Middle Park Stakes winner Spanish Express, who was soon exported to Japan and sired the local 1976 1,000 Guineas and Oaks winner Teitaniya. Another son of Sovereign Path, Sovereign Edition, was third in that Middle Park Stakes, and has done well on export to stand at stud in New Zealand.

Sovereign Path's best racecourse representative was the filly Humble Duty (seven-length winner of the 1,000 Guineas but died prematurely), but his sons have done well at stud, notably Supreme Sovereign (sire of 1,000 Guineas winner Nocturnal Spree), Town Crier (sire of Cry of Truth, leading two-year-old filly of 1974) and most particularly Wolver Hollow (won three races in England for £34,227, including the Eclipse Stakes, and sire of Wollow, winner of nine races for £200,790, including the 2,000 Guineas, and Eclipse Stakes on a technical disqualification; and Furry Glen, winner of the Irish 2,000 Guineas).

With the promising stallion Warpath also to represent him, Sovereign Path seems sure of retaining his place as Grey Sovereign's most influential son. As for daughters of the last-named, they have also produced a distinguished list, being responsible for such as Stintino (Prix Lupin, exported to Japan); My Swanee (hardy winner of 17 races); Mabel (Yorkshire Oaks, dam of Yorkshire Oaks and Park Hill Stakes winner May Hill); Parsimony (July Cup); Recupere (French Gold Cup); Mummy's Pet (useful sprinter); Prince Tenderfoot (Coventry Stakes) and Cranberry Sauce (best three-year-old mile and a quarter filly of her year).

BAYTOWN (1925)

To return to Nimbus, his dam Kong was by Baytown, a tough customer who raced for four seasons and won ten events value £9,034, including the Irish 2,000 Guineas and Derby. He went to stud in England as a six-year-old but did not do well during the ten seasons or so he held court. He was by Achtoi (1912), who won four races value £1,644 and was third in the St. Leger. Achtoi spent some 20 seasons at stud in Ireland, during which he sired winners who netted about £107,000 in stakes between them on the Flat, and in addition got many winners under National Hunt Rules. His Flat-race progeny were not of the highest class but they were game, hardy, genuine animals who could be relied upon to do their best. Stallions who get this class of stock are not too numerous these days.

In addition to Baytown, Achtoi sired Arctic Star (11 wins value £4,714 including the Cesarewitch, a gelding); Nitsichin (nine wins value £8,515 including the Cesarewitch, Jockey Club Cup and Irish Oaks, no use as a broodmare); Cariff (£6,468 including the Irish 2,000 Guineas; at stud in Ireland with indifferent results); Achtenan (£4,000; died after a few seasons at stud in Ireland), and Poor Man (£5,750; sired a few winners in Ireland before being sent to Russia in 1936).

Achtoi was by the Ascot Gold Cup winner Santoi (1897), who provides another case in which a breeder came off badly, as he was sold for 190 guineas as a yearling and secured £11,265 in stakes besides earning a great deal in covering fees. He was at stud for 21 seasons in Ireland and although he never quite reached the top of the tree, his name was third on the list of sires of winners in 1913. He was a very bad-tempered horse both in training and when at stud but no fault could be found with his gameness.

Santoi's best son was the Ascot Gold Cup winner Santorb, who was not foaled until 1921, so was 24 years his sire's junior. Santorb only spent four seasons at stud in England before export to Hungary but he left an imprint on Turf history by siring Rockliffe, dam of Rockfel.

OLD PARENTS' SUCCESSES

Achtoi's dam Archray (1894) was bred in New Zealand. She was by Martini-Henry (1880), who was the first good racer sired by Musket (1867) after his arrival there. Musket was foaled in England and his breeder thought so little of him in his early days that he planned to shoot him. Before this was carried into effect the breeder himself died and the horse was reprieved. He won nine races including the Ascot Stakes before going to stud in England, where he met with limited patronage but sired the 2,000 Guineas winner Petronel (1877). Whilst Petronel was still a yearling Musket was sold to New Zealand, where he was at first used with half-bred mares. He was soon promoted, with wonderful results to his breed throughout the world as he sired Carbine, Trenton and others of note.

I have just mentioned that Santorb was Santoi's best son and 24 years his junior. Here is another case, for Musket died aged 19, the year after his best get Carbine was born. A study of the Stud Book brings to light many instances of a parent, male or female, producing their best in old age. It will be remembered that Kizil Kourgan was 19 years old when Ksar was foaled, Pearl Cap 16 when she threw Pearl Diver, Plucky Liège 23 at the time of Bois Roussel's birth, Cherry Tree 17 before his only good winner Cherimoya appeared, Hyperion 20 years older than Aureole, Hurry On 20 years older than Precipitation, Carbine 18 years older than his only Classic winner Spearmint, and Admiral Drake 21 years older than Phil Drake.

I can offer no explanation to this. It would be easy enough to understand in the case of a parent who has worked his or her way up during their stud career so as to be mated with superior partners in old age. But this is not so usually. In fact an old sire gets fewer mates than one in the prime of life, whilst the tendency is for an old mare who has not previously foaled anything of merit to be put to less expensive, and so probably inferior stallions. I have touched on the subject before and will only repeat that these results clearly prove that parents who have reached an age which many breeders would regard as too old, are in fact fully capable of producing their best.

The question is a vast one, and one of the things concerning which I think an extensive scientific research into the history of the Stud Book would produce useful results. Opportunity, frequency of matings of old parents compared to the others, class of partners, average results from the matings of parents both old and young, etc., would require most careful consideration. There are obviously far more mares and stallions at stud between the ages of five and 15 than from 16 upwards, so the young ones must produce more good winners, but it is the percentages balanced against opportunity which is the relative point. In this respect fashion often plays its part in determining opportunity, and it is current practice among stallion owners not to persevere with older stock that have failed to prove their worth, replacing such sires with horses who have shown their capabilities on the racecourse and trusting that they can repeat the trick at stud.

THE WAR DANCE (FRENCH) SIRE LINE

Baytown's dam Princess Herodias won one small race value £83 and bred no other winners. She was by the French-bred Poor Boy (1905), who won nearly £4,000 in stakes but lived up to his name during the 15 or so seasons he held court at stud in England. He was by the French Derby, 2,000 Guineas and Grand Prix de Paris winner, and three times French champion sire, Perth (1896). This is an almost exclusively French sire line, having its origin in Galopin's son Galliard (1880), who won the 2,000 Guineas and other good races. He rendered inestimable service to English and Irish breeders by siring Black Duchess, whose issue included Bay Ronald, Blandford and a host of important horses, but left no son to carry on his sire line at home. This honour was reserved for his French-bred son War Dance (1887), who in addition to siring Perth made a mark on the Stud Books of the world by siring Roi Herode's dam Roxelane.

Perth got the French Derby winner Alcantara II (1908), who had a long and successful stud career, though curiously enough, his two sons who maintained his line, Kantar and Pinceau, were both foaled in the same year, 1925. Kantar went to America as an 11-year-old but before leaving his native shores sired the French St. Leger winner Victrix (1934). Victrix among others got the 1946 Grand Prix de Paris winner Avenger and the German 2,000 Guineas winner Logiste (1940), who went to stud in France in 1948. Victrix is also accepted as the sire of Bastia, a maiden on the racecourse but dam of Right Royal V (winner of eight races for £102,357, including the French Derby and 2,000 Guineas, and King George VI and Queen Elizabeth Stakes; sire of the Irish Sweeps Derby winner Prince Regent, French 1,000 Guineas winner Right Away, Italian Derby winner Ruysdael, and Hardwicke Stakes winner Salvo in a top-class stud career that ended prematurely with his death at the age of 15). Bastia is also the dam of Neptunus, winner of the French 2,000 Guineas. Pinceau sired the French Derby and St. Leger winner Verso II.

GALOPIN (1872)

Far and away the most influential blood from Galopin comes through his unbeaten son St. Simon, who was foaled a year later than Galliard. There can be no doubt that no horse in the past 100 years has had a greater effect on the thoroughbred breed the world over than Galopin, so I will mention a few facts about him. He was a brown foaled in 1872 by the 2,000 Guineas winner Vedette (1854), who was not a great sire of winners. Vedette had a long innings at stud but his name rarely appears amongst the list of the 20 leading sires of winners of his

period, although he also got the Goodwood Cup winner Speculum (1865), who
founded an important sire line carried on to Santoi and his male descendants in
one stirp and to Sunstar, Mon Talisman, Admiral Drake, Phil Drake etc. in
another.

VEDETTE AND FLYING DUCHESS (1853)

Vedette's status as a sire declined to such a degree that when he was 17 years old
he fetched only 42 guineas when sold at auction—the year before his great son
Galopin was foaled. Galopin's dam Flying Duchess never advanced much beyond
selling-plate class during her racing days and had a most unsatisfactory record at
stud until she was 19 years old, as may be judged by the fact that at that age she
was sold with a foal at foot (Galopin) by auction for 100 guineas the pair. She was a
well-bred mare being by The Flying Dutchman, who won the Derby, St. Leger,
Ascot Gold Cup and later went to France to found the important Dollar sire line,
whose chief scion of recent years has been Tourbillon. Flying Duchess was a
half-sister to Besika, dam of the 2,000 Guineas winner Moslem, who secured this
race only after a dead heat with the subsequent 1,000 Guineas, Oaks and St. Leger
winner Formosa, followed by a walk-over on the withdrawal of that filly. I have
just referred to the many cases of parents producing their best in their old age and
it will be noticed that Galopin's sire was 18 years and his dam 19 years his senior,
and that neither had a satisfactory previous stud career.

Galopin was sold again as a yearling for 520 guineas. He raced for two seasons
and won ten of his 11 starts, being third in the other. His great triumph was in the
Derby of 1875, which he won with the utmost ease, but he was not entered in either
the 2,000 Guineas or St. Leger. Galopin was not raced as a four-year-old as his
owner had a weak heart and could not stand the excitement of running a horse of
this class. This precaution was very wise, as a few years later, when Galopin's son
Galliard won the 2,000 Guineas, the nervous strain was too great for Galopin's
owner Prince Batthyany, who had a heart attack and died in the Newmarket
racestand. Galopin went to stud as a four-year-old at a fee of 100 guineas, but
received very poor patronage. Considering his racing merit this is surprising and
was to some extent due to a series of articles in "The Sporting Times" decrying the
Blacklock blood to whom he was inbred. His short pedigree, which shows he had
seven great-grandparents instead of eight, reads:

		VOLTIGEUR. *2*		VOLTAIRE by BLACKLOCK
	VEDETTE. *19*	(Br. 1847)		MARTHA LYNN
	(Br. 1854)			BIRDCATCHER
		MRS. RIDGEWAY. *19*		
GALOPIN. *3*		(B. 1849)		NAN DARRELL
(Br. 1872)		THE FLYING DUTCHMAN. *3*		BAY MIDDLETON
		(Br. 1846)		BARBELLE
	FLYING DUCHESS. *3*			VOLTAIRE by BLACKLOCK
	(B. 1853)	MEROPE. *3*		
		(B. 1841)		VELOCIPEDE'S DAM

GALOPIN POORLY PATRONIZED

In his first five seasons at stud Galopin did not average more than ten or 12 mates a year, in spite of the fact that prior to his fourth season his fee was reduced to 50 guineas. Most of his partners during this period belonged to his owner and he rarely got more than one or two public mares. Luckily, amongst his second crop of foals was a very good filly called Corrie Roy, who won the Jockey Club Cup, Great Ebor, Cesarewitch, Alexandra Plate (Ascot), Manchester November Handicap and Goodwood Stakes. She was soon followed by Galliard, St. Simon, etc., and quickly they put Galopin's name on the map as a stallion, but it was not until he was an 11-year-old that he worked his way to near the top of the tree. Later he was champion sire three times.

I think the lesson to be learnt from this story is that even now writers in the Press should be extremely careful about adversely criticizing horses unless their proven shortcomings fully justify it. A young stallion's chances to make good at stud are to a large extent dependent on public support, and as no man can say with certainty whether a horse will do well or otherwise at stud, until he has proved himself, it is extremely unwise to crab him unjustifiably. Incidentally, I believe at one time Galopin was offered for sale to the German Government but they refused to buy him as they did not like his make or shape, which shows how foolish it is to be hidebound to mere good looks.

POOR BOY (1905)

To return to Poor Boy, he was a half-brother to the wonderfully hardy French filly Punta Gorda, who during four seasons in training faced the starter 73 times and won 25 races value 370,000 francs. Moreover she made her mark at stud as she was the grandam of the Cambridgeshire winner and a good winner in France Palais Royal II, and of Bois Josselyn. Palais Royal II spent about ten years at stud in France and then went to Belgium. Whilst in France he sired the dams of Nikellora (French 1,000 Guineas, Oaks and Arc de Triomphe), and Souverain (Grand Prix de Paris, French St. Leger, Ascot Gold Cup etc.). Punta Gorda was later exported to America.

PRINCE METEOR (1926)

Baytown's grandam Queen Herodias by The Tetrarch never ran and produced four winners of lowly status and also Prince Meteor by Flying Orb, whose merit cannot be accurately determined. He raced only as a two-year-old and only in Ireland. He won all his three races with the greatest ease and was without doubt head and shoulders above the others of his generation in Ireland. He then broke down and was sold to an English stud for, I believe, £10,000—a big price considering the world economic conditions then prevalent and that he was far from a fashionably-bred horse.

He went to stud as a four-year-old and in his first season sired Diana Buttercup, who won six of her eight races in her first season, but beyond that he was a total stud failure—possibly on account of the poor patronage he received. He might have done better in Ireland, where his racing reputation was very high, but on transfer to England breeders took that with a grain of salt.

HAINAULT (1914)

Clang, the grandam of Nimbus, started her racing career well by winning an event worth £586 the first time she appeared on a racecourse, but never scored again. At stud she produced five winners—including one over fences at Aintree—with a marked tendency towards hardiness rather than class. She was by Hainault, who won six races value £1,382 but was possibly a better racehorse than these figures indicate. He sustained a knee injury as a yearling and could not be trained as a two-year-old. He managed to run 13 races of moderate class as a three- and four-year-old and was only once unplaced, but all through his racing career he was continually lame from his knee injury, so did well all things considered. He died after six seasons at stud in Ireland, during which he sired the winners of 202 races value £61,281. He was by Swynford out of Bromus and so a half-brother to Phalaris.

FORFARSHIRE (1897)

Vibration by Black Jester, the great grandam of Nimbus, did not win, whilst at stud she foaled four unimportant winners. The next in his tail female line, Radiancy by Sundridge, won the Coventry Stakes at Ascot and was second, beaten a short head, in the 1,000 Guineas. Her dam Queen Elizabeth was a sister in blood to Forfarshire, the best two-year-old of his generation who later proved a very versatile horse, being placed in such diverse races as the Ascot Gold Cup (2½ miles), Goodwood Stewards' Cup (6 furlongs), Jockey Club Stakes (1¾ miles) and Royal Hunt Cup (7 furlongs 166 yards). In all he won about £8,000 in stakes. He was at stud in England for many years but was not a success.

Nimbus is inbred at his fourth remove to the mare Bromus.

CHAPTER 24

1950 – 1952
Success of Late Foals

1950

1 M. Boussac's ch c Galcador (Djebel – Pharyva).
2 W. Woodward's b c Prince Simon (Princequillo – Dancing Dora).
3 Lady Zia Wernher's ch c Double Eclipse (Hyperion – Doubleton).
    25 ran. Time: 2 min 36.8 sec

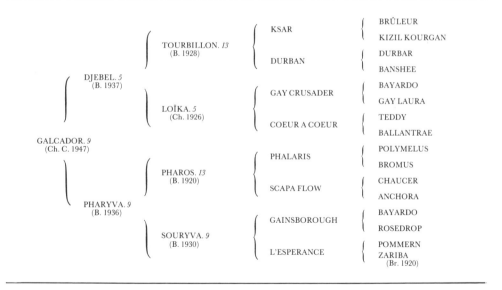

		KSAR	BRÛLEUR
	TOURBILLON. *13*		KIZIL KOURGAN
	(B. 1928)	DURBAN	DURBAR
DJEBEL. *5*			BANSHEE
(B. 1937)		GAY CRUSADER	BAYARDO
	LOÏKA. *5*		GAY LAURA
	(Ch. 1926)	COEUR A COEUR	TEDDY
GALCADOR. *9*			BALLANTRAE
(Ch. C. 1947)		PHALARIS	POLYMELUS
	PHAROS. *13*		BROMUS
	(B. 1920)	SCAPA FLOW	CHAUCER
PHARYVA. *9*			ANCHORA
(B. 1936)		GAINSBOROUGH	BAYARDO
	SOURYVA. *9*		ROSEDROP
	(B. 1930)	L'ESPERANCE	POMMERN
			ZARIBA
			(Br. 1920)

PREVIOUSLY I HAVE drawn attention to the many cases of the successes of the produce of old parents, and therefore naturally the long period of years covered in a few generations of the pedigree. Galcador affords an illustration of the opposite side of the picture as his great-great-grandam Zariba was born in 1920, so his four generations can be squeezed into 27 years. I do not think there is any particular

317

lesson to be learnt from this and merely draw attention to the fact as it is most unusual.

The Derby was Galcador's only appearance in England in a season when French-bred horses won over £200,000, and it was all the more galling for British breeders because he was not even his owner M. Boussac's first choice for the race. Pardal should have carried the famous Boussac colours at Epsom, but was injured just before the race and Galcador substituted. In his native France Galcador ran in only one Classic, the 2,000 Guineas, in which he was second to another Boussac horse Tantieme, but he scored in two other races bringing his earnings to 1,245,482 francs in France and £17,010 in England. At stud he made no impact and was eventually exported to Japan.

DJEBEL (1937)

Galcador's sire Djebel was a small, hardy horse who raced for four seasons and was probably at his best as a five-year-old, when he ran seven times without defeat. In all he won 15 races including both the English and French 2,000 Guineas and the Arc de Triomphe. It was intended that he should run for the English Derby of 1940 but owing to the general war situation it was not possible to transport him from France. I have little doubt, however, that he was considerably superior to the winner Pont l'Eveque. Instead, Djebel's astute owner, appreciating the danger, sent him to the south of France out of harm's way. There he remained until things had settled down and he resumed racing in the middle of October that year. It was then he scored in the French 2,000 Guineas, the race having been postponed from its normal date. In all his victories Djebel netted 4,153,760 francs in France and £6,115 in England, plus about 200,000 francs in place money.

PLACE MONEY IN A HORSE'S EARNINGS

Incidentally, I think it is a pity that in Britain we generally exclude place money when reckoning a horse's earnings, except as I discussed in Chapter 6, in determining the winnings of a stallion's progeny. So far as I am aware we are the only country in the world which still adheres to this, to my mind, outmoded practice. In days long since passed, when the owner of a placed horse received a purely nominal sum, things were different. Then the amount involved ranged down to £5 or sometimes less, and even in the Derby at the turn of the century the second-place money was £300.

Prince Simon's earnings for running second in the Derby under discussion came to £2,013, yet he won only two races value £2,565, which is all he is credited with winning according to our system of compiling Turf statistics, and which does not adequately reflect the fact he was reckoned in the three-year-old Free Handicap of that season to be the best of his generation trained in England by no less than half a stone.

Examples from more recently, when the level of place money has been adjusted so as to provide adequate rewards for horses not quite capable of winning the most important races, are equally illuminating. As a five-year-old Welsh Pageant won the Lockinge Stakes and Hungerford Stakes to earn £9,767, but he was also placed third in the Eclipse Stakes and Champion Stakes, bringing his owner a further £7,363 but not necessarily enhancing his racecourse reputation in terms of financial reward. Similarly, six wins in England credited to Our Mirage earned a total of £14,670, but his three placings in this country—second in the St. Leger, third in the King George VI and Queen Elizabeth Stakes, and fourth in the Derby—were together worth £24,390. And while Giacometti owed the greater part of his winnings of £58,458 to success in the Champion Stakes, he was also good enough to collect £45,633 in place money, chiefly from seconds in the 2,000 Guineas and St. Leger and a third in the Derby.

Rightly or wrongly, in a rough and ready way, a horse's racing ability is assessed on the amount of stakes he wins, and the exclusion of place money often creates a false impression. In the case of Our Mirage, his English winnings of £14,670 could place him on a par with a second-class handicapper for whom there are now numerous opportunities to win a decent prize, but this is not a correct assessment of Our Mirage's worth. Of course, it is possible for the inclusion of place money to create its own false impression, if for instance a horse picks up a good prize for finishing a distant third in a three-horse race, which is not unknown. But by and large the inclusion of place money would give a far better picture of a good-class horse's worth. If the rest of the world uses the system, why should we be out of step and run the risk of creating the impression our horses are not so good as they really are?

DJEBEL'S GET

Djebel went to stud as a six-year-old in 1943, and his impact was immediate. He was easily France's leading first-season sire in 1946, when his son Clarion headed the local Free Handicap after winning the Grand Criterium, and that year his daughter Djerba won the Cheveley Park Stakes in England. Thereafter Djebel consolidated his position, and before his death in 1958 he was leading sire in France four times.

His most important winners were: My Babu (£29,830 including the 2,000 Guineas); Hugh Lupus (£15,232, Champion Stakes and Irish 2,000 Guineas); Djelfa (French 2,000 Guineas); Coronation V (French 1,000 Guineas and Arc de Triomphe, beaten a neck for the English Oaks); Arbar (Ascot Gold Cup, second in the St. Leger); Djeddah (six races in France, Eclipse Stakes and Champion Stakes in England); Montenica (French Oaks); Djebellica (Irish Oaks); Argur (Eclipse Stakes); Apollonia (French 1,000 Guineas and Oaks); and Djelal (Prix Lupin, died as a four-year-old).

Some of Djebel's sons were not top class on the racecourse but still made their mark at stud, notably Djefou (whose principal win was in the Grand Prix de Marseilles), sire of Rapace (French Derby) in his first crop, and later Puissant Chef (French St. Leger); and Le Levandon (winner of the Portland Handicap at Doncaster), sire of Le Levenstell. However, the major influences for maintaining Djebel's success have been Clarion, My Babu, Hugh Lupus and Djeddah.

Clarion, who sired Sena II, dam of the French 1,000 Guineas winner La Sarre, was responsible for a single top-class son, Klairon, winner of six races including the French 2,000 Guineas, and a powerful influence at stud before his death in November 1977, when his produce had won more than 650 races worldwide for £1.25m. Klairon had an unusually large number of locations at stud, starting in France in 1957, being found at three separate studs in England between 1963 and 1974, and ending his life in Ireland. From his second crop in France he sired Monade (Oaks), and from his second in England came Luthier (a leading three-year-old in France, and champion sire in France in 1976, when his produce Riverqueen—the French 1,000 Guineas winner—Ashmore and Tip Moss filled the first three places in the Grand Prix de Saint-Cloud); Lorenzaccio (winner of seven races for £87,936, including the Champion Stakes), and D'Urberville (an exceptionally fast sprinter but infertile as a stallion and subsequently gelded). Among the fillies sired by Klairon were Altissima, winner of the French 1,000 Guineas; Rose Dubarry, unbeaten in three starts at two and third in the 1,000 Guineas; and Peace, who won the Blue Seal Stakes and bred three top-class runners in Intermission, Peaceful and Quiet Fling.

My Babu sired seven crops in England who won 240 races for almost £180,000, and included most notably Our Babu (four wins, £22,080, including 2,000 Guineas, retired to stud in America and moderately successful before being sent to Ireland in 1963 but made little impact and was sent to Japan); and Milesian (four wins, £8,195, a better sire than racer, sire of 293½ winners of £277,340 at the time of his death in 1971, including Falcon, a very fast two-year-old and promising sire; Partholon, winner of the Ebor Handicap, leading sire in Japan; Ionian, second in the 2,000 Guineas beaten a neck, exported to Japan; and Scissors, disqualified in the Timeform Gold Cup). In 1956 My Babu was bought for 600,000 dollars and sent to America, where he continued his success, his best produce being the Kentucky Derby second Crozier (10 wins, 641,733 dollars). My Babu was also maternal grandsire of Gamely, champion American three-year-old filly of 1967, winner of 16 races for 574,961 dollars, and dam of the English and Irish-raced Cellini.

Hugh Lupus was the sire of Pourparler (1,000 Guineas) and Hethersett, who will be discussed in relation to his Derby-winning son Blakeney. Djeddah, sire of winners in Europe and America, will be best remembered for two successful broodmares—Lalun (winner of the Kentucky Oaks and dam of Mill Reef's sire Never Bend) and Breath O'Morn (dam of the Kentucky Derby winner Proud Clarion). Another son of Djebel worth considering for his stud success abroad is Argur, who became leading sire in Argentina, where his produce included South America's leading money-spinner Arturo A.

Djebellica, one of Djebel's best racemares, proved her worth at stud as the dam of Cambremont (French 2,000 Guineas) and Torbella (Dewhurst Stakes, dam of the Sussex Stakes winner Carlemont), and grandam of Bon Mot III (Arc de Triomphe, sire of French Gold Cup winner Lassalle) and Bonami (the best French two-year-old of 1970 who broke a leg in his first race at three). Other top-class racers produced by Djebel mares included the Eclipse Stakes winner Javelot; smart sprinter French Plea; useful filly Pugnacity (dam of Derby runner-up Relkino); and the half-brothers Aegean Blue (Chester Cup) and Khalekan (second in Irish St. Leger).

LOIKA (1926)

Djebel was by Tourbillon out of Loïka by the English Triple Crown winner Gay Crusader. A fact which often escapes comment is that Loïka, when carrying Djebel, was entered for the Newmarket December Sales of 1936 but did not change hands. How different the course of Turf history would probably have been if an English or Irish breeder had bought her—or for that matter if M. Boussac had allowed Djebel to fall into German hands at the time of the invasion with the chance of non-recovery. During her racing days Loïka scored in only one small two-year-old race value 17,150 francs. At stud in addition to Djebel she produced Hierocles (six wins value 1,115,590 francs; sired some winners in France before export to America); Imperator (two wins); Phidias (six wins value 1,043,200 francs; sent to Argentina); Djask (six wins value 803,340 francs; a leading sire in South Africa); Nokka (winner and dam of winners), and Typhao (one win in France).

BALLANTRAE (1899)

Loïka's dam Coeur a Coeur by Teddy did not win and produced in all six winners of moderate ability, the most successful being Xudan (eight Flat and two hurdle races). Coeur a Coeur's dam Ballantrae, who was bred in England, was a different proposition as she won over £4,000 in stakes including the Cambridgeshire of 1902 and the Criterion Stakes the previous year. At stud she produced Mediant (Goodwood Stewards' Cup and other races, and also a good winner in America). Ballantrae was also the ancestress of Swinging (12 wins in America); Equipoise (raced 51 times in six seasons, won 29 and placed on 14 other occasions, netting in all 338,610 dollars; died after four seasons at stud but sired Kentucky Derby and Preakness Stakes winner Shut Out); Seabiscuit (failed to win in his first 17 starts in claiming races; raced until he was seven years old winning 27 events and about 428,000 dollars; a disappointing stallion), and other good horses. This is a female line which has met with great success in both France and America.

PHARYVA (1936)

Galcador's dam Pharyva by Pharos did not win but produced a string of winners including Galcador and his half-sisters Galagala (French 1,000 Guineas and 1,080,407 francs) and Windorah (1,026,507 francs in France and the Criterion Stakes in England), besides others of lesser merit. Galcador's grandam Souryva by Gainsborough was a non-winner and bred one unimportant winner. The next in the tail female line, L'Esperance, won one race in France and bred four winners of no account.

Galcador's great-great-grandam Zariba won 13 races value 458,975 francs, and was second in the French Oaks. At stud she produced Goyescas, Abjer, Corrida and Goya II. As we have seen on page 291 the line traces to Fairy Gold (1896) by Bend Or, who produced the great American racer and sire Fair Play, and is one of the most influential in the whole Stud Book.

Galcador is inbred to Bayardo at his fourth remove.

1951

1　J. McGrath's br c Arctic Prince (Prince Chevalier – Arctic Sun).
2　Lord Milford's b c Sybil's Nephew (Honeyway or Midas – Sybil's Sister).
3　F. W. Dennis's ch c Signal Box (Signal Light – Mashaq).
　　33 ran. Time: 2 min 39.4 sec

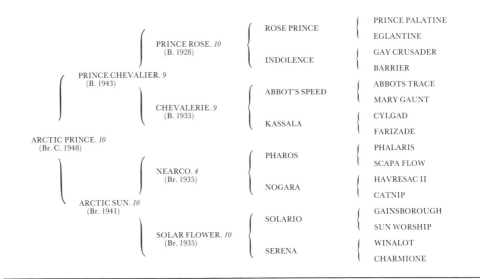

Arctic Prince ran twice as a two-year-old, winning a minor race value £207. The following season he was unplaced in the 2,000 Guineas, won what was regarded as a below-average Derby with the greatest ease by six lengths, and broke down in his next race, the King George VI and Queen Elizabeth Stakes. His victories were worth £19,593. He retired to stud in Ireland as a four-year-old at a fee of 400 guineas, and at the end of the 1956 breeding season was sold for a reputed 900,000 dollars to America, where his owner Mr. Joseph McGrath had also sent Nasrullah.

Arctic Prince's best get before export was the Eclipse Stakes winner Arctic Explorer (five wins for £14,677), who was sent to Australia as a stallion at the end of his four-year-old career and stood there from 1958 to 1964, until he crossed to New Zealand. Arctic Explorer's best son in Australia was Tobin Bronze, winner of the Caulfield Cup, third in the Washington International Stakes and sent to stud in Kentucky.

Arctic Prince was also responsible for Exar (Goodwood and Doncaster Cups, and a good winner in Italy where he retired to stud); Avon's Pride (Cesarewitch, at stud in South Africa), and Collyria (Park Hill Stakes); while he is the maternal grandsire of Approval (Observer Gold Cup); Luciano (German Derby and St. Leger); that fine filly Park Top; and the Derby winner Santa Claus. While in England Arctic Prince also sired Snow Cat, a fairly useful half-brother to the Oaks winner Carrozza who took advantage of ground and weight conditions to become the only horse to beat Alcide in 1958 and then emerged as a leading sire on export to Argentina. In America Arctic Prince sired numerous winners of no great merit, with the exception of the useful handicapper Parka, though it was during his sojourn in the States that he sired Black Prince II, third in the Derby after being bought cheaply as a foal in America.

PRINCE CHEVALIER (1943)

Arctic Prince's sire Prince Chevalier won the French Derby and six other races, and was second in the Grand Prix de Paris, French St. Leger and Arc de Triomphe. Bought as a stallion to stand in England, he commenced his duties in 1947—so Arctic Prince came from his first crop—and died of a heart attack in 1961. Three of his sons—Charlottesville, Court Harwell and Doutelle (see page 279)—have had particular importance at stud; his daughter Royal Danseuse won the Irish 1,000 Guineas, and another of his fillies La Paiva produced the outstanding racer Brigadier Gerard. Charlottesville, winner of six races including the French Derby, who went to stud in the year his sire died, will be discussed in detail in relation to his Epsom Derby-winning son Charlottown.

Court Harwell (1954) was just below the highest class, winning the Jockey Club Stakes but running his best race when second in the St. Leger. Unfortunately, he spent only three seasons at stud in Ireland before being exported to Argentina, where he became leading sire. From his first two crops in Ireland came Christmas Island, winner of the Irish St. Leger, and from his third Meadow Court, whose success in the Irish Sweeps Derby and King George VI and Queen Elizabeth Stakes largely made Court Harwell leading sire in 1965. Meadow Court had the misfortune to meet Sea-Bird II in the Epsom Derby, and was given little chance to prove his worth at stud in Europe, being shipped off to Canada after a brief spell in Ireland.

PRINCE ROSE (1928)

Prince Chevalier's sire Prince Rose was bred in England and sold to Belgium as a foal for 260 guineas. He turned out to be about the best racer in Europe of his generation, winning 14 of his 17 starts value 1,647,350 Belgian francs and 483,000 French francs. I would not go so far as to say that he was actually the best as most of his racing was done in Belgium, but he beat such as Pearl Cap, Tourbillon, Shred (second in French Derby) and Orpen (second in English Derby) at one time or another.

When he arrived in Belgium as a foal he was "naturalized". This was a system, since discontinued, whereby an imported horse could race at level weights with Belgian-bred horses provided his owner guaranteed never to export him, except under special licence to run in a specific race abroad and to return immediately. Imported horses who were not naturalized had to carry a 6lb penalty in all races in Belgium. The objective of the scheme was to encourage the importation of promising youngsters who, if they turned out well, would of necessity be available in due course to Belgian breeders for the general benefit of bloodstock in that country.

I do not know of any horse to gain international fame, except Prince Rose, who was tied by these conditions. He accordingly went to stud in Belgium in 1933 as a five-year-old, and although a few French and other breeders sent him good-class mares, his chances of making good as a top-grade sire were greatly hampered both by the scarcity of mates of the requisite calibre in Belgium and by the very low prices for bloodstock prevailing there. In fact at one time it looked as if a very good horse was likely to be wasted at stud so far as world interests were concerned. Luckily some agreement was arranged with the Belgian authorities and in 1938 Prince Rose was moved to France where he remained until his death in 1944.

In spite of such humble beginnings and a generally difficult background, Prince Rose did much more than become champion sire in both France and Belgium in 1946. He helped to put into motion three sire lines with enormous international importance. That of Prince Chevalier has been examined, but equally notable are those of his sons Princequillo (1940) and Prince Bio (1941).

Princequillo's career began in the same way as his sire's, in that he was foaled in England, was exported (to America) and began in humble surroundings (in his case claiming races). However, Princequillo progressed, numbering America's two major stamina tests, the Saratoga Cup and Jockey Club Gold Cup, among his 12 wins, though they seemed to count for little when he retired to stud at the lowly fee of 250 dollars a mare, since he received only 17 visitors in his first season. Then, however, he was moved to Mr. "Bull" Hancock's famous Claiborne Farm Stud, and as a result received more mares of much better quality. Princequillo's role as a sire of broodmares who struck gold when mated with Nasrullah and his sons was noted in Chapter 5, but Princequillo was a useful stallion in his own right, twice being leading sire of winners in America to the eight times he led the list of broodmare sires. His best get were Hill Prince (Preakness Stakes); Tambourine II (Irish Sweeps Derby); the exasperating Prince Simon; two good fillies who won the Coaching Club American Oaks, How and Cherokee Rose; and Round Table (43 wins for 1,749,869 dollars, leading sire of winners in America in 1972, and sire of the important European winners Baldric II, Apalachee, Cellini and Targowice).

Princequillo also sired Prince John, who won three races as a two-year-old, including America's richest event the Garden State Stakes, for a total of 212,818 dollars. Prince John did not race again after breaking a bone in his foot and was retired to stud as a four-year-old in 1957. He has sired Stage Door Johnny (Belmont Stakes, sire of the record weight-carrying Cesarewitch winner John Cherry); Magazine (CCA Oaks); Typecast (Man O'War Stakes); and the dams of Riverman (French 2,000 Guineas) and Caracolero (French Derby). In 1973 Prince John was responsible for the champion American two-year-old colt Protagonist, while his son Speak John sired the equivalent two-year-old filly Talking Picture.

Whereas Princequillo's fame is founded in America, Prince Bio made his name through his exploits in France, where he won their 2,000 Guineas in record time and became leading sire of winners in 1951. His most notable get were Sicambre (French Derby and Grand Prix de Paris); Le Loup Garou (French Gold Cup); Northern Light (Grand Prix de Paris); Le Petit Prince (French Derby); Sedan

(Italian Derby); Alibella (Italian 1,000 Guineas and Oaks); Pres du Feu (French 2,000 Guineas); Rose Royale II (1,000 Guineas and Champion Stakes), and Khairunissa (smart sprinter and dam of the French 2,000 Guineas winner Kalamoun).

It was through Sicambre, and the less-successful racer Prince Taj, that Prince Bio enhanced the Prince Rose sire line.

Sicambre, who died in 1975 at the age of 27, was a first-rate racehorse who won from a mile to 15 furlongs, and he passed on his racing ability to such as Sicarelle (Oaks); Pepin le Bref (second in Grand Prix de Paris and third in French Derby); Shantung (third in Epsom Derby); Spalato (Austrian Derby); Celtic Ash (Belmont Stakes, sire in America, England and Japan); Tiziano (Italian St. Leger); Ambergris (Irish Oaks, second in 1,000 Guineas and Oaks); Hermieres (French Oaks); Belle Sicambre (French Oaks); Cambremont (French 2,000 Guineas); Diatome (Washington International Stakes, second in French Derby, Grand Prix de Paris and Grand Prix de Saint-Cloud); Phaeton (Grand Prix de Paris); and Roi Dagobert (Prix Lupin). Sicambre was also the maternal grandsire of Sea-Bird II (see page 396) and La Lagune (Oaks). Shantung has since emerged as a sire of Classic-winning stock, through his produce Lacquer (Irish 1,000 Guineas), Full Dress II (1,000 Guineas) and Ginevra (Oaks), as well as the smart winners Felicio (Grand Prix de Saint-Cloud) and Saraca (Prix Vermeille), and is beginning to make a name as a potentially good broodmare sire, notably through the St. Leger winner Bruni.

Prince Taj (1954), a beautifully-bred colt the first foal of a sister to Nasrullah, was one of Prince Bio's least successful sons on the racecourse, with two small wins in France and ten unplaced outings in America to his credit, but his earnings did not do justice to his ability, at least not in Europe, where he was third in the French 2,000 Guineas and fourth in the Arc de Triomphe. He retired to stud in France in 1960 and was sent to America six years later. His most important produce in Europe are Rajput Princess (French 1,000 Guineas); La Sarre (French 1,000 Guineas); Astec (French Derby); Taj Dewan (Prix Ganay, second in 2,000 Guineas, Champion Stakes and Eclipse Stakes); Prince Tady (smart two-year-old in Italy); Reltaj (leading two-year-old in Spain); and Tidra (second in French Oaks and dam of the Grand Prix de Paris winner Tennyson). It is a pity he was not allowed to continue his career in Europe, for his export to Florida provided no such comparable success.

ROSE PRINCE (1919)

Prince Rose's sire Rose Prince was bred in France where he won five good races in addition to the Cesarewitch and Queen Alexandra Stakes in England. He had a long stud career alternated between France and England, mostly the latter, but apart from Prince Rose cannot be accounted a great success as a stallion. His son Mousson (1934) won seven races value 845,700 francs but proved a poor sire in France.

Rose Prince was by the dual Ascot Gold Cup and St. Leger winner Prince Palatine out of Eglantine by Perth. Eglantine was a winner in France and produced about ten unimportant winners. Her dam Rose de Mai (1900) won the French 1,000 Guineas, Oaks, etc., and set in motion a succession of winners which exists today, quite apart from the Prince Rose line. Rose de Mai bred, apart from Eglantine, the fillies Mauve (dam of the French Gold Cup winner Gris Perle) and Menthe. The last-named had a daughter called Melanie, who at the age of 14 is reputed to have been sold to the French army after being barren for three successive years. But before that fate befell her Menthe foaled the winner Agathe, who in turn bred the French Oaks winner Aglae Grace, dam of the Arc de Triomphe winner Soltikoff and smart miler Aigle Gris, and grandam of the French 2,000 Guineas winner Red Lord. Another offspring of Aglae Grace— whose sire Mousson was by Rose Prince—was Aglae, who finished fourth in the French Oaks but lost second place only in a finish of necks. In her turn Aglae bred the 1975 French Derby winner Val de l'Orne.

Rose de Mai's dam May Pole won the French 1,000 Guineas, whilst the next in the tail female line Merry May was full sister to Miss Mabel, whose son Trayles won the Ascot Gold Cup and Goodwood Cup.

INDOLENCE (1920)

Prince Rose's dam Indolence won one race value £267 and produced two other winners of no importance, but was a full sister to Hellespont (1921), grandam of the Derby and St. Leger winner Airborne.

To turn to the distaff side of Prince Chevalier's pedigree, his dam Chevalerie won two Flat and two hurdle races value 21,820 francs in all, so she was no great racemare. At stud she also produced Legende II, who after winning seven races value 398,390 francs and being placed in 11 other events came to England, where she foaled the Jockey Club Stakes winner Mister Cube, who scored five wins in all worth £7,715, and was sent to Argentina.

Chevalerie's sire Abbot's Speed won 8½ races value £10,806, including the Kempton Park Jubilee Handicap twice. He went to stud in France as a six-year-old in 1929 and disappeared during the German invasion, but up to that time had not shown himself to be an important sire. He was by Tracery's son Abbot's Trace (1917), who won six races value £5,473 and during his ten seasons at stud in England sired the winners of approximately 300 races worth £150,000. Although Abbot's Trace was never a very popular stallion, his comparatively early death was, I think, a loss to breeders, as he had the ability to sire stock who were better racers than he was himself, for example Abbot's Speed; Doctor Dolittle (£10,606); Supervisor (£6,879); The Macnab (£6,072); Roral (£5,959); O'Curry (£7,048); Sunny Trace (£6,872) and Justice F (the equivalent to about £19,000 in America).

Mary Gaunt by John O'Gaunt, the dam of Abbot's Speed, never ran but produced nine winners in all, who won 28½ races value £18,483 between them, whilst his grandam Quick won a small race and bred Mushroom (1908), who won £5,738, including the City and Suburban, but was a stud failure at home although he achieved some success after export to Belgium; and Kerasos (1917), who won six races value £2,979 and went to South Africa. Quick was by Cherry Tree, whose curious history will be found on page 222.

PAPPAGENO II (1935)

Prince Chevalier's grandam Kassala ran only twice and won both times. She produced five winners, including Pappageno II who was by Prince Rose and therefore brother in blood to Prince Chevalier. Pappageno II raced for four seasons in England, winning the Manchester November Handicap and six other races to the total value of £4,596. As a six-year-old he went to stud in Ireland at a fee of £9 a mare. By the end of 1953 his stock had won about 160 races worth under £30,000. At the end of that season he thus had ten crops of foals of racing age. Even allowing for wartime stakes values for the first two batches, an average winning return of £3,000 per crop cannot be regarded as other than indifferent. But in 1954 he suddenly jumped from his moderate status to a position of considerable importance, for that year alone his stock won over £15,000, and in 1955 some £22,726. It will be noted that he was 19 years old before he made his name as a sire, but unfortunately he died after the 1955 stud season.

When he was about 12 years old he was acquired by a syndicate and I think that his history demonstrates the good results of syndication. If he had remained in private ownership, it is more than probable that indifferent results in the early part of his stud career would have caused him to have been starved of mares. But members of the combine availed themselves of their free nomination rights, so except for one season (1953) Pappageno II always had a full complement of mates.

I am inclined to think there is a considerable wastage of good blood caused by the present practice of hastily deserting stallions (and mares for that matter) who do not show immediate results, and transferring patronage to the latest "new boy" just out of training. But I fear that it cannot be helped in these days of commercial breeding. It is no use sending to the sales a yearling who is by a stallion with a poor record. Nevertheless there are any number of parental stock who after a slow beginning have achieved high repute later in life.

Pappageno II's best get was the champion sprinter Pappa Fourway, who incidentally brought his breeder only £150 as a yearling, but before export to American won 12 races worth £9,891, including eight off the reel in 1955 when he was unbeaten. Some of Pappageno II's stock were remarkable for their hardiness and soundness, as for example Pappatea, who raced on the Flat for five seasons and won the Northumberland Plate and other useful races, and Tintinnabulum, who ran 85 times up to the age of 11, won 21 and was placed in half the rest, in spite of the fact he was unsound of wind from his three-year-old days.

CYLGAD (1909)

Kassala's sire Cylgad was a useful but not quite first-class racer who went to stud at a fee of 98 guineas as a four-year-old. After about ten seasons in England he was sent to France. His best get was the moderate Ascot Gold Cup winner Tangiers, but he cannot be accounted a sire above the average. He was by Cyllene out of Gadfly by Hampton. In addition to Cylgad, Gadfly produced Willia (1910), the dam of Parth, who won six races in England and was third in the 1923 Derby, besides securing the Arc de Triomphe. Parth did well at stud in France as a stallion without reaching the top class. Gadfly's dam Merry Duchess won the City and Suburban and was a half-sister to Your Grace, the grandam of Our Lassie (Oaks) and Your Majesty (a St. Leger winner who made good when sent to Argentina as a stallion after six seasons at home with poor results).

ASTERUS (1923)

Prince Chevalier's great-grandam Farizade by Sardanapale was a winner in France and third in the French Oaks. At stud she foaled four winners of moderate class. The next in the female line, Diavolezza (1911), won the French 1,000 Guineas and produced four winners, including Celerina (won about 630,000 francs and dam of the 1943 Grand Prix de Paris winner Pensbury, who broke a leg and never went to stud). Diavolezza was a half-sister to Astrella, dam of the famous Asterus, who won 486,855 francs in France, including the French 2,000 Guineas, besides the Royal Hunt Cup and Champion Stakes in England.

Asterus went to stud in France in 1928, being M. Boussac's first stallion, and died in 1939, but not before he had headed their list of sires of winners and exercised great influence on international bloodstock breeding. In view of his importance, especially to the Boussac stud and as a sire of successful broodmares, it is worth setting out his important produce in some detail:

Abjer: see page 276.

Astrophel: seven wins value 631,039 francs; a leading sire in France where his best get was the Grand Prix de Paris and French Oaks winner Bagheera.

Atys: an indifferent racer but sire of the Ascot Gold Cup winner Pan II, who produced the Grand Prix de Paris winner Altipan but after two seasons at stud in France was sold to Australia for £12,000.

Merry Boy: although not much of a racer, sired in France the English 2,000 Guineas winner Thunderhead II but achieved little of note on transfer to Ireland.

Formasterus: a good winner in France and a leading sire in Brazil.

Kipling: winner of the Durban July Handicap; champion sire in South Africa; died in 1953 after 11 seasons at stud.

Adargatis: winner of French Oaks; dam of Ardan, who scored 15 times including the French Derby, at stud in France and America, and sire of Hard Sauce (sire of Derby winner Hard Ridden); also dam of Pardal, who will be discussed in relation to his Derby-winning son Psidium.

Sanaa: dam of the French 1,000 Guineas winner Esmeralda, dam of Coronation V.

Djezima: bred eight winners; dam of Priam II, who won ten races value 3,928,950 francs and did well as a stallion in France and America; also dam of Djeddah (see page 319), and grandam of Corejada (French 1,000 Guineas and Irish Oaks).

Astronomie: one of the world's finest broodmares, whose produce included Arbele II (though not a Classic winner, one of the best fillies of her time in France); Arbar (Ascot Gold Cup); Caracalla (unbeaten winner of the Ascot Gold Cup, Arc de Triomphe and Grand Prix de Paris but a stud failure); and Marsyas II (four times winner of the French Gold Cup, along with the Goodwood Cup and Doncaster Cup).

Orlamonde: dam of the St. Leger winner Scratch II, who went on to success at stud in Argentina after three seasons in France, where Dushka (French Oaks) was his most notable produce.

Thaouka: dam of the St. Leger winner Talma II.

Asbestos II: three times champion sire in South Africa.

Astrologer: winner of the French Gold Cup.

Tant Mieux: winner of £10,200; best two-year-old of his year in England.

Jock II: good winner and stallion in France, but died after one season at stud in England; sire of Sunny Boy III (see below).

Dadji: French Gold Cup and 943,635 francs.

SUNNY BOY III (1944)

It appeared at one time that the Asterus male line in Europe would peter out in the top bracket, but in 1954, almost out of the blue, Sunny Boy III emerged as a first-class sire. He was not a great winner, although he had some very good form to his name, beating in one race Avenger and in another running second to Souverain, both winners of the Grand Prix de Paris.

Sunny Boy III went to stud as a five-year-old in 1959 at a fee of 200,000 francs, with the proviso that in the case of barrenness mares would be qualified for another nomination the following season at half fee. Amongst his second crop was the Oaks winner Sun Cap, and Sica Boy, whose success in the French St. Leger and Arc de Triomphe put Sunny Boy III at the head of the list of winning sires for 1954. Unfortunately, the Asterus male line has now resumed its precarious position, for although Sunny Boy III also sired the French Derby winner Tamanar, Midnight Boy II, who dead-heated with Saint Crespin III for the Arc de Triomphe but lost his share of first prize on an objection, and the subsequent stallions Soleil Levant and Yorick, none made much impact, outside Sica Boy's son Negresco, several times leading sire in Poland.

The importance of Sunny Boy III is now to be seen as the sire of Vali, dam of Val de Loir (French Derby, third in the Arc de Triomphe, sire of Classic winners Tennyson, La Lagune, Chaparral and Val de l'Orne, as well as the Arc de Triomphe runner-up Comtesse de Loir) and Valoris (Oaks and Irish 1,000 Guineas); and grandam of Roi Lear (French Derby). Such are the vicissitudes of bloodstock breeding!

The tail female line of Asterus and Prince Chevalier traces to Fairy Gold, whose details can be found on page 291.

ARCTIC SUN (1941)

Arctic Prince's dam Arctic Sun by Nearco raced only as a two-year-old and only in Ireland, where she proved herself the best juvenile of her generation, winning five of her seven starts and £2,438. On retiring to the paddocks she also bred the winners Arctic Blue (grandam of the hardy Lockinge Stakes winner Bluerullah) and Windsor Sun, as well as Arctic Prince's unraced sister Certosa (dam of the useful two-year-old Tulyartos, who finished second in the Eclipse Stakes and was exported to Japan).

Arctic Sun's dam Solar Flower by Solario was a first-class racer and brood-mare. She won four races value £10,243, including the Coronation Stakes at Ascot, besides being third in both the 1,000 Guineas and Oaks. At stud she produced five other foals including Peter Flower (£11,408; spent three seasons at stud in England and sent to America in 1954) and Solar Slipper (£5,822 and third in the St. Leger; went to stud in Ireland in 1950, and sire of the Irish Derby winner and Epsom Derby second Panaslipper; sent to America in 1956; also maternal grandsire of Derby winner Royal Palace).

The next in the tail female line Serena did not win and bred six other winners but none in any way comparable to Solar Flower. She also foaled the unraced Arctic Star (1942), sire of the Derby third Roistar but whose chief successes came after his death in 1958, through Fidalgo (Chester Vase and Irish Derby, second in the Epsom Derby and St. Leger; sired the Ebor Handicap winner Twelfth Man and the useful middle-distance horse Chicago before export to Japan in 1971), and Arctic Storm (Irish 2,000 Guineas and Champion Stakes; sire of many winners but none of outstanding merit). Arctic Star also sired Arctic Time (fourth in the King George VI and Queen Elizabeth Stakes; sire of Irish St. Leger winner Arctic Vale; exported to South Africa as a 16-year-old), and Arctic Slave (a good racehorse when in the mood and later a leading National Hunt sire), and was maternal grandsire of the Irish 1,000 Guineas winner Royal Danseuse.

Serena was by Winalot who as we have already seen was by Son-in-Law and won £8,964 before a fairly successful stud career. Winalot's dam Gallenza did not win but produced six other winners of no special merit except Perce-Neige (1916).

Perce-Neige won two small races and bred the Oaks winner Rose of England and Winterhalter. The latter won the Coronation Cup and six other races, but was a disappointing sire at first in England and later in France. Rose of England was the dam of the St. Leger winner Chulmleigh, who went to Argentina after seven seasons in England; British Empire, winner of £5,623 and a leading sire in Argentina; Coastal Traffic, whose stud career encompassed Ireland, France, America and France again; and Faerie Queen, winner of the Scottish Derby, and grandam of the Derby runner-up Arabian Night.

Arctic Prince's great-great-grandam Charmione by Captivation won a small race and foaled three winners, including Lady Maderty. She was sold in 1935 as a 13-year-old to Russia for 180 guineas. Lady Maderty did not win a great amount—£1,373—but was one of the fastest of her time. She produced only three live foals, all of whom won.

1952

1   H.H. Aga Khan's br c Tulyar (Tehran – Neocracy).
2   Mrs. J. V. Rank's b c Gay Time (Rockefella – Daring Miss).
3   F. Dupre's b c Faubourg II (Vatellor – Fast Lady).
      33 ran. Time: 2 min 36.4 sec

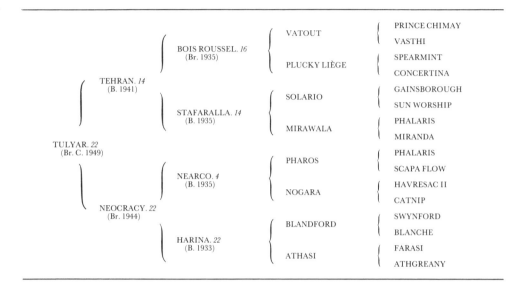

LATE FOALS

A first foal, Tulyar was not born until May 12. Curiously, his sire Tehran was also a late foal, not making his appearance until May 14. Bearing in mind the handicaps they suffer from, which are four-fold, late foals have a very respectable record as racehorses. In addition to Tulyar and Tehran they include Polemarch (June 3, St. Leger); Cyllene (May 28, Ascot Gold Cup); Pan II (May 26, Ascot Gold Cup); Gold Bridge (May 26, champion sprinter and good sire); Pont l'Eveque (May 25, Derby); Bebe Grande (May 23, won eight races value £18,586 as a two-year-old); Trigo (May 23, Derby and St. Leger); Umidwar (May 20, Champion Stakes, etc.); Wilwyn (May 20, won 21 races value £12,940 and 32,500 dollars); Brown Betty (May 15, 1,000 Guineas); Adam's Apple (May 10, 2,000 Guineas); Hurry On (May 7, never beaten); Ki Ming (May 6, 2,000 Guineas); Neasham Belle (May 6, Oaks); Toboggan (May 5, Oaks); Windsor Slipper (May 4, unbeaten Irish Triple Crown winner); Tiberius (May 4, Ascot Gold Cup); Hypericum (May 3, 1,000 Guineas); Solario (May 1, St. Leger and Ascot Gold Cup); Saltash (May 1, Eclipse), etc.

In my view the handicaps late foals suffer from are as follows: first, a great numerical inferiority, since it is plain that far more foals must be born in the four months January to April than in May. Second, late foaling is far more common amongst mares of doubtful antecedents than amongst better-class ones. It is rare for the owner of a mare likely to breed Classic winners to allow her to be covered in

June. He usually prefers to accept her as barren for the season and to try for an early foal the following year.

This is quite understandable as once a mare has entered a cycle of late foalings she will probably continue the practice until eventually her foals are born altogether too late to be of use for Flat racing. Theoretically as the period of gestation is usually taken as about 11 months it ought to be possible to peg a mare back to earlier and earlier foaling each season, but in effect this does not work out. In my experience many mares carry their foals longer than a bare 11 months, and then there are nine days to wait after foaling before even first service can be carried out. If the mare does not hold to this—which is quite common—the few days' leeway made up will be lost and in all probability a later foal than ever will be the result the following season. On the other hand amongst mares in a lower category many owners consider that it is better to have a late foal than no foal at all and consequently are keen to have them mated at a very late date.

A third handicap is that the late foal cannot reasonably be expected to match in physical development as a two-year-old or early three-year-old, rivals who are three or four months his senior. At the commencement of the racing season an early-born two-year-old will be some 28 months old whilst a late one barely 24 months old—four months in two years is a lot to give away.

A fourth handicap is that a late foal sent to the yearling sales will appear under-developed compared to his elders and so has a reduced chance of falling into ownership which will give him the best possible opportunities of making good on the Flat at home. Amongst the late-maturing and particularly the 'chasing type of stock I think probably the late foal is at no disadvantage, except for the fact that if he is intended for sale as a yearling he may have a less attractive appearance at that age than his elders.

To return to Tulyar, he ran six times as a two-year-old winning on two occasions. The Jockey Club Handicapper assessed his merit in the Free Handicap as 19lb below the best of his generation (Windy City). The following season provided a different story as Tulyar ran seven times without defeat, winning amongst other races the Derby (for the fifth and last time by his owner), St. Leger, King George VI and Queen Elizabeth Stakes and Eclipse Stakes. He did not run for the 2,000 Guineas.

Tulyar's nine victories earned £76,417, at that time the largest amount won by any horse in the history of the English Turf, beating the record of £57,455 held by Isinglass since before the turn of the century. Isinglass (1890) won the Triple Crown, Ascot Gold Cup, Eclipse Stakes, etc., but it must be remembered that in his day there was no King George VI and Queen Elizabeth Stakes, and that the Classic races then carried stakes of only about a third of the 1950s. The Derby then was worth about £7,000 and the St. Leger £5,000. It is also interesting to note how values have rocketed even in the short time since Tulyar. There is now no equivalent race to the Henry VIII Stakes Tulyar won on his first outing as a three-year-old, but a horse emulating the rest of his victories in 1976—the Ormonde Stakes, Lingfield Derby Trial, Derby, Eclipse Stakes, King George VI and Queen Elizabeth Stakes and St. Leger—would have netted a few pence short of £297,475, almost four times the amount Tulyar earned in his career.

It was intended that Tulyar would stay in training as a four-year-old, but in the winter of his Classic year he was sold by the Aga Khan to the Irish National Stud for £250,000—a world record price for a thoroughbred at that time—and the Irish Government decided not to race him again, although he was a perfectly sound horse. The racing public would have liked to have seen more of him, but the decision was probably a correct one as he retired when his reputation was at a peak, and if he had suffered defeat as a four-year-old, breeders' interest in him might have cooled. As it happened, the field for the 1953 Ascot Gold Cup was of poor quality and barring accidents Tulyar could hardly have lost it, thus repaying over £10,000 of his purchase price. But the Irish Government could not foresee this.

He took up stud duties as a five-year-old in 1954 at a fee of 600 guineas, with the proviso that in the case of a mare being barren who was owned by an Irish citizen half the fee was returnable. However, a year later he was sold to an American syndicate for £240,000, and delivery was effected after the 1956 stud season. Unhappily, after arrival in the States he contracted such a serious illness that he could not be used in 1957. It cannot be said how much this affected him, but he was a disappointing sire who produced fewer than the average number of foals—only 217 were named in his 14 American crops—before he died in 1972, though like others before him he has gained posthumous fame through his brood-mares.

Tulyar's best European winners were Ginetta (French 1,000 Guineas), Fiorentina (Irish 1,000 Guineas) and Tulyartos (second in Eclipse Stakes after a useful two-year-old career; sent to Japan), whilst in America he was notable for Castle Forbes (joint top two-year-old filly in 1963 and winner of eight races) and Mako (Steeplechaser of the Year in 1966). Castle Forbes became the dam of Irish Castle, a top-class two-year-old who in his first crop sired the Kentucky Derby and Belmont Stakes winner Bold Forbes. Another daughter of Tulyar, Tularia, became the dam of What a Pleasure, who sired the Kentucky Derby second Honest Pleasure, winner of seven Grade One races in 1975 and 1976, and his full brother the good two-year-old For the Moment. In Europe Tulyar mares who have made their mark include Ginetta (dam of Grand Roi, who sired the Norwegian St. Leger winner and useful English handicapper Royal Park); Gracious Me (dam of the Arc de Triomphe runner-up and prolific place-money earner On My Way); Tina II (dam of the Ebor Handicap winner Ovaltine, and grandam of three fast fillies in Mange Tout, Hecla and Rose Dubarry); and Rose of India (grandam of the French Derby winner Hard to Beat).

Mention should be made of Tulyar's son Menelek, who after a hard-working racing career became a legend for his output as a stallion. Foaled in 1957, he ran for five seasons, won 11 of his 40 races for £5,983 and was placed 14 times. He showed his best form at a mile and a half, and went to stud in Ireland in 1964 at a fee of £48 after being sold for 3,000 guineas at the previous Newmarket December Sales. He became a leading National Hunt sire, his winners including Rag Trade (Grand National) and April Seventh (Whitbread Gold Cup), and at the age of 19 in 1976, the year he had 70 live produce, was credited with having covered 135 mares.

TEHRAN (1941)

Tulyar's sire Tehran won six races value £7,258, including a wartime St. Leger. He was placed in the Derby, 2,000 Guineas and Ascot Gold Cup. Syndicated in 1945 for a total of £100,000, he went to stud the following year, and as a result of Tulyar's success was champion sire in 1952—half Tulyar's winnings alone would have guaranteed him superiority over the second-best stallion, Hyperion, and Tehran's other 13 individual winners contributed only £9,004 out of a total of £84,177. That, however, was just about the only high spot in Tehran's career, though he did sire Mystery IX (Eclipse Stakes; sent to South Africa); Tabriz (useful two-year-old who broke a bone in his foot and could not race at three; sire of 2,000 Guineas winner Taboun); Raise You Ten (Goodwood Cup and Doncaster Cup; sire of Goodwood Cup winner Girandole, and good sire of National Hunt horses including Cheltenham Gold Cup winner Ten Up); Dawn Watch (Ayr Gold Cup); and Kurdistan (moderate racehorse in England but four times leading sire of jumpers in New Zealand).

Tehran was by the French-bred English Derby winner Bois Roussel out of Stafaralla by Solario. Stafaralla won two races value £3,035 and bred at least nine winners. In addition to Tehran she was the dam of Noorooz (seven wins value £6,687 including the Ebor Handicap; to stud in America) and the moody Anwar (2½ wins value £3,012; at stud in England and Ireland; sired many winners but none of outstanding merit, though is the maternal grandsire of Irish Oaks winner Pampalina). Stafaralla is also the grandam of the French 1,000 Guineas winner Toro.

MIRAWALA (1923)

Stafaralla was out of Mirawala, who won four races worth £1,956 and foaled six winners, the best of which was Sind, who scored twice in England to the tune of £3,366 and was second in the Grand Prix de Paris. Sind went to stud in France in 1937 and later moved to Argentina where he did well.

Mirawala was bred by Lt. Col. Giles Loder in Ireland and sold by him for 1,500 guineas as a yearling at the 1924 Dublin Sales. The purchaser re-sold the filly a month later to the Aga Khan for 2,900 guineas, so making a swift profit on the deal. She is the tail female ancestress of innumerable good horses, in addition to those mentioned above, such as Argur (£12,367, including the Eclipse, third in the 2,000 Guineas); Holmbush (£6,267, including the Jockey Club Stakes), etc.

Mirawala's dam Miranda, who was a full sister to Pretty Polly, won two races value £439 and produced seven winners who secured between them 15 races worth £10,600. The major subscriber to these figures was the Irish Derby and Manchester November Handicap winner King John (£5,201), who went to New Zealand but was not successful.

ADMIRATION (1892)

Miranda's dam Admiration raced on the Flat 18 times between the ages of two and four. She was a most indifferent performer who failed to score on English courses but won two events worth £150 in all in Ireland. She then ran unsuccessfully under Irish National Hunt Rules. She came from a family with a very poor record. Her dam Gaze ran only once, and that in a selling race, and she produced nine foals but only two winners. Between them they secured six races worth £772. Gaze's dam Eye-Pleaser was first covered as a two-year-old and produced eight foals, who between them won one solitary race value £102. Moreover, for 50 years prior to Pretty Polly's birth the line had not thrown a single horse of class.

However, at stud Admiration proved an outstanding success, producing nine winners from her 13 foals, and I think there have possibly been more good winners of recent years in various parts of the world who trace in tail female to her than to any other broodmare of her period. Her own nine winners won between them 42 races value £52,484.

PRETTY POLLY (1901)

Incomparably the best of Admiration's winners was Pretty Polly, who won 22 races and garnered £37,297. Her victories included the 1,000 Guineas, Oaks and St. Leger. She was not entered for either the 2,000 Guineas or Derby. She was beaten only twice—once in France and once in the Ascot Gold Cup—and in both she was second. She lived until she was 30 years of age and bred ten foals. Four of these were winners on the Flat who between them secured 11 races worth £6,107, and a fifth, called Tudor King (1920), won a small hurdle race. Pretty Polly was by Gallinule whose dam Moorhen also had a most unusual history and background as I have related on page 193.

Others of Admiration's progeny were Admiral Hawke (£7,197; sent to Germany; full brother to Pretty Polly); Admirable Crichton (£2,036; sent to France after a few seasons in Ireland; by Isinglass so brother in blood to Pretty Polly); Coriander (£839; sired some winners of moderate class in Ireland); Veneration II (£444; dam of four winners who won £18,589, including the disqualified Derby winner Craganour); Adula (£3,080; had only three foals before dying; all won and secured between them stakes worth £9,778); Miranda (see above); and Addenda (£442; dam of four winners of £1,740 in stakes between them at home, and Silvius, second in the Melbourne Cup and a useful winner and sire in Australia).

For 13 consecutive seasons Admiration never missed having a foal, and her descendants in tail female were chronicled in Chapter 9, where it can be seen how her influence extends to the modern day. At this point it is worth referring to Brigadier Gerard, the most recent outstanding member of the family and whose breeding was the climax of Mr. John Hislop's ambition.

A celebrated amateur rider and journalist, Mr. Hislop nurtured the vow to himself that one day he would own a mare from the Pretty Polly family, and true to his word he bought Brazen Molly (by Horus out of Molly Adare, by Phalaris out of Molly Desmond, by Desmond out of Pretty Polly) just after the Second World War. A half-sister to the Goodwood Cup winner Fearless Fox, Brazen Molly was never broken and was barren at the time of her purchase for 400 guineas—but she did trace to Pretty Polly. She threw four winners for Mr. Hislop, including the 2,000 Guineas runner-up Stokes, before what proved to be her last foal, the filly La Paiva. And in 1968 La Paiva foaled Brigadier Gerard, winner of his first 15 races (including the Middle Park Stakes, 2,000 Guineas, Champion Stakes and Eclipse Stakes), second in the Benson & Hedges Gold Cup, and winner of his last two races, including the Champion Stakes, for a total of £253,024 in win and place money. This is one example in which long-term planning paid off handsomely—but it is not always so in the world of thoroughbred breeding!

NEOCRACY (1944)

Tulyar's dam Neocracy by Nearco was sold as a yearling to the Aga Khan, for whom she won two races value £2,562 as a two-year-old. At stud in addition to Tulyar she bred Tarjoman, who won a useful race in France; Cobetto, whose success in the Princess of Wales's Stakes at Newmarket earned his owner a caution from the Stewards about the future running of his horses, after he had shown vastly improved form in blinkers for the first time; and Saint Crespin III, who was not at his best when third in the Derby but made up for that misfortune by winning the Eclipse Stakes and Arc de Triomphe, the latter on an objection after dead-heating with Midnight Sun II. Seldom fully fit as a three-year-old, Saint Crespin III was lame after the Arc, later damaged his shoulder and did not run again. He retired to stud in Ireland, and it is a curious feature of his career that his best produce were fillies—Dolina (Italian Oaks), Casaque Grise (Prix Verme-ille), Mige (Cheveley Park Stakes), Altesse Royale (1,000 Guineas, Oaks and Irish Oaks), Crespinall (Nassau Stakes) and Shebeen (Cumberland Lodge Stakes and Princess Royal Stakes)—with only the Derby runner-up Shoemaker rising into the top class amongst his colts.

Altesse Royale pulled off a remarkable Classic treble in 1971, and so enabled Saint Crespin III to take third place in the list of winning sires, with 13 individual winners contributing just £7,833 to his total, and Altesse Royale providing the other £75,685. By then, however, Saint Crespin III had followed the bloodstock drain to Japan.

Success with his fillies suggests that Saint Crespin III's name might yet live on as a sire of winning broodmares, and Alvertona (dam of the prolific Flat and National Hunt winner Alverton) and Cutle (dam of the Irish 2,000 Guineas winner Sharp Edge) have done a good job in founding the theory.

Tulyar's grandam Harina won the Imperial Produce Stakes worth £4,128. The only winner to her credit is Neocracy, but she produced Kyanos, dam of Ocean Sailing (a good staying filly who won six races worth £3,678); Daemon (Chester Vase); Twinkling Eye (£429), and Straight Way (four wins in Ceylon). Harina was a full sister to the Derby and St. Leger winner Trigo.

Tulyar was inbred to Phalaris at the fourth remove.

CHAPTER 25

1953 – 1957
Winners of the Mid-1950s

1953

1 Sir Victor Sassoon's b c Pinza (Chanteur II – Pasqua).
2 H.M. The Queen's ch c Aureole (Hyperion – Angelola).
3 Prince Said Toussoun's b c Pink Horse (Admiral Drake – Khora).
   27 ran. Time: 2 min 35.6 sec

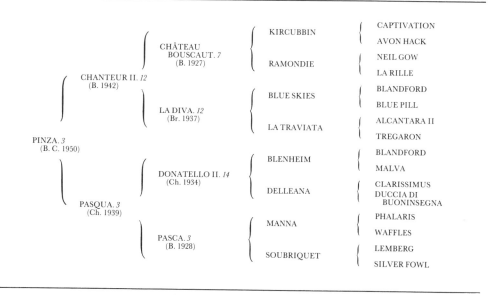

PINZA, A VERY strongly-built horse, was an exceptionally good racer. He was actually bred by the famous trainer Fred Darling, but it was Mrs Morris of the Banstead Manor Stud who had the honour of being responsible for Pinza's production. At the 1949 Newmarket December Sales Mrs Morris sold the ten-year-old mare Pasqua in foal to the Banstead Manor stallion Chanteur II for 2,000 guineas to Mr. J. A. Dewar, acting for the absent Mr. Darling, who was not enamoured of the mare when he saw the substance behind the name, and exactly a year later she was back at Newmarket, being sold for only 560 guineas to an Argentinian breeder. Shortly after Pasqua's departure, her yearling by Chanteur II went to the Newmarket Sales and made 1,500 guineas to the bid of Sir Victor Sassoon, who named the colt Pinza.

He won two races as a two-year-old and was allotted 9st 2lb in the Free Handicap, top weight on 9st 7lb being the subsequent 2,000 Guineas winner Nearula. During the winter Pinza slipped on the road and injured an elbow joint. This took a long time to mend satisfactorily and it was not until March of his three-year-old days that he could resume training, too late to get him ready for the 2,000 Guineas. He made his first appearance on a racecourse that season in the middle of May when he won the Newmarket Stakes. This success was followed by victories in the Derby (a first winner at the 28th attempt for the newly-knighted champion jockey Sir Gordon Richards) and the King George VI and Queen Elizabeth Stakes. He broke down in his preparation for the St. Leger and was syndicated as a stallion at a capital valuation of £200,000. During his racing career he won in all five races worth £47,401, and can be counted as one of the best Derby winners this century, which makes his failure at stud all the more disappointing, especially after a fine start which in 1957 saw him leading first-season sire, largely through Pinched (Royal Lodge Stakes) and Pin-Wheel (Imperial Produce Stakes). However, Pinza's standing soon went into decline, interrupted only by the occasional above-average exception, such as Pindari (third in the St. Leger, won King Edward VII Stakes and Great Voltigeur Stakes), Pinturischio (ante-post favourite for the Derby but believed to have been doped before the race which he had to miss); and Violetta III (dead-heated for the Cambridgeshire; dam of the Irish 1,000 Guineas winner Favoletta). Mares by Pinza have also produced Jimmy the Singer (Goodwood Stewards' Cup) and Princely Son (Vernons Sprint Cup).

CHANTEUR II (1947)

Pinza's sire Chanteur II was bred in France. A plain, coarse but very tough horse, he won seven races value 3,156,150 francs in his native land and three worth £9,019 in England including the Coronation Cup. He was second in the Grand Prix de Paris and also second twice in the Ascot Gold Cup. He was bought as a stallion for England and commenced duties as a six-year-old in 1948. Pinza was thus from his second crop, and the latter's success in 1953 enabled Chanteur II to head the list of winning sires that year. Chanteur II sired two other Classic winners—Cantelo (St. Leger) and Only for Life (2,000 Guineas in 1963, the year after his sire died)—while his daughter Patti was beaten a head in the St. Leger. His other important winners included the sisters Bracey Bridge (Ribblesdale Stakes, Park Hill Stakes) and Romp Home (Great Metropolitan; dam of the Doncaster Cup winner Biskrah); Best Song (Manchester November Handicap);

and Chatsworth (seven races for £9,024; a leading sire in Australia and New Zealand; sire of AJC Derby winner Royal Sovereign and New Zealand Oaks winner Blyton).

As a sire of broodmares Chanteur II distinguished himself through Chorus Beauty (dam of the Oaks second Windmill Girl and grandam of the Derby winners Blakeney and Morston); Warning (dam of the Royal Lodge Stakes winner Escort and the triple Champion Hurdle winner Persian War); Cavatina (dam of the Doncaster Cup winner and St. Leger runner-up Canterbury); and Somnambula (dam of the Champion Stakes winner Hurry Harriet).

CHATEAU BOUSCAUT (1927)

Chanteur II was by the French Derby winner Château Bouscaut, who died some months before Chanteur II was foaled. He was a consistent racer whose victories also included the French Gold Cup, and accumulated 1,943,291 francs in stakes before retiring to stud in 1932. Thus he had a comparatively short career as a stallion.

Among others he got were the French 1,000 Guineas winner Longthanh; the St. Leger second Château Larose (sent to Argentina); the Irish 2,000 Guineas and Derby winner The Phoenix; Châteauroux (1,080,000 francs and sire of winners of nearly 30m francs); Macaron (759,815 francs; to stud in France); and Patchouly (four wins in France, three wins in England; third in the Cambridgeshire; to stud in France) etc.

Most important of these was The Phoenix, winner of five races and essentially a source of speed at stud, where he sired the Irish 1,000 Guineas winners Morning Wings and Lady Senator, as well as Victorina (Goodwood Stewards' Cup), Rising Flame (Craven Stakes; a leading sire in Japan), and one of the exceptions to the rule Prince Hansel, winner of the Doncaster Cup. Mares by The Phoenix produced Lorenzaccio (Champion Stakes); the brothers Balidar (smart sprinter; sire of the 2,000 Guineas winner Bolkonski) and Balliol (Cork & Orrery Stakes); the brother and sister Kamundu (Royal Hunt Cup) and Somersway (Ayr Gold Cup); and High Line (twice winner of the Jockey Club Cup; sire of the Lancashire Oaks winner Centrocon).

KIRCUBBIN (1918)

Château Bouscaut, who never quite reached the top of the tree as a stallion, was by the Irish St. Leger winner Kircubbin, who was sold to France for £4,000 as a three-year-old and proceeded to win six more races value 436,825 francs besides valuable events in Spain. He was at stud in France for about a dozen seasons, until his death in 1937, with excellent results. He was their champion sire in 1930 and for some years was near the head of that compilation. His best performer in England was the 1,000 Guineas winner Mesa (1932). Kircubbin's stock secured over 330 races, which must be considered as very good bearing in mind that he did not have a long career.

CAPTIVATION (1902)

Kircubbin was by Captivation, who raced only once and then without success. Captivation went to stud in Ireland as a four-year-old at a fee of £5 a mare. To start with he was an Irish registered stallion, the counterpart of the English premium stallion, and covered all kinds of mares. He gradually improved his status and before he died in his 28th year had sired the winners of some 236 races on the Flat at home, in addition to some good horses under National Hunt Rules.

The average value of races won by Captivation's stock was very low, and his name lived on mostly through his daughters, who included Dark Eyes (a half-bred mare; dam of Soloptic, who won the Irish 1,000 Guineas and Oaks and £6,125, and of Sweet Wall, who won 13 consecutive races value £1,952; grandam of Hyacinthus, one of the best English two-year-olds of 1940 and sire of numerous winners, none of outstanding merit, before his death in 1973; also grandam of Solerina, who won eight races value £3,712 including the Goodwood Stewards' Cup); Casquetts (grandam of the unbeaten Grand Prix de Paris winner and great French sire Pharis II, and ancestress of the Ascot Gold Cup winner Elpenor, and of Nirgal, winner of 14 races in high class in England and France who went to stud in America and sired Li'l Fella, successful in 22 of his 73 outings from the age of two to seven, and Nail, a smart two-year-old who won five races for 239,930 dollars); Charmione (grandam of Solar Flower and third dam of Solar Slipper); Captive Princess (grandam of the 1,000 Guineas and Oaks winner Exhibitionnist); and Capdene (dam of Diomedes who won 16½ races value £11,320 and was the fastest horse of his time in England but proved a very disappointing sire; grandam of that excellent 'chaser Prince Regent, who secured 21 events under both rules and £8,712 in stakes including the Cheltenham Gold Cup and Irish Grand National).

Incidentally, Prince Regent affords an illustration of the slow build-up of blood lines which sometimes takes place. His sire My Prince was racing before the First World War, whilst he himself was competing after the Second World War. My Prince first saw a racecourse in 1913, whilst his son's last appearance was 36 years later, in 1949. In fact, it might be said they belonged to two different epochs and almost two different worlds as far as the way of life of their human connections was concerned.

Considering his humble start at stud, Captivation achieved very good results. He was by the Ascot Gold Cup winner and very fine sire Cyllene, out of Charm (1888) by St. Simon. Charm was a full sister to the 1,000 Guineas and Oaks winner Amiable (1891).

KIRCUBBIN'S BROTHERS AS HUNTER SIRES

Kircubbin's dam Avon Hack (1907) produced six other winners who won 22 races value £8,468 between them. The best of these was Ardavon, winner of 11 events worth £3,934, who became an English premium stallion. Others who achieved similar status were Santavon (who won £1,793; went to stud in Ireland) and Ballynahinch (who won £2,087). The vast difference in the stud performances of the brothers and half-brothers is interesting, since one, Kircubbin, became a champion sire in France, while the others were hunter sires in England and Ireland. There was not a great deal to choose between the value of stakes won by each of the four when racing at home, but Kircubbin's successes in France put an entirely different complexion on his ability. One wonders what his fate would have been had he broken down at the end of his three-year-old days. Would he also have followed the almost family tradition of becoming a hunter sire, or would he have made good as a bloodstock stallion?

RODOSTO (1930)

Château Bouscaut's dam Ramondie (1920) won four races in France and produced three other winners. The best of these was Rodosto, who won nine races including both the English and French 2,000 Guineas. He went to stud in France in 1935 and moved to Argentina at an advanced age. His French-bred stock won roughly 270 races worth 45m francs. The best of these was probably Dogat (1940), who won the French 2,000 Guineas and 8½ other races to the tune of 2,396,470 francs. He retired to stud in his native land in 1944, was brought to England eight years later and died in 1958. His most important produce to run in Britain were Damremont (winner of 15 races for £9,951; sire of several good-class handicappers) and Trouville (winner of the Cork & Orrery Stakes; died at the age of 13; sire of Red Slipper, an Irish-trained colt who must have set a record of sorts by winning three races, including the Prix du Moulin, in France in 1963, and went on racing as a full horse until the age of 11).

Ramondie also produced Rovigo (Grand Prix de Vichy and other good races) and Rodna (dam of Rondo II, who won 2,206,350 francs in France and over a dozen races under National Hunt Rules in England before becoming a premium stallion).

NEIL GOW (1907)

Ramondie's sire Neil Gow, by Marco, won £25,771, including the 2,000 Guineas, but died after eight seasons at stud, though not before he had made a comfortable impact on the Stud Books of the world. His son Re-echo won eight races worth £6,376 including the Cambridgeshire and then went to Argentina, where he was a great success as a stallion. However, Neil Gow's name lives on mainly through his daughters, who were excellent broodmares, although limited in numbers.

Amongst the most notable were Best Wishes (grandam of the Ascot Gold Cup winner Felicitation, who was sent to Brazil after eight stud seasons at home); Perce-Neige (see page 330); Herself (dam of the Oaks winner Chatelaine, and ancestress of the Grand Prix de Paris winner Avenger, and the Cambridgeshire winner Raymond, who went to Japan); Nilghai (dam of the Northumberland Plate winner Leonard); Roxana (dam of Soliped, the best two-year-old of his generation in Ireland); Santa Cruz (dam of seven winners of £12,858; grandam of Heliopolis, winner of £14,792 and twice leading sire on export to America, and of 1,000 Guineas winner Tide-way and 1,000 Guineas and Oaks winner Sun Stream; also ancestress of Eclipse Stakes winner Gulf Stream, who after winning £10,537 went to Argentina); and Thistle (dam of Fohanaun who won £10,245 and went to stud in France).

Neil Gow's name never appeared as champion maternal grandsire but the great influence exercised by his comparatively few daughters shows that they were of exceptional merit. Château Bouscaut's grandam La Rille (1908) produced in addition to Ramondie the French 1,000 Guineas winner Rebia, Enfilade, who won 16 races in America, and others of lesser merit.

BLUE SKIES (1927)

Chanteur II was the first foal of his dam La Diva, a small winner by Blandford's son Blue Skies. The last-named was born in France, where he won four races value 245,065 francs. He was a good but not great performer, and his stud career was at about a corresponding level—useful but no more. He reached about sixth or seventh on the French list of leading sires of winners. His dam Blue Pill was a winner and also produced the French Oaks winner Perruche Bleue (1929) and a very good filly called Cerulea. Blue Pill was a half-sister to Scammonee, grandam of Bagheera, who won the Grand Prix de Paris and French Oaks and ran second to Supertello in the Ascot Gold Cup.

Chanteur II's grandam La Traviata was a winner and bred several winners, including Pantalon, who took the Grand Criterium and Grand Prix de Marseilles before becoming a leading sire in Argentina. She was by the French Derby winner and good sire Alcantara II.

DELLEANA (1925)

Turning to the distaff side of Pinza's pedigree, he was the sixth foal of his dam Pasqua, who failed to win and previously had produced the small winner Petros, who earned £599 from two wins on the Flat before being gelded and put to National Hunt racing. Petros was sold for 185 guineas in 1955, and won one novice hurdle worth £102 that year and three novice 'chases worth a total of £374 the following year, before breaking down in a three-horse race at Hexham in May 1956.

Pasqua was by Donatello II, whose history I outlined on page 217. Donatello II's dam Delleana won eight races including the Italian 1,000 and 2,000 Guineas, and bred seven winners. Of these, second in importance to Donatello II was the Italian 1,000 Guineas and Oaks winner Dossa Dossi, successful in seven races and dam of five winners, the best being De Corte (Polish Derby and St. Leger) and Dagherotipia (Italian 1,000 Guineas, grandam of Psidium), as well as Daniela da Volterra (dam of the Italian St. Leger winner Derain). Delleana also produced Donata Bardi (dam of Degas, who won about 12m lire in stakes; grandam of Favorita, winner of six races as a two-year-old in England, and I Titan, winner of the Queen's Vase but died at three years); and Donatella (top two-year-old of her year; dam of ten winners including the Italian Derby and St. Leger winner Daumier, and the Italian St. Leger winner De Dreux). This is the famous 'D' family that served the Dormello Stud well, though it should be pointed out that the last Classic winner from this source was the 1967 Italian 1,000 Guineas and Oaks winner Dolina (whose great-grandam was Dossa Dossi), and that she came from a branch established in France and was a dozen years distant from De Dreux's exploits in Italy. The family seems to be in a parlous state to say the least.

Delleana was by Clarissimus, who was bred in England and won the 2,000 Guineas. He spent three seasons at stud at home and then went to France, where he sired a lot of winners before dying in 1933. His best son was Nino, who won the French Gold Cup, 1,586,930 francs in stakes, and later became a prominent stallion. The daughters of Clarissimus were particularly good broodmares. Amongst them were the dams of two Grand Prix de Paris winners, Pharis II and the filly Crudite, whilst the latter's half-brother Brantome had an unbeaten certificate in France. Brantome was by Bend Or's son Radium (1903), who won the Goodwood Cup, Doncaster Cup, etc., but was not a great sire, out of the unbeaten 1,000 Guineas winner Quintessence (1900).

DUCCIA DI BUONINSEGNA (1920)

Donatello II's grandam Duccia di Buoninsegna was bred in Ireland and sold to Italy as a yearling for only 210 guineas. She won the Italian Oaks and seven other races, and though producing only three foals to live, became a great foundation mare in the Tesio stud. She was by Bridge of Earn out of Pretty Polly's daughter Dutch Mary, dam of six winners including Christopher Robin and Spelthorne (Irish St. Leger), who both went to Australia as sires, and Kopje, dam of the Grand Prix de Paris winner Cappiello.

PASCA (1928)

Pinza's grandam Pasca won two races value £500 in all. At stud she bred Pasch (£19,030, including the 2,000 Guineas and Eclipse Stakes; died after one season at stud); Château Larose (second in the St. Leger; by Château Bouscaut so closely related to Pinza; sent to Argentina); and Pascal (£4,539; died as a four-year-old in training). Pasca was by the 2,000 Guineas and Derby winner Manna out of Soubriquet, a half-sister to the Derby and Oaks winner Fifinella. Details of the family history will be found on page 185.

Pinza was inbred to Blandford at the fourth remove.

1954

1 R. S. Clark's ch c Never Say Die (Nasrullah – Singing Grass).
2 J. E. Ferguson's b c Arabian Night (Persian Gulf – Faerie Lore).
3 Sir Percy Lorraine's b c Darius (Dante – Yasna).
     22 ran. Time: 2 min 35.8 sec

		PHAROS	PHALARIS
	NEARCO. *4*		SCAPA FLOW
	(Br. 1935)	NOGARA	HAVRESAC II
NASRULLAH. *9*			CATNIP
(B. 1940)		BLENHEIM	BLANDFORD
	MUMTAZ BEGUM. *9*		MALVA
	(B. 1932)	MUMTAZ MAHAL	THE TETRARCH
NEVER SAY DIE. *1*			LADY JOSEPHINE
(Ch. C. 1951)		MAN O'WAR	FAIR PLAY
	WAR ADMIRAL. *11*		MAHUBAH
	(Br. 1934)	BRUSH UP	SWEEP
			ANNETTE K.
SINGING GRASS. *1*		VATOUT	PRINCE CHIMAY
(Ch. 1944)			VASTHI
	BOREALE. *1*		SIR GALLAHAD III
	(Ch. 1938)	GALADAY II	SUNSTEP

Never Say Die was conceived in Ireland a few months before his sire Nasrullah was exported to America. His dam Singing Grass was boarded in Ireland for her first two years at stud, but after being covered by Nasrullah she was repatriated and produced Never Say Die in Kentucky. Never Say Die came to England as a yearling and in due course showed he was a horse for the big occasion, since his three wins from 12 races included the Derby (in which 18-year-old Lester Piggott became the youngest jockey to win the race this century) and St. Leger (in which Charlie Smirke deputized for the suspended Piggott), and earned £31,146. His other win was as a two-year-old, when he was successful once from six outings, being rated 18lb inferior to the best of his generation in the Free Handicap. As a three-year-old he was unquestionably the best horse of his age in England, but he was not kept in training after the St. Leger, his aged owner taking the view that he wanted to see his horse's progeny race and that he should be retired to stud as quickly as possible.

It was arranged that Never Say Die would stand his first season in Newmarket at a fee of 500 guineas, and in May 1955 his owner gave the horse to the British Government, free, gratis and for nothing, with the one condition that each season ten nominations should be earmarked for Irish breeders. The horse was valued at 750,000 dollars and the deed of gift meant he became an English National Stud stallion. Sadly, his owner died in December 1956 at the age of 76, before his fine horse could fulfil his intentions.

Never Say Die, a strongly-made horse with lop ears, was Nasrullah's best racehorse in England and made an auspicious start at stud, with Never Too Late II (1,000 Guineas and Oaks) and Die Hard (second in St. Leger, won Ebor Handicap) in his second crop; Sostenuto (Ebor Handicap) in his third; and the Derby winner Larkspur in his fourth. But after Larkspur's somewhat fortuitous success had put Never Say Die at the head of the sires' list in 1962, his star began to wane. When he was put down at the age of 24 in 1975, in which year he had covered only three mares because of deteriorating health, his produce had won 309 races in Britain for a total of £400,527.

Though a hard puller and a difficult ride on the racecourse, Never Say Die showed none of his sire's temperament at stud, where one of his least attractive traits was to produce stock that split pasterns or had weak bones. He might well have sired another Derby winner in General Gordon, who won the Chester Vase on only his second racecourse appearance but broke a leg on the gallops six days before Epsom, where the stable ran the inferior Pretendre to be a neck second to Charlottown.

Mellay and Immortality were among sons of Never Say Die who did not reach the racecourse because of injury. Mellay (who was out of Meld and therefore a half-brother to Charlottown) went to New Zealand, became champion in the 1972–73 season but died in 1974 at the age of 13. Immortality (whose grandam was La Troienne) went to stud in Ireland in 1961 but was sent to Argentina in 1967, the year his daughter Fleet won the 1,000 Guineas.

Immortality, who began his career at a fee of £48, became the only unraced horse this century to sire an English Classic winner, and his case raises the question of such stallions. The best example of this type of success is provided by Alibhai, who boasted an exemplary pedigree, since he was by Hyperion out of Teresina, a mare who won the Goodwood Cup, was placed in the Oaks, St. Leger and Eclipse Stakes, bred eight winners and became one of the Aga Khan's foundation mares. Alibhai was bought as a yearling in England, showed great promise in trials in America but injured his sesamoids and was saved for stud. His best get included Determine (Kentucky Derby); Your Host (winner of 12 races from 23 runs; sire of Kelso, the world's leading money-spinner and American Horse of the Year five times) and Traffic Judge (third in Preakness Stakes; sire of smart American sprinter Delta Judge, himself sire of the very fast Swing Easy, winner of the King's Stand Stakes and Nunthorpe Stakes). Alibhai also figures in the pedigrees of Oaks winner Long Look (whose grandam Secret Meeting was by Alibhai) and Dahlia (whose maternal grandsire Honey's Alibi was by Alibhai).

Since opportunity is one of the guiding factors in stud success, it is obvious that a stallion who never got on to a racecourse has an uphill task from the start, with his pedigree the only "asset" by which breeders can judge his potential. British breeders, and American come to that, have so much choice when deciding which stallion to use, that they have little need to gamble on an unraced sire when they can take up an option on a horse who boasts both acceptable pedigree and creditable racecourse performances. This probably explains why overseas breeders, lacking the same opportunity, are not so reluctant to take on an unraced stallion, providing he has the right pedigree. Mellay is such an example.

Mellay is not the only son of Never Say Die to have prospered abroad. Battle-Waggon, a modest winner but a son of the Oaks winner Carrozza, also did exceptionally well in New Zealand, where Philoctetes (eight races for £20,739, including the Northumberland Plate) was sent. Sostenuto was a success in Australia, and Casabianca (Royal Hunt Cup) made a name for himself in South Africa. However, it was the Japanese who seemed to want first call on Never Say Die's sons, claiming Larkspur; Die Hard (consistently among his new land's top sires); Never Beat (a moderate racehorse but half-brother to the St. Leger winner Hethersett; leading sire in Japan three times up to 1975); Cipriani (fifth in the Derby; sire of a Japanese Derby winner) and Philemon (a good-class stayer and half-brother to the leading Scandinavian sire Carnoustie; little patronized but with his last crop in England the sire of Coventry Stakes winner Whip It Quick, a subsequent success in Germany).

The name of Never Say Die is clearly going to be found in many overseas pedigrees for years to come, and though his success as a sire of winners declined in later years, his broodmares helped to arrest the graph. Among the useful winners produced by Never Say Die mares are Moulines (French 1,000 Guineas); Attica Meli (whose grandam was the 1,000 Guineas winner Honeylight; won Park Hill Stakes, Yorkshire Oaks and Doncaster Cup); Mistigri (Irish St. Leger); Celina (Irish Oaks); the half-sisters Julie Andrews and Little Audrey who both won the South African Oaks; Sky (Magyar Derby) and Without Fear (out of Never Too Late II; leading sire in Australia after only a modest racing career in France).

NASRULLAH (1940)

Never Say Die's sire Nasrullah was bred in Ireland by the Aga Khan, for whom he won five races worth £3,348, doing all his racing at Newmarket because of wartime restrictions. As a two-year-old he was rated one pound behind the top weight Lady Sybil (also by Nearco) and by coincidence was the same margin from the best as a three-year-old, when he won the Champion Stakes, ran third in the Derby but was unplaced in the 2,000 Guineas and St. Leger. He was a very temperamental horse—a trait hardly helped by being virtually confined to barracks at Newmarket, where he many times refused to participate in gallops—and he was inclined to transmit this defect to his stock when he went to stud. However, it did not prevent him from becoming one of the greatest sires of the last 35 years, and one whose influence will be felt at least until the end of this century.

From the first crop to last, Nasrullah was a success. Starting stud duties at the age of four, he stood for one season in England at a fee of 198 guineas, after which he was bought by Irish breeder Mr. Joseph McGrath for 19,000 guineas, and finally went to America in 1950, when he was purchased by a syndicate for a reported 340,000 dollars. In his first crop in Ireland he sired Nathoo (£11,269 including the Irish Derby) and Noor (four wins worth £4,705 and third in the Derby and Eclipse Stakes; eight races worth 365,940 dollars in America), and in his last he sired Never Say Die and Princely Gift (nine races worth £6,673 including Challenge Stakes, and Portland Handicap under 9st 4lb). In his first crop in America he sired Nashua (22 wins from 30 starts worth 1,288,565 dollars for world-record earnings by an individual horse); and a month before he died in May 1959 he covered the mare Lalun, the resulting foal being Mill Reef's sire Never Bend. Between these five horses there were many more of high merit.

Amongst his most important Irish-sired winners were Nearula (£27,350; including the 2,000 Guineas); Musidora (£26,798 including the 1,000 Guineas and Oaks); Belle of All (£25,264 including the 1,000 Guineas); Golestan (£5,108 in England and 831,420 francs in France, where he was second in the 2,000 Guineas); Zucchero (ten wins value £15,788; a highly temperamental horse who would have won more but for his habit of refusing to race), and Orgoglio (a hardy horse who won 9½ races worth £11,854 from 39 outings).

After his departure to America, Nasrullah was responsible for only a few racecourse successes in Britain, the chief ones being Red God (won Richmond Stakes and beaten a head in Champagne Stakes at two years; repatriated to America and won four small races); Bald Eagle (won three of his six races in England for £5,886 but never fulfilled his promise; repatriated to America at four years and won nine races for 676,442 dollars, including Washington International Stakes twice); and Nasram II (French-trained winner of three races, and emulated Never Say Die as a horse for the big occasion by gaining his one win from eight outings at four years in the King George VI and Queen Elizabeth Stakes).

However, in his adopted America Nasrullah quickly proved that his son Nashua was not a flash in the pan. He sired more than 100 stakes winners in the States, and was leading sire of winners five times—the first horse to become champion sire on both sides of the Atlantic. Further indication of his prowess comes from the fact that he was in the top ten sires in Britain or America 17 times in all.

In America his chief winners, apart from those already mentioned, were Bold Ruler (won 23 out of 33 races for 764,204 dollars, including the Preakness Stakes); Jaipur (won ten out of 19 races for 618,926 dollars, including the Belmont Stakes); Nadir (11 wins, 434,316 dollars, including America's richest race the Garden State Stakes; best two-year-old of 1957); Nasrina (seven wins, 171,000 dollars, best two-year-old filly of 1955); Leallah (ten wins, 152,784 dollars, best two-year-old filly of 1956); On-and-On (12 wins, 390,718 dollars) and Fleet Nasrullah (11 wins, 223,150 dollars).

With so much top-class talent to represent him, it would have been a surprise if Nasrullah's stock had not been capable of gaining further honours in the bloodstock world, but the extent of that success is still remarkable. Nasrullah's influence as a sire of sires can be largely divided into two sections—those at stud in Britain and Ireland, who in general were a source of speed; and those in America, who by and large produced better middle-distance performers. Nasram II, his most successful son to stand in France, slotted into the latter category by siring the French 2,000 Guineas winner Zug and the German Derby winner Athenagoras.

NASRULLAH'S SONS IN BRITAIN & IRELAND

Nasrullah's influence on the stallion lists of Europe devolved largely on three of his lesser winners—Red God, Grey Sovereign and Princely Gift. Like Grey

Sovereign, who was discussed in Chapter 23, Princely Gift was successful in sprint races—though he stayed seven furlongs well enough to win the Hungerford Stakes—and as a result did not receive the financial reward his talents deserved. Whatever his exact status on the racecourse Princely Gift became a better sire than racer, and when he was put down at the age of 22 in 1973 his produce had won 383 races for £431,668. They included an impressive array of sons who showed top-class racing form before retiring to stud, such as:

Faberge II (second in 2,000 Guineas; sire of Italian Derby winner Gay Lussac, short-head Derby runner-up Rheingold, and Champion Stakes winner Giacommeti; sent to Japan in 1970); Floribunda (five races, £6,668, including Nunthorpe Stakes; sire of good sprinters Florescence and Porto Bello; sent to Japan); Tesco Boy (see page 139); King's Troop (five races, £6,690, including Royal Hunt Cup; a better sire than racer, and sire of Irish 2,000 Guineas winner King's Company and useful racers Caspoggio in Italy and Lear Jet in France before being sent to New Zealand); Frankincense (six races, £18,596, including Lincoln Handicap); So Blessed (six races, £17,086, including July Cup and Nunthorpe Stakes; sire of the hardy filly Duboff, winner of nine races as a three-year-old including Sun Chariot Stakes); Tribal Chief (very fast two-year-old winner; sent to Japan in 1975); Realm (five races, £17,733, including Diadem Stakes and Challenge Stakes; leading first-season sire in 1975); and Sun Prince (three races, £32,240, in England, including Coventry Stakes and St. James's Palace Stakes, and valuable Prix Robert Papin in France).

Red God, who unlike Princely Gift inherited a fair measure of the temperament associated with Nasrullah, was also a better sire than racehorse. He was almost wholly associated with speed on his return from America to stand at stud from 1960, and his best get include:

Jacinth (top-rated two-year-old of 1972; winner of Cheveley Park Stakes and Coronation Stakes, and over £44,000); the brothers St. Alphage (ten races, £7,017; sire of champion sprinter Sandford Lad) and Yellow God (five races, £29,047, including Gimcrack Stakes, and sire of similar winner Nebbiolo); Red Alert (five races, £19,413, including Goodwood Stewards' Cup as a three-year-old carrying 9st 2lb); and Green God (six races, £16,217, including Vernons Sprint Cup; disqualified in Nunthorpe Stakes). Although Jacinth and Yellow God were second in the 1,000 Guineas and 2,000 Guineas respectively, it was not until the age of 22 that Red God sired a Classic winner, with Red Lord taking the French 2,000 Guineas of 1976.

NASRULLAH'S SONS IN AMERICA

From relatively few opportunities in Britain and Ireland, Nasrullah did exceptionally well as a sire of sires; from his increased band in America he fashioned an influence that can only be described as phenomenal. Reference to the list of leading sires of winners in America, and to the record of Kentucky Derby winners amply illustrates the point.

Nasrullah himself was leading sire in 1955, '56, '59, '60 and '62; his son Bold Ruler eclipsed all other 20th Century stallions by leading the list eight times, in 1963, '64, '65, '66, '67, '68, '69 and '73; TV Lark (by Nasrullah's son Indian

Hemp) followed in 1974; and What a Pleasure (by Bold Ruler) was leading sire in 1975 and '76. In 22 seasons from 1955 to 1976 the Nasrullah male line was at the top of the sires' list 16 times—and that in a country where annual foal production reaches almost 30,000. As a further indication of Nasrullah's influence, there were over 700 stallions most regularly used in America in 1976, and 20 per cent of them stemmed from his male line.

In the Kentucky Derby, America's most prestigious race, Nasrullah was not directly responsible for a winner, but his sons and grandsons have provided ample compensation, as the following table shows:

1968   Forward Pass (on a disqualification) by On-and-On, by Nasrullah.
1970   Dust Commander, by Bold Commander, by Bold Ruler, by Nasrullah.
1973   Secretariat (won Triple Crown), by Bold Ruler.
1974   Cannonade, by Bold Bidder, by Bold Ruler.
1975   Foolish Pleasure, by What a Pleasure, by Bold Ruler.
1976   Bold Forbes, by Irish Castle, by Bold Ruler.

A further sidelight is the financial attraction of Nasrullah's sons. When Nashua—who won a record 752,550 dollars in a season—was sold for 1,251,200 dollars in 1955, he was at the centre of the first million-dollar syndication. Soon to follow were Nasrullah's sons Bald Eagle (1.4m dollars), Never Bend (1,225,000 dollars) and Fleet Nasrullah (1,050,000 dollars).

Nashua, who won the Preakness Stakes and Belmont Stakes but was beaten by Swaps in the Kentucky Derby, never quite made it to the top of the American sires' tree—one second was his highest placing—but he did consistently well, especially with his fillies, of whom the most notable included the sisters Shuvee (16 wins between five furlongs and two miles, from 44 races, for 890,445 dollars, including Frizette Stakes at two years, CCA Oaks at three years, and champion handicap mare at four and five) and Nalee (eight wins for 141,631 dollars; dam of seven consecutive winners including Irish St. Leger victor Meneval, and two other CCA Oaks winners Bramalea—dam of the Derby winner Roberto—and Marshua).

Bald Eagle, who went to stud in America in 1961 and was sent to France 11 years later, made less impact, his most notable winner being the filly Too Bald (13 races, 174,722 dollars, and dam of the Grand Prix de Paris and French St. Leger winner Exceller). More important, and with especial interest to British breeders, were Jaipur, who sired Amber Rama, an outstanding sprinter and the first horse to break 60 seconds over Ascot's five furlongs; and Nantallah, who belied his early shortcomings of winning four races for a total of only 17,825 dollars. To matings with the Gold Bridge mare Rough Shod, Nantallah sired Ridan (11 wins, 635,074 dollars, second in Preakness Stakes and third in Kentucky Derby; sired the winners of over 2.7m dollars at stud in America from 1964 before being sent to Ireland in 1973 but died in 1976; sire of French St. Leger winner Busiris and very fast two-year-old filly Brave Lass); Moccasin (champion two-year-old filly of 1965; dam of Apalachee, Britain's leading two-year-old of 1973, winner of the Observer Gold Cup, third in 2,000 Guineas); Lt. Stevens (good-class racer; sire of Prix Robert Papin winner Sky Commander); and Thong (dam of the high-class Thatch, winner of July Cup and Sussex Stakes among seven wins from nine starts;

Lisadell, winner of the Coronation Stakes; and King Pellinore, second in the St. Leger and Irish Sweeps Derby). This highly successful mating has become well-regarded in Britain, for obvious reasons, and further honour has been accorded to the aforementioned Rough Shod through her daughter Gambetta (by My Babu), who to a mating with Nasrullah himself produced Zonah (dam of the Observer Gold Cup winner Take Your Place), and to a mating with Nasrullah's son Bold Ruler produced Gamely (whose first foal was the Dewhurst Stakes winner Cellini, third behind Apalachee in the 1973 Free Handicap).

BOLD RULER (1954)

Gamely, champion filly at three years, leading handicap mare at four and five years, and winner of 16 races for 574,961 dollars, was one of a host of tip-top horses produced by Bold Ruler, who himself was foaled within half an hour and a few yards of another great American racer Round Table. The odds against such an eventuality must be enormous!

Bold Ruler topped the sires' list for the first time when his initial crop was three years old, and by the time he was humanely destroyed in July 1971, as a result of cancer, his 276 named foals had produced 165 individual winners of 806 races worth more than 13m dollars, though it is an interesting point that he sired only one—the Kentucky Oaks victress Lamb Chop—to win an American Classic in his lifetime. Then in 1973 his son Secretariat carried off the American Triple Crown for the first time since Citation in 1948. So the theory that Bold Ruler was capable of siring only very fast two-year-olds who did not train on was exposed as groundless.

Between Lamb Chop's championship year of 1963 and Gamely's final season of 1969, sons and daughters of Bold Ruler took top honours 11 times. In 1964 Bold Lad (U.S.A.) (see page 250) and Queen Empress were the leading two-year-old colt and filly respectively in America; in 1966 Successor (seven wins for 532,254 dollars) and Bold Lad (Ire) (see page 250) were the top two-year-old colts in America and Britain respectively, and Bold Bidder (13 wins, 478,021 dollars) was champion handicap horse. In 1967 Bold Ruler was again responsible for America's leading two-year-old colt and filly, in Vitriolic and Queen of the Stage respectively, and Gamely began her three championship seasons.

Eventually along came Secretariat (1970), who proceeded to set records in almost every department during the course of his 16 wins from 21 starts, once being disqualified. He was voted American Horse of the Year at both two and three years, a rare honour indeed; and in the course of winning the Triple Crown he broke the track record in both the Kentucky Derby and Belmont Stakes. He also won the Man O'War Stakes in record time, and established a world best for nine furlongs in another race. His three-year-old earnings of 860,404 dollars was a record, and so too was his syndication as a stallion for 6,080,000 dollars. Incidentally, he was syndicated before he ran as a three-year-old, so it is anybody's guess what American breeders might have been asked to pay if those arrangements had been delayed until the end of his second and final season. No doubt they would have paid up regardless!

However, once Secretariat did go to stud, anyone with an interest in him could not help but realize their assets. The first mares in foal to him came on the market in 1974 and top price was 385,000 dollars for Artists Proof (whose foal was subsequently named Dactylographer and raced in Britain). In 1975 his first foals were sold and included the world record price of 250,000 dollars, which was eclipsed the following year when a colt out of Trevisana made 370,000 dollars. Also in 1976 his first yearling went up for public auction and the world record was more than doubled when a half-brother to Dahlia (later named Canadian Bound) made 1.5m dollars—a figure that makes a mockery of the average small breeder's dealings.

It is too early to judge Secretariat's stud achievement, but in terms of conformation, racing performance and pedigree, he had a head start over most stallions, and there was the evidence of Bold Ruler's other sons at stud to provide further encouragement. Yet if Secretariat does not match what is expected of him, he will not be the first stallion to do so.

The table of Kentucky Derby winners on page 349 reveals the influence of Bold Ruler's sons, who provided four winners as well as the runner-up Honest Pleasure (by What a Pleasure by Bold Ruler). In addition, Bold Forbes won the Preakness Stakes, as did Master Derby (by Dust Commander by Bold Ruler). Dust Commander was also the second leading sire of two-year-olds in 1976, behind another Bold Ruler son Raja Baba, whilst What a Pleasure was leading sire both of winners of all ages and of two-year-olds in 1975. Mention must also be made of Bold Ruler's sons Boldnesian (who won the Santa Anita Derby among only five outings and sired Wing Out, winner of 500,000 dollars, and also Bold Reasoning, winner of eight races between three and four years, sire of the leading two-year-old of 1976 Seattle Slew in his first crop, but died in 1975 after one more crop); and Reviewer (sire of the CCA Oaks winner Revidere, and the ill-fated champion three-year-old filly of 1975 Ruffian).

Bold Ruler's influence has been felt less keenly in Europe than in America, though the two Bold Lads have played an important part in establishing their sire's worth, and Bold Bidder offers plenty of hope for the immediate future through the young stallions Mount Hagen (winner of the Prix du Moulin and three other races for £64,870) and Auction Ring (July Stakes). It should not be assumed that Bold Ruler's stock cannot adapt to European conditions; the likelier explanation for their comparatively lower success is that American breeders have held on to the best stallions and only occasionally has top-class blood been allowed to cross the Atlantic. Even the English National Stud appeared to find this out with the Preakness Stakes runner-up Stupendous, who was imported as a four-year-old in 1967 but five years later left for Japan branded a failure. This is the danger breeders face when attempting to jump on a particular foreign stallion's band-waggon; the risk of being landed with another country's cast-offs at an exorbitant price must always be borne in mind.

MUMTAZ BEGUM (1932)

Reverting to Nasrullah, he was by Nearco out of Mumtaz Begum by Blenheim. Mumtaz Begun, a half-sister to Mah Mahal, the dam of Mahmoud, won two races value £461 before producing eight winners of 22 races worth £11,501 in all, and was an important source of speed. Her best winner was Rivaz, a sister to Nasrullah who won the Queen Mary Stakes and became the dam of Spicy Living (ten wins in America worth 251,205 dollars) and Palariva, a top-class sprinter. Palariva is the dam of Atrevida (third in the Irish 1,000 Guineas; dam of the very useful two-year-old Habat, who won five races for £49,636); Khairunissa (a smart French two-year-old; dam of the French 2,000 Guineas winner Kalamoun and the useful sprinters Aslam and Karayar); and Zahedan (winner of the National Stakes at two years; exported to Australia).

Among Mumtaz Begum's other winners were Nizami II (a brother to Nasrullah; two wins; sent to America); Darbhanga (six wins worth £1,915; a leading sire in Sweden); Dodoma (two wins value £230; dam of Diableretta, eight wins worth £14,492 including the Queen Mary Stakes, dam of the French 1,000 Guineas winner Ginetta, and of Jambo, who won £3,281 and was sent to Australia); and Malindi (won over five furlongs as a two-year-old but did not train on, dam of Prince Taj, see page 325).

Mumtaz Begum also produced the non-winner Sun Princess, who foaled Royal Charger; Lucky Bag (five wins, £2,739, sent to New Zealand); Tessa Gillian (five wins, £6,662, second in 1,000 Guineas, dam of the Gimcrack Stakes winner Test Case, and grandam of the useful sprinter D'Urberville) and three other winners.

ROYAL CHARGER (1942)

Royal Charger, by Nearco, was closely related to Nasrullah, and the pair had very similar careers. Royal Charger won six races worth £3,426 and was third in the 2,000 Guineas; and at the end of his racing days he went to stud in Ireland, being sold to the Irish National Stud for 50,000 guineas. He commenced stallion duties as a five-year-old, and was sold to America, to stand in Kentucky, after the 1953 stud season. His maximum fee in Ireland was £250, but on transfer to America his services were available by private contract only, beginning I am advised at the equivalent of £3,570 a mare.

Though he never emulated Nasrullah, Royal Charger did more than satisfactorily, and he has established a strong male line. His important successes in Britain were Happy Laughter (£26,908, including the 1,000 Guineas); Royal Challenge (£6,225, one of the best two-year-olds in England in 1953); Royal Serenade (ten wins, £10,971, in England where he was leading sprinter; moderately successful at stud in America); Sea Charger (£8,914, including the Irish 2,000 Guineas and St. Leger); Royal Duchy (four wins, £3,926, as a two-year-old when the best of her generation in Ireland); Gilles de Retz (2,000 Guineas), and Royal Palm (useful sprinter and sire of 630 winners of £448,000, including Royal Avenue and Young Christopher).

Easily Royal Charger's most successful son in America was Turn-To (1952), who was foaled in Ireland but was sent to America as a yearling. He won six races from eight starts at two and three (including one on a disqualification) for 280,032 dollars, and sired First Landing (19 wins, 779,557 dollars, leading two-year-old of 1958), Hail to Reason (nine wins, 328,434 dollars, leading two-year-old of 1960) and Sir Gaylord (ten wins, 237,404 dollars). Sir Gaylord sired the Derby winner

Sir Ivor, and Hail to Reason was responsible for Roberto; they will be discussed in more detail later.

WAR ADMIRAL (1934)

Never Say Die's dam Singing Grass was bred in America but raced in the north of England, winning seven events worth £1,875 in all. Their average value was low but she was a game, consistent sort whose distance was a mile and a quarter. Her first produce was a filly called Banbox (by Combat), who won a £100 race in Ireland, and next came her Derby-winning son. She was by War Admiral, who stood no more than 15.2 hands high, faced the starter 26 times, won 21 events, and was placed on four more occasions, earning 273,240 dollars. His successes included the American Triple Crown of Kentucky Derby, Belmont Stakes and Preakness Stakes, as well as the Jockey Club Gold Cup, etc.

He proved a first-grade stallion, being champion sire in 1945 when his daughter Busher (15 wins for 344,035 dollars) was Horse of the Year in America, and leading sire of broodmares in 1962 and 1964. Best known for his fillies, War Admiral produced the dams of Iron Liege (Kentucky Derby) and Buckpasser, as well as Never Say Die. Busanda, dam of Buckpasser, was a useful filly in her own right, winning ten races for 182,460 dollars including the Saratoga Cup twice, and also bred Bupers (seven wins, 221,688 dollars, including the Futurity Stakes at two years) and Bureaucracy (11 wins, 156,635 dollars). Buckpasser himself was one of America's best horses of the century, his 25 wins (including 15 in succession) for 1,462,014 dollars putting him into fifth place in the list of world all-time money-earners. His success was all the more creditable as he could not run for the Classics because of injury. Syndicated for a record 4.8m dollars, he sired a champion two-year-old filly in each of his first two crops—Numbered Account and the ill-fated Le Prevoyante—and also produced L'Enjoleur, the champion colt in Canada. Buckpasser's best-known racer in Britain was Swingtime, who dead-heated for the Diadem Stakes, won the Cork & Orrery Stakes outright, and was sold in November 1975 for 132,000 guineas, an Irish auction record for a horse in training.

War Admiral's further contribution to international racing came through another daughter Searching (1952), a high-class winner of 25 races in America whose grandam was La Troienne. Searching has bred three notable fillies— Admiring, a good-class stakes winner; Affectionately, champion racemare and winner 28 times for over 500,000 dollars, and dam of Personality, Horse of the Year in 1970 when he won the Preakness Stakes; and Priceless Gem, a useful two-year-old who won seven races but is better known as the dam of the top-notch filly Allez France.

MAN O'WAR (1917)

War Admiral's sire Man O'War, who died at the age of 30 in 1947, must be regarded as near the best and possibly actually the best horse bred in America this century. He changed hands as a yearling for only 5,000 dollars, and raced as a two-and three-year-old only. He contested 21 events, won 20 of them and was second in the other. His earnings on the Turf came to 249,465 dollars, an American record in the days of limited stake values. He was champion sire in America in 1926, and his best get were the Belmont Stakes winners American Flag and Crusader; the Kentucky Derby winner Clyde van Dusen; the CCA Oaks winners Florence Nightingale, Edith Cavell and Bateau; and War Admiral.

Man O'War's success was all the more remarkable because he was virtually a private stallion, his owner Samuel D. Riddle rarely allowing him to serve any but his own mares. Since Mr. Riddle possessed neither an outstanding nor a prolific band of mares, Man O'War's services were restricted to approximately 25 a season—with the consequence that future generations have been left to speculate on what he would have achieved had he been permitted the usual complement of top-class mares which undoubtedly would have been sent to him under normal circumstances.

Probably Man O'War's best-known get to race in Britain was the 'chaser Battleship, who won the 1938 Grand National. All kinds of horses have won the world-famous 'chase but few can have been further removed from the accepted pattern of a National horse than Battleship, who stood no more than 15.2 hands high. He was Flat-race bred, his dam Quarantaine being also the dam of the French Oaks winner Quoi, and he was an entire. After his racing days were over he went to stud in his native America and was a very good sire of 'chasers.

FAIR PLAY (1905)

Man O'War's sire Fair Play was a good though not outstanding two-year-old, but was the best of his age the following season in America except for the unbeaten Colin. He was then sent to England, turned sour and failed to win a race. On return to America he became a most successful stallion, being champion sire on three occasions. Fair Play's dam was Fairy Gold by Bend Or.

THE FAIR PLAY SIRE LINE

The War Admiral–Man O'War–Fair Play stirp is an exclusively American branch of the Matchem sire line. The Triple Crown winner West Australian (1850), who was sent to France after six seasons at stud in England, sired amongst others at home Solon (1861) and Australian (1858). Solon went to stud in Ireland and sired unbeaten Barcaldine, the tail male ancestor of Hurry On, whilst Australian went to America early in his life. Australian bred three winners of the Belmont Stakes—Joe Daniels, Springbok, and Spendthrift—of which Spendthrift was the most significant, being the sire of Hastings, who also won the Belmont Stakes and got Fair Play.

The Matchem line is commonly spoken of these days in Europe as the Hurry On line, but in America it is the Fair Play line. Curiously enough, although for a long time past Americans have been buying good-class stallions of other sire lines, they never acquired a Hurry On-line horse of class, in spite of the fact that various of his sons were leading stallions in Britain, Argentina, Australia, Italy, France, New Zealand and South Africa. In contrast, the import of Fair Play-line sires to Britain after Relic's arrival in 1951 has had notable effect in establishing the standing of this branch of the Matchem line in Europe.

Relic (1945), by War Relic by Man O'War, was brought from France and sired the brothers Buisson Ardent and Venture VII, as well as Polic (sire of the useful sprinter and promising sire Polyfoto in a brief career before export to Japan), Pieces of Eight and the prolific broodmare Relance III (dam of Reliance II, Match III and Relko). Most significant of these as a sire was Buisson Ardent, who died in December 1963 after only seven covering seasons but produced Roan Rocket (St. James's Palace Stakes and Sussex Stakes), Ardent Dancer (Irish 1,000 Guineas), Atherstone Wood (Irish 2,000 Guineas) and Silver Shark (ten races, £73,825, including Prix du Moulin and Prix de l'Abbaye). Roan Rocket has made a useful if not outstanding start at stud, and Silver Shark stayed long enough in Ireland before departure for Japan in 1972 to sire the Irish 2,000 Guineas winner Sharp Edge.

Immediate prospects for the Fair Play branch of the Matchem line in Europe depend largely on Roan Rocket, Polyfoto and Sharp Edge, and while it may not reach Classic-winning standard on more than rare occasions, it seems safe to suggest it will remain a prolific source of winners.

THE BONNIE SCOTLAND SIRE LINE

War Admiral's dam Brush Up (1929) was a non-winner who bred five other winners but none of special merit. She was by Sweep (1907), a first-class stallion particularly of broodmares, and a member of another exclusively American sire line. It is a branch of the Eclipse male line emanating from the English-bred Bonnie Scotland (1853), who won a couple of races in his native land and in America sired Bramble (1875).

Bramble got Ben Bush (1893), who sired two important horses in Sweep and Broomstick (1901). On this side of the Atlantic Sweep's name is most familiar as the sire of Dustwhirl, dam of Whirlaway (1938), who won 32 races and 561,161 dollars but did not come up to expectations as a sire, first in America and later in France where he died in 1953; whilst Broomstick got the English 2,000 Guineas winner Sweeper II.

In the reverse direction Bonnie Scotland traces to Eclipse through his son Pot-8-o's, who got Waxy, the sire of Whalebone, who in turn sired Sir Hercules, male ancestor of Blandford, Tracery and Bend Or; Camel, ancestor of Hermit, Hampton and Musket; Defence, from whom descend the Monarque line; and a horse called Waverley, who sired the 1838 St. Leger winner Don John. The last-named got Iago (1843), and there this small stirp in Europe petered out. But Iago sired Bonnie Scotland, whose dam was that great broodmare Queen Mary (1843), dam of the Derby and Oaks winner Blink Bonny (dam of Derby and St. Leger winner Blair Athol), and ancestress of innumerable high-class horses.

ANNETTE K (1921)

War Admiral's grandam Annette K did not win but produced three winners. Far and away the best was War Glory, who won 11 races value 55,050 dollars and sired some winners but was not a stallion of standing. Annette K was conceived in England but foaled in America. She was by Harry of Hereford (1910), a full brother to Swynford but a non-winner, who spent six seasons at stud in England and then went to France. In common with the majority of horses lacking racing merit he was a stud failure although he sired a few winners.

BOREALE (1938)

Never Say Die's grandam Boreale by Vatout was bred in France and won one race in America as a three-year-old. She bred at least six winners in various parts of the world, the most notable apart from Singing Grass being Blow Wind Blow (1951), who won six races worth £2,196 as a two-year-old but failed in all 11 outings the next year, and after raising only 1,850 guineas at the 1954 Newmarket December Sales, was gelded and had his attention turned to National Hunt racing.

Boreale's dam Galaday II was bred in America, where she won seven races in addition to one in England. She was also third in the Kentucky Oaks. At stud she produced four winners; of these Easter Day and Marie Galante scored in America, whilst Trois Pistoles and Galatea II secured their successes in England. Galatea II, who won the 1,000 Guineas, Oaks and £16,131 in all, was incomparably best of the four.

SIR GALLAHAD III (1920)

Galaday II's sire Sir Gallahad III won ten races in France including the local 2,000 Guineas and the Lincolnshire Handicap in England. He was also third in the French Derby and second in the French St. Leger. After retiring to stud in France in 1925, he was bought by Arthur Hancock for 125,000 dollars a couple of years later and became one of the first syndicated stallions in America, where he proved to be a sire of exceptional merit, topping the list for winners in 1930, '33, '34 and '40, and for broodmares no less than 12 times.

His son Gallant Fox (1927) won 11 races and 328,165 dollars (a world record at the time), including the Triple Crown. He started well at stud, with the Triple Crown winner Omaha (a flop at stud) and Flares, who came to England and won the 1937 Champion Stakes and 1938 Ascot Gold Cup, but was not much good as a stallion in America and went to Canada. But Gallant Fox could not maintain the standard and by the time of his death in 1954 at the age of 27, he was reckoned only a modest success.

However, other sons and grandsons of Sir Gallahad III have done well in America, notably Fighting Fox (nine wins) and Roman (18 wins). Fighting Fox was the sire of Crafty Admiral (18 wins, 499,200 dollars, and sire of the smart French-trained two-year-old Neptune II, himself the sire of French 2,000 Guineas winner Neptunus; and Admiral's Voyage, winner of 12 races for 455,879 dollars and beaten a nose into second place in the Belmont Stakes); whilst Roman sired Hasty Road (14 wins for 541,402 dollars including the Preakness Stakes).

Fighting Fox was also responsible for one of the few branches of the Sir Gallahad III male line to become established in Britain, through his son Fighting

Don, who was imported from America to stand at stud in Ireland. Fighting Don produced few top-class racers among his several winners but achieved notable success in the Nunthorpe Stakes through his sons Althrey Don (exported to Australia) and Forlorn River (who himself sired a Nunthorpe winner in Rapid River, as well as the fast filly Melchbourne).

Sir Gallahad III was by Teddy out of Plucky Liège, and so half-brother to the Derby winner Bois Roussel. Further details of the family were noted on page 254.

SUNSTEP (1915)

Never Say Die's great-great-grandam Sunstep by Sunstar was bred in England and sent to America as a foal. There she produced several winners including Sun Spot, whose son Gallant Sir won 16 races and 115,965 dollars. Sunstep was a half-sister to the 1923 Irish Derby winner Waygood (£8,128), who also went to America and fared moderately well as a stallion. Their dam Ascenseur (1911) was a non-winner by Eager (1894), a great sprinter who won over 20 races including the Wokingham Stakes, Portland Plate, Rous Memorial Stakes (twice) etc. Eager was not a great stallion but got plenty of winners including Electra (1906), who secured the 1,000 Guineas. Ascenseur's dam Skyscraper (1900) won £4,631 and was third in the 1,000 Guineas and Oaks. She was out of Chelándry.

1955

1 Mme. L. Volterra's b c Phil Drake (Admiral Drake – Philippa).
2 J. McGrath's ch c Panaslipper (Solar Slipper – Panastrid).
3 Alice Lady Derby's ch c Acropolis (Donatello II – Aurora).
    23 ran. Time: 2 min 39.8 sec

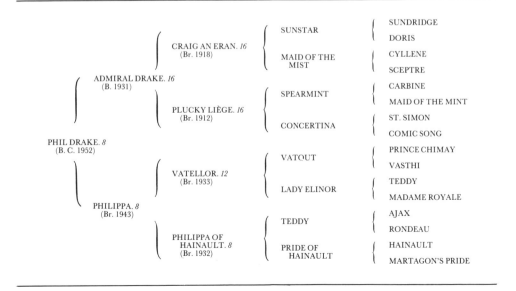

Phil Drake did not run as a two-year-old and appeared on a racecourse only twice prior to his Derby victory. After his success he returned to his native France and won the Grand Prix de Paris, thus being the third, and last horse this century to complete the double, after My Love and Spearmint. Phil Drake came to England again in the middle of July for the King George VI and Queen Elizabeth Stakes but ran unplaced. Training troubles prevented him from racing again and he retired to stud in France in 1956. Before his death in 1964 at the age of 12 he sired only one top-class racer, the French-trained Epsom Derby second Dicta Drake (1958).

Dicta Drake won the Grand Prix du Printemps and Grand Prix de Saint-Cloud as a three-year-old and the Coronation Cup at four, after which he was syndicated at 2,000 guineas a share to stand at stud in England from 1962. He was not a success as a stallion, failing to get a horse as good as himself, and before export to Japan in 1970 his best get were Dan Kano (Irish St. Leger), Star Ship (Ribblesdale Stakes, Lancashire Oaks) and the prolific handicap winner Chinatown.

ADMIRAL DRAKE (1931)

Phil Drake's sire Admiral Drake won five races value 1,413,344 francs including the Grand Prix de Paris. He was second in the French 2,000 Guineas and third in the French Derby, but finished last in the Epsom Derby. He went to stud in France as a five-year-old in 1936 and died there in 1952, the year Phil Drake, who was 21 years his junior, was foaled.

Admiral Drake got nothing of merit in his first few seasons and was a 14-year-old before his first good horse—Mistral (French 2,000 Guineas and died after about six seasons at stud in France)—showed his ability on a racecourse. Thereafter Admiral Drake got Monsieur l'Amiral (a hardy horse who raced from three to seven years, winning some 520,800 francs in France and £8,404 in England, including the Cesarewitch, Goodwood Cup etc.; at stud in England and Ireland but of little account); Amour Drake (won races worth 3,783,377 francs in France, including the 2,000 Guineas; second in English Derby; to stud in Ireland); Alindrake (won eight good-class races in France and England including the Queen Alexandra Stakes; to stud in France); Royal Drake (won six good races in France and England; second in English Derby; to stud in France); and Pink Horse (useful two-year-old winner in France; third in English Derby; to stud in France).

The only one of these sons of Admiral Drake to make much impact as a stallion was Amour Drake, though several of his progeny were temperamental and his influence in the top grade was restricted to Sarcelle, the best two-year-old filly of 1956 who failed to train on. Amour Drake was exported to Peru in the mid-1960s and his star has since flickered only briefly as the sire of the filly Château, whose son Davy Lad won the Cheltenham Gold Cup.

Admiral Drake's most lasting influence has been as the sire of two important broodmares—Source Sucree (1940), dam of Turn-To; and Toute Belle II (1947), dam of Hard Ridden.

Admiral Drake was by Craig an Eran out of Plucky Liège and so half-brother to the English Derby winner Bois Roussel. Further details of the family history have already been noted.

Phil Drake's dam Philippa by Vatellor won two races worth 265,400 francs and produced a foal who died; Bozet who won three races value 5,192,750 francs including the Prix Morny; La Perie who won one race in France in 1954, and Phil Drake. Her dam Philippa of Hainault by Teddy won four races including the French Free Handicap and bred six winners of which the best was Mistralor (six wins and 6,370,900 francs).

Phil Drake's great-grandam Pride of Hainault by Hainault was bred in Ireland. She never ran and was exported to France in 1929 as a five-year-old. There she foaled seven winners who won over 20 races between them. Amongst these was Mercia, dam of the French Derby winner Le Petit Prince, who was sold by auction in Buenos Aires in May 1955 for about £25,000—which seems a most reasonable price for a young French Derby winner, especially as he was by the French 2,000 Guineas winner and champion sire of 1951 Prince Bio. Mercia, by Teddy and so full sister to Philippa of Hainault, was foaled in 1931. She was thus 20 years old when she produced Le Petit Prince and had never previously given birth to a horse of comparable ability. Another of Pride of Hainault's foals, but a non-winner, was Sea Pride II, who produced six winners in various parts of the world including Marshal at Arms (1950), winner of five races worth £2,072 in England.

The next mare in Phil Drake's tail female line, Martagon's Pride (1917), won two races and bred three winners of humble status, the best being the selling plater Marlborough, who won £1,128 at home and also no less than 27 races in Belgium. Martagon's Pride was by Desmond's son Lomond (1909), who won £6,689 and was at stud in Ireland for ten seasons. He was then sent to Argentina, where he died at the age of 21. His Irish-sired stock won about 175 races value some £68,000 on the Flat, whilst he got a fair number of winners under National Hunt Rules.

Phil Drake's further female line goes back to Fleur de Marie (1885), whose descendants include Sainte Nitouche (Coronation Stakes, Ascot); Chosroes (foaled in 1919, won £8,215, a stud failure); Champ de Mars (14 wins value £10,455); Torelore (£7,912, sent to Poland after nine seasons at stud in England with poor results); Dansellon (£2,419, second in the 1917 Derby; died as a five-year-old); and the remarkably robust Donzelon (1921), who won a good two-year-old race worth £1,600 and continued racing on the Flat until he was a six-year-old, winning in all £3,744. As a seven-year-old Donzelon was transferred to 'chasing and raced under this code until he was a 14-year-old, scoring 26 times. He thus had 13 seasons in training and was a winner every year except two.

Phil Drake is inbred at the third and fourth remove to Teddy.

1956

1 P. Wertheimer's b c Lavandin (Verso II – Lavande).
2 R. B. Strassburger's ch c Montaval (Norseman – Ballynash).
3 J. McGrath's br c Roistar (Arctic Star – Roisin).
   27 ran. Time: 2 min 36.4 sec

	PINCEAU. *67* (B. 1925)	ALCANTARA II	PERTH TOISON D'OR
VERSO II. *3* (B. 1940)		AQUARELLE	CHILDWICK TEMESVAR
	VARIÉTÉ. *3* (Ch. 1924)	LA FARINA	SANS SOUCI II MALATESTA
LAVANDIN. *4* (B. C. 1953)		VAYA	BEPPO WATERHEN
	RUSTOM PASHA. *2* (B. C. 1927)	SON-IN-LAW	DARK RONALD MOTHER-IN-LAW
LAVANDE. *4* (Br. 1936)		COS	FLYING ORB RENAISSANCE
	LIVIDIA. *4* (B. 1927)	EPINARD	BADAJOZ EPINE BLANCHE
		LADY KROON	KROONSTAD NEEDLE ROCK

Lavandin's success marked the climax of foreign domination in the Derby between 1947 and 1956, his being the sixth overseas win in ten years. The same year French-trained horses won the St. Leger, Oaks, Ascot Gold Cup, Coronation Cup and Eclipse Stakes. Such events usually go in cycles, and I think the reasons for such concentration at this time could be found in opportunity and economics, rather than in any sudden deterioration of our bloodstock. During this period the English Classics were worth a great deal more to the owners of the winners than the corresponding events in France, as the figures for 1950 and 1955 show:

	1950		1955	
	ENGLAND £	FRANCE £	ENGLAND £	FRANCE £
1,000 Guineas	10,895	2,156	10,653	6,146
2,000 Guineas	12,982	2,148	13,666	6,471
Derby	17,010	5,602	18,702	17,268
Oaks	13,509	4,298	14,078	10,277
St. Leger	13,959	2,702	13,478	7,038
Total for all Classics (5)	68,355	16,906	70,577	47,150
Average value per Classic	£13,671	£3,381	£14,115	£9,430

It will be observed that in 1955 the average value of an English Classic was 50 per cent higher than for a French one, whilst a few years earlier, and still within the ten-year period under consideration, the figures hardly bear comparison. Is it to be wondered that the French left no stone unturned to scoop our pools?

Gradually, however, the French set about increasing their levels of prize money, especially in the top grade, and aided by breeders' premiums and improved added money derived from the Totalisator, they made their own Classics more attractive. The following figures, representing the position 20 years on—1976 being used because the French Oaks was abandoned by rioting stable lads the previous year—tell their own story:

	1970		1976	
	ENGLAND £	FRANCE £	ENGLAND £	FRANCE £
1,000 Guineas	21,015	24,286	39,447	34,965
2,000 Guineas	28,295	23,929	49,581	34,965
Derby	62,311	90,955	111,825	110,835
Oaks	31,319	69,993	50,117	86,205
St. Leger	37,082	34,881	53,638	51,370
Total for all Classics (5)	180,022	244,044	304,608	318,340
Average value per Classic	£36,004	£48,808	£60,921	£63,668

It can be seen that by 1970, though the English Classics had been increased in total value by 155 per cent, the French improvement was 417 per cent and enabled them significantly to overtake the values for the English Derby and Oaks and to bring about a virtual parity with the other three. In the years after 1970 England took steps to redress the balance and by 1976 something approaching equality had been reached, but the fact remains that the position compared with 1955 shows a vast improvement in French values, and helps to explain why the years since Lavandin's success have not been so fruitful for the French—not necessarily because of any significant change in bloodstock standards, but because of different circumstances of opportunity and economics.

LAVANDIN (1953) and VERSO II (1940)

Lavandin—the last horse wearing bandages to win the Derby—ran only once more after his Epsom success, and a badly cut knee sustained in the Grand Prix de Paris ended his racing career. He started at stud in France in 1957 at a fee of £300 but was not a success and went to Japan in 1963. His best get in Europe were Blabla (French Oaks) and Mehari (useful stayer who like Lavandin's brother Lavarede won the Prix Kergorlay). Lavandin was also the maternal grandsire of Realm (winner of five sprint races and leading first-season sire in 1975).

Lavandin's sire Verso II won the French Derby and St. Leger and Arc de Triomphe. He went to stud in 1945 and died some ten years later. Probably next in merit to Lavandin was the latter's full brother Lavarede, who ran Sicambre to a neck in the Grand Prix de Paris, whilst his daughter Vice Versa was third in the

English Oaks. Verso II's best-known get in England, prior to Lavandin, was Osborne (1947), a winner in France as a three-year-old who came to England in 1951 and really blossomed as a seven-year-old, when he won the Doncaster Cup and Goodwood Stakes. He later went to stud in Australia.

Verso II's best claims to fame as a broodmare sire come through Almora (dam of the French Gold Cup winner Azincourt), Anguar (dam of the French Derby winner Le Fabuleux, who stood for six seasons in his native country before being exported to America), Limicola (a fairly useful stayer who bred seven winners from eight foals, including the Derby runner-up Pretendre), and Barbara Sirani (the best Italian two-year-old of 1955 and winner of the Italian St. Leger; dam of the Italian 1,000 Guineas winner Bronzina).

PINCEAU (1925)

Verso II was by Pinceau, a good winner in France besides being second in both the Grand Prix de Paris and French St. Leger. He had a long stud career in his native land, being champion sire in 1943, largely through Verso II's efforts. In all his stock won some 65m francs and it must be remembered that five crops of them were racing for the very low stakes prevailing during the war.

Pinceau was by Alcantara II (1908) and so a member of the War Dance sire line. Pinceau's dam Aquarelle was a good winner in France and amongst others produced the French Oaks winner Aquatint II, a full sister to Pinceau. Aquarelle also foaled La Brume, whose son Brumeux won the Jockey Club Cup etc., and went to stud in England. Aquarelle was by Childwick out of Temesvar, a sister in blood to Semendria and half-sister to the Grand Prix de Paris and French Derby winner Ragotsky. Semendria won the French 1,000 Guineas, Oaks, Grand Prix de Paris and was one of the great fillies of 20th Century French Turf history. Her produce were not much good as racers, but her granddaughter La Bidouze foaled the French St. Leger winner and highly successful sire Biribi (1923).

I have referred to Verso II's maternal relations on page 253.

EPINARD (1920)

Lavandin's dam Lavande won three races worth just under 100,000 francs. On retiring to the paddocks she produced, in addition to Lavandin and Lavarede, the Portland Handicap (Doncaster) winner Le Lavandou, who went to stud as a six-year-old in Ireland in 1950, but died after seven covering seasons. He was chiefly a source of speed and bred three horses that each won six races as two-year-olds—Nonchalance, Galley Hill and Vandoulay. However, his most influential product was Le Levenstell (see page 270). Le Lavandou's most notable broodmares reverted to type—Lavant breeding two July Cup winners in Lucasland and So Blessed, and Street Song being responsible for the very fast early two-year-old Porto Bello (who as a sire produced the first two in the 1975 Ayr Gold Cup, Roman Warrior and Import, after the latter had won the Goodwood Stewards' Cup).

Lavande was by the Eclipse Stakes winner Rustom Pasha out of Livadia, who did not race but produced four winners of moderate attainments. Livadia was by the brilliant Epinard, who won 11 races in France and the Goodwood Stewards' Cup. His stud career was probably to some degree hampered by frequent changes of location. In 1926 he was in America, whilst two years later he returned to his native land. In 1930 he made another excursion to America, only to return in 1932 to France, where he died in 1942. His best-known get in Europe was the French and English 2,000 Guineas winner Rodosto. Theirs is a branch of the Dollar sire line which came down through Cambyse's son Callistrate (1890) via Gost (1898) and Badajox (1907) to Epinard.

Epinard's dam Epine Blanche (1913) was bred in France but conceived in America, being one of Rock Sand's daughters sired before his return to Europe. She was out of an American mare named White Thorn (1904), whose sire Nasturtium, a brilliant two-year-old, changed hands at that age for the equivalent of £10,000—a very big price for those days. He was then sent to England to run for the 1902 Derby but went wrong in his wind. His stud career in America was somewhat disappointing.

KROONSTAD (1900)

Lavandin's great-grandam Lady Kroon was bred in England and proved incapable of winning or even being placed in selling plates. She was put to stud as a three-year-old and later sent to France, where she produced three winners, the best of whom was Koenigsmark II, who won 4½ Flat races and three over jumps. Lady Kroon's sire Kroonstad was neither a high-class racer nor a successful sire, but he was a fine old warrior. He raced on the Flat for eight consecutive seasons, starting 91 times, winning 22 events and netting £7,714 in stakes. Some of his races were low-class handicaps and even selling plates, although as a three-year-old he won the Ascot Derby, but his toughness and gameness were greatly to be admired. As a ten-year-old he went to stud in England at a fee of nine guineas a mare and lived until he was 27 years old. He did not attract much patronage as a stallion and most of his mates were of modest quality, but he managed to sire about 30 winners and his stock were like himself—hardy, honest and game. His best-known get was Querquidella (1913), who produced that consistent and popular gelding Brown Jack.

Lavandin's great-great-grandam Needle Rock (1915) was also a daughter of Rock Sand conceived in France but foaled in England. She never ran but produced the 2,000 Guineas winner and Derby third Diolite. The next in female ascent Needlepoint could manage to win only one maiden two-year-old race and bred no winners. The line traces to Dee (1874) by Blair Athol, whose descendants include unbeaten Cavaliere d'Arpino (1926).

1957

1 Sir Victor Sassoon's ch c Crepello (Donatello II – Crepuscule).
2 J. McShain's ch c Ballymoss (Mossborough – Indian Call).
3 S. Niarchos's br c Pipe of Peace (Supreme Court – Red Briar).
    22 ran. Time: 2 min 35.4 sec

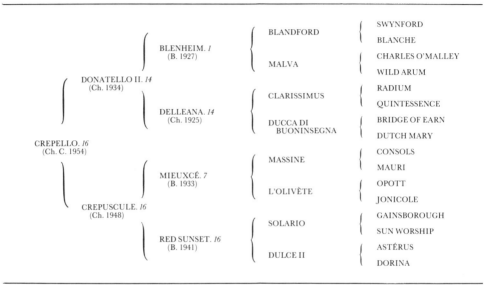

A powerful, heavy-topped horse, Crepello showed remarkable ability as a two-year-old for one so stoutly-bred, winning the Dewhurst Stakes on the last of his three outings and earning 9st in the Free Handicap. He made his first appearance as a three-year-old in the 2,000 Guineas and won; his next was in the Derby and he did not race again. Such a brief career was unfortunate but made his successes all the more creditable, for the distance was too short and the ground too firm for him to be at his best in the Guineas, and though he acted round Epsom his make and shape made it likely he would be better suited by a more galloping track. He never got the chance to prove the point, however, being controversially withdrawn from the King George VI and Queen Elizabeth Stakes on the morning of the race, and finally breaking down in a gallop three weeks after the Ascot race. He was promptly retired to stud in Newmarket for the 1958 season, having won three of his five races for £34,201.

His stud career can perhaps be best described as chequered, and on balance his top-class fillies outnumbered the colts, with the result he is now an important sire of broodmares. As an indication of his ups and downs, he was fifth leading sire with his third crop, but three seasons later, in 1966, he managed only one winner, and then in 1967 he reached third place on the list, largely through the exploits of his best son Busted (1963), who came into his own as a four-year-old and won the Eclipse Stakes and King George VI and Queen Elizabeth Stakes.

Crepello's other useful sons included Credo (Chester Cup); The Bo'sun (Blue

Riband Trial and City & Suburban Handicap); Soderini (Hardwicke Stakes, second in King George VI and Queen Elizabeth Stakes and Coronation Cup); Candy Cane (Ballymoss Stakes in Ireland, fourth in Arc de Triomphe); Linden Tree (Observer Gold Cup, second in Derby); and Pentland Firth (third in Derby). It will be seen that there are no Classic-winning colts among this list—though Busted was undoubtedly up to that standard—and while such as Soderini (in Germany), Minera (Argentina) and Crest of the Wave (New Zealand) have done well as stallions abroad, only Busted has kept the name of Crepello to the fore in this country. Indeed, Busted has done better than his own sire, since among his top-grade winners are two in the Classics—Bustino (£145,858 including the St. Leger) and Weavers' Hall (£67,757 including the Irish Sweeps Derby). Busted, who in 1976 covered 64 mares, the most by any of the leading sires, is still going strong; Weavers' Hall had his first runners in 1977, and Bustino was represented for the first time in 1979. It is on this trio that Crepello's slim male line depends.

However, as a sire of fillies Crepello is soundly established, his Classic winners in this group being Caergwrle (1,000 Guineas); Celina (Irish Oaks); Crepellana (French Oaks) and Mysterious (1,000 Guineas and Oaks). In addition his daughters The Creditor and Cranberry Sauce were in the highest bracket of their sex as three-year-olds; Cursorial and Mils Bomb won the Park Hill Stakes; Pink Gem dead-heated for the Park Hill and won the Cheshire Oaks, which was also won by Yelda and Milly Moss; and Lucyrowe won the Coronation Stakes.

Most notable of Crepello's daughters at stud has been Bleu Azur (1959), a winner of two small races who bred Royal Saint (Fred Darling Stakes); Altesse Royale (1,000 Guineas, Oaks and Irish Oaks); Yaroslav (Royal Lodge Stakes); and Imperial Prince (second in the English and Irish Derbys). Other mares by Crepello to produce Classic winners are Zest (dam of the Oaks heroine Ginevra, who was sold to Japan for 106,000 guineas in 1972) and Marlia (dam of the French 1,000 Guineas winner Mata Hari); while The Creditor has produced the good sprinter Abwah and the Dante Stakes winner Owen Dudley, both now at stud.

Crepello had to be put down in October 1974 because of old age. A month earlier his grandson Bustino had won the St. Leger; a month later, at the end of the racing season, he himself stood tenth in the list of leading sires for stakes won and top of the leading sires of broodmares. That adequately seemed to sum up his standing—a sound sire of winners, with the occasional top-class colt and a host of good fillies who hopefully will maintain his name as a stallion for several years to come.

CREPUSCULE (1948)

Crepello's dam Crepuscule won a mile and a quarter maiden race worth £276 as a three-year-old and then was promptly packed off to stud. She can be regarded as one of the best broodmares of the post-war era, since her first foal was the 1,000 Guineas winner Honeylight and her second Crepello. Thus she created a record for British bloodstock as the first mare to breed Classic winners with her first two foals. Several British mares have foaled Classic winners in consecutive years— such as Absurdity (dam of the 1,000 Guineas and Oaks winner Jest and the St.

Leger winner Black Jester); Galicia (dam of the St. Leger winner Bayardo and the Derby winner Lemberg); Devotion (dam of the 1,000 Guineas and Oaks winner Thebais and the 1,000 Guineas winner St. Marguerite) etc.—but all had earlier progeny.

It would have been amazing had Crepuscule improved on that Classic record, but she did enhance her reputation, as among her foals after Crepello were Night Court (winner of the Ebbisham Stakes and dam of four minor winners); Dark Alley (who had a high reputation but broke a leg at exercise as a three-year-old before he could run); Crepina (a maiden but dam of the Yorkshire Cup winner Rangong), and Twilight Alley (bred like Crepello by a son of Donatello II, in his case by Alycidon). Twilight Alley, a massive horse who had only four races in his career, won the Ascot Gold Cup as a four-year-old but broke down in his next race, the King George VI and Queen Elizabeth Stakes. He retired to stud but has not been a success. Honeylight bred Come on Honey, dam of the fine staying filly Attica Meli (seven races including Yorkshire Oaks, Park Hill Stakes, Geoffrey Freer Stakes and Doncaster Cup).

MIEUXCE (1933)

Crepuscule was by Mieuxcé, who won five races in France including the Derby and Grand Prix de Paris, and was then bought for stud in England. He must be regarded among those stallions—like Straight Deal, Combat, Vimy, Worden II and Big Game—who disappoint as sires of winners but excel as sires of brood-mares. Indeed, Mieuxcé's list of broodmare successes was enormous, and it is an indication of his influence that he was last represented with a winner in Britain as late as the 1974 season—41 years after his birth.

His best winners were Commotion (Oaks) and Feu du Diable (2,000 Guineas), whilst Paddy's Point and Stokes finished second in the Derby and 2,000 Guineas respectively. Quantity was the keynote to Mieuxcé's success as a broodmare sire, but apart from the offspring of Crepuscule, he was also responsible for the dams of the financially successful Major Portion (six races, £24,358, including the Middle Park Stakes and St. James's Palace Stakes, and second in the 2,000 Guineas; successful sire); Marguerite Vernaut (Champion Stakes); and Saucy Kit (Champion Hurdle).

Mieuxcé was by the one-time French champion sire Massine, whose racing successes included the Ascot Gold Cup and important races in his native country, and whose achievements at stud were of the highest order. In addition to Mieuxcé he got the French Derby and Grand Prix de Paris winner Strip the Willow; the French St. Leger winner Laeken (who as a French National Stud horse sired that unusually good English hurdler Sir Ken); Maravedis (who got the Grand Prix de Paris, Ascot Gold Cup and French St. Leger winner Souverain); the French Gold Cup winner Chaudiere; the French 1,000 Guineas and Oaks winner Feerie, and so on. Massine's daughters produced many winners but, on the whole, he was perhaps a better sire of winners than of broodmares—in complete contrast to his son Mieuxcé.

Massine, who died in 1939, was by Consols (1908), who was by either St. Bris (1893) or more probably by Florizel II's son Doricles (1898), who after winning the English St. Leger went to stud in France. It will be remembered that Florizel II was by St. Simon and a full brother to the Derby winners Persimmon and Diamond Jubilee.

LA CAMARGO (1898)

Massine's dam Mauri (1909) was by Ajax out of La Camargo by Childwick. La Camargo was a wonderful filly who won top-class races in France at two, three, four, five and six years. In all she ran some 32 times and scored on 24 occasions, including successes in the French 1,000 Guineas, Oaks and Gold Cup.

Mieuxcé's dam L'Olivète, who won three races in France, was by the French Gold Cup second Opott (1910). This horse made but little imprint on the Stud Book, but he was a stoutly-bred horse, being by Maximum II, who came from France to win the 1903 Ascot Gold Cup, out of Oussouri by the Doncaster Cup winner Chesterfield out of Reve d'Or, who secured a number of races including the 1,000 Guineas and Oaks.

RED SUNSET (1941)

Returning to the distaff side of Crepello's pedigree, his grandam Red Sunset won one race worth £235 and produced four winners at home and one in France. Most money came via Mieux Rouge but as her racecourse earnings amounted to only £828, it is clear the others were of limited ability. Red Sunset was by Solario out of Dulce II, who won three races in France and ran second in their 1,000 Guineas before breeding four winners in England and Ireland, including Mehmany (£2,764). Dulce II was a full sister or half-sister to the important French stallion Astrophel; the English Derby second Fox Cub; Dorinda, who was placed in both the French 1,000 Guineas and Oaks; and Donnemarie II (by Donatello II), who went to America where her son Stepfather started 64 times and won 217,425 dollars before retiring to stud in California.

Crepello's fourth dam Dorina was successful six times including the French Oaks, whilst she was second in their 1,000 Guineas.

CHAPTER 26

1958 – 1963
Aloe, Pretty Polly and Sceptre

1958

1 Sir V. Sassoon's b c Hard Ridden (Hard Sauce – Toute Belle II).
2 F. N. Shane's b c Paddy's Point (Mieuxcé – Birthday Wood).
3 Mme. A. Plesch's ch c Nagami (Nimbus – Jennifer).
   20 ran. Time: 2 min 41.2 sec

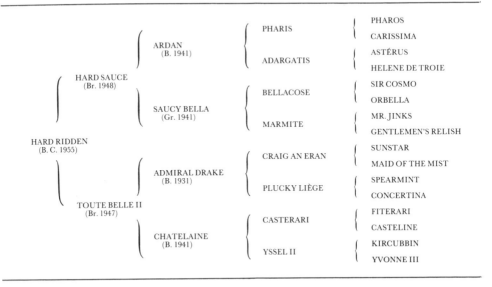

HARD RIDDEN'S SUCCESS was remarkable on several counts. He was only the second Irish-trained winner (after Orby in 1907); he was ridden by 51-year-old Charlie Smirke, and he had been bought for only 270 guineas by his owner, Sir Victor Sassoon, from Lady Lambert at the Dublin Yearling Sales. Described as "a lean, rangy colt, not impressive in looks", he baffled several observers with his win at Epsom, since his sire was a sprinter.

He served to emphasize that the influence of the dam can be just as important as that of the sire. He also bore out the views on inheritance of speed and stamina that Professor Robertson expressed around 1910, namely that "the mating of a sprinter with a stayer may result in anything from one end of the scale to the other".

Hard Ridden ran once as a back-end two-year-old, was beaten a short head first time out at three and then trotted up in the Irish 2,000 Guineas. He won the Derby decisively by five lengths from Paddy's Point. Sixth to the year-older Ballymoss in the King George VI and Queen Elizabeth Stakes, he was found to be lame after the race and did not run again.

He retired to stud with two wins worth £23,741, and for his first season, 1960, stood at a fee of 300 guineas. Exported to Japan in 1967, whereupon he became one of that country's leading sires, he cannot be judged anything other than disappointing over here. His best winners sired in Ireland were Hardicanute and Giolla Mear, along with the French Champion Hurdle winner Hardatit, though he sired plenty of minor winners, the majority of those that were successful at three years or more being at a mile and a quarter or further.

Hardicanute won three races for £31,938, including the Champagne Stakes and Timeform Gold Cup at two years, on his only starts, and at stud he has sired Hard To Beat (French Derby) and Hardiemma (dam of Shirley Heights). Giolla Mear won four races for £17,256, including the Irish St. Leger, and on retiring to stud in 1970 he emerged as a sire of useful National Hunt stock.

HARD SAUCE (1948)

The cause of surprise at Hard Ridden's Derby success was Hard Sauce, owned by Sir Victor Sassoon and never raced beyond six furlongs. Hard Sauce's victories included the July Cup and Challenge Stakes among six wins worth £5,097. Retired to stud in 1952, he got Hard Tack, a useful but temperamental sprinter in the same crop as Hard Ridden. Had the critics been able to see into the future, they might not have been so perplexed by the 1958 Derby result, for although Hard Sauce sired Bleep-Bleep (Nunthorpe Stakes) and Gazpacho (Irish 1,000 Guineas), he was also responsible for several useful middle-distance winners, such as Horse Radish (Northumberland Plate) and Saucy Kit (Champion Hurdle), not to mention Bonjour (third in the French Derby) and Miss Onward (Japanese 1,000 Guineas and Oaks).

Gazpacho went on to produce the stayer Cumbernauld, and mares by Hard Sauce also bred the sprinter Manacle and the Ascot Gold Cup winner Lassalle. Meanwhile, Hard Tack at stud was responsible for Lock Hard (Grand Handicap de Deauville) and Windfield Lily (sprinter who bred the Irish St. Leger winner Conor Pass), as well as the top-class milers Right Tack and Sparkler. Right Tack won eight races, worth £57,406, including the 2,000 Guineas, Irish 2,000 Guineas, St. James's Palace Stakes and Middle Park Stakes. Sparkler won 13 races, worth £107,361, including the Prix du Moulin, Lockinge Stakes and Queen Anne Stakes.

Ardan, sire of Hard Sauce, was an influence for stamina since he won the French Derby and Prix de l'Arc de Triomphe, and was disqualified after finishing first in the Grand Prix de Paris.

SAUCY BELLA (1941)

Saucy Bella, the dam of Hard Sauce, won two small races, one at two years and one at three. She bred four winners at up to a mile, the only other of importance being Sweet Pepper (by Nasrullah), dam of three winners including Game Bird, whose son Birdbrook was a tough individual that won 16 races.

Marmite, the dam of Saucy Bella, won two races over five furlongs as a two-year-old. At stud she bred six winners, including Soupçon, who barely stayed five furlongs as a two-year-old but who won two races and bred three sprint winners; and Madrilene, a Court Martial mare who ran only as a two-year-old and won two of her four races, both over five furlongs. Madrilene bred five winners, including the smart miler Petite Marmite (dam of the Prix Vermeille winner Paulista) and the less able Camenae, who followed her one win, in maiden company over 14 furlongs, by breeding the 2,000 Guineas winner and successful sire High Top. Madrilene also bred Fran, a non-winner and badly left in the last of her four runs in selling plates but the dam of Tudor Music (six wins, £23,602, including the Gimcrack Stakes and July Cup).

Gentlemen's Relish stayed better than her daughter Marmite, for her 4½ wins, worth £2,841, included the Bessborough Stakes and a share of the Ebor Handicap. At stud her two other winners in Britain were Goudyswell, and Marmite's sister Raven Locks, who was the dam of Star Lyon (eight races, £5,269, and placed in several good handicaps including the Ebor), Periwig (three races, £3,744), Lindrick (three races and fourth in the 2,000 Guineas) and Aggravate (three races, £5,667, including the Park Hill Stakes).

TOUTE BELLE II (1947)

Bred by M. Leon Volterra, Toute Belle II was weeded out of his string and bought cheaply by Lady Lambert. She ran only once, was not an immediate success at stud and went up to the Newmarket December Sales of 1953, where she failed to make her reserve. She stayed on to be covered by Hard Sauce, and the produce was her second living foal, Hard Ridden (her first foal to live ran under NH Rules). Before Hard Ridden was a year old, Toute Belle II was back at the December Sales (1955), where she was sold for 160 guineas. The following year, in foal to Patton, she was sold to a breeder in Peru for £700.

Toute Belle II was by Admiral Drake out of Chatelaine, a mare who won one race at two years. Chatelaine bred one winner, Le Bourgeois, a stayer that was successful in France and England and ran third in the Ascot Gold Cup, before she was sent to Canada.

Chatelaine's sire Casterari ran second in the Arc de Triomphe and Prix du Cadran, while her dam, Yssel II (by Kircubbin), bred eight winners including the tough stayer Vatelys (Prix Gladiateur and French Champion Hurdle).

ALCIDE (1955)

The 1958 Derby is remembered as much for Alcide's absence as Hard Ridden's victory. Having won the Chester Vase and then the Lingfield Trial by 12 lengths, Alcide was favourite for Epsom, but he missed the race because he was "nobbled" in his stable. He came back to win the Great Voltigeur Stakes by 15 lengths and the St. Leger. At four years he won three races before being held up in his work and beaten inches by Wallaby II in the Ascot Gold Cup, after which he won the King George VI and Queen Elizabeth Stakes over a mile shorter.

Alcide retired after winning eight races worth £56,042, and cannot be regarded as an outstanding stallion, with several of his produce showing temperamental defects. His best winners were Oncidium (Jockey Club Cup, Coronation Cup; leading sire in New Zealand); Approval (Observer Gold Cup); Atilla (Vaux Gold Tankard, Grosser Preis von Baden); Alignment (Ebor Handicap); Remand (unbeaten two-year-old including Royal Lodge Stakes; Chester Vase; fourth in Derby); Sea Anchor (Doncaster Cup) and Grey Baron (Goodwood Cup, Jockey Club Cup).

It was not long before the connections of Alcide gained compensation.

1959

1 Sir H. de Trafford's b c Parthia (Persian Gulf – Lightning).
2 G. A. Oldham's b c Fidalgo (Arctic Star – Miss France).
3 Baron G. de Rothschild's b c Shantung (Sicambre – Barley Corn).
   20 ran. Time: 2 min 36.0 sec

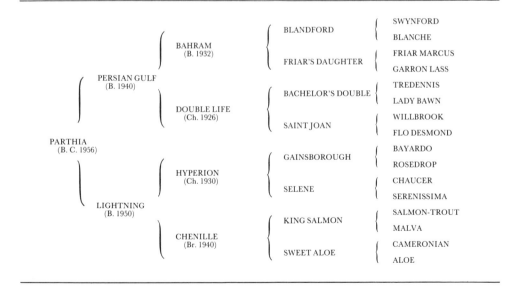

Parthia put the Derby record straight for his connections after their disappointment with Alcide, and following the latter's St. Leger win, he provided a hitherto-overshadowed branch of the Aloe family with its second classic winner in successive years.

Two late-season races as a two-year-old revealed Parthia to be a colt with classic potential, for in the second he finished less than three lengths third to Billum in the Dewhurst Stakes. He won his three races before the Derby, but was progressively less impressive, giving the impression he was short of pace when winning the Dee Stakes at Chester and the Lingfield Trial. Connections believed he was idle, not one-paced, and he was worked in blinkers before Epsom, where he led two furlongs out and drew clear close home to beat Fidalgo by a length and a half, with Shantung unlucky in running and a strong-finishing third.

Parthia missed the Great Voltigeur Stakes because of coughing, and though odds-on for the St. Leger, he was only fourth to the filly Cantelo, with Fidalgo again second. Parthia was found to have cut a hind leg after the St. Leger and did not run again that season. He stayed in training and won two races including the Jockey Club Cup, but was beaten by Petite Etoile in the Coronation Cup, and ran sixth in the King George VI and Queen Elizabeth Stakes, where his appetite for racing seemed to have disappeared.

A handsome, powerful, deep-bodied colt, he retired with six wins worth £43,786 to his credit, and was the first Derby winner since Airborne (who was by Persian Gulf's half-brother Precipitation) not to have a French or Italian sire or paternal grandsire.

Syndicated to stand at stud in Newmarket from 1961, he was exported in 1968 to Japan, where he died in 1970. A stayer who was not patronized for speed, his best produce in Britain were all fillies—Sleeping Partner (Oaks); Parthian Glance (Yorkshire Oaks, Park Hill Stakes, Ribblesdale Stakes); Every Blessing (Princess Elizabeth Stakes); Parsimony (Cork and Orrery Stakes, July Cup; unlike most of the others a sprinter, which is probably attributable to her dam, Money For Nothing, by Grey Sovereign); and Sentier (Horris Hill Stakes).

PERSIAN GULF (1940)

Parthia's Derby success enabled his sire Persian Gulf to emulate his half-brother Precipitation (by Hurry On), sire of the 1946 winner Airborne. Persian Gulf retired to stud in 1946 after a war-time racing career in which he earned £2,123 in winning stakes, principally from the Coronation Cup as a four-year-old. His best winners before Parthia were Zabara (1,000 Guineas) and Zarathustra (Irish Derby, Ascot Gold Cup), while later his daughter Plaza, a miler, produced the St. Leger winner Intermezzo.

Persian Gulf's immediate family has been discussed in relation to Airborne.

LIGHTNING (1950)

Lightning won the Sandwich Stakes at Ascot as a two-year-old, her only season to race. She bred one foal before Parthia, the filly Media (also by Persian Gulf), who won a Windsor nursery in 1957 but failed to win over middle distances as a three-year-old, and at stud bred two minor winners in Britain.

Lightning bred four further winners: Rainstorm (by Premonition), three races on the Flat including the Newbury Autumn Cup, and over hurdles and fences; Monitor (by Worden II), one maiden race for £884; Village (by Charlottesville), one maiden race for £345 and dam of the useful handicapper Yamadori; and the filly Flash of Light (by Charlottesville), one maiden race for £207.

CHENILLE (1940)

Lightning's dam and grandam never ran. Chenille, her dam, was bought at the 1948 Newmarket December Sales by trainer Marcus Marsh on behalf of Sir Humphrey de Trafford for 3,100 guineas to fulfil a nomination to Hyperion (which eventually produced Lightning). Chenille was bought in foal to Borealis, the produce being Papillio, winner of six races, £3,696½, including the Goodwood Stakes.

Papillio was Chenille's second winning produce, following the Hamilton Park maiden-race winner Proscenium (by Wyndham). After foaling Lightning, she returned to Hyperion, the produce being Thunder, winner of two races, and dam of Lionhearted, who managed to remain a maiden despite finishing second in the Irish Sweeps Derby, and the winning staying handicapper Thundridge.

Chenille's dam Sweet Aloe (by Cameronian) bred one winner, Pontypridd (by Pont l'Eveque), who according to a contemporary account "beat six bad horses in a £138 plate for maiden fillies over 1½ miles at Folkestone". Pontypridd bred Mr. McTaffy, Jock Scot's first successful progeny, who won several races on the Flat and under NH Rules from the age of two to 11 and was still running at 14; and Miss McTaffy, a useful staying filly whose seven wins included the Great Metropolitan, and dam of six winners.

ALOE (1926)

Sweet Aloe had not hitherto made much impact in supplying big-race successes for the family of her dam Aloe, whose importance as a broodmare was charted in Chapter 9. Aloe's success at the paddocks easily outweighed her significance on the racecourse, since she was unraced at two years, and as a three-year-old failed to win in ten races, when though second in the Nassau Stakes and third in the Falmouth Stakes, she seemed to run out of enthusiasm when dropped in class towards the end of the season.

A half-sister to the Ascot Gold Cup winner Foxlaw, Aloe bred five winners. It was from the best and most productive of these, Feola, that the family's most important winners were descended, and these will be discussed in the next section in connection with St. Paddy's sire Aureole, whose grandam was Feola.

1960

1   Sir V. Sassoon's b c St. Paddy (Aureole – Edie Kelly).
2   Sir R. Brooke's ch c Alcaeus (Court Martial – Marteline).
3   E. R. More O'Ferrall's b c Kythnos (Nearula – Capital Issue).
        17 ran. Time: 2 min 35.6 sec

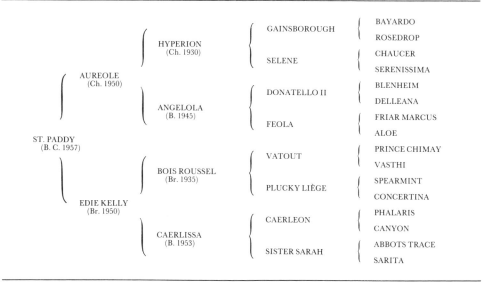

St. Paddy was given 8st 13lb in the Free Handicap after his two outings as a two-year-old, when he was a beaten favourite on his debut but made no mistake in the Royal Lodge Stakes, which he won impressively by five lengths from four others. His first race at three years was in the 2,000 Guineas, where he finished sixth, six lengths behind the winner Martial, and he went for the Derby after winning the Dante Stakes easing up by three lengths.

Favourite at Epsom was the French colt Angers, who shattered his near foreleg not long after halfway. St. Paddy, always a free-running horse who was a handful for his jockey in his early days, took some restraining but put up a smooth display to beat Alcaeus by three lengths. St. Paddy jarred himself at Epsom, then had the cough and was beaten a neck on his return for the Gordon Stakes at Goodwood. He won the Great Voltigeur Stakes and the St. Leger.

As a four-year-old St. Paddy won his first three races—a small event at Sandown; the Hardwicke Stakes from the St. Leger second and third, Die Hard and Vienna; and the Eclipse Stakes, in which he made all and broke the course record. He seemed to be more impressive with each race, but proved to be a horse who looked to have more in hand than was the case, and in his last three races, though beating a small field for the Jockey Club Stakes, he was beaten by Right Royal V in the King George VI and Queen Elizabeth Stakes and by Bobar when odds on for the Champion Stakes.

A handsome horse who grew stronger and more muscular, St. Paddy retired having won nine races worth £97,193, and was syndicated to stand his first season at stud in 1962. He has been a sound but not outstanding sire.

His most successful produce was Connaught, winner of seven races worth £69,212, who though useful as a three-year-old, when he won the King Edward VII Stakes (by 12 lengths) and Great Voltigeur Stakes (on a disqualification) and was second in the Derby, improved with age. He won the Prince of Wales Stakes (by five lengths in record time) as a four-year-old, and again as a five-year-old, when he was unbeaten and broke another record when winning the Eclipse Stakes.

St. Paddy's other notable produce were Parnell, 14 races worth £50,475, including the Prix Jean Prat twice, Irish St. Leger, Queen's Vase, Jockey Club Cup, second in the Prix du Cadran, King George VI and Queen Elizabeth Stakes and Washington International as a four-year-old; Sucaryl, three races worth £28,148, including News of the World Stakes, second in Irish Sweeps Derby, exported to New Zealand; Patch, two races in England worth £16,533, including Great Voltigeur Stakes, and one in Italy, second beaten a head in French Derby; St. Chad, five races worth £14,206, including Wills Mile, Jersey Stakes and Hungerford Stakes, stud in Ireland before being exported to Brazil in August 1976 and sired the very useful sprinting filly Street Light, and the tip-top miler Court Chad; St. Pauli Girl, second in 1,000 Guineas and Oaks; and Welsh Saint, seven races worth £8,538 in Britain and 31,650 francs in France, promising sire.

ANGELOLA (1945)

St. Paddy avenged the Derby defeat of his sire Aureole, who was the first foal of the useful middle-distance racemare Angelola, unraced at two years but the winner at three of the Lingfield Oaks Trial, Yorkshire Oaks, Princess Royal Stakes and Newmarket Oaks for £6,198, and second in the Epsom Oaks.

Angelola bred four winners, but while Aureole amassed £36,226, the other trio managed a total of £1,904. They included Aureole's sister Angel Bright, who won the Lingfield Oaks Trial but did nothing in the Oaks itself and after an unsuccessful spell at stud was sold at the 1960 Newmarket December Sales for 22,000 guineas and went to America.

FEOLA (1933)

Angelola represented the most successful branch of the Aloe family, being a daughter of Feola, whose ten living foals included seven winners and have made her one of the most influential mares this century.

Bought on behalf of King George V as a yearling for 3,000 guineas, Feola was unplaced six times as a two-year-old but at three years won two races worth £1,267. Her victories came after the death of her owner and she carried the colours of Lord Derby. She was also second in the 1,000 Guineas and third in the Oaks.

Feola's first foal was Foretaste, who won one small race before being exported to France, where she bred the mare Windy Cliff, dam of the Ascot Gold Cup winner Lassalle. Feola's other winning produce included Kingstone (nine wins, £6,391) and the influential mares Above Board, Knight's Daughter and Hypericum.

Knight's Daughter was among the best two-year-old fillies of her generation and won three races. Exported to America in 1952, she became the dam of Round Table (43 wins, 1,749,869 dollars, including American Derby; sire of 2,000 Guineas and Champion Stakes winner Baldric II); Monarchy (seven wins, 85,737 dollars, including Arlington Lassie Stakes); and Love Game (minor winner in England; in America dam of the stakes winner Road House II).

Hypericum (by Hyperion) won three races, including the 1,000 Guineas, and bred six winners. Her winners included Restoration (King Edward VII Stakes, second in Eclipse Stakes, stud failure in Argentina); Belladonna (one race, second in Newmarket Oaks, exported to Italy and dam of Ben Marshall, whose ten wins before export to Japan included the Italian St. Leger); Prescription (exported to Chile and grandam of the Chilean Oaks winner Nisapur and Chilean 2,000 Guineas winner Nagvilan); and Highlight (two races at 1½m, dam of five winners including the 1,000 Guineas and French Oaks winner Highclere).

Above Board won two races, the Cesarewitch and Yorkshire Oaks, and bred four winners, two of little consequence and two out of the top drawer. Her top-notchers were Doutelle (seven wins, £10,400, died young but sired the Eclipse winner Canisbay) and Above Suspicion (two wins, St. James's Palace Stakes and Gordon Stakes, for £9,226).

Feola was also responsible for Starling, who was unplaced on the racecourse but made her mark at stud in Argentina, where to matings with the local Derby winner Seductor she produced Sideral (nine wins, three times leading sire and four times leading broodmare sire in his native country), Siderea (best mare of her generation, six wins including Argentinian Oaks) and Sagitaria (top two-year-old of her generation).

EDIE KELLY (1950)

St. Paddy is from the family of Admiration, which as was explained in Chapter 9 is probably too remote to be worth retaining the reference nowadays. Even the identification with Admiration's daughter Pretty Polly seems too indistinct, because of the various branches. St. Paddy represents the branch emanating from Pretty Polly's daughter Molly Desmond—though before he had run there was no great encouragement from this fact to suggest that the undistinguished immediate family of his dam would be capable of producing a top-class winner.

In two seasons and 14 races Edie Kelly, St. Paddy's dam, managed to win once, in an apprentice plate worth £133 at The Curragh. Sold for 1,750 guineas when her racing days were over, she passed into the hands of St. Paddy's breeder, Sir Victor Sassoon, for 3,500 guineas in 1955. Her first foal Ben Beoch (by Golestan) won four small sprints worth a total of £884 and was sold for 170 guineas as a five-year-old, in the year that St. Paddy won the Derby. In between, Edie Kelly's second foal Fighting Edie (by Guersant) had failed to win but was placed at two and three years.

St. Paddy signalled the turning-point in Edie Kelly's fortunes, though there were ten years between the Derby winner and the next-best produce from his dam. That was Parmelia (by Ballymoss), who won the Ribblesdale and Park Hill Stakes for £9,420, and was second in the Irish Oaks. The following year Edie Kelly foaled Parmelia's brother Balios, who won the Ascot Stakes. Between

St. Paddy and Parmelia, Edie Kelly threw four winners of no great consequence.

Fighting Edie deserves further mention, since though incapable of winning a race, she proved herself at stud, where she is the dam of Felicio II and Formentera. Felicio II won the Grand Prix de Saint-Cloud and Prix Jean de Chaudenay, and was second, beaten half a length by Royal Palace, in the King George VI and Queen Elizabeth Stakes. Formentera (by Ribot) was a very moderate racemare but her first foal was Flying Water, winner of the 1,000 Guineas.

While Edie Kelly managed only one small win, her dam Caerlissa was never even put into training, so badly deformed was one of her feet. She was virtually given away by her breeder, Lt. Col Giles Loder, who sold her as a seven-year-old uncovered mare for 55 guineas. What a bargain she proved, since she bred eight winners. Only one of them won at further than a mile. Her most notable produce other than Edie Kelly were The Web (Anglesey Stakes and three other races for £3,717, sent to America and won there) and Who You (five races including Gosforth Park Cup).

SISTER SARAH (1930)

Caerlissa was the first of 20 foals produced by Sister Sarah, who had raced only as a two-year-old, winning twice over five furlongs. She was not covered as a three-year-old, but more than made up for that before she died six hours after foaling her last produce, Welsh Abbot, in 1955.

Welsh Abbot was Sister Sarah's eighth winner and her best since Lady Sybil 15 years earlier. Lady Sybil (by Nearco) won six races and was the leading two-year-old filly of 1942. She became the dam of Count Rendered (useful middle-distance horse and champion sire in New Zealand), his sister Reckless Lady (winner and dam of the useful sprinter Klondyke Bill), Nepthos (eight races and leading sire in Greece) and Esquire Girl (three wins, £1,236, dam of eight winners from ten foals including Lucyrowe, six wins, £22,202, including Coronation Stakes and Sun Chariot Stakes).

Welsh Abbot (by Abernant) won five races worth £6,883, including the Portland Handicap, and was a fairly useful sprint sire. Sister Sarah's other winners included Black Peter (by Blue Peter), whose five wins included the Jockey Club Stakes, and Lady Angela (by Hyperion), winner of one race worth £276 and better known as the dam of 13 winners from 15 foals in North America including Nearctic (21 wins, 152,384 dollars, Canadian Horse of the Year 1958, sire of Northern Dancer).

To a further mating with Nearco, Sister Sarah produced Sybil's Sister, another whose importance emerged when she went to the paddocks, where she bred seven winners, including Sybil's Nephew (Newmarket St. Leger and five other races; leading sire in South Africa) and Sybil's Niece (three wins, £4,135, including Queen Mary Stakes; dam of Grundy's sire Great Nephew).

Sarita, Sister Sarah's dam, produced one other winner, The Dolphin, whose two victories earned £654. Sarita, a half-sister to the Irish Classic winners Spike Island and Zodiac, won once over a mile for £445 from her seven races, but was last of seven in the Park Hill Stakes, where she probably did not stay the trip. Sarita was by Swynford out of Molly Desmond.

1961

1   Mme A. Plesch's ch c Psidium (Pardal – Dinarella).
2   Mme L. Volterra's b c Dicta Drake (Phil Drake – Dictature).
3   Mrs C. Iselin's ch c Pardao (Pardal – Three Weeks).
    28 ran. Time: 2 min 36.5 sec

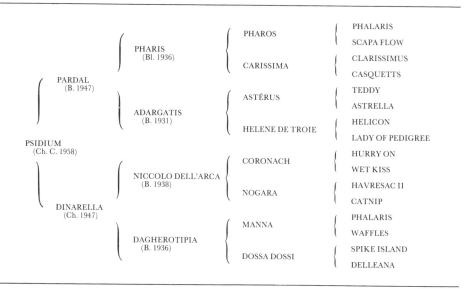

Psidium was a remarkable Derby winner in that he was trained in England, was bred by a Hungarian in Ireland from an Italian mare, and was ridden at Epsom by a Frenchman. Added to that, he was the second 66-1 winner of a Classic that year, following Rockavon in the 2,000 Guineas.

Psidium had seven races as a two-year-old and may have been flattered on 8st 4lb in the Free Handicap, 17lb behind the top horse, the filly Opaline II, after beating three previously unraced rivals in the Duke of Edinburgh Stakes and finishing second in the Horris Hill and Dewhurst Stakes.

As a three-year-old, Psidium finished a close third in the Kempton 2,000 Guineas Trial, fourth in the Prix Daru, and 19th of 22 in the 2,000 Guineas itself. He went to Epsom having been beaten by eight of the field, and ridden by his seventh different jockey. It was not easy to explain his Derby win, especially as he came from last place six furlongs out, but he was running over a mile and a half and on firm ground for the first time. Having beaten Dicta Drake by two lengths, he never got the chance to show whether this was a fluke or not, since in a gallop prior to the Gordon Stakes he pulled up lame with tendon trouble and never ran again.

A tall, strong, well-made horse, he retired to stud at Newmarket in 1962 at a fee of 500 guineas. His two wins had been worth £37,048. Before export to Argentina in 1970, Psidium had only one produce that won more than £10,000 in a season. That was Sodium, winner of the St. Leger and Irish Sweeps Derby in 1966, when his three victories were worth a total of £96,232, and helped put Psidium top of the sires' list with earnings of £100,927—an indication of the false position a stallion can get in this particular league table. One swallow does not make a summer, and one winner does not make a stallion!

PARDAL (1947)

The 1961 Derby was a triumph for Psidium's sire Pardal, who was also responsible for the third horse, Pardao, winner of four races in England for £10,922, including the Jockey Club Cup. Pardao went to America, where he won two races worth 84,400 dollars, before being brought back to stand at stud in Newmarket from 1964. Two of his best produce, Sovereign and Moulton, were trained by Harry Wragg, trainer of Psidium. Sovereign won the Queen Mary Stakes and three other races at two years when she was the leading filly, and the Coronation Stakes at three. Moulton, who was out of a half-sister to Sovereign's dam, won six races for £97,041, including the Benson and Hedges Gold Cup and was second in the Eclipse Stakes. Pardao's only other produce of any consequence was Spoiled Lad, winner of the Blue Riband Trial Stakes and Extel Handicap before his export to Australia.

Pardal, owned and bred by M. Boussac, did not win until the age of four but showed decent form just below the top class. He ended his racing career with six wins, worth £10,171 and 579,775 francs, including the Prince of Wales's Stakes and Jockey Club Stakes, before being bought for £50,000 to stand at stud in England.

Pardal's best produce before Psidium and Pardao were Chantelsey (second 2,000 Guineas), Firestreak (City & Suburban, sire), London Cry (Cambridgeshire) and Eudaemon (Gimcrack and Champagne Stakes). He later sired two Ascot Gold Cup winners in Pardallo (who followed Psidium to Argentina and has done well at stud there) and Parbury. Pardal died in 1971.

Pardal was a brother to Ardan, their dam being the Astérus mare Adargatis, who beat the French 1,000 Guineas winner Mary Tudor II when she won the French Oaks in 1934. Ardan, six years older than Pardal, won 15 races, ranging from the Prix Robert Papin as a two-year-old to the French Derby, Arc de Triomphe and Coronation Cup, as well as being disqualified from the Grand Prix de Paris. As a sire in France and America, to where he was exported in 1949, he was responsible for Hard Sauce (sire of Hard Ridden), Dacia (Italian Oaks and Gran Criterium), Damaka (Grand Prix de Deauville) and Miss Ardan (Canadian Oaks).

HELENE DE TROIE (1916) and LA TROIENNE (1926)

Adargatis was out of Helene de Troie, dam five years earlier of an even more notable daughter in La Troienne (by Teddy), who bred ten winners from 14 foals and founded one of the most spectacularly successful families in American racing history.

Bred by M. Boussac, La Troienne was thought good enough to run in the French 1,000 Guineas but was unplaced, and the best she did when sent to be trained at Newmarket was a second and third in small races over seven furlongs and five furlongs. She was covered by Gainsborough in her first season at stud, 1930, and at that year's Newmarket December Sales was sold for 1,250 guineas to the British Bloodstock Agency, acting for Col. Edward Riley Bradley, who founded one of the most famous breeding establishments in Kentucky, Idle Hour Farm (later Darby Dan Farm).

The Gainsborough foal did not live, but in America La Troienne's ten winners included five in stakes races. She did particularly well to matings with her owner's stallions Black Toney, Bubbling Over and Blue Larkspur.

To Black Toney, La Troienne bred:

Black Helene: La Troienne's first foal conceived in America; a small but very able racemare who won the Coaching Club American Oaks, Florida Derby and American Derby;

Bimelech: won 11 races, unbeaten as a two-year-old, surprisingly beaten in the Kentucky Derby but won the Preakness and Belmont Stakes; not outstanding at stud but did sire Be Faithful, the grandam of Never Bend;

Big Hurry: speedy and won four races; dam of 12 winners including Searching, who won 25 races for 327,381 dollars and became the dam of six winners including Affectionately (28 races, 546,660 dollars, and dam of the Preakness Stakes winner Personality) and Priceless Gem (seven races, sold for record broodmare price of 395,000 dollars in 1970, the year her daughter Allez France was born).

To Bubbling Over, the Kentucky Derby winner, La Troienne bred Baby League, who won once and bred nine winners, including Busher (champion filly in 1945, 13 wins, including Hollywood Derby, for 334,035 dollars; dam of Jet Action, 308,225 dollars) and Striking (three wins, dam of Glamour, six wins and dam of eight winners including the St. Leger winner Boucher, and Intriguing, dam of the 1971 champion American filly Numbered Account, whose earnings of 446,595 dollars were a record for that age and sex). Baby League was also the dam of the unraced filly La Dauphine (by Princequillo), who bred the very useful two-year-old Bucks Nashua and the Gordon Stakes winner Guillaume Tell.

To Blue Larkspur, La Troienne bred:

Belle of Troy: unraced; dam of Cohoes (13 wins, 210,850 dollars, sire of Quadrangle, ten races, 559,386 dollars, including Belmont Stakes) and Immortality (unraced; sire of 1,000 Guineas winner Fleet before export to Argentina);

Big Event: won two small races; dam of six winners from eight foals and the unraced Blackball, dam of Malicious (14 wins, 317,237 dollars) and The Axe (two wins in England including Imperial Stakes, and 13 wins in America, 393,391 dollars);

Businesslike: non-winner; bred eight foals, all winners, including Busanda, ten races, 182,460 dollars, dam of Buckpasser (25 wins, 1.46 million dollars, one of the best American racers since the war) and Bupers (seven wins, 221,688 dollars);

Belle Histoire: non-winner; dam of Royal Record II, 12 wins in America, stood first season at stud in Britain in 1964 but exported to Japan in 1968, sire of Seventh Bride (Princess Royal Stakes, dam of Oaks winner Polygamy).

DINARELLA (1947)

Dinarella, the dam of Psidium, was the first mare bought by the Plesch family. She had won one small race at two years and been second twice and fourth in the Italian Oaks when she was bought from the Dormello Stud in Italy in 1955 for almost £3,000. She was in foal to Tornado and the resulting produce, named Thymus, won the French 2,000 Guineas before being exported to South Africa. After Psidium, Dinarella bred Rosanella II (by Alizier), a stayer who won three races in France in 1956, and Notonia (by Never Say Die), a maiden-race winner over a mile and a half.

Dinarella was one of seven winners bred by Dagherotipia (by Manna) but none matched her achievements. Dagherotipia emulated her dam Dossa Dossi by winning the Italian 1,000 Guineas, as well as two other races, and was one of five winners bred by her dam, who was a half-sister to Donatello II. The career of Delleana, dam of Dagherotipia and Donatello II, was discussed in relation to the 1953 Derby winner Pinza.

Psidium was inbred to Phalaris at the fourth remove. His Derby was the second in as many years for the Pretty Polly family, since Duccia di Buoninsegna, fifth dam of Psidium, was a granddaughter of Pretty Polly.

1962

1   R. Guest's ch c Larkspur (Never Say Die – Skylarking)
2   M. Boussac's b c Arcor (Arbar – Corejada).
3   Mme. L. Volterra's b c Le Cantilien (Norseman – La Perie).
      26 ran. Time: 2 min 37.6 sec

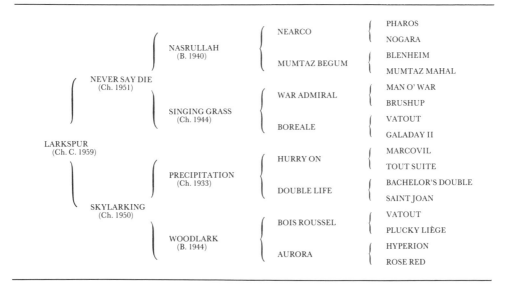

Though it was not obvious at this time, a new era was about to dawn for the Derby and British bloodstock breeding, since Larkspur provided a first success in the race for his trainer Vincent O'Brien, and he had been expensively bought by an American owner. The combination of O'Brien, American interests and money was soon to assume greater significance.

The O'Brien stable ran two in the 1962 Derby—Sebring, who had been bought for 13,000 guineas as a yearling, and Larkspur, a January foal who cost 12,200 guineas at the same stage of his career. Larkspur, a neat, strong colt, was one of the smallest Derby winners at 15.2½ hands. He was also unusual for the fact he had his first race over five furlongs in a race worth £200 to the winner at The Curragh in May of his first season. He was beaten there but reappeared to win over seven furlongs at Leopardstown before running third, beaten less than a length, in the National Stakes, and seventh in the Timeform Gold Cup.

Before the Derby Larkspur was beaten over seven furlongs but then won over a mile and a half at Leopardstown. His preparation for Epsom was held up after he developed a thoroughpin (an inflammation of the flexor tendon sheath over the hock) and only on his owner's insistence did he run in the English Derby and not the French equivalent. The Derby he did win has been described as "no more completely unsatisfactory" race, since of the 26 that ran, seven fell and six more were hampered in a pile-up six furlongs out.

Larkspur recovered after the favourite Hethersett had fallen in front of him, and after leading two furlongs out won comfortably by two lengths. There was an inquiry into the falls, but no-one was found to be at fault. However, the statement added: "The stewards regret that such a large number of horses not up to Classic standard were allowed by their owners and trainers to start." Larkspur never won again.

He suffered another thoroughpin before the first-ever Irish Sweeps Derby, in which he finished fourth on firm ground, after which he was kept off the course for two months with sore shins. He was a short-head second in the Blandford Stakes, but only sixth in the St. Leger won by Hethersett.

Larkspur stayed in training at four years but did not run again, and was syndicated to stand at stud in Ireland from 1964, his three wins having been worth £38,080. In December 1967 an offer of £100,000 was accepted and he left for Japan. From four crops in Ireland he sired the winners of 79 races and £100,664, the best being Crazy Rhythm (Ebor Handicap) and Wenona (Blandford Stakes).

Being by Never Say Die, Larkspur became the first Derby winner to be sired by a similar winner since Mahmoud (by Blenheim) in 1936.

SKYLARKING (1950)

Larkspur was the third of nine winners bred by Skylarking, though only two of the others won more than £1,000 in a season. They were Rising Wings (by The Phoenix), winner of two races in Ireland as a two-year-old and fourth in the Irish Oaks, and Ballymarais (by Ballymoss), winner of three races for £7,995, including the Dante Stakes, and fourth in both the Irish Sweeps Derby and Doncaster St. Leger before export to Japan. Rising Wings went on to produce Meadow Pipit, who won both her races, one at two years and one at three, for £825, and was the dam of Meadowville (six wins, £17,661, including Great Voltigeur Stakes, Jockey Club Stakes, and second in St. Leger) and Nuthatch (Nijinsky Stakes).

Skylarking had raced in Lord Derby's colours and won three races up to a mile and threequarters at three years, when she was also third in the Park Hill Stakes. She was sold at that year's Newmarket December Sales for 4,000 guineas to Mr. Philip Love, who usually sold the colts from his Irish stud and retained the fillies for racing. No-one can say that Skylarking owed him a penny, for her own produce, and those of her daughters, regularly found their way into the top-priced yearlings sold in Ireland. The following details show how the commercial breeder can turn the right pedigree into a goldmine—until, of course, either fashions change, or the pedigree is shown to be less successful than was at one time hoped.

Produce of Skylarking:

YEAR SOLD	NAME	PRICE GNS.	
1957	Rousseau's Dream	9,500	Dublin sale top
1959	Vimy filly	5,900	
1960	Larkspur	12,200	Dublin sale top
1961	Clear Air	4,000	
1963	Ballymarais	8,000	
1964	Charlottesville Flyer	16,400	Dublin sale top
1965	Larkhill	15,800	Dublin sale top
1966	Phoenix Bird	13,000	Dublin sale top
1968	Mr. Mixer	13,000	Dublin sale top
1969	Charlottesville colt	3,400	

Produce of Rising Wings:

1961	Arise	6,500	
1963	Menloe	12,500	7th highest of yr.
1966	Golden Gripper	7,400	
1971	Old Connell	3,600	
1972	Sea Kestrel	7,800	
1973	Petingo filly	6,000	
1974	Captain's Wings	3,000	sold privately

Produce of Meadow Pipit:

1968	Meadowville	10,500	
1969	Simead	7,100	
1971	Meadow Lady	6,400	
1974	Young Pip	13,000	

NB: Rising Wings herself was sold, in foal to Sea Hawk II, for 7,200 guineas in 1970. The resulting foal was Sea Kestrel.

By way of coincidence in discussing money and breeding, the 1953 December Sales produced another bargain. Two lots before Skylarking, from the same draft of four three-year-old fillies submitted by Lord Derby, was Barley Corn, a well-bred maiden out of the dam of the St. Leger winner Herringbone and the

Derby third Swallow Tail. Barley Corn had had five races over two seasons when trained in the north of England, but even modest company had been too good for her and she was left at the start on her final outing. Pedigree alone carried her up to 4,100 guineas at Newmarket, where she was bought by the Curragh Bloodstock Agency and sent to France for the Rothschild family.

In France, Barley Corn bred:

Dame d'Atour: unraced first foal; dam of Roi Dagobert (top two-year-old of his year and best three-year-old) and Ruta (sprint winner in Ireland; dam of French Derby and Arc de Triomphe winner Sassafras);

Shantung: unlucky third in the Derby; sire;

Imberline: third in Oaks; dam of four good-class winners including Percale (dam of Paysanne, who dead-heated with San-San for Prix Vermeille);

Diagonale: dam of Djakao (Grand Prix de Deauville; third in French Derby; sire).

It must be a toss-up which was the better bargain, Skylarking or Barley Corn. And one can only guess whether the fortunes of Lord Derby's bloodstock, now sadly only a shadow of their former glories, would have been saved by the retention of either or both these subsequently most successful mares.

WOODLARK (1944)

Skylarking's dam, Woodlark, won one small race, worth £207, over six furlongs as a two-year-old. She did not reappear until October at three years, when she showed only modest form in three races. At stud, before being sent to America in 1956, she bred three winners apart from Skylarking, the best and most successful being Admiral Byrd, whose seven wins were gained from one mile to two miles. Woodlark also bred Sylvan, a winner and dam of the sisters Beaufront (Lingfield Oaks Trial) and Muscosa (dam of the Irish Sweeps Derby third Wedding Present).

Woodlark, by Bois Roussel, was a very well-bred filly, since her dam, Aurora, also bred the brothers Alycidon and Acropolis, Agricola (winner of the Newmarket Stakes and a great success at stud in New Zealand) and Borealis (whose seven wins included the Coronation Cup). Aurora, who won one seven-furlong maiden event for £363 and finished second to Galatea II in the 1,000 Guineas, also bred Woodlark's sister Amboyna, a non-winner but grandam of the Belmont Stakes winner and successful sire Celtic Ash.

Aurora was from the first crop of Hyperion, out of Rose Red, the daughter of Marchetta whose influence has been chronicled in Chapter 9.

Larkspur was inbred at the fourth remove to Vatout.

1963

1   F. Dupre's b c Relko (Tanerko – Relance III).
2   Sir Foster Robinson's ch c Merchant Venturer (Hornbeam – Martinhoe).
3   J. R. Mullion's b c Ragusa (Ribot – Fantan II).
      26 ran. Time: 2 min 39.4 sec

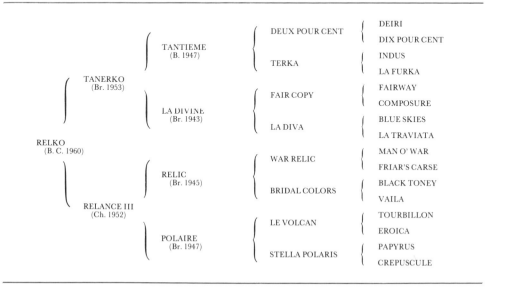

			DEIRI
		DEUX POUR CENT	
	TANTIEME		DIX POUR CENT
	(B. 1947)		INDUS
		TERKA	
TANERKO			LA FURKA
(Br. 1953)			FAIRWAY
		FAIR COPY	
	LA DIVINE		COMPOSURE
	(Br. 1943)		BLUE SKIES
		LA DIVA	
RELKO			LA TRAVIATA
(B. C. 1960)			MAN O' WAR
		WAR RELIC	
	RELIC		FRIAR'S CARSE
	(Br. 1945)		BLACK TONEY
		BRIDAL COLORS	
RELANCE III			VAILA
(Ch. 1952)			TOURBILLON
		LE VOLCAN	
	POLAIRE		EROICA
	(Br. 1947)		PAPYRUS
		STELLA POLARIS	
			CREPUSCULE

Relko was a highly-strung horse who improved physically through his career. He won his first two races at two years, but his fourth place in the Grand Criterium and second in the Prix Thomas Bryon showed him to be some way behind the best of his age. He came into his own as a three-year-old and was unbeaten in his first four races, including the French 2,000 Guineas, the Epsom Derby (where he swept into the lead two furlongs out and beat Merchant Venturer by six lengths) and French St. Leger (from a field that included the French Derby and Grand Prix de Paris winner Sanctus). Relko sweated up when only sixth in the Arc de Triomphe, by which time he was probably over the top.

Despite this record of success, there were two elements of controversy about Relko's three-year-old career. There was the inquiry into his dope test after the Derby, when a non-normal nutrient was found present in the urine but not the saliva, as a result of which the Jockey Club urged that no action be taken. And he fell lame on the way to the start of the Irish Sweeps Derby, whereupon he had to be withdrawn, though he was recovered the following day.

Relko ended his racing career with three unbeaten races as a four-year-old in the Prix Ganay, Coronation Cup (where the soft ground was against him) and Grand Prix de Saint-Cloud. One of the best racers of post-war years, he retired having won nine races, worth £149,136, and after Lord Sefton purchased a half-share in him, he came to England to stand his first season at stud in 1965. Ironically, his best produce proved to be those that raced in Italy or France.

The best were Breton (Grand Criterium and Prix de la Salamandre; died aged six after siring the St. Leger and Grand Prix de Paris runner-up Secret Man and the useful miler Monsanto); Master Guy (Prix Jean Prat); Mia Pola (Prix la Rochette) and Tierceron (Gran Premio d'Italia, Italian St. Leger and Gran Premio del Jockey Club).

Relko's best produce to win in Britain were Relay Race (Hardwicke Stakes), Ranimer (Sun Chariot Stakes), Relkino (second in the Derby), Freefoot (John Porter Stakes, third in the Derby), Reload (Park Hill Stakes) and the good handicappers Royal Prerogative and Royal Echo. Relko also bred the smart mare Mariel, who won three races including the Pretty Polly Stakes and bred the Yorkshire Oaks winner Sarah Siddons.

TANERKO (1953)

Tanerko, the sire of Relko, was unraced at two years and showed top-class form at three over a mile and a half. That year he won four races including the Prix Noailles and Prix Lupin, was third in the French Derby and Arc de Triomphe, and was also beaten only a neck by Midget in the seven-furlong Prix de la Fôret. Tanerko stayed in training for two more years, and won the Prix Ganay and Grand Prix de Saint-Cloud each season, as well as two other races.

Put to stud in 1959, he never reached the leading sire's position on the Flat, though he was once top over hurdles and once top for broodmares in France. Relko was his best produce. Tanerko's other classic winners were White Label V (Grand Prix de Paris; disappointing sire in France and Italy); Traverline (Spanish Derby) and Caracol (German 2,000 Guineas, Grosser Preis von Baden). He also bred Djakao (sire of the Grand Criterium winner Mariacci) and Orvilliers (French Champion Hurdle; sire).

Tanerko was out of the unraced mare La Divine, who bred three winners to staying sires before Tanerko. They were Lucinda (won four small races from five furlongs to seven); Clair Soleil (won over two miles on the Flat in France and won the Champion Hurdle and Triumph Hurdle in England) and Don Carlos (won a small race over a mile and a half). After Tanerko, La Divine bred three more winners.

La Divine was by Fair Copy, winner of eight middle-distance races for £10,500. He was at stud in France, England and Sweden, and his best-known offspring were Orfeo (Grand Prix de Paris) and Sayani (Cambridgeshire; successful sire in France). Fair Copy was out of Composure, a half-sister by Buchan to Selene, Tranquil and Bosworth. Composure won three small races and bred four winners, the others including Complacent, third dam of Park Top's sire Kalydon.

La Diva, the dam of La Divine, was a minor winner who also bred Chanteur II, sire of Pinza (see Chapter 25).

TANTIEME (1947)

Tantieme, the sire of Tanerko, showed the same highly-strung nature that Relko was to display. It almost certainly explained why the spare, light-framed Tantieme never showed his best form when he travelled to England. He was beaten only once in England, when almost seven lengths third to Supreme Court in the King George VI and Queen Elizabeth Festival of Britain Stakes, but he had only a head to spare over Coronation V in the Queen Elizabeth Stakes at Ascot and struggled to beat Saturn by less than a length when odds-on for the Coronation Cup. Those efforts were not comparable with the superlative form he showed in France, where he won four out of five races, including the Grand Criterium, as a two-year-old, the French 2,000 Guineas and Arc de Triomphe as a three-year-old, and the Arc again as a four-year-old.

He was a noted influence for stamina at stud, where he sired the sister and brother La Sega (French 1,000 Guineas and Oaks) and Danseur (Grand Prix de Paris), as well as Oakville (a useful stayer in England and successful sire in New Zealand where he sired the smart 'chaser Grand Canyon), Tantaliser (dam of the St. Leger winner Provoke), Brioche (Hardwicke Stakes, Great Voltigeur Stakes and Yorkshire Cup) and Agio (German St. Leger).

Tantieme also teamed up with the mare Relance III (dam of Relko) to produce Match III and Reliance II, of which more shortly. Tantieme died in March 1966.

Terka, dam of Tantieme, failed to win but produced six other winners. She was out of La Furka, a mare who won nine races.

DEUX POUR CENT (1941)

Tantieme was the sole top-class winner sired by Deux Pour Cent before departing for Czechoslovakia in 1954. Deux Pour Cent is credited with winning the 1943 Grand Prix de Paris at 28-1, but Ardan was first past the post in a time that has not been beaten since only to be disqualified and placed third. Deux Pour Cent was also the sire of Xora, who bred the French Derby winner Tapalque, and two unsound horses in Trenel, who went wrong after being beaten half a length in the French Derby, and Calchaqui, who did not even get to that stage after advertising his Classic claims by winning the Prix Noailles.

Deux Pour Cent's sire, Deiri, sired two "genuine" Classic winners in Pirette (French Oaks) and Bey (French Derby). Pirette became the dam of Primera (Ebor Handicap, sire of two smart fillies in the King George VI and Queen Elizabeth Stakes winner Aunt Edith and the Oaks winner Lupe).

RELANCE III (1952)

Relko gave the family of Sceptre its first Derby success, and also brought into the story for the first time one of the finest mares of the 20th Century, Relance III.

Winner four times from six races as a two-year-old, and of three races at three years including the Handicap Optional, Relance III produced three top-class racers from the six consecutive winners she bred between 1957 and 1962.

Relance III's second foal was Match III (by Tantieme), winner of seven races, worth £110,815. Unraced at two years, Match III won three races at three including the French St. Leger and was second in the French Derby behind Right Royal V. At four years Match III won four of his seven races including the Grand Prix de Saint-Cloud from Exbury, the King George VI and Queen Elizabeth

Stakes from the St. Leger winner Aurelius, and the Washington International Stakes.

Syndicated in shares of £6,000 to stand at stud in England from 1963, Match III died in September 1965 after completing only three covering seasons. His produce included Ovaltine (Ebor Handicap, Goodwood Cup), Palatch (Yorkshire Oaks, Musidora Stakes; dam of the Great Voltigeur Stakes winner Patch), Torpid (Jockey Club Stakes; second French St. Leger), Photo Flash (second 1,000 Guineas) and Murrayfield (nine wins in England and Italy, £25,546, including Coventry and Solario Stakes).

Relko was Relance III's fourth foal; her sixth, on a return visit to Match III's sire Tantieme, was Reliance II, who won five races in France worth the equivalent of £152,435. Like Match III, Reliance II was unraced at two years but as a three-year-old he carried all before him, winning the French Derby, Grand Prix de Paris and French St. Leger—until he met Sea-Bird II. Then, Reliance II had to concede defeat, going down by six lengths in the Arc de Triomphe.

Following the death of Match III, Herbert Blagrave bought a half-share in Reliance II for him to stand at stud in England after a four-year-old career. In the event Reliance II did not race at four years and started at stud in 1967. Though he stood in England, the majority of his stakes earnings came from France, where his best produce, Recupere, won eight Pattern races including the Prix du Cadran. Recupere also won in America where he retired to stud. Reliance II's best produce trained in England were Consol (Geoffrey Freer Stakes) and Proverb (Goodwood Cup, Doncaster Cup, Chester Vase).

The fact that Relance III threw her best produce as her second, fourth and sixth foals, to either Tantieme or his son Tanerko, was coincidence, for her other foals were not entirely devoid of merit. Her first foal, Dalida, won three races in France; her third, Desirade, a sister to Match III and Reliance II, won two races in France and bred Khadine, a good two-year-old in England in 1969; and her fifth, Bing, won three races in France and was second in the Criterium de Saint-Cloud before a stud career in England.

RELIC (1945)

Relance III was from the first crop in France of her sire Relic, who showed good form in his native America up to seven furlongs, winning five of his seven races and finishing second in the others for 69,275 dollars, including the Hopeful Stakes. He spent one season at stud in America before being exported to France in 1951, and left in 1957 for England, where he died in December 1970. He proved one of the principal means of bringing the influence of Man O' War to Europe.

Relic's French-trained produce included Blockhaus (12 wins, including Prix Perth and Prix d'Ispahan); Mincio (French 2,000 Guineas); Buisson Ardent (French 2,000 Guineas, Middle Park Stakes, third in English 2,000 Guineas, sire in Ireland); the top-class sprinting sisters Texana (winner of all 11 races at two years in 1957 including the Prix d'Arenberg and Prix de l'Abbaye) and Texanita (won eight out of 11 at two years including Prix d'Arenberg and Prix de l'Abbaye, and won Abbaye again at three years) and Venture VII (Middle Park Stakes, Sussex Stakes, St. James's Palace Stakes; second in 2,000 Guineas).

Relic also produced Olden Times (17 wins in America for 603,875 dollars); Hecuba (second in 1,000 Guineas), and his best son trained in Britain, Pieces of Eight (three wins, £56,054 including Eclipse Stakes and Champion Stakes; at stud in America and England and sired winners in America, France, Italy and Britain but none up to his own class).

Relic's sire War Relic won nine races at three years. His best winners at stud were Intent (317,775 dollars, and sire of Intentionally, 652,258 dollars); Battlefield (474,727 dollars, and sire of Yorktown, 156,335 dollars) and Missile (193,369 dollars). War Relic was out of the American stakes winner Friar's Carse, who was also the fourth dam of Sword Dancer (15 wins, 829,610 dollars, Horse of the Year at three in 1959 when won the Jockey Club Cup and was second beaten a nose in the Kentucky Derby; sire of Damascus, 21 wins, 1,185,421 dollars including Preakness Stakes, Belmont Stakes, Woodward Stakes and American Derby).

Relic's dam was the winning mare Bridal Colours, whose own dam Vaila won over five furlongs in England as a two-year-old and was exported to America as a four-year-old after being sold for 80 guineas at the 1915 Newmarket December Sales.

Vaila also bred Blossom Time, dam of the noted American horse Blue Larkspur, a good two-year-old who missed out on the Kentucky Derby when favourite but won the Belmont Stakes and Arlington Classic. At stud Blue Larkspur got 44 stakes winners, and his fillies were more noteworthy than his colts.

Mention of Blue Larkspur's influence when mated with La Troienne was made in connection with Psidium. Blue Larkspur's further importance can be traced to his daughter Ampola and the three useful sprinters she bred—Takawalk II; Polamia (dam of the French 1,000 Guineas winner Right Away II; the French 2,000 Guineas winner Grey Dawn, who beat Sea-Bird II in the Grand Criterium; and the useful filly Mia Pola); and Sly Pola (dam of Green Valley, the record-priced yearling at Deauville in 1968 when she fetched 410,000 francs, unraced but dam of the Observer Gold Cup and French 2,000 Guineas winner Green Dancer, her first foal).

Blue Larkspur also sired Bloodroot (third dam of Mill Reef's sire Never Bend), Bleebok (third dam of Roberto) and Our Page (third dam of Nijinsky). His best sons were Blue Swords (second to Count Fleet in the Kentucky Derby and Preakness Stakes; sire of the Preakness Stakes winner Blue Man and Hail to Reason's dam Nothirdhance), and Revoked (good two-year-old; sire of three top-class winners in Furl Sail, Rejected and Reneged).

POLAIRE (1947)

Relance III was out of Polaire, winner of six races at three years and a most successful broodmare. Her six winners also included Midnight Sun II, who dead-heated with Saint Crespin III for the Arc de Triomphe at 50-1 but lost his share of the race for badly bumping his rival and was placed second (he was also second to Bald Eagle in the Washington International); and Esquimau III, winner of the Prix Maurice de Nieuil.

To a repeat mating with Relic the year after Relance III was born, Polaire bred Polic, a smart sprinter at two years when his four wins included the Prix d'Arenberg. Polic won over a mile at three, when he also ran fifth in the King's Stand Stakes. He became the sire of Polyfoto, who was remarkable in that his temperament improved as he got older. He had great trouble at the start as a two-year-old, despite which he won twice, but when he again became bothersome he was sent to France and proceeded to win his last three races from stalls, including the Prix d'Arenberg. He wore blinkers at two years but they were left off the following season when he won the Prix de Saint-Georges and Nunthorpe Stakes before being sent to America, where he won at up to a mile. At stud in Ireland (from 1970) and France, Polyfoto sired the top-class sprinter Bay Express (six wins, £29,048, including King's Stand and Nunthorpe Stakes).

Polaire was by the sprinter Le Volcan out of Stella Polaris, a winner at two years who to a mating with Prince Bio bred the Grand Prix de Paris winner Northern Light, who also ran second in the St. Leger. Stella Polaris, whose dam Crepuscule (1925) was by Galloper Light and should not be confused with the dam of Crepello, had as her fourth dam Sceptre.

CHAPTER 27

1964 – 1967
Extremes of Breeding

1964

1   J. Ismay's b c Santa Claus (Chamossaire – Aunt Clara).
2   C. Engelhard's b c Indiana (Sayajirao – Willow Ann).
3   L. Gelb's b c Dilettante II (Sicambre – Barbizonette).
17 ran. Time: 2 min 41.98 sec

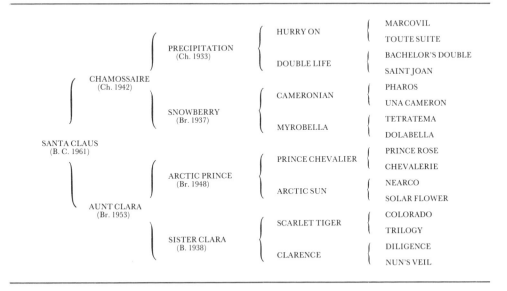

SANTA CLAUS, WHOSE Derby success was a triumph for the small breeder, became winter favourite for Epsom on the strength of his two runs as a two-year-old, both in Ireland where he was trained. A remote fifth in the Anglesey Stakes, he won the National Stakes by eight lengths from Mesopotamia and a good field, of which nine out of 12 were winners.

He had only one race before the Derby as a three-year-old, winning the Irish 2,000 Guineas effortlessly by three lengths. At Epsom he was ridden with great confidence by Breasley—the Australian jockey's first Derby winner in 13 attempts and at the age of 50—and at halfway there were only three behind him. At Tattenham Corner there were eight in front and he had 12 lengths to make up, but after gaining ground from three furlongs out, he led well inside the last 100 yards and won by a length from Indiana.

In his next two races Santa Claus was ridden by his work-rider Burke. He won the Irish Sweeps Derby by four lengths, but on hard going was beaten two lengths by the front-running Nasram II when given far too much to do in the King George VI and Queen Elizabeth Stakes. Jarred up after Ascot, he missed the St. Leger, and in his next race, the Arc de Triomphe, was second, beaten threequarters of a length by Prince Royal II.

The Arc was Santa Claus's last race, and he retired with four wins worth £138,127. Soon after the Irish Sweeps Derby he was syndicated at £10,000 per share, putting a valuation of £400,000 on him, and he retired to the Airlie Stud in Ireland, where his dam, Aunt Clara, and one of his half-sisters, Clara Rebecca, also eventually found themselves.

The valuation at which Santa Claus retired was in sharp contrast to that which obtained when he was sold first as a foal and then as a yearling. Bred at the Rugby stud of Dr. Frank Smorfitt, who had two mares at the time, Santa Claus fetched 800 guineas as a foal at the Newmarket December Sales when bought by the Irish breeder Mr. A. N. G. Reynolds, who re-sold him the following year as a yearling, at Newmarket in September, for 1,200 guineas to the BBA, acting for Mr. John Ismay.

Santa Claus died in February 1970—on Friday the 13th!—after five seasons at stud, during which time he bred one Classic-winning son, Reindeer, and one Classic-winning daughter, Santa Tina. Reindeer, from his sire's first crop, won six out of 11 races, for £37,756, including the Irish St. Leger and Prix Kergorlay; he was third in the Irish Sweeps Derby, and after one season at stud in Ireland was exported to New Zealand. Santa Tina—a sister to Reindeer—won five of her six races, including the Irish Guinness Oaks and was sold to America for a near-six-figure sum.

Other produce of note by Santa Claus were Yaroslav (unbeaten in two races at two years; hard to keep sound and did not win, though showed good-class form, in five races over the next two seasons); Sleat (Sun Chariot Stakes), and Bonne Noel (Ebor Handicap; stud in Ireland).

CHAMOSSAIRE (1942)

Santa Claus was born when his sire, Chamossaire, was 19, while his dam, Aunt Clara, was eight. Chamossaire was about to go into dramatic decline, since in his last three crops—the ones after Santa Claus—he numbered only nine foals. Described as "a thorough-going stayer with little or no speed in his make-up", he had sired predominantly stayers, and had had few good fillies. In financial terms his greatest success came towards the end of his career. He also bred two Irish

Derby winners in Chamier and Your Highness, the St. Leger winner Cambremer, and Le Sage (eight wins including Sussex Stakes). By the end of 1963 his progeny had won £190,341 and he was never higher than sixth on the list of leading sires. Santa Claus changed all that, and in 1964 Chamossaire was leading sire. Also that year, in the November, he died, at which point his progeny had won £326,848 (from 152 winners of 360½ races), of which £132,103 came from Santa Claus in 1964. The figures indicate that breeders can take too much notice of statistics and not enough of the facts behind them if they are not wary.

Chamossaire, by Precipitation, was out of Big Game's half-sister Snowberry, who won the Queen Mary Stakes and was rated the second best filly of her age in 1939.

Snowberry bred nine winners in all, including the useful miler Massif and the useful stayer Jardiniere (Northumberland Plate). Her daughter White House won a mile and a half maiden worth £207 and then bred Hopeful Venture (7 wins, £70,906, including Grand Prix de Saint-Cloud, Hardwicke Stakes; second in St. Leger; at stud in England from 1969 until exported to Australia in 1975). Snowberry also bred the maiden Ariana, second dam of Snow Knight.

AUNT CLARA (1953)

Though the first three dams in Santa Claus's pedigree did not win (and two did not race), he came from a female family that had rare bursts of respectability, of which he provided one. But when the family was apparently upgraded and became more fashionable, the prices paid for its members which appeared on the open market increased, but their success did not necessarily follow suit. There is a lesson in that for all who dabble in bloodstock breeding.

After being unraced as a two-year-old, Aunt Clara was sent to the 1955 Newmarket December Sales by her owner Miss Dorothy Paget and was bought by Dr. Smorfitt for 130 guineas. She ran three times as a three-year-old for her new owner but appeared totally void of ability in her races at five and six furlongs. Retired to stud, her first foal was Clara Rebecca (by Neron), a poor maiden who was beaten in a seller before being sold for 300 guineas at Newmarket in October 1961 as a three-year-old. Clara Rebecca bred The Real McCoy, a modest winner of two small races in Ireland, and Nevada, who won one hurdle race.

Aunt Clara's second foal was Millden (by Dumbarnie), who was bought by Lord Derby for 4,300 guineas as a yearling but did not race as a two-year-old and had already been gelded when he went to that year's Newmarket December Sales and was sold for 220 guineas, in the same batch that included Bow Tie, subsequently the leading sprinter in Scandinavia after his sale for 1,100 guineas. Millden himself developed into a decent sprinter at three years, and eventually won four of his 60 races, his best form being at six furlongs and his last success coming in a seller.

Santa Claus was Aunt Clara's fourth foal, being followed by the filly Clarity (by Panaslipper), who was sold as a yearling in 1953 for 4,200 guineas and sent to Spain. After the exploits of Santa Claus, Clarity was brought back to England, where as a three-year-old she won two races at up to a mile and a half, and then bred four winners on the Flat as well as the four-mile Warwick National 'Chase winner Clarification.

Evidence that the Derby win of Santa Claus had upgraded the family of Aunt Clara came in 1965, when her fifth foal, to be named Saint Christopher (by Aureole), came up for sale as a yearling. Saint Christopher was in a batch submitted by the better-known Airlie Stud, which had bought Aunt Clara, in foal to Aureole, in a private deal before Santa Claus won the Irish 2,000 Guineas. Saint Christopher made 25,000 guineas! Third twice in three races as a two-year-old, Saint Christopher won his last two races from four outings at three. In 1969 the next yearling out of Aunt Clara offered on the open market, a colt by Sea Hawk II called Skyhawk, fetched 31,000 guineas, and went on to win one small race over a mile and a half as a three-year-old.

SISTER CLARA (1938)

Aunt Clara's dam, Sister Clara, had an even less distinguished early brush with the world of racing, since she was sold as a yearling for a mere 20 guineas. She had extremely bad forelegs and never raced, but to show that conformation is but one consideration when fillies come to be sold, she gained from the exploits of her half-sister Sun Chariot to such extent that when she was sent to the 1944 Newmarket December Sales, as an unraced mare who had not bred a winner, she made 11,000 guineas to the bid of Miss Paget. The same owner bought Sister Clara's yearling that year, Sarah Clara, for 6,600 guineas.

Sister Clara went on to breed seven winners, though none was above modest class and none won more than £800 in any one season. Her two most successful daughters (until Aunt Clara bred Santa Claus) were Sarah Clara, dam of Salute (13 wins, £3,647, mainly at 1m or 1¼m), and Mary Clare, dam of Fulshaw Cross (three wins, £3,549, at 1m).

Sister Clara was by Scarlet Tiger, a son of the 2,000 Guineas winner Colorado and half-brother to the Oaks winner Light Brocade. Scarlet Tiger, third in the St. Leger and fourth in the Derby, both won by Hyperion, retired to stud at a fee of £99 but made little mark and was sent to Argentina.

CLARENCE (1934)

Like Sister Clara, her dam Clarence did not race, but she was easily the most distinguished of the first three dams in Santa Claus's pedigree, because she bred the triple Classic winner Sun Chariot.

Leased for her racing career to King George VI, Sun Chariot won eight races for £9,209, including the 1,000 Guineas, Oaks and St. Leger, and can be regarded as one of the best fillies of this century. She bred seven winners, including Blue Train (see page 257); Gigantic (Imperial Produce Stakes; top sire in New Zealand); Landau (six wins, £6,103, including Sussex Stakes, but was not always reliable; sold to Australian stud for 20,000 guineas at end of three-year-old days); and Pindari (four wins, £18,456; including King Edward VII Stakes, Great Voltigeur Stakes; third in St. Leger; little impact at stud and was sent to India in 1970). Sun Chariot, who died in 1963, bred only one winning filly, Persian Wheel, who won a minor race at a mile, finished fourth in the 1,000 Guineas, and bred two minor winners.

Clarence bred three other winners, who between them won a total of £675. They were the sisters Golden Coach and Calash (by Hyperion), and the colt Howdah (by Big Game), who was beaten at odds on first time out but managed to win at 5-2 on next time and was exported to America. Golden Coach struggled to beat some poor-class fillies when 9-2 on for a maiden race, while Calash won at 4-1 on first time out but then failed at 5-2 on next time. It was from examples like these that the family of Clarence—Sun Chariot apart—got a reputation for doubtful soundness, either physical or mental.

Calash made a name for herself as a broodmare with three winners that included the Oaks winner Carrozza (three wins, £22,952) and Snow Cat (three wins, £3,633, including Rous Memorial Stakes; a leading sire in Argentina). Carrozza was modestly successful at stud in her own right, as the dam of three winners in Britain who each won one race and totalled stakes of £1,262, but two of those winners went on to better things. Battle-Wagon became a top sire in New Zealand, and Camilla Edge was the dam of Calvello (14 races in Italy from three years to six), Bonne Noel (five wins, £5,747, including Ebor Handicap) and Tender Camilla (four wins in Ireland at two years including Railway Stakes).

When it was decided to sell all the National Stud mares in 1964, Carrozza made 20,000 guineas and the following year she left for America. The foal she was carrying at that time was Carromata (by St. Paddy), who managed one third place from 11 starts in America but as her third foal produced Matahawk, winner of the Grand Prix de Paris.

One of Calash's non-winners, The Chase, a filly who was blind in her near eye and would probably have won a small race had she exerted herself, became the dam of Arnica (Prix Fille de l'Air).

Clarence was by the middle-distance winner Diligence, who was exported to Russia in 1935, out of Nun's Veil, who ran only as a two-year-old. Nun's Veil won three races in 1932 worth £2,297 including the Caterham Stakes on Derby day, and was second when a short-priced favourite for the New Stakes at Royal Ascot, where the winner was a horse called Hyperion! Nun's Veil, who was out of Swynford's dam Blanche, bred two minor winners.

1965

1 J. Ternynck's ch c Sea-Bird II (Dan Cupid – Sicalade).
2 M. Bell's ch c Meadow Court (Court Harwell – Meadow Music).
3 L. Freedman's br c I Say (Sayajirao – Isetta).
      22 ran. Time: 2 min 38.41 sec

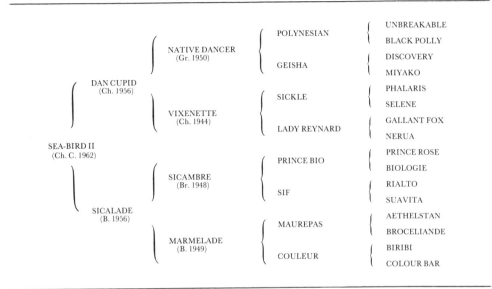

Santa Claus had come from a line of mares with little to recommend them on the racecourse; Sea-Bird II was even more bereft in this direction, since not one of the first five dams in his pedigree won under rules on the Flat. Sea-Bird II represented an act of faith in the family on behalf of French owner M. Jean Ternynck, who was rewarded with one of the best horses in post-war years, and one whom several observers regard as the best Derby winner in that time.

Unraced until September as a two-year-old, Sea-Bird II won his first two races by a short neck and then was beaten by his more precocious stablemate Grey Dawn in the Grand Criterium. It was as a three-year-old that Sea-Bird II came into his own, totally dominating his generation in Europe and winning all five races: the Prix Greffulhe by three lengths, the Prix Lupin by six, the Derby by two, the Grand Prix de Saint-Cloud by two and a half, and the Arc de Triomphe by six.

He won the Derby without coming off the bit, and the runner-up Meadow Court went on to win the Irish Sweeps Derby and King George VI and Queen Elizabeth Stakes. The Arc field was as strong as any assembled, and Sea-Bird II made them look ordinary, with the French Derby winner Reliance II second and Diatome, who went on to win the Washington International, five lengths away third.

Sea-Bird II ended his career in the Arc and retired having won seven races for £65,301 and 2,268,649 francs, a total of £225,000 in stakes.

Before the Arc, a syndicate of American breeders agreed to lease Sea-Bird II for five years, at a sum reported to be £95,000 a year, and he entered stud in America in 1966. The contract was renewed for a further two years before he was returned to France late in 1972 for the next covering season. Before he could stand his first season in Europe, Sea-Bird II contracted colitis and in April 1973 he died, with seven crops to represent him.

The interest of the Americans in Sea-Bird II marked a new departure in European breeding; the mighty dollar was about to be used to tempt some of the best racers across the Atlantic, whether it was in the interest of the horse or not. Sea-Bird II, for instance, did not seem the ideal type to produce horses for the American style of racing, which depends more on fast-maturing runners able to show their best on tight tracks and from sprints to ten furlongs.

Sea-Bird II got the reputation of being a "failure" at stud, but in the first place he could not be expected to sire more than a handful of horses anywhere near his own brilliance, and also he had to overcome the apparent disadvantage of standing in a country that could not bring out the best in him. Looked at in this light, Sea-Bird II did well, with several very good produce to his credit and one that was brilliant.

An interesting aspect of Sea-Bird II's career is the influence of first-crop sires. His sire Dan Cupid was from the first crop of Native Dancer; Sea-Bird II was from the first crop of Dan Cupid; and in Sea-Bird II's own first crop came his best son Gyr, who won four races from the seven he contested. Gyr was a nervous horse but he had such a reputation that Etienne Pollet, who also trained Dan Cupid, delayed his retirement to train him. Gyr rewarded him by winning the Prix Hocquart and Grand Prix de Saint-Cloud, and finishing second to Nijinsky in the Derby. A major share in him to stand at stud in England gave him a valuation of £500,000 but within a couple of years of his arrival in 1971 he was off to France.

Gyr was out of a top-class Italian mare called Feria, winner of the Italian Oaks and St. Leger, but was bred in America, where Sea-Bird II had the advantage of being put to a series of superbly-bred mares. The result was a series of other top-line winners such as Dubassof (11 wins, 304,889 dollars, including American Derby); Sea Pigeon (prolific winner on the Flat and over hurdles in Britain); Kittiwake (smart filly in America with 18 wins for 338,086 dollars); Little Current (Preakness Stakes, Belmont Stakes); Sea Break (second in Observer Gold Cup); Great Heron (second in Irish 2,000 Guineas) and Guillemot (third in Irish Sweeps Derby).

Above all, Sea-Bird II will be remembered for the filly Allez France, who has had few peers among her sex since the Second World War. She was bred in America and bought there, and won 13 of her 21 races for £493,100, second only in European stakes to another filly Dahlia, whom Allez France met six times and beat six times. Allez France won the French 1,000 Guineas and Oaks, the Arc de Triomphe, Prix Vermeille, and Prix Ganay twice. She was second in another Arc, and the only blemish on her career was that she was not at her most fortunate in England, where she was twice second in the Champion Stakes.

DAN CUPID (1956)

Sea-Bird II's sire, Dan Cupid, was more precocious than his best son, as evidenced by the fact that his trainer, who did not normally introduce his Classic horses until at least August as a two-year-old, first raced him in June. Dan Cupid won three races at two years and finished second in the Middle Park Stakes and third in the Prix Morny. Unplaced in the English 2,000 Guineas and Derby, he was second, beaten a neck by Herbager, in the French Derby, after which his form tailed off.

An enigmatic racehorse, Dan Cupid carried on this trait at stud, which he entered in 1961, producing Sea-Bird II in his first crop and a handful of other smart racers, including Miss Dan (best three-year-old filly of her year in France; won Prix Fille de l'Air and Prix Kergorlay; second in Prix Vermeille and Washington International; third in Arc); Silver Cloud (Grand Criterium, third in French Oaks); Gift Card (seven wins, including Prince of Wales Stakes; second in French 2,000 Guineas), and Dankaro (Prix Lupin; second in French Derby; third in King George VI and Queen Elizabeth Stakes).

NATIVE DANCER (1950)

Dan Cupid's sire, Native Dancer, was a massive, powerfully-built horse who won all but one of his 22 races despite soundness problems. He was a brilliant racehorse and earned 230,495 dollars, a world record at the time, from an unbeaten run of nine races at two years. As a three-year-old he was beaten once in ten starts, being hampered early and failing by a neck to peg back Dark Star in the Kentucky Derby. He won the other Classics, the Preakness Stakes and Belmont Stakes, as well as the American Derby, Arlington Classic and Wood Memorial, and at four years he won his only three races.

Retired to stud in 1955 at a fee of 20,000 dollars, he died in November 1967 from the effects of a serious operation. In Europe, apart from Dan Cupid, his best winners were Hula Dancer (brilliant filly who won 1,000 Guineas, Grand Criterium and Champion Stakes); Secret Step (very speedy filly who won July Cup, King's Stand Stakes and King George Stakes); Takawalk II (four wins in France including Prix Djebel and Prix Saint-Georges), and Hul a Hul (fastest two-year-old trained in Ireland in 1964).

Native Dancer's biggest earners in America were Kauai King (nine wins, 381,397 dollars, including Kentucky Derby and Preakness Stakes; disappointing sire in America, England and Japan); Protanto (eight wins, 322,085 dollars); Native Charger (five wins, 278,893 dollars, including Florida Derby; sire of Forward Gal, 12 wins for 438,933 dollars and leading two-year-old filly of 1970, and Summer Guest, 11 wins for 372,900 dollars including CCA Oaks); Dancer's Image (12 wins, 236,636 dollars, disqualified for positive dope test after Kentucky Derby); and Native Street (ten wins, 236,808 dollars, including Kentucky Oaks in same year that Kauai King won Kentucky Derby).

One of Native Dancer's less successful sons, Raise a Native, has done best at stud and he promises to be a potent force in international bloodstock breeding. His earnings of 45,955 dollars came from being unbeaten in four races at two years from three furlongs to five and a half furlongs. After breaking down, he was syndicated for the equivalent of £1,089,000 and went to stud in 1964 as a three-year-old. Two of his best produce came when he was mated to the Royal Charger mare Gay Hostess, who foaled Majestic Prince (won Kentucky Derby and Preakness Stakes from Arts and Letters, who beat him in the Belmont Stakes, after which Majestic Prince went wrong with splint trouble), and Crowned Prince (fetched world-record price of 510,000 dollars (£212,000); topped English Free Handicap after winning Champagne and Dewhurst Stakes, but suffered from soft palate and flopped at three years). Another son of Raise a Native to break down was Exclusive Native, who was retired after winning the Arlington Classic at three years.

Sons of Native Dancer have not done particularly well at stud in Europe, as evidenced by Kauai King. Takawalk II and Gala Performance (nine wins, 143,855 dollars) stood in Ireland with limited success. Atan, who won his only race as a two-year-old before he too broke down, stood for a short time in Ireland before export to Japan, and left one notable offspring in Sharpen Up (Middle Park Stakes; sire).

The best produce of one of Native Dancer's sons to race in Europe, apart from Crowned Prince, were Lianga (by Dancer's Image), whose 11 wins in three seasons included the July Cup, Prix de l'Abbaye and Vernons Sprint Cup, and her brother Garda's Revenge, a very useful two-year-old trained in Ireland in 1975.

The best-known mare in America produced by Native Dancer was Natalma, whose three wins earned 16,015 dollars before she bred Northern Dancer (14 wins, 580,806 dollars), Native Victor (18 wins, 71,400 dollars) and Regal Dancer (13 wins, 51,500 dollars).

Native Dancer's sire, Polynesian (by Unbreakable), was a tip-top racer who won the Preakness and Withers Stakes, and set four track records in a career total of 27 wins from 58 races in four seasons. Native Dancer was his best son at stud, but he also sired Imbros (sire of Native Diver, the winner of 1,026,500 dollars) and Polly's Jet, whose son Turbo Jet won 242,448 dollars.

Mares by Polynesian included Banquet Bell, dam of the Kentucky Derby winner Chateaugay and the best handicap mare of 1962 Primonetta. Banquet Bell was also the second dam of Little Current (by Sea-Bird II).

Native Dancer was out of Geisha, a mare who managed to win 4,120 dollars as a three-year-old. She bred four winners from eight foals. Her daughter Orientation bred four stakes winners and was the grandam of Morris Dancer, a popular horse who held the record for earnings by a gelding trained in Britain.

Geisha, one of three foals, all winners, bred by the five-race two-year-old winner Miyako, was by Discovery, winner of 27 races for 198,287 dollars, many times carrying big weights. Discovery's best son was Find (winner of 803,615 dollars), but he is better known as a sire of broodmares, having produced the dams of Bold Ruler (764,204 dollars), Intentionally (652,258 dollars) and Hasty Road (541,402 dollars).

VIXENETTE (1944)

Vixenette, the dam of Dan Cupid, ran only at two years and was never placed. She was by Hyperion's half-brother Sickle, a good two-year-old who went on to run third in the 2,000 Guineas and became a leading sire in America.

Lady Reynard, the dam of Vixenette, was by the American Triple Crown winner Gallant Fox, who sired a similar American champion in Omaha, as well as the Ascot Gold Cup and Champion Stakes winner Flares.

SICALADE (1956)

Sea-Bird II's dam, Sicalade, had a chequered career. She raced only as a two-year-old, dead-heating for second place on the second of her only two starts. She bred three foals, of whom two ran, but then was so badly kicked that she had to be put down, ending her life being sold for £100 as butcher's meat. Her first foal never ran; her second was Sea-Bird II; and her third, Syncom (by Beau Prince II), won three races.

Sicalade was by the French Derby and Grand Prix de Paris winner Sicambre, an outstanding sire whose dam Sif was unraced at two years, won two races at three and one at four, and was fourth in the French Oaks. Sicambre's sister, Senones, won two races at Saint-Cloud as a three-year-old, including the Prix Penelope, in which she beat the subsequent French Oaks winner Douve. Senones was the dam of Snob, who won the Prix de la Fôret, finished fourth in the Arc de Triomphe, and sired Goodly (French Derby), Glaneuse (Prix de Malleret, Gran Premio del Jockey Club in Milan) and Batitu (Prix de la Salamandre) from his first two crops but was exported to Japan after seven seasons at stud in France.

Sif also bred Fantine (by Fantastic), dam of Free Man, winner of the French 2,000 Guineas, second to Sicambre in the French Derby, and sire of Free Ride (Prix Ganay, fourth in the Arc) and Royal Lady (Belgian Derby).

Sif was by the middle-distance winner Rialto, winner of 17 races from 37 starts and sire of Wild Risk, out of the Alcantara II mare Suavita.

MARMELADE (1949)

Marmelade, the dam of Sicalade, had one outing, unplaced, as a three-year-old, and bred three foals that raced. They were Sicalade, Marmite (placed on the Flat and over jumps) and Magnae (winner of three races in the French provinces). Marmelade was by the Grand Prix de Paris and Prix du Cadran winner Maurepas out of the Biribi mare Couleur, breeding which produced for Sea-Bird II's owner his first Classic winner, the 1,000 Guineas heroine Camaree.

Couleur, the dam of Marmelade and Camaree, failed to win on the Flat but was successful over hurdles at Nice and was placed over fences. She was bought in 1946 by M. Ternynck, in foal to Maurepas with a filly foal by the same sire. The filly foal was Camargue II, who won the Prix de Malleret and bred Dalaba (Prix Royallieu); and the foal born in 1947 was Camaree, who as well as her English Classic won three races in France. Couleur went on to breed only one other winner on the Flat, called Conakry, and a steeplechase winner called Caddor, before she died in 1957.

Couleur was by the Prix Royal Oak winner Biribi out of the Colorado mare Colour Bar, another mare with an interesting history. Colour Bar ran in sellers in her only season on the racecourse as a two-year-old, and her form went from moderate to worse, though she did go on to win under Pony Club Rules. As a four-year-old she was sold, covered by Cyclonic, at the 1934 Newmarket December Sales for the surprisingly high price of 750 guineas to Major D. McCalmont.

Colour Bar bred Quarteron (three wins, £1,690, including Woodcote Stakes) for her new owner, but as a six-year-old she was passed on to France, where she bred Tetrabar (three wins at two years including Grand Criterium d'Ostende) and Le Bosc Giard (four wins in England, £1,256, and three in France) as well as Couleur. Colour Bar, whose dam Lady Disdain ran twice and bred Crosspatch (Molecomb Stakes), was presumed killed in the fighting around Normandy during the Second World War.

1966

1   Lady Zia Wernher's b c Charlottown (Charlottesville – Meld).
2   J. A. C. Lilley's ch c Pretendre (Doutelle – Limicola).
3   E. B. Benjamin's bl c Black Prince II (Arctic Prince – Rose II).
       25 ran. Time: 2 min 37.63 sec

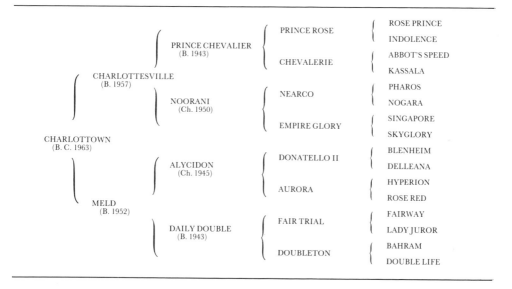

If the best in breeding is to mate a Derby winner with an Oaks winner—as some would suggest—the next best would be a French Derby winner with an Oaks winner, as happened with Charlottown, who if any horse can be said to have been bred to win a Derby, is the one. And it did not take him long to show he was out of the ordinary, winning his first race, the Solario Stakes, by eight lengths. Thereafter it got a little harder, though after scrambling home at odds-on for a small race at Ascot and winning the Horris Hill Stakes convincingly, he was allotted 9st 9lb in the Free Handicap, only 5lb below the top.

At three years he ran five times, won twice and was second in the other three, being given too much to do when beaten by Black Prince II in the Lingfield Derby Trial on his only outing before Epsom. In the Derby, with a change of jockey, he turned the tables on his Lingfield conqueror, despite having to be re-shod during the preliminaries. He got up to beat Pretendre by a neck with Black Prince II five lengths away third.

In his last three races in his second season Charlottown came up against Sodium, who beat him in the Irish Sweeps Derby and St. Leger, but in between was beaten by him in a small race at Newbury.

At four years Charlottown beat Salvo narrowly in the John Porter Stakes, and won the Coronation Cup after the field had crawled for five furlongs. His only bad race was his third and last as a four-year-old, when he was sixth in the Grand Prix de Saint-Cloud. He was syndicated at 10,000 guineas a share, giving him a total value of £420,000, and retired to stud in Newmarket with a career record of seven wins for £101,210.

Unimpressive to look at, Charlottown never filled out and remained rather narrow and sparely-made. Nor did he fulfil expectations at stud and in 1976 he became the first Derby winner to be exported to Australia. His best produce while he stood in England were Charling (Lingfield Derby Trial); Lottogift (Cambridgeshire); Roman Blue (Italian 1,000 Guineas and Oaks) and Tomatin (Warren Stakes; to stud in Tasmania).

One point to be borne in mind in assessing Charlottown is that he won his Derby under unusual circumstances, since 1966 was the year no French-trained horse ran in England after the Guineas meeting, because of swamp fever (equine anaemia) restrictions imposed by the Ministry of Agriculture.

CHARLOTTESVILLE (1957)

Charlottesville, the sire of Charlottown, was not an easy horse to train, and the fact he won five races off the reel as a three-year-old, including the French Derby and Grand Prix de Paris, was a tribute to his trainer. Charlottesville did not run again after finishing sixth in the Arc de Triomphe, and retired to his owner's stud in Ireland having won six races for the equivalent of £75,507. His syndication at 8,000 guineas a share gave him a European record valuation of £336,000.

His first crop raced in 1964, the best being Carlemont (Sussex Stakes and £18,191); Charlottown was from his second crop. His other important winners included Varinia (Lingfield Oaks Trial; third in Oaks); Stratford (Gran Premio di Milano, Grosser Preis von Baden, Milan Gold Cup twice); Canterbury (Doncaster Cup; second in St. Leger; sire in Brazil); Bonconte di Montefeltro (Italian 2,000 Guineas and Derby, Gran Premio d'Italia); Meadowville (Great Voltigeur Stakes, John Porter Stakes; second in Irish Sweeps Derby, St. Leger and Irish St. Leger); Biskrah (Doncaster Cup); Jefferson (seven wins in France; sire) and Selhurst (Hardwicke Stakes).

Charlottesville died suddenly after a heart attack in February 1972, at the age of 15.

Prince Chevalier, sire of Charlottesville, was also responsible for the sires of the second and third in the 1966 Derby, since Pretendre was by Doutelle, and Black Prince II by Arctic Prince.

Charlottesville was the second living foal of the Nearco mare Noorani, whose first produce was Sheshoon (by Precipitation). A non-runner at two and only moderately successful in the top class at three, Sheshoon came into his own as a stayer at four, when he won the Ascot Gold Cup from Exar and Le Loup Garou after being beaten a short head by the latter in the French equivalent. Sheshoon also won the Grand Prix de Saint-Cloud and Grosser Preis von Baden over middle distances, but was regarded mainly as a long-distance horse, which probably explains why his syndication was fixed at a total value of £140,000, while Charlottesville's was £336,000, and why Charlottesville always received the more choice patronage after they retired to stud in 1961.

Sheshoon spent all his career at the Limestone Stud in England, but his biggest successes were in France, where Sassafras, Pleben and Samos III all won the St. Leger.

Sassafras won six races for £249,544, including the Arc de Triomphe (where he beat Nijinsky) and French Derby and St. Leger. A promising sire, he got the French St. Leger winner Henri la Balafre in his first crop.

Samos III, as well as the French St. Leger, won the Prix Gladiateur, finished second in the Ascot Gold Cup, and went to stud in Argentina. Pleben won the Grand Prix de Paris, as well as the French St. Leger.

Sheshoon's other top-class winners included Society (Anglesey Stakes; exported to America where he won ten races for 131,784 dollars) and Stintino (Prix Lupin; third in Derby). Ironically his best son trained in England was his one English Classic winner, Mon Fils, who won the shortest-distance Classic, the 2,000 Guineas.

Sheshoon's best fillies were Vela and Oak Hill, both winners of the Criterium des Pouliches for two-year-olds, and Riverside, who won the Prix de Royallieu and bred Riverqueen (French 1,000 Guineas, Prix Saint-Alary and Grand Prix de Saint-Cloud).

EMPIRE GLORY (1933)

Noorani, the dam of Sheshoon and Charlottesville, was acquired by the Aga Khan as a foal in 1950, among a batch of 54 bloodstock and the Phantom House Stud at Newmarket which he purchased from Mr. Wilfred Harvey. The transaction also included Noorani's dam Empire Glory, then aged 17.

Empire Glory was fairly successful as a two-year-old, when she won the five-furlong Prendergast Stakes worth £1,265 at Newmarket, after finishing third in the Imperial Produce Stakes. Empire Glory bred four winners, the best being Fair Fame, the top two-year-old filly of 1943, when she won four races for £2,539 including the Queen Mary and Cheveley Park Stakes.

Empire Glory was by the St. Leger winner Singapore out of Skyglory, who also produced Glorious Devon (Park Hill Stakes, Yorkshire Oaks and Manchester

November Handicap; dam of Devonian, sire of winners including Vintage, Queen Mary Stakes and July Cup) and Peter the Great (won over hurdles; sire of Grand National winner Russian Hero). Skyglory was out of a sister to Rabelais, and was half-sister to Fair Simone (seven wins; dam of Ascot Gold Cup second and sire Finglas).

MELD (1952)

Returning to the distaff side of Charlottown's pedigree, he was out of Meld, a superb racemare and a good-looker too. Meld won five races worth £43,051, and was unbeaten at three years in the 1,000 Guineas, Oaks, St. Leger and Coronation Stakes.

She bred four foals before Charlottown—the sister and brother Intaglio and Carrara (by Tenerani), Lysander (by Nearco) and Mellay (by Never Say Die)—which in achievement were about as far removed from a Derby winner as one could get.

Intaglio won a mile and a half maiden race in Ireland on her only start, and bred two winners, the better being Hants, a filly who won four races at around ten furlongs as a three-year-old. Intaglio also bred Seedling, who was unplaced in her only race but bred the Criterium de Saint-Cloud winner Easy Regent.

Carrara won one four-year-old novice hurdle, but failed to add to that success in five subsequent seasons with five different trainers. Lysander did less well, being a temperamental handicapper who was placed six times from seven starts without winning, and went to Australia as a stallion. Mellay did not even get on to the racecourse, because of injury, but he redeemed himself when exported to New Zealand, where he was most successful at stud, two of his best produce being Trelay (23 wins, and broke down in the Wellington Cup after seeming to have the race won) and Princess Mellay (18 wins, including New Zealand Oaks and New Zealand Cup twice, each time beating Trelay).

Meld continued to be sent to high-class stallions after producing Charlottown, and seemed to have produced another fine colt in Donated (by Princely Gift), who finished second in the Coventry Stakes but then broke a leg on the gallops in August of his first season and had to be destroyed.

Meld bred two further winners in Britain to Charlottesville—Canasta Girl, who failed to live up to her early promise but did win a small race over ten furlongs and bred a winner; and Scarletville, who won one race over a mile and a half. Meld also bred Ragotina (by Ragusa), foaled when her dam was 21, and after splitting a pastern at the age of two, won two races the following year.

Several members of this family seemed not so honest as Meld had been. Carrara was not reliable; Canasta Girl did not maintain her progress; Scarletville was once tried in blinkers with no success, and Ragotina once gave the impression of having put little heart into a finish. It was a trait that could also be spotted in other members of the female family, especially those from Duplicity, a half-sister to Meld's dam Daily Double.

Meld was far and away the best of five winners bred by Daily Double, who also produced Chione (two wins including Galtres Stakes; dam of three winners including Warren Stakes winner Castro) and Jekyll (a brother to Meld; one-paced stayer who won two races up to 18 furlongs). Daily Double herself was a successful racemare, winning four races for £1,653, the last when she upset by a head the odds laid on the temperamental Royal Charger when they were the only runners for the six-furlong Challenge Stakes at Newmarket. Daily Double ended her racing career when seventh of 34 behind Sayani, carrying 8st 1lb, in the Cambridgeshire. Earlier she had distinguished herself by getting loose before the 1,000 Guineas and bolting back into Newmarket!

DOUBLETON (1938)

Doubleton, an unraced sister to Persian Gulf and half-sister to Precipitation, bred five winners, though there were 11 years between the fourth (Double Eclipse) and the fifth (Double Exposure). Double Eclipse was the best of these and has been discussed in relation to Airborne, where Doubleton's dam, Double Life, was also dealt with.

Doubleton's daughter Duplicity won two races for £1,438 and bred seven winners, including Duplation (vicious and unmanageable as a two-year-old, he benefited from corrective treatment long enough to win the £3,461 Lingfield Derby Trial but refused to do his best in the Derby, refused to start at all in his next attempt, and never ran again); Dual (three wins, £4,177, including Solario Stakes; sire of mainly moderate stock); Duplex (won small maiden event; fourth in Eclipse Stakes), and Sonsa (Ebbisham Stakes, £2,550, on debut; failed to win again but showed useful form in top class).

Sonsa went on to produce six winners on the Flat and a decent winning hurdler called Dogged (a misnomer if ever there was one for certain members of this family, if not for those who followed its breeding). The best of Sonsa's foals were Sunspeck (two wins, £4,938, including Ebbisham Stakes like her dam, but behaved badly at start there; also left at start in 1,000 Guineas and refused to take part in Coronation Stakes, the last of her three races at three years; dam of four winners) and Inventory (ten wins including Newbury Autumn Cup; also won under National Hunt Rules but not always so genuine as his long-serving record might suggest).

Doubleton's daughter Didima won the second of her two races at two years, a small event at Newmarket for £435; ran only twice at three years; failed to breed a winner in Britain, and her daughter Double Charm bred only one foal before dying of grass sickness, that foal being the Goodwood Cup and Yorkshire Cup winner Sagacity.

1967

1 H. J. Joel's b c Royal Palace (Ballymoss – Crystal Palace).
2 C. Engelhard's b c Ribocco (Ribot – Libra).
3 M. Sobell's ch c Dart Board (Darius – Shrubswood).
  22 ran. Time: 2 min 38.30 sec

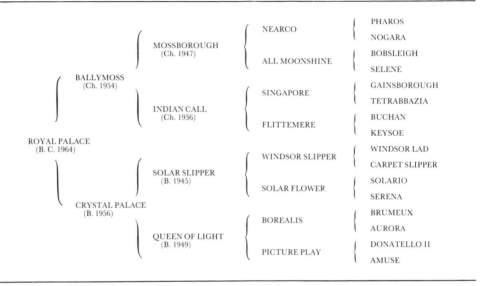

Royal Palace bridged the gap between Mr. Jim Joel's two Classic winners, the first being Picture Play, who won the 1,000 Guineas in 1944. He also brought into the Derby story one of the best female families of the century.

Royal Palace won the last two of his three races at two years, ending with a length and a half victory in the Royal Lodge Stakes, after losing six lengths at the start, which earned him second place in the Free Handicap on 9st 4lb, 3lb behind the more-precocious Bold Lad.

As a three-year-old Royal Palace might have won the Triple Crown but for a mishap. He had the speed to win the 2,000 Guineas, where he got up close home to beat Taj Dewan, and he won the first Derby dispatched from starting stalls smoothly by two and a half lengths from Ribocco. Royal Palace had to miss the St. Leger after knocking a joint the weekend before he was due to reappear in the Great Voltigeur Stakes, and in his only other race that season was beaten by Reform and Taj Dewan in the ten-furlong Champion Stakes.

The decision to keep him in training as a four-year-old paid off handsomely with five wins, though the last was gained at the expense of an injury. The most important of these were the last two, in the Eclipse Stakes, where he met the Derby winner Sir Ivor and his old rival Taj Dewan and got up in the last strides to beat the latter, with Sir Ivor less than a length away third; and the King George VI and Queen Elizabeth Stakes, where he got home by half a length from Felicio and the previous year's Arc de Triomphe winner Topyo.

Royal Palace, an attractive, shapely horse, returned from his final race lame in his near foreleg with what was later diagnosed as a badly torn suspensory ligament. He retired with nine wins from 11 races and stakes of £163,951, a record for a British-based horse in training.

He stood his first season at stud in 1969, having been taken on by a partnership of his owner Mr. Joel, Lady Macdonald-Buchanan and Lord Howard de Walden. His earliest years as a stallion were most disappointing, the only traces of class among his offspring coming from the fillies Escorial (Musidora Stakes) and Antrona (Prix de Malleret).

BALLYMOSS (1954)

Royal Palace's sire, Ballymoss, was among the best Derby runners-up, since he went on from Epsom to win the Irish Derby and Doncaster St. Leger as a three-year-old, and the Coronation Cup, Eclipse Stakes, King George VI and Queen Elizabeth Stakes and Arc de Triomphe at four. He was a top-class horse who stayed well but also had a turn of speed, which he turned to such advantage that he retired to stud in Ireland in 1959 having won seven races for £60,726 and 45,510,700 francs (£107,166). In contrast to his own sire, Mossborough, who got offspring better than himself, Ballymoss at stud did not emulate Ballymoss on the racecourse.

Royal Palace was easily the best son of Ballymoss, whose best daughters were the Irish Oaks winners Merry Mate and Ancasta, Parmelia, who echoed her sire by improving with time and distance to win the Ribblesdale Stakes and Park Hill Stakes, and Miba (Princess Elizabeth Stakes).

Ballymoss has emerged as a significant sire of broodmares, his daughters having produced such as Levmoss (eight wins, £132,951, including the rare Ascot Gold Cup–Arc de Triomphe double); Stage Door Johnny (Belmont Stakes); Sweet Mimosa (French Oaks); Blue Cashmere (seven wins, £32,582, including Nunthorpe Stakes and Ayr Gold Cup); Mil's Bomb (Lancashire Oaks, Park Hill Stakes) and African Dancer (Park Hill Stakes).

Indian Call, by Singapore, bred Ballymoss when she was 18. She had shown no ability on the racecourse but made up for that at stud, with eight winners of 28 races. The best before Ballymoss was Guide (by Chamossaire), who won eight staying races for £5,399 and is best known as a National Hunt stallion.

FLITTEMERE (1926)

Flittemere, the dam of Indian Call, had a career of contrasts. She did not race at two years but made up as a three-year-old with 12 races, of which she won three for £1,173, including the Yorkshire Oaks (in which Aloe was sixth) and then an apprentice event. In 1929 she was sent to the Newmarket December Sales and fetched top price of 3,500 guineas among a large draft submitted by Lord Derby. That also proved to be one of the highest prices at the whole sale, but ten years later she was back at the same venue, having passed into the hands of Lord Glanely, and made only 10 guineas. At the same sale Indian Call, then a three-year-old, did slightly better and was sold for 15 guineas to the Irish breeder Mr. Richard Ball, for whom she eventually bred Ballymoss.

Flittemere bred no winners in Britain, her only success being Grand Flit, successful 15 times in Belgium.

Early interest in Flittemere at the sales arose from her pedigree, since she was by Buchan out of Keysoe, the St. Leger winner who herself was by Swynford out of the Oaks winner Keystone. Keysoe bred one other winner besides Flittemere— Caissot, winner of three races for £4,355, of which more than half came from the Prince of Wales Stakes at Ascot and a further £830 from a walk-over in the Newmarket St. Leger. Keysoe had only four live foals in eight years at stud, and the only filly was Flittemere.

MOSSBOROUGH (1947)

Mossborough, the sire of Ballymoss, was a good racehorse just below top class. In 16 races he won five times for £4,606, and was at his best as a four-year-old, when he won the Churchill Stakes and was second, beaten half a length, in the Eclipse Stakes. Retired to stud in 1952, he did honourable service until his death in July 1971. He was champion sire in 1958, when Ballymoss was the prime contributor, and was undoubtedly a better sire than racehorse.

He was responsible for one superb filly in Noblesse, who won four races for £46,443, including the Oaks and Timeform Gold Cup, and became the dam of three winners. Also by Mossborough were Yelapa (Grand Criterium); Cavan (Belmont Stakes); Morecambe (11 wins, £24,151, including Cesarewitch and Ebor Handicap); Craighouse (Irish St. Leger); Birdbrook (16 races, £9,453; sire); Anticlea (Italian 1,000 Guineas and Oaks); and Spartan General (useful stayer; important National Hunt stallion).

Mares by Mossborough have done well, chiefly by producing Exbury (seven wins, £156,784, including Arc de Triomphe and Coronation Cup); Sagaro (Ascot Gold Cup), Pandora Bay (Ribblesdale Stakes); Current Coin (Cork and Orrery Stakes); Popkins (Princess Elizabeth Stakes, Sun Chariot Stakes) and Peace (Blue Seal Stakes).

Mossborough's dam, All Moonshine, was by Bobsleigh out of Hyperion's dam Selene. All Moonshine won a race worth £315 over seven furlongs on her debut at Newmarket as a three-year-old, but showed nothing in her other races and may not have been entirely genuine. Mossborough was her second foal, and she produced five winners from 11 foals before being exported to America in 1960.

All Moonshine's first foal, Eyewash, won two races including the Lancashire Oaks, and bred 11 winners in Britain. They included Sijui (two wins, including Fred Darling Stakes; dam of seven winners) and two more top-class fillies in Collyria (£5,835, including Park Hill Stakes; dam of three winners) and Varinia (£3,436; third in Oaks). Eyewash also bred Visor, a plating-class maiden who went to America in 1956 but was repatriated to produce four winners in Britain, the best being Raise You Ten (four wins, £11,352, including Doncaster Cup, Goodwood Cup). Visor's other winners included Aranda, who was twice successful in Ireland for a total of £266 but failed to breed a winner in Britain, though her daughter Arandena, despite being useless on the racecourse, bred the Italian Derby winner Ardale.

CRYSTAL PALACE (1956)

The distaff side of Royal Palace's pedigree has served the Joel interests well for many years, and Royal Palace's dam Crystal Palace played her part on the racecourse by winning five of her nine races, for £5,817, at up to ten furlongs. She was beaten whenever she tackled the highest company, but showed her merit by winning the Falmouth and Nassau Stakes.

Royal Palace was her fourth foal. The first did not win; the second was Crystal Glass, a filly who won four races in Ireland as a three-year-old for a total of £895; and the third died as a yearling.

After Royal Palace, Crystal Palace bred five more winners, including two colts just below the highest class—Prince Consort (three wins, £5,284; third in St. Leger) and Selhurst (four wins, £15,618, including Hardwicke Stakes)—and the filly Glass Slipper (second in Musidora Stakes).

Crystal Palace was by Solar Slipper, a son of Windsor Slipper, whose dam Carpet Slipper was one of eight winners bred by Simon's Shoes, though she added only £323 to the total of over £10,000 won in stakes by her dam's produce. Carpet Slipper also bred the sisters Godiva (1,000 Guineas and Oaks but died without issue) and Nova Puppis (unraced; dam of six winners including the very useful two-year-old Novarullah).

QUEEN OF LIGHT (1949)

Crystal Palace's dam, Queen of Light, showed her best racecourse form at a mile, over which she won the Falmouth Stakes at Newmarket, as her daughter was to do seven years later. Queen of Light also won two races at two years to earn a total of £4,780. She bred five winners, the first four coming in her first four years at stud.

The first was Picture Light, a filly by Court Martial who won four races for £3,776, including the Prince of Wales's Stakes. She was the dam of nine winners who were more noted for outstanding stamina than outstanding ability, though the majority were above the average expected from a racehorse. They included Illuminous (Sandleford Priory Stakes); Photo Flash (second in 1,000 Guineas); Welsh Pageant (11 wins, £41,234, including Queen Elizabeth II Stakes, Lockinge Stakes twice; third in 2,000 Guineas; successful sire); and Miss Pinkie (Argos Star Fillies' Mile).

The second was Chandelier, a filly by Goyama who showed good form but also a measure of temperament and ended her racing career with a single win for £967. Mr. Joel sold her as a back-end three-year-old for 6,300 guineas and she proceeded to throw three winners including Crocket (seven wins, £19,321; leading two-year-old of 1962 when unbeaten; sire of Oaks runner-up Frontier Goddess, Queen Mary Stakes winner Farfalla, and good sprinter Burglar before export to Japan in 1973); and Crystal Light, who like her dam had good two-year-old form but appeared not to train on, and became the dam of Headlamp (a very useful two-year-old in Ireland who later won in America) and Street Light (very useful sprint filly who won the Prix Meutry).

Queen of Light's third foal was Crystal Palace; her fourth was Ancient Lights, a colt by Supreme Court who won the Dewhurst Stakes on his only outing at two years. He had training problems and ran only once at three, finishing second in the Dante Stakes, and three times more at four. He was exported to Argentina.

Borealis, the sire of Queen of Light, was by Brumeux out of Aurora, and therefore a half-brother to the brothers Alycidon and Acropolis, and to Woodlark, grandam of Larkspur.

Borealis won seven races over three seasons for £3,635, including the Coronation Cup. He was second in the St. Leger, but was not entered for the 1944 Derby, which was a pity because there was form that year which suggested he might have won it. He was only moderately successful at stud, with such as Double Bore (Goodwood Cup), Magnetic North (Irish St. Leger), Northern Gleam (Irish 1,000 Guineas), Herero (German Derby) and Stella Polaris (third in Oaks) as his most important offspring to race. Borealis, who was exported to Sweden in 1961 and died there three years later, did notably well as a broodmare sire with Northern Hope, dam of Agreement (seven wins, £13,483, including Doncaster Cup twice); Stella Polaris, dam of Discorea (Irish Oaks); and Highlight, dam of Highclere (1,000 Guineas, French Oaks).

PICTURE PLAY (1914)

Picture Play, the dam of Queen of Light, became Donatello II's first Classic winner when she beat a modest field for the 1,000 Guineas. She broke down in her next race, the Oaks, when her near-fore suspensory ligament gave way in an incident similar to that which befell her great-grandson Royal Palace. Picture Play had won her first two races as a two-year-old before disappointing in the Cheveley Park Stakes.

Picture Play bred seven winners, beginning with the most prolific, Full Hand, winner of six races for £2,679 and a gelding bred to stay well but who did best when allowed to bowl along in front and consequently a miler.

She bred another successful filly, apart from Queen of Light, in Red Shoes, whose one win for £2,198 was the family's almost customary one in the Falmouth Stakes. Red Shoes was also second in the Yorkshire Oaks, and bred five winners, including West Side Story (two wins, £4,923, Nell Gwyn Stakes and Yorkshire Oaks; short-head second in Oaks, third in 1,000 Guineas) and Pembroke Castle (five wins, £8,997, second in Hardwicke Stakes).

Picture Play's daughter Love Parade, a twin, bred the prolific Flat and hurdles winner Mayfair Bill, and her last winning produce, Promulgation, proved to be financially the most successful. Promulgation won the National Breeders Produce Stakes and Richmond Stakes for £9,544 as a two-year-old before being exported to America, where he won one of his four starts at the age of five to increase his tally by 3,300 dollars. He was retired to stud and has appeared as the maternal grandsire of Noble Decree, the Observer Gold Cup winner and 2,000 Guineas runner-up.

Picture Play's first success, in 1943, was the first to be credited to her dam Amuse, who was 16 at the time and a latecomer in other ways. Amuse's first outing on the racecourse was not until October of her three-year-old season, when she was second, after which she ran only once more, in the King's Stand Stakes the following year.

Apart from Picture Play, who was her only winner, Amuse bred Queen's Pleasure, dam of three winners including Royal Pardon, a successful sire in South Africa.

Amuse was by Phalaris out of Gesture, dam of three winners and herself a daughter of Absurdity, who won two small races and bred eight winners. Absurdity's produce included the brother and sister Absurd (Middle Park Stakes; leading sire in New Zealand) and Jest (1,000 Guineas and Oaks; dam of the Derby winner Humorist); and Black Jester (St. Leger, Sussex Stakes).

CHAPTER 28

1968 – 1972
The American Influence

1968

1   R. Guest's b c Sir Ivor (Sir Gaylord – Attica).
2   H. J. Joel's b c Connaught (St. Paddy – Nagaika).
3   A. J. Struther's b c Mount Athos (Sunny Way – Rosie Wings).
      13 ran. Time: 2 min 38.73 sec

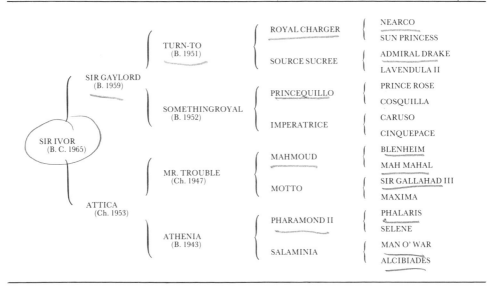

		TURN-TO (B. 1951)	ROYAL CHARGER	NEARCO
				SUN PRINCESS
	SIR GAYLORD (B. 1959)		SOURCE SUCREE	ADMIRAL DRAKE
				LAVENDULA II
SIR IVOR (B. C. 1965)		SOMETHINGROYAL (B. 1952)	PRINCEQUILLO	PRINCE ROSE
				COSQUILLA
			IMPERATRICE	CARUSO
				CINQUEPACE
	ATTICA (Ch. 1953)	MR. TROUBLE (Ch. 1947)	MAHMOUD	BLENHEIM
				MAH MAHAL
			MOTTO	SIR GALLAHAD III
				MAXIMA
		ATHENIA (B. 1943)	PHARAMOND II	PHALARIS
				SELENE
			SALAMINIA	MAN O' WAR
				ALCIBIADES

SIR IVOR HERALDED a run of four Derby wins in five years by horses bred in North America and owned by American residents. He was bought as a yearling for 42,000 dollars (then the equivalent of £15,272) at the Keeneland Sales and proved the outstanding two-year-old trained in England, Ireland and France. After finishing sixth on his July debut, he was unbeaten in three races, the last a comprehensive defeat of Pola Bella and Timmy My Boy in the Grand Criterium at Longchamp. His two other wins were in Ireland, where he headed the Free Handicap. Had he been placed in the French Free Handicap he would have topped that too, since Pola Bella made second spot, 1lb below the top.

As a three-year-old Sir Ivor had the unusually large number of nine races. He won the Guineas Trial at Ascot before the 2,000 Guineas itself, where he did not have a hard race to win by a length and a half from Petingo. He started 5-4 on for the Derby, for which in physical appearance he looked head and shoulders above the rest, and in the race, after making a steady run to close on the leaders, he produced a burst of speed which took him past Connaught inside the final furlong and he won by a length and a half, after being eased close home.

Sir Ivor lost his next four races after the Derby. It was a mystery how he was beaten by Ribero in the Irish Sweeps Derby; he was a close third to Royal Palace in the Eclipse Stakes; he was second in his preparation race for the Arc de Triomphe, and he beat all but the brilliant Vaguely Noble in a strongly-contested race for the Arc. Sir Ivor returned to winning form on his last two outings, in the Champion Stakes and Washington International.

A big, strong, handsome horse as a three-year-old, Sir Ivor retired at the end of his second season with eight wins for £116,787 in England and Ireland, 876,100 francs in France, and 100,000 dollars in America. His owner, Mr. Raymond Guest, who had won the race with Larkspur and seven years later owned the Grand National winner L'Escargot, left Sir Ivor to stand for two seasons, 1969 and 1970, at stud in Ireland on generous terms which reflected his feeling for European racing. From 1971 Sir Ivor stood at Claiborne Farm in Kentucky.

His crops born in Ireland included two Derby seconds: Cavo Doro (also won Ballymoss Stakes and Royal Whip) and Imperial Prince (also second in Irish Sweeps Derby and Benson and Hedges Gold Cup). Sir Ivor's Irish-bred produce also included two good fillies in Istiea (Lancashire Oaks) and Northern Princess (Ribblesdale Stakes), and the useful colt Sir Penfro (five wins; fourth in Irish Sweeps Derby).

His first crop in America, foaled in 1972, produced his best winner to date, the filly Ivanjica, who at three years won the French 1,000 Guineas and Prix Vermeille, and at four years recovered her form in the autumn to win the Arc de Triomphe from the St. Leger winner Crow and the French Derby winner Youth.

Several more produce foaled in America found their way to Europe, and his reputation as a top-class sire of fillies was enhanced by Miss Toshiba (Pretty Polly Stakes at The Curragh); Val's Girl (Pretty Polly Stakes at Newmarket; second in Oaks); Cloonlara (unbeaten two-year-old of exceptional speed including winning Phoenix Stakes from odds-on Godswalk); and I've a Bee (third Irish Oaks).

His best colts from this period were Sir Wimborne (unbeaten two-year-old when awarded Royal Lodge Stakes on a disqualification); Malinowski (Craven Stakes; second in Dewhurst Stakes in lightly-raced career) and Sir Penfro's brother Padroug (Acomb Stakes).

RACING TWO-YEAR-OLDS AT A MILE

The emergence of Sir Ivor and Vaguely Noble as the star two-year-olds of 1967, and their subsequent performances in the top middle-distance races of Europe, provided evidence of the value of staging valuable late-season races over a mile for two-year-olds. Sir Ivor won the French Grand Criterium; Vaguely Noble won the more recently introduced Observer Gold Cup, formerly the Timeform Gold Cup and more recently the William Hill Futurity.

The French have long acknowledged the value of putting on a good-class race for those two-year-olds which by their nature cannot be expected to show the early maturity that wins races over five and six furlongs before the autumn. The Grand Criterium was first run in 1853. Its record since the war and up to Sir Ivor's success was one of uninterrupted high class, franked by several Classic winners in their second seasons. It clearly did no harm to Rigolo, Tantieme, Sicambre, Apollonia, Tyrone, Bella Paola, Right Royal V, Hula Dancer, Neptunus and Soleil, who between 1947 and 1965 won one or more European Classics after winning the Grand Criterium.

The Grand Criterium served its purpose as a top-class race in its own right, worthy of a valuable place in the two-year-old calendar, and also pinpointed horses likely to do well in the Classic generation. Despite this record there was much opposition to the idea of a high-class mile race being given a prize relative to its status in England, and it was only thanks to the persistence of Mr. Phil Bull and the foresight of Doncaster racecourse that the Timeform Gold Cup was introduced in October 1961.

By the time Vaguely Noble won the Observer Gold Cup—which replaced the Timeform Gold Cup in 1965—the critics had been made to think again, since two of the six winners, Noblesse and Ribocco, had won a Classic in this country, and another, Pretendre, had been beaten a whisker in the Derby. Vaguely Noble put the seal on the sense of staging such a race, and nothing has happened since, with the race undergoing another change of backer and title when it became the William Hill Futurity in 1975, to upset the theory.

SIR GAYLORD (1959)

Sir Gaylord, the sire of Sir Ivor, was a good two-year-old in America, where he won six races and five and six furlongs, including the Sapling Stakes. He kept his form as a three-year-old, winning four races up to nine furlongs from four starts, but while favourite for the Kentucky Derby he broke down and had to be taken out of training. He retired to stud in 1963 with earnings of 237,404 dollars from his ten wins. He was exported to stand in France from 1973.

Sir Gaylord had a most uneven stud career. The best horses in his first three crops, apart from Sir Ivor, were Village Square (useful up to a mile in France) and Gay Matelda (very smart juvenile in America over an extended mile; third in CCA Oaks; nine wins, 409,945 dollars). Then, in 1966, came the smart winner and phenomenally successful sire Habitat, and after that very little apart from Delmora (Prix de la Salamandre, Prix du Moulin; second in Cheveley Park Stakes).

Habitat was bought in America for 105,000 dollars (then £38,182). He did not race at two, but at three was the best miler in Europe, winning four of the most important races over the trip in his last five races. In all he won five races for the equivalent of £40,840, including the Prix du Moulin and Lockinge Stakes. Bought for £400,000 to go to stud, he quickly made his mark, becoming second leading sire in 1975 and the leading sire of two-year-olds with each of his first three crops, though events showed he was capable of producing horses able to train on to become decent three-year-olds.

Habitat's first crop was highlighted by Habat (five wins, £44,972; leading English-trained two-year-old; won Middle Park Stakes, Norfolk Stakes) and Bitty Girl (joint leading English-trained two-year-old filly; won four races at five furlongs including Queen Mary and Lowther Stakes, for £12,305).

His second crop included Hot Spark (three wins, £14,684; Flying Childers Stakes, Palace House Stakes; second in King's Stand Stakes); Rose Bowl (five wins, £61,943, including Champion Stakes, Queen Elizabeth II Stakes twice); Roussalka (seven wins, £45,017, including Nassau Stakes twice, Coronation Stakes); and Steel Heart (four wins, £49,229, including Middle Park Stakes, Gimcrack Stakes).

Habitat's third crop included Hittite Glory (three wins, £31,637, including Middle Park, Flying Childers Stakes) and Flying Water (French-trained; three wins in England, £43,310, including 1,000 Guineas and Champion Stakes).

TURN-TO (1951)

Just as Sir Gaylord had to be retired prematurely from racing, so did his sire Turn-To, an Irish-bred colt who was imported into America as a foal by Mr. Claude C. Tanner and on his death was sold to A. B. Hancock for 20,000 dollars. As a two-year-old Turn-To won the richest American juvenile race, the Garden State Stakes, and on the strength of three straight wins in his second season was made favourite for the Kentucky Derby. Just as Sir Gaylord had to be retired as early favourite for America's premier Classic, so did Turn-To, since he bowed a tendon and never raced again. One of America's best horses up to a mile, he retired having won six races from eight starts, for 280,032 dollars.

He proved a first-class sire, and in his first crop produced First Landing, the top two-year-old of 1958 and winner of 19 races in all, worth 779,577 dollars, including the Champagne Stakes, Hopeful Stakes, Garden State Stakes (like his sire), and Saratoga Special (on a disqualification). First Landing went on to sire Riva Ridge (Kentucky Derby and Preakness Stakes).

Turn-To also sired Hail to Reason, (top two-year-old of 1960; nine wins, 328,434 dollars, including Hopeful Stakes; sire of Roberto) and Captain's Gig (eight wins, 205,312 dollars; two seasons at stud in America and two in Ireland before his premature death). Turn-To himself died in 1973.

Source Sucree, the dam of Turn-To, won once at two years in France, where she was foaled four years after her dam's export in 1936. Source Sucree bred nine winners, the most noteworthy apart from Turn-To, her sixth foal, being Cagire II (four wins, £13,204, including Ormonde Stakes; only moderately successful at stud before export to America in 1953, the same year as his dam's export there); Sourcillon (nine wins in France) and Black Brook (eight wins in France; dam of winners).

LAVENDULA II (1930)

Lavendula II, dam of Source Sucree, was a mare culled from the Lord Derby stable in 1933, when Mr. Benjamin Guinness bought her for 750 guineas and sent her to stud in France. She had won three races in France as a two-year-old but ran without success in two handicaps in England the following summer. At stud she bred six winners from ten offspring, and her influence can be said to have been exerted worldwide.

Lavendula II's son Ambiorix headed the French Free Handicap after winning the Grand Criterium in 1948 and the following year won the Prix Lupin and was second to Good Luck in the French Derby. Ambiorix did well on his export to stand at stud in America, where he was leading sire in 1961. His daughter Fantan II bred Ragusa. He died in 1975.

Lavendula II also bred Babiste (22 wins and good sire in Belgium) and Singadula (dam of Singlspieler, the top three-year-old colt in Germany in 1944 when he won the local 2,000 Guineas and Derby), as well as two important mares in Source Sucree and Perfume II.

The unraced Perfume II bred seven winners, including Sayani (four wins as three-year-old, including Cambridgeshire under 9st 4lb; leading sire in France in 1953 when his daughter La Sorellina won the French Oaks and Arc de Triomphe, beating her half-brother Silnet a short head in the latter; leading sire in Brazil in 1960 after his death); My Babu (earlier named Lerins; won $10\frac{1}{2}$ races, £29,830, including 2,000 Guineas; sire), and Marco Polo II (three wins, £2,949; sire in New Zealand of two Melbourne Cup winners, Polo Prince and Macdougal, and the New Zealand St. Leger winner Bali Ha'i, who came to England as a present to the Queen Mother and won the Queen Alexandra Stakes).

Two of Lavendula II's winning fillies, apart from Source Sucree, are worth recording. They are Vertige, who did not race at two years because of an accident and after winning a mile and threequarters maiden race at three years was sold for 7,800 guineas at the 1954 Newmarket December Sales, and Virelle, who won a small race in France. Vertige bred six winners, including Romantica (five wins, £5,254, came into her own in the autumn of her three-year-old career over middle distances, including Princess Royal Stakes; dam of Ebbisham Stakes winner Ileana). Virelle became the grandam of the Irish Sweeps Derby winner English Prince.

SOMETHINGROYAL (1952)

Sir Gaylord's dam, Somethingroyal, ran only once, unplaced, as a two-year-old. She more than made up for that lack of success at stud, where she bred four top-class racers among the eight winners she had from 13 foals. Apart from Sir Gaylord, they were First Family (by First Landing), who won 188,040 dollars and finished third in the Belmont Stakes; and two produce by Bold Ruler—Syrian Sea (six wins, 178,245 dollars, including Selima Stakes and Colleen Stakes; second in National Stallion Stakes) and Secretariat (Triple Crown).

Somethingroyal's dam, Imperatrice, was a high-class racer who won 11 times for 37,255 dollars, including the New Zealand Oaks and Test Stakes. She bred ten winners, the most successful being Squared Away (31 wins, 255,145 dollars, mainly in handicaps) and Scattered (four wins, 80,275 dollars, including Pimlico Oaks and CCA Oaks).

The family of Somethingroyal and Imperatrice goes back to Cinq à Sept, who produced Assignation, the dam of Cinquepace, in turn the dam of Imperatrice. It was Cinq à Sept who introduced the family to America, when she was exported there as a five-year-old in 1929, the year after she had finished her racing career by winning the Ebor Handicap under 8st 4lb. The previous year Cinq à Sept had won the Irish Oaks and Park Hill Stakes.

As well as Assignation, who bred seven winners, Cinq à Sept bred Twilight Tryst (fifth dam of the 2,000 Guineas winner Nonoalco) and Gentle Tryst (whose daughter Up the Hill is third dam of the Preakness Stakes winner Elocutionist, and fourth dam of Alleged, runaway winner of his only two races at two years in Ireland).

Cinquepace, who was unraced, bred only two foals, both winners.

ATTICA (1953)

Turning to the distaff side of Sir Ivor's pedigree, his dam Attica won five races at around a mile at three and four years of age, and was placed in two stakes races at 8½ furlongs. From five foals before Sir Ivor she bred four winners, including Sir Ivor's brother, Young Noble, and Greek To Me (38,532 dollars).

Attica, who was by the stayer Mr. Trouble, third in the Kentucky Derby and Belmont Stakes won by Middleground, was out of Athenia, a good racemare who won ten races for 105,170 dollars, including the valuable Ladies Handicap over a mile and a half at Belmont Park. Athenia, who was also second in the Kentucky Oaks, produced 11 foals, nine of which were winners.

Athenia was by Hyperion's half-brother Pharamond II out of Salaminia, probably the best daughter of Man O'War to race. Salaminia won two top-class races, the Alabama Stakes and the Ladies Handicap. Salaminia was a half-sister to Menow, sire of Tom Fool, who retired with 21 wins for 570,165 dollars. The best two-year-old colt of 1951 and Horse of the Year in 1953, Tom Fool put himself among the leading American sires with the great Buckpasser, Tim Tam (Kentucky Derby) and many more. Menow also sired Capot (Belmont and Preakness Stakes).

The dam of Salaminia and Menow was Alcibiades, the Kentucky Oaks winner. The dam line was originally introduced to America in 1923, six years before the sire line was first seen there, when Regal Roman was exported from England as a two-year-old after being bought by the father of Sir Ivor's breeder, Mrs Alice Headley Bell, for 560 guineas. Regal Roman became the dam of Alcibiades.

1969

1   A. M. Budgett's b c Blakeney (Hethersett – Windmill Girl).
2   P. G. Goulandris's ch c Shoemaker (Saint Crespin III – Whipcord).
3   Comtesse de la Valdene's br c Prince Regent (Right Royal V – Noduleuse).
    26 ran. Time: 2 min 40.30 sec

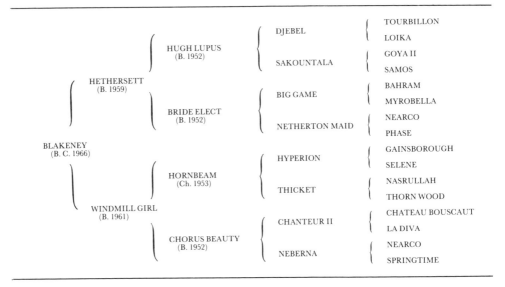

		DJEBEL	TOURBILLON / LOIKA
	HUGH LUPUS (B. 1952)		
		SAKOUNTALA	GOYA II / SAMOS
HETHERSETT (B. 1959)			
		BIG GAME	BAHRAM / MYROBELLA
	BRIDE ELECT (B. 1952)		
		NETHERTON MAID	NEARCO / PHASE
BLAKENEY (B. C. 1966)			
		HYPERION	GAINSBOROUGH / SELENE
	HORNBEAM (Ch. 1953)		
		THICKET	NASRULLAH / THORN WOOD
WINDMILL GIRL (B. 1961)			
		CHANTEUR II	CHATEAU BOUSCAUT / LA DIVA
	CHORUS BEAUTY (B. 1952)		
		NEBERNA	NEARCO / SPRINGTIME

Blakeney, the one English-bred Derby winner in the five-year period under discussion in this chapter, could have been bought for 5,000 guineas as a yearling, but he failed to make his reserve at the sales and stayed with his owner-breeder-trainer Arthur Budgett, who eventually sold two quarter shares to patrons and kept the other half for himself.

As befitted a horse who was bred for stamina rather than speed, Blakeney was lightly-raced as a two-year-old, winning the second of his two races, the seven-furlong Houghton Stakes, from 26 others.

He won only one race as a three-year-old, the Derby, where the fact there were 26 runners, twice the previous year, reflected the feeling there was no outstanding horse among that generation. Blakeney got a smooth run on the inside and won by a length from Shoemaker, with Prince Regent third after being given too much to do by his French jockey. In his other races at three years Blakeney was second in the Lingfield Derby Trial, and then after Epsom was fourth to Ribofilio (who had been fifth at Epsom) in the Irish Sweeps Derby, fifth to Intermezzo (eighth at Epsom) in the St. Leger, where he did not get a trouble-free run, and ninth of 24 in the Arc de Triomphe won by Levmoss.

At four years Blakeney showed his superiority over his own generation at a mile and a half, despite a modest start to the season in which he was fifth in the Jockey Club Stakes and a narrow winner of the moderately-contested Ormonde Stakes. After finishing half a length second to Precipice Wood in the Ascot Gold Cup, Blakeney was two lengths second to Nijinsky in the King George VI and Queen

Elizabeth Stakes, and was the first older horse home when fifth to Sassafras in the Arc.

At the end of his four-year-old career Blakeney was retired to the National Stud, which for £250,000 had bought a half-share in him following the King George VI and Queen Elizabeth Stakes. He was the first Derby winner to be purchased for the National Stud under a new policy, formulated in 1963 when responsibility for its operation and control was taken over by the Horserace Betting Levy Board. Instead of being funded by government contributions, the Stud took on its own identity and accounts under the Levy Board banner, and thus is now national in name only. It also acquired a new policy, by which the practice of maintaining approximately 15 mares was discontinued, and the whole emphasis was placed on stallions.

The National Stud also found new premises, its previous operations at Gillingham and West Grinstead being moved to Newmarket, where in 1964 a long lease was taken on the Jockey Club's Bunbury Farm and work on a new stud began. In 1966 the existing National Stud stallions Never Say Die and Tudor Melody moved in; Stupendous was bought from America in 1967; Hopeful Venture joined in 1969; and Blakeney followed in 1971.

A neat, attractive horse, if on the small side, Blakeney made the headlines with his first crop, which contained Juliette Marny, the Oaks and Irish Oaks winner. It was a notable achievement, since not only did he get a Classic winner in his first crop, Juliette Marny was his first Classic runner, and her dam, Set Free, was the first mare he officially covered at stud.

Blakeney originally showed a marked reluctance for his new duties, and it was not until March 11 that he covered the first mare on the list of 35 booked to him. That was Set Free, and after his suspicious start, Blakeney got 31 of the 33 mares safely in foal.

His second crop contained the Lingfield Derby Trial winner Norfolk Air.

DERBY WINNERS AND THE ASCOT GOLD CUP

Blakeney's appearance in the Ascot Gold Cup was highly unusual. The increasing tendency has been to downgrade staying races, and Blakeney was the first Derby winner to take the field for the Ascot Gold Cup since Ocean Swell won the race 25 years earlier. Ocean Swell himself was the first to bring off the Epsom–Ascot double since Persimmon in 1896–97, so it can be seen that the trend away from long-distance events is not exactly recent. However, it has been accelerated in recent times, when victory in any race beyond a mile and a half has seemed to spell disaster for a potential stallion. That is a short-sighted policy and has been discussed already. Sadly, it will probably be as many years as passed between Ocean Swell and Blakeney before another Epsom Derby winner ventures so far as the Ascot Gold Cup, more's the pity.

HETHERSETT (1959)

Blakeney's victory brought some degree of reflected compensation to the name of his sire, Hethersett, who was the chief casualty in the falls of the 1962 Derby.

Hethersett had won two of his three races before Epsom and returned to win the Great Voltigeur Stakes and St. Leger before finishing second to Arctic Storm in the Champion Stakes. At four years his three races were at a mile and a half, and he finished second in the Jockey Club Stakes and Coronation Cup, and fifth in the Hardwicke Stakes.

He had a sadly short career at stud, retiring in 1964 but dying in 1966 before his first crop got on to the racecourse. His first crop included Dalry (Duke of Edinburgh Stakes); Heathen (third Dewhurst Stakes), and Hibernian (Beresford Stakes, Irish Cambridgeshire under 9st as three-year-old). His second crop, as well as Blakeney, featured Harken (Chesterfield Cup), and his final crop included Highest Hopes (five wins, including Prix Vermeille); Rarity (four wins; short-head second to Brigadier Gerard in Champion Stakes), and Hazy Idea (Clarence House Stakes; dam of Hittite Glory).

Hethersett was by Hugh Lupus, whose racing and breeding career has been noted, out of the Goya II mare Sakountala, who was foaled in France and began life under the name Amortisseur. Though she won once, she had an undistinguished racing career, and after being brought to Britain in 1952, she bred only one other winner here, Consistency, successful in a three-year-old maiden plate worth £203 at Limerick Junction.

Sakountala had a better pedigree than racing record. Her dam Samos was second in the French Oaks and won the Arc de Triomphe in 1935, the year the first three all were fillies. Samos also won the Prix de la Rochette, then run over two and threequarter miles, and at stud bred Marveil II, winner of the two-mile King George VI Stakes.

BRIDE ELECT (1952)

Major Lionel Holliday, one of Britain's largest owner-breeders in post-war years, never won the Derby, but Blakeney's success can be directly traced to his influence. He bred Hethersett, Blakeney's sire, from a family that emanated from Phase, one of the foundation mares at his stud, and he bred Windmill Girl, Blakeney's dam, from another line which he introduced to start the Cleaboy Stud concern.

Bride Elect, dam of Hethersett, won the Queen Mary Stakes on her second start and was rated the second best filly behind Gloria Nicky, who beat her in the Cheveley Park Stakes, in the Free Handicap. As a three-year-old Bride Elect showed she had made no progress and had only one race, when last of 12 in the 1,000 Guineas.

At stud Bride Elect bred 11 winners from 11 runners she had out of 12 live foals. The best after Hethersett were Proud Chieftain (three wins, £7,515, including Magnet Cup; second in Eclipse Stakes) and Royal Prerogative (seven wins, £21,991, including William Hill Gold Cup). Bride Elect was also the dam of Never Beat, who won one race, worth £402, at Ripon, but became champion sire in Japan.

Bride Elect was a daughter of Netherton Maid, whose relevance was discussed in relation to her dam Phase, by Windsor Lad.

WINDMILL GIRL (1961)

Just as Blakeney could have been bought quite cheaply as a yearling, so could his dam, Windmill Girl, though she did change hands early in her life for not a lot of money. As a foal she accompanied her dam Chorus Beauty to the 1961 Newmarket December Sales. Chorus Beauty fetched 700 guineas, but her foal, Windmill Girl, failed to make her reserve of 1,000 guineas, eventually being bought at that price by Arthur Budgett with the intention of sending her on to the yearling sales the following year. When that time came, Windmill Girl had injured herself. She was reserved for the 1962 December Sales but still showed a trace of the injury, and in consequence failed to make a reserve of 5,000 guineas—the price at which Blakeney was also led out unsold in the same ring five years later.

Windmill Girl went into training with Budgett and she proved a top-class filly, winning two races for £9,936 at three years, including the Ribblesdale Stakes, but running her best race when she finished fast to be beaten two lengths into second place behind Homeward Bound in the Oaks. Windmill Girl also finished third in the Irish Oaks.

At stud she bred six foals in the seven years before her death in November 1972, when she apparently slipped up while galloping in a paddock and fractured her skull.

Blakeney was her first foal. She bred three other winners: Morston, who gave the family and her breeder immortality by winning the Derby; and the moderate winners Alderney (by Alcide) and the filly Cley (by Exbury). Windmill Girl's other produce were Derry Lass (by Derring-Do), who could not be trained because of a hip injury, and Mendham (by Relko), a highly-regarded colt who despite his reputation failed to win in three attempts against maiden class.

HORNBEAM (1953)

Hornbeam, the sire of Windmill Girl, was a most able stayer who won 11 races, worth £10,634, from a mile to two miles, including the Great Voltigeur Stakes. He was also second in the St. Leger, Doncaster Cup and Ebor Handicap. He was retired to stud in 1959, and before being exported to Sweden in 1966 his story was unremarkable, his best produce being Merchant Venturer (Dante Stakes; second to Relko in Derby); Ostrya (Ribblesdale Stakes; second in Park Hill Stakes), and Windmill Girl.

As sometimes can occur, as soon as Hornbeam was exported, things started to happen for him. The same year that Blakeney won the Derby, Hornbeam's son Intermezzo won the St. Leger, and his daughter Queen Bee, who preceded him to Sweden in utero, won eight races including that year's Swedish Oaks. Then came Iskereen (Pretty Polly Stakes at The Curragh; second in Irish Oaks) and Hardiesse (four wins from seven starts including Cheshire Oaks).

While in Sweden, Hornbeam became leading sire both there and in Denmark, with several Classic winners, and he also emerged as a strong influence as a broodmare sire. Apart from Windmill Girl, his important broodmares included Hornpipe (dam of Protagonist, the top American two-year-old of 1973); Great

Occasion (dam of the Dewhurst Stakes winner Lunchtime); La Milo (dam of the Washington International winner Admetus) and the unraced Bean Feast (whose son Brook became a top-class miler in Italy but is best remembered for the one of his 12 victories in 20 races which resulted from three horses that finished in front of him in the Queen Anne Stakes being disqualified).

Hornbeam was the first of five winners bred by his dam Thicket, who won one maiden race worth £414 but finished last when she was tried in the higher class of the Heathcote Stakes at Epsom and Falmouth Stakes at Newmarket. Her other winners included Tarbert Bay (three races, £1,306; dam of the Irish St. Leger winner Craighouse and the Newbury Autumn Cup winner Ruantallan) and High Trees (ten wins, including nine in one season, £4,146; three wins in America).

Thorn Wood, the dam of Thicket, won two small races up to a mile but was accused of being fainthearted and at three years she ran her two best races when ridden by an apprentice who did not carry a whip. She bred four equally modest winners in England—as well as one in America—the most successful being Buffalo-Thorn, a selling plater whose seven wins earned £1,790, before Thorn Wood was exported to America.

Thorn Wood was by Bois Roussel out of Point Duty, a Grand Parade mare who won three races for £2,932, including the Falmouth Stakes. Point Duty bred five winners, including Traffic Light (three wins, £7,036, including Park Hill Stakes, Coronation Stakes; dam of Jockey Club Cup winner Amber Flash who in turn bred the Oaks winner Ambiguity), and Fair Ranger (won one race, £214; dam of seven winners including Kipling, who upset odds laid on St. Paddy in Gordon Stakes). Point Duty also bred the maiden Sun Helmet, who became the dam of Woodburn (£6,315, Cesarewitch, Yorkshire Cup) and Sanlinea (three wins, £2,872; third in St. Leger; dam of the smart colt Amerigo who on export to America as a three-year-old won 12 races for 419,171 dollars).

CHORUS BEAUTY (1952)

Chorus Beauty, the dam of Windmill Girl, won two races for £1,230, and also finished third in the Ribblesdale Stakes and Lancashire Oaks, and fourth in the Cesarewitch.

She had two foals of racing age before Windmill Girl but neither was any good on the Flat, though Thornfield did manage to win a hurdle race. When Chorus Beauty was sent up to the 1961 Newmarket December Sales, barren and with the filly foal Windmill Girl at foot, she changed hands for 700 guineas. She failed to get in foal again and was exported to Sweden in 1967.

Neberna, dam of Chorus Beauty, won one small race over a mile and was placed second six times. Such a run of placings may have been responsible for doubts about her generosity being expressed, though it may have been that she needed firm ground to give her best.

Neberna bred four winners on the Flat, the best being the thorough stayer Philos (eight wins, £6,321, including Warren Stakes and Bentinck Stakes at Goodwood) and the unbeaten, speedy two-year-old filly Phantom Star (three wins, £1,882).

It was Neberna's dam, Springtime, a half-sister to Blue Peter, who brought the family into the Holliday Stud, when she was purchased privately and cheaply from Lord Rosebery. Springtime bred 12 winners, including Neberna's sister Nelia, who ran third in the Oaks and produced four winners in England including Penitent (a gelding whose 20 wins spanned the years 1951 to 1961 and earned £11,065, including the Old Newton Cup and five wins from eight starts at the age of 11); Crotchet (winner of the Lingfield Oaks Trial, second in the Park Hill Stakes, and dam of the Blue Seal Stakes winner Ballette), and Trimmer (winner of five races for £5,933 before being exported to Colombia).

Blakeney was inbred to Nearco at the fourth remove.

1970

1 C. Engelhard's b c Nijinsky (Northern Dancer – Flaming Page).
2 W. Guest's ch c Gyr (Sea-Bird II – Feria).
3 G. A. Oldham's b c Stintino (Sheshoon – Cynara).
    11 ran. Time: 2 min 34.68 sec

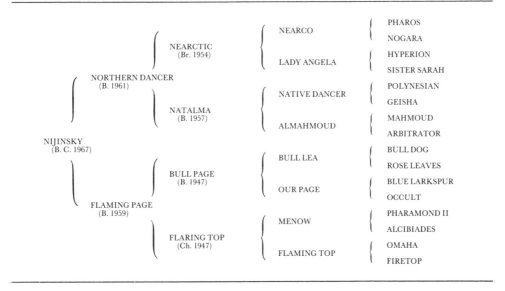

Not only was Nijinsky the first Canadian-bred to win the Derby, he was the first to complete the Triple Crown of 2,000 Guineas, Derby and St. Leger since Bahram in 1935. He may well also be the last to do this, since because of injury, alternative engagements or prejudice against stayers, relatively few Derby winners run in the St. Leger these days.

When sold as a yearling, Nijinsky broke the Canadian record at 84,000 dollars (then £32,307), but such were his exploits that his year-younger brother Minsky fetched 140,000 dollars (£53,846) as a yearling. It is remarkable that the sire and dam of Nijinsky and Minsky could have been bought for a total of 45,000 dollars. Both Northern Dancer and Flaming Page were bred by Mr. E. P. Taylor; both went to the Canadian Yearling Sales, and both failed to reach their reserves—Northern Dancer at 25,000 dollars and Flaming Page at 20,000 dollars.

In most ways Nijinsky was the perfect racehorse: he was a well-grown individual, good-looking and a fluent mover. As a two-year-old he was unbeaten in five races, and his one appearance in England, for the Dewhurst Stakes, so impressed the Jockey Club Handicapper that he was placed top of the Free Handicap, 2lb above the next horse.

Nijinsky kept his unbeaten record for six more races at three years, adding the Irish Sweeps Derby and King George VI and Queen Elizabeth Stakes (where he beat Blakeney) to the Triple Crown. He was beaten in his last two races, finishing second to Sassafras in the Arc de Triomphe and second to Lorenzaccio in the Champion Stakes.

So surprising was it to find Nijinsky beaten that reasons other than ability, or lack of it, had to be put forward for the phenomenon, especially since the two horses that beat him had not previously suggested they were good enough to topple what many regarded as a horse to rank with Sea-Bird II. Nijinsky's trainer was convinced he was not at his best in the Arc because a severe attack of ringworm in August meant he was not as fit as he could have been in the St. Leger, and the race took more out of him than should have been the case, which in turn affected him in the Arc, where also he was given a lot to do by his jockey. There was probably a more simple explanation for the Champion Stakes—he was over the top in his sixth race of the year outside Ireland, where he was trained.

Nijinsky retired at the end of his three-year-old career with a record of 11 wins from 13 races for total earnings of £246,132 in England and 480,000 francs in France. When he began stud duties in Kentucky in 1971, his valuation was 5.44 million dollars (£2,248,000) based on the American system of 32 shares, at 170,000 dollars each, of which his owner retained ten.

His earliest crops at stud tended to lack his own precocity; they were more slow maturing and less speedy, but nonetheless he made a flying start with two Classic winners in his first crop, born in 1972. They were Green Dancer (Observer Gold Cup at two years; French 2,000 Guineas and Prix Lupin at three) and Caucasus (four wins from five races at three years including Irish St. Leger; five wins at four and five in America). Also among Nijinsky's first crop was Quiet Fling (second to Caucasus in Irish St. Leger; won Coronation Cup, John Porter Stakes); and the second included African Dancer (Park Hill Stakes and Cheshire Oaks) and Bright Finish (Jockey Club Cup and Yorkshire Cup).

NORTHERN DANCER (1961)

Northern Dancer, the sire of Nijinsky, was a first foal who after failing to make his yearling reserve went into training in Canada, where he won five of his first seven races as a two-year-old. Sent to race in America, he won twice more in his first season, and at three years again won seven of his nine races, including the Kentucky Derby, Preakness Stakes, Queen's Plate in Canada, Florida Derby, Flamingo Stakes and Blue Grass Stakes. He was third in the Belmont Stakes on his only attempt at a mile and a half.

Northern Dancer retired to stud for the 1965 season with 14 wins from 18 races, worth 580,806 dollars, and commanded a nomination fee of 10,000 dollars. His first crop included Viceregal (Canadian Horse of the Year in 1968, when he was unbeaten in his eight races as a two-year-old); Nijinsky was from his second crop. His third crop, in 1968, included Northfields, a half-brother to Habitat, and Fanfreluche, one of the best racemares of her generation in North America. Northfields won seven races and 195,071 dollars before retiring to stud in Ireland, where he sired in his first crop Oats (third in Derby) and Northern Treasure (Irish 2,000 Guineas; third in Irish Sweeps Derby). Fanfreluche won 11 races for 238,688 dollars including the Alabama Stakes, and her first foal, L'Enjoleur, was the second highest-weighted two-year-old in the American Free Handicap of 1974.

Northern Dancer's fourth crop included Lyphard, who won six races, including two over ten furlongs but was best at a mile or thereabouts, where he won the Prix Jacques le Marois (from the 2,000 Guineas winner High Top) and the Prix de la Fôret. He was not an easy ride, and English racegoers remembered him most for disgracing himself in the Derby, where he failed to negotiate Tattenham Corner in the accepted manner. He made a sensational start to his stud career in France, being the leading first-season sire in both Britain and France in 1976, when Durtal won the Cheveley Park Stakes and Pharly emulated his sire by winning the Prix de la Fôret. The signs are that with Nijinsky, Northfields and Lyphard to represent him, Northern Dancer will become a most important international influence.

NEARCTIC (1954)

Northern Dancer's sire, Nearctic, was one of the best horses bred in Canada, where he raced from two to five years, completed 47 races and won 21, for a total of 152,384 dollars. His best year was as a four-year-old, when his nine wins netted 95,000 dollars and he was voted Canadian Horse of the Year. He died in 1973.

Northern Dancer was his best offspring, but he was also responsible for Icecapade (13 wins, 256,468 dollars, including four stakes); Cold Comfort (15 wins, 319,022 dollars); Cool Reception (second in Belmont Stakes); and Nonoalco (seven wins and equivalent of £148,255, including 2,000 Guineas; also won Prix Morny and Prix de la Salamandre and second in Grand Criterium at two years; stud in France for one season before moving to Ireland in 1976).

Nearctic, who was by Nearco, was out of the Hyperion mare Lady Angela, financially the least successful among eight winners of 30 races bred by Sister Sarah, since her sole victory, in an Epsom maiden plate at three years, was worth £276. Five years later, in 1952, she was bought for 10,500 guineas by the British Bloodstock Agency on behalf of Mr. E. P. Taylor and went to Canada. At that time she had bred one winner in England—Lady Mills, who won over five furlongs as a two-year-old—and her next three foals followed suit after her export, though none was of any great account. At the time of her sale Lady Angela was in foal to Nearco (sire of two of her winning foals) and she was bought on the understanding she would go back to that source. The foal she was carrying was named Empire Day, who won three races but died in 1961 after a few seasons at stud, and the result of her return was Nearctic.

In all, Lady Angela was the dam of 13 winners from 15 foals. Her others included Choperion (five wins, 76,450 dollars); Lady Victoria (four wins, 31,428 dollars; dam of 11-race winner Canadian Victory and 12-race winner Canadian Prince, as well as Northern Taste, who was by Nearctic's son Northern Dancer, cost 100,000 dollars as a yearling and won five good-class races in France including the Prix de la Fôret); and Countess Angela (by Bull Page, as was Nijinsky's dam Flaming Page; won three small races; bred four winners including Titled Hero, winner of 16 races for 214,690 dollars including the Queen's Plate).

NATALMA (1957)

Natalma, the dam of Northern Dancer, was bought as a yearling by Mr. E. P. Taylor for 35,000 dollars and showed good-class form, winning three races. But when being prepared for the Kentucky Oaks, she chipped a bone in her knee. Though the stud season was almost over, she was retired immediately, visited Nearco and bred Northern Dancer to a mating on June 28. She later bred Native Victor (18 wins, 71,400 dollars) and Regal Dancer (13 wins, 51,500 dollars).

Natalma was by Native Dancer out of the Mahmoud mare Almahmoud, whose four wins at two and three years included the Colleen Stakes. Almahmoud bred five winners, the best being Cosmah, winner of nine races for 85,525 dollars and a significant broodmare. Cosmah bred Tosmah (23 wins, 612,588 dollars; voted best two-year-old and three-year-old of her generation and best handicap mare of 1964) and Father's Image (six wins, 173,318 dollars; second in Arlington–Washington Futurity when beaten half a length by Buckpasser).

Almahmoud's dam, Arbitrator, was unraced. She bred six winners, the most successful being Burra Sahib (19 wins from two years to ten years for 71,765 dollars).

FLAMING PAGE (1959)

Turning to the distaff side of Nijinsky's pedigree, his dam Flaming Page, as has been recalled, failed to make her reserve as a yearling. She did actually change hands but her owner asked Mr. Taylor to take her back because she had a slight injury to a hind leg. Mr. Taylor agreed, and can hardly have regretted the decision, since Flaming Page proved one of the best fillies to race in Canada, with four wins for 108,836 dollars, including the Queen's Plate and Canadian Oaks, and second place in the Kentucky Oaks.

Flaming Page's stud career started on a poor note, since she produced dead twins to Northern Dancer. Then came a filly by Victoria Park, called Fleur, who won three races at two and three, worth 9,235 dollars. Her second live foal was Nijinsky.

After Nijinsky, she bred his brother Minsky, who showed promise at two years by winning the Railway Stakes and Beresford Stakes and was second in the Observer Gold Cup. He won two races at odds-on at three years before running fourth of six in the 2,000 Guineas, after which he was returned to Canada and won four further races before he was eventually exported to Japan. Flaming Page had a bad birth after Minsky and was barren for the next two years, so her stud record rested on two colts and a filly.

Flaming Page's sire, Bull Page, was one of the least successful sons of Bull Lea, who sired three Kentucky Derby winners in Citation (Triple Crown), Hill Gail and Iron Liege, as well as high-class winners in Coaltown and Mark-Ye-Well.

Flaming Page's dam, Flaring Top, was responsible for introducing the family into the Taylor stud, being purchased after a racing career which brought three small wins from 14 outings at two and three years for 5,020 dollars. Flaring Top bred 11 foals, all winners, of which four won stakes races and Flaming Page was easily the best.

Flaring Top was by Menow—whose half-sister Salaminia produced Sir Ivor's grandam Athenia when mated with Menow's sire Pharamond II—out of Flaming Top (an irritating similarity in names for students of bloodstock). Flaming Top, who bred eight winners, ran only three times, all as a two-year-old, and never reached the frame. The best of her produce to race was Doubledogdare, the top filly of her generation in both 1955 and 1956 and winner of 13 races for 258,206 dollars.

While Flaring Top did not win, neither did her dam Firetop, who herself was out of a mare that failed to score in three seasons. Firetop redeemed herself by breeding 17 foals, 11 winners and two in stakes races.

1971

1 P. Mellon's b c Mill Reef (Never Bend – Milan Mill).
2 Mrs. D. McCalmont's ch c Linden Tree (Crepello – Verbena).
3 E. Littler's b c Irish Ball (Baldric II – Irish Lass).
    21 ran. Time: 2 min 37.14 sec

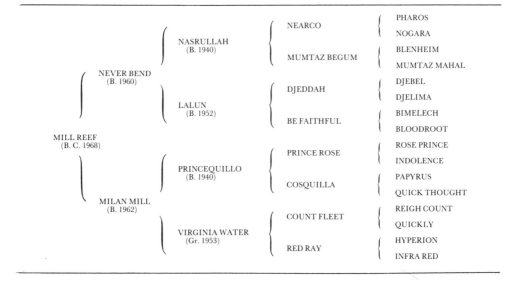

Mill Reef never looked back once he had beaten the 9-2-on chance Fireside Chat in his first race, at Salisbury in May as a two-year-old. He ended his first season with five wins, including the Coventry Stakes by eight lengths, the Gimcrack Stakes by ten lengths and the Dewhurst Stakes, and met his sole defeat when second to another outstanding youngster, My Swallow, in the Prix Robert Papin, where a short head separated them.

As a three-year-old Mill Reef again had six races and again was beaten in only one, when second to Brigadier Gerard in the 2,000 Guineas. He beat Linden Tree by two lengths in the Derby, Caro by four lengths in the Eclipse Stakes, Ortis by six lengths in the King George VI and Queen Elizabeth Stakes, and Pistol Packer by three lengths in the Arc de Triomphe. The official handicapper put him 4lb in front of Brigadier Gerard at the end of the season—despite the result of their only meeting—but others were not so willing to separate their achievements at different distances.

At four years prospects of a meeting between Mill Reef and Brigadier Gerard were high, but went unfulfilled. Mill Reef won the Prix Ganay by ten lengths but had to struggle to beat Homeric by a neck in the Coronation Cup. It transpired that he was not to run again. He was thought to have a virus infection and missed the Eclipse Stakes; he sustained a swollen hock and missed the Benson and Hedges Gold Cup, and finally, on August 30, 1977, came a complicated fracture of the near foreleg, in which damage to the cannon bone was the worst aspect.

Had the accident happened to any other horse than a valuable stud prospect the chances are he would have been put down, so great would have been the expense for the owner and possible discomfort for the horse. But by virtue of his status Mill Reef was spared. In a six-hour operation at his trainer's yard he had a steel plate put into the leg, and he was in plaster for six weeks. The operation was the first of its kind to be done in England, though there was a successful precedent in America, where Hoist the Flag was saved for stud after breaking a hind leg.

Mill Reef recovered so well that he was able to join the band of stallions at the National Stud in 1973, and his original book of 21 mares became 23. His American owner made six nominations available to British breeders, which resulted in a ballot among 85 applications. They paid 25,000 dollars (approximately £10,600) for the first five with a live-foal concession, and the sixth came without that proviso. Permanent syndication was then arranged whereby 41 shares were distributed at £50,000, for a total valuation of £2,050,000. Nine shares bought by the Levy Board for use by British breeders were allocated by ballot each year at a fee of £15,000.

A neat colt, strong and compact, Mill Reef retired from racing with 12 wins and two seconds from 14 races, having earned £300,422, a European record.

He covered his first mare seven and a half months after his accident, and when his first crop made their debut in 1976 as two-year-olds, his five individual winners earned a total of £6,791 from seven wins. The previous year four members of his first crop had been sold as yearlings, and they ended in the top 23 of the year's sales—a filly out of Prudent Girl for 31,000 guineas; a colt out of Hecla for 75,000 guineas; a filly out of Dress Uniform for 50,000 guineas, and a colt out of Lalibela for 202,000 guineas.

RECORD-PRICED YEARLINGS

Mill Reef's son out of Lalibela, a smart racemare who won the Cheveley Park Stakes but showed no form beyond sprint distances and had previously bred three winners, was called Million. The two lots nearest to him in the 1975 sales chart made 127,000 guineas and 75,000 guineas, and he comfortably broke the record for a yearling sold at public auction in Britain. Be My Guest, the colt who fetched 127,000 guineas, held the record, but only briefly.

The record of Million has no place in this discussion, since he started his racing career outside its scope. But the portents for his proving worth the money were not good, at least judged on previous record holders. The following are the record-priced yearlings of the 20th Century, which opened with Childwick having topped the list since his sale for 6,000 guineas in 1891.

1900: 9,100 guineas—Cupbearer—never won in three seasons.

1900: 10,000 guineas—Sceptre—bought from the same batch as Cupbearer, having been sent up by the executors of the late Duke of Westminster, and proved as successful as Cupbearer was disappointing, by earning £38,283 and making a name as a broodmare.

1919: 11,500 guineas—Westward Ho—proved one of his dam Blue Tit's least successful produce, winning two races for £1,024, including the Great Yorkshire Stakes, at three years.

1920: 14,500 guineas—Blue Ensign—a half-brother to Westward Ho, also bought by Lord Glanely, but even less successful, being unplaced on his only outing, in the Craven Stakes, and ending his days as a sire at 9 guineas.

1936: 15,000 guineas—Colonel Payne—Miss Dorothy Paget's purchase, a half-brother to Orwell, managed to win two races, including the ironically-named County Moderate Plate at Bath, for a grand total of £344.

1945: 28,000 guineas—Sayajirao—one of the rare successes, after Dante's brother had been bought on behalf of the Maharajah of Baroda, since he won two Classics, went some way towards recouping his purchase with wins to the value of £19,343, and did well at stud.

1966: 31,000 guineas—Rodrigo—never even saw the racecourse, and may have retained some of his value as a result since he was sold for stud purposes to Japan.

1967: 36,000 guineas—Democratie—the first filly on the list since Sceptre, and enhanced her standing as Fleet's full sister by winning four races in France and finishing fourth in the local 1,000 Guineas.

1968: 37,000 guineas—Entrepreneur—a three-parts brother to Molvedo, who fell about four stones short of his relative's class and recouped only £797 of his purchase price before being sold to go to stud in Australia.

1969: 51,000 guineas—La Hague—a sister to Fleet and Democratie who matched neither in performance, winning one small race in France.

1970: 65,000 guineas—Cambrienne—half-sister to the Sussex Stakes winner Carlemont, she had four races in Ireland and won one of them as a two-year-old for £730.

1971: 117,000 guineas—Princely Review—took the record into six figures but left his owner Sir Douglas Clague with a massive deficit, reduced only by one win in a Salisbury handicap worth £663 before being sold to stud in Australia. He was out of Review, and therefore half-brother to Democratie and La Hague.

1975: 127,000 guineas—Be My Guest—sold in Ireland, and therefore the first record-holder not to have passed through the hands of Tattersalls at either Doncaster or Newmarket. He held the record for only a few weeks before Million came along.

Twelve record-holders, and only three—Sceptre, Sayajirao and Democratie—came anywhere near justifying their high prices. At first sight, that might seem scant reward for such financial bravery, but a success rate of 25 per cent is probably as much as can be expected, remembering that money alone, and its expenditure, cannot guarantee success in this particular branch of sport. Considering the relatively small number of horses bred each year which make the racing grade, let alone win a race, 25 per cent success is about as much as anyone can hope for. Unfortunately, those who spent vast sums on horses which proved the age-old theory that a second dip into the genetic well need not be as fruitful as the first might not agree.

It is also important to appreciate the extra strain placed on the high-priced individual. Only a minority of horses ever win a race; but there are also only relatively few high-priced yearlings, so the latter, by virtue of their status alone, begin their careers as a minority aiming to join a further minority. The odds are stacked against them from the start, and are probably further lengthened by the fact that they will usually start by being tried against the highest class. No-one who has laid out a fortune wants to start in the bottom grade, though the odds are that that is where they will end.

NEVER BEND (1960)

Mill Reef's sire, Never Bend, was a top-class two-year-old in America, where he headed the Experimental Free Handicap after seven wins from ten races. His two-year-old earnings of 402,969 dollars were then a world record, and in all he won 13 of his 23 races for 641,524 dollars, including the Futurity Stakes and Champagne Stakes at two and the Flamingo Stakes at three. He was second to Chateaugay in the Kentucky Derby and third in the Preakness Stakes but missed the Belmont Stakes because of an ankle injury.

Never Bend's few produce that raced in Europe showed him to be an influence for stamina as well as speed. In addition to Mill Reef he sired Riverman (five wins including French 2,000 Guineas; second in Champion Stakes and third in King George VI and Queen Elizabeth Stakes; stud in France) and J. O. Tobin (top English-trained two-year-old of his year, winning three races including the Champagne Stakes; third in Grand Criterium; sent to America as three-year-old). Never Bend's reputation in his native America was not so high.

Never Bend was the product of a covering between his dam, the Djeddah mare Lalun, and Nasrullah a month before the latter died in May 1959. Lalun was a smart racemare who won five races and 112,000 dollars, including the Kentucky

Oaks, and was second in the CCA Oaks. She bred at least five winners, the best after Never Bend being Bold Reason, who won the Hollywood Derby, American Derby and Travers Stakes, and finished third in the Kentucky Derby.

Lalun was out of Be Faithful, a mare by the Belmont Stakes and Preakness Stakes winner Bimelech. Be Faithful won 14 races, worth 189,040 dollars, and raced until she was five. Her best wins came in handicaps, and she was placed several times in stakes races. Her eight foals produced four winners, of which Lalun was easily the best.

Be Faithful, who became the grandam of the useful filly Artists Proof, a winner of four races as a two-year-old and later third in the Kentucky Oaks, was one of four stakes winners among the eight successful produce of Bloodroot, a granddaughter of Sunshot whose racing record contained eight wins and a second in the CCA Oaks. Bloodroot's other stakes winners were Ancestor (26 wins, 237,956 dollars, up to the age of ten); Bric a Brac (13 wins, 103,225 dollars, including San Juan Capistrano Handicap) and Bimlette (Frizette Stakes).

MILAN MILL (1962)

The distaff side of Mill Reef's pedigree has to be traced back four generations to find a winning filly, since among his first three dams, two did not race and the other did not win. The unsuccessful member of the group was Mill Reef's dam Milan Mill, a Princequillo mare who managed two minor placings at two years. She produced one foal before Mill Reef, his brother Milan Meadow, a gelding who won two small races in America at four years after injury had kept him off the course.

Following Mill Reef, she foaled the unraced filly Millicent (by Cornish Prince); a filly by Jacinto who had to be destroyed; Mille Fleurs (by Jacinto), a filly who won one small race but showed above-average form at up to a mile; and Memory Lane (by Never Bend), Mill Reef's sister who won two races including the Princess Elizabeth Stakes.

Virginia Water, the dam of Milan Mill, could not be trained. She bred four winners, two in England and two in America. Goose Creek won four races for £2,883 and was second in the Royal Lodge Stakes, after which he returned to America, won over hurdles, and at stud on his owner's farm bred Red Reef and Aldie, winners on the Flat in England, as well as the high-class two-mile steeplechaser Tingle Creek. Goose Creek's half-sister Berkeley Springs won three races, worth £10,723, including the Cheveley Park Stakes. She was second in the 1,000 Guineas and Oaks but ended her career running over six furlongs. Virginia Water's winning offspring in America were Carter's Creek (three wins, 12,310 dollars, died at three years) and Fountain Hill (five races, 25,265 dollars).

Count Fleet, Virginia Water's sire, won ten of his 15 races at two years and was placed in the rest. As a three-year-old he won all six starts, including the Triple Crown of 1943—the Kentucky Derby by six lengths, the Preakness Stakes by eight lengths and the Belmont Stakes by 25 lengths. He did not race after the Belmont Stakes because of injury.

At stud Count Fleet was responsible for Count Turf (Kentucky Derby) and

One Count (Belmont Stakes, Horse of the Year), but he became even better known for his daughters, who produced the likes of Kelso (leading money-spinner in America); Lucky Debonair (Kentucky Derby); Prince John (Garden State Stakes; sire); Lamb Chop (Kentucky Oaks) and Quill (champion two-year-old filly of 1958).

RED RAY (1947)

Mr. Paul Mellon, owner-breeder of Mill Reef, laid the foundation for that success when he bought Red Ray as an unraced two-year-old at the dispersal of the late Lord Portal's horses in training at Newmarket in July 1949. She made the second-highest price of the sale, 12,000 guineas, and Mellon's faith in going above his intended limit of £10,000 was not immediately rewarded. Red Ray did not race, and she died only four years after her purchase, leaving two surviving foals, Claret (a gelding who won two races) and Virginia Water (her third and last foal).

It is an interesting sidelight to see what happened to the rest of the horses that went with Red Ray to Newmarket in July 1949 for one of the most important dispersal sales of the time. A total of 14 lots changed hands for a combined bill of 71,850 guineas at an average of 5,132 guineas. Only one purchase, Burnt Brown, did himself justice on the racecourse; only Red Ray is remembered with any significance today, and by virtue of being the highest-priced lot, Hedgerow was the biggest flop.

HORSE	SALE PRICE GNS.	RECORD
Laverstoke (3ycolt)	4,800	Half-brother to Airborne; won one race, £276.
Meadow Mist (3yc)	3,200	Half-brother to Vilmorin; won 4½ races, £1,905.
Lone Planet (3yc)	2,200	Won one race, £276.
Burnt Brown (3yc)	11,500	Won seven races, £8,299, at three, four and five, including City & Suburban; second beaten short head by Peter Flower in Champion Stakes; third Hardwicke Stakes.
Red Ray (2yfilly)	12,000	As above.
Sailor's Knot (4yc)	1,100	Not win Flat or hurdles; died at six.
Northern Star II (3yg)	2,300	Not win Flat or hurdles.
Yelix (3yg)	1,350	Not win Flat or hurdles.
Fairshot (3yf)	2,200	Won one race, Leicester Oaks, £805, on only run for Sir Victor Sassoon; at stud founded a family of platers.
Fairock (2yf)	1,050	Not win and no winning produce in England.
Hedgerow (2yc)	17,000	Top-priced lot and half-brother to Vilmorin and Meadow Mist; won one race, £207 maiden event, and not run again after finishing 11th of 19 in 2,000 Guineas; sold at December Sales for 110 guineas.
Main Road (2yc)	1,250	Won one race, £332.
Bowline (3yf)	3,400	Exported by BBA.
Grecian Flower (2yf)	8,500	Half-sister to Airborne and Laverstoke; won one race, £207; bred one minor winner.

Infra Red, the dam of Red Ray, was one of 19 consecutive foals produced by the mare Black Ray, who was bred by Mr. J. B. Joel and owned by the American, Marshall Field. Black Ray won one of her two races as a two-year-old and was covered for the first time in 1922. She bred ten winners, including Jacopo (four wins, £6,673; top of the Free Handicap after winning the Windsor Castle Stakes; a leading sire in America); Foray (eight wins, £9,674; top of Free Handicap), and Eclair.

Eclair won seven races for £5,975, including the Leicester Oaks and Lingfield Autumn Oaks. She bred Khaled (unbeaten in three races at two years in England; second in 2,000 Guineas; successful sire in America and best known for Kentucky Derby winner Swaps), and Lady Electra (Mr. Phil Bull's foundation mare who won ten races including the Lincolnshire; grandam of Gimcrack Stakes winner Eudaemon).

Infra Red was a useful racemare, winning two races, the Great Surrey Foal Stakes and Princess Elizabeth Stakes for £2,756. She bred four winners: Magic Red, her first foal and a good sprinter who proved a fairly successful sire, including the Grand National winner Red Alligator among his victories; and three fillies. The female winners from Infra Red were: Riding Rays (dam of six winners; grandam of the American Derby winner Dubassof and the Royal Hunt Cup winner Regal Light; third dam of the Gran Premio del Jockey Club winner Glaneuse); Red Briar (dam of the 1,000 Guineas runner-up Big Berry, and Middle Park Stakes winner and 2,000 Guineas and Derby third Pipe of Peace); and Excelsa (winner of six races in Ireland; dam of Goodwood Cup and Doncaster Cup winner Exar, and Wichuraiana, dam of the 2,000 Guineas and Eclipse Stakes winner Wollow).

Unlike Red Ray, who was sold publicly on Lord Portal's death, Infra Red passed into the hands of her trainer Boyd-Rochfort, but she was barren five years running and was put down in 1954.

1972

1 J. W. Galbreath's b c Roberto (Hail to Reason – Bramalea).
2 H. Zeisel's b c Rheingold (Faberge II–Athene).
3 V. Hardy's b c Pentland Firth (Crepello – Free For All).
  22 ran. Time: 2 min 36.09 sec

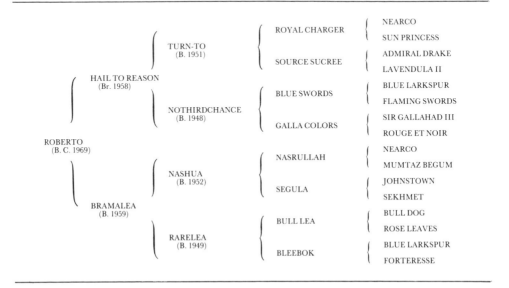

		ROYAL CHARGER	NEARCO
	TURN-TO		SUN PRINCESS
	(B. 1951)	SOURCE SUCREE	ADMIRAL DRAKE
HAIL TO REASON			LAVENDULA II
(Br. 1958)		BLUE SWORDS	BLUE LARKSPUR
	NOTHIRDCHANCE		FLAMING SWORDS
	(B. 1948)	GALLA COLORS	SIR GALLAHAD III
ROBERTO			ROUGE ET NOIR
(B. C. 1969)		NASRULLAH	NEARCO
	NASHUA		MUMTAZ BEGUM
	(B. 1952)	SEGULA	JOHNSTOWN
			SEKHMET
BRAMALEA		BULL LEA	BULL DOG
(B. 1959)			ROSE LEAVES
	RARELEA	BLEEBOK	BLUE LARKSPUR
	(B. 1949)		FORTERESSE

A well-made colt, though neither impressive nor imposing, Roberto was a particularly good mover who was evidently ideally suited by firm ground, though he was able to act in any conditions. His career can best be described as chequered, and at least one aspect revealed a disturbing feature that was in danger of starting a trend for the worst.

Bred by his owner, Roberto was unbeaten in his first three races in Ireland before finishing fourth to Hard to Beat in the Grand Criterium at Longchamp, where he was given a lot to do after failing to hold his place soon after halfway. The Irish handicapper put him at the top of his listings on 10st, 8lb above the next best.

At three years he had two distinguished achievements to his name—he won the Derby under an inspired ride from Lester Piggott who forced him home by a short head from Rheingold in the first short-head verdict in the race since 1909, and even more noteworthy he became the only horse to beat Brigadier Gerard in 18 races. On balance of form Roberto was not three lengths in front of Brigadier Gerard, the distance by which he beat the four-year-old when making all the running on fast ground in York's Benson and Hedges Gold Cup. Roberto won one other race at three years, the Vauxhall Trial Stakes in Ireland; he finished second to High Top in the 2,000 Guineas, and was twice beaten in France after his York success, ending the season by running seventh in the Arc de Triomphe. He was also 12th in the Irish Sweeps Derby, for which he started favourite.

Kept in training as a four-year-old, Roberto ran three times. He won the

Coronation Cup in a canter, finished second to Ballymore when odds-on for the Nijinsky Stakes, and ended his career by running badly in the King George VI and Queen Elizabeth Stakes. His third season was also notable for three races in which he did not run.

Roberto went to France for the Prix Ganay but was withdrawn because of an injury sustained on the crossing. Odds-on ante-post for the Eclipse Stakes, he was pulled out on the morning of the race because the ground "had good and soft patches and is loose and false" said trainer O'Brien. And after warning that Roberto would be taken out of the Benson and Hedges Gold Cup if rain made the going soft or false, O'Brien did just that on the day of the race, when he thought the course unsafe for racing after rain two days before had ended a very long spell of hot, dry weather.

The defection of Roberto at Sandown Park and York was to be regretted, since the racing public did not see the contests they had been promised overnight. No-one will argue that owners are entitled to their opinion, and that their first consideration must be to their horses. But in these days of enormous syndications, it would be easy to create an atmosphere in which defeat is regarded as dangerous. It has nothing to do with the horse physically; more it is concerned with financial arrangements and the possible loss of revenue brought about by defeat. If a horse does not run, he cannot be beaten, so connections are tempted to pull out, rather than risk defeat and the possibility of their investment being seen to be slightly less valuable than it might have been. Such a situation can give rise to false reputations and unwarranted values.

In the case of Roberto and the two races he missed because his connections did not want him to run under certain ground conditions, there were plenty of horses, valued equally highly by their connections, who did run at Sandown Park and York, and the majority were beaten. No-one complained unduly that their investment had been ruined.

In September of his third season Roberto pulled a ligament on the gallops and was retired to stand at his owner's stud in Kentucky from 1974. Syndicated in 32 shares of 100,000 dollars each, he realized a total valuation of the equivalent of £1,254,900. His first crop were two-year-olds in 1977.

HAIL TO REASON (1958)

injury prone

Hail to Reason, the sire of Roberto, was unusual in that he raced for the first time on January 21, two months before he was actually two years old. That year, 1960, he was America's top two-year-old, with nine wins from 18 races, including the Hopeful Stakes and Sapling Stakes, and he topped the Experimental Free Handicap by 4lb. But in the September of his two-year-old career he fractured both sesamoids in his near-fore fetlock joint, and his racing life was over. His sire Turn-To broke down at three years, as did Turn-To's son Sir Gaylord.

Hail to Reason soon made his mark at stud, and in 1970 ended Bold Ruler's seven-year spell as leading American sire. In Hail to Reason's first two crops he sired three particularly good mares: Straight Deal (21 wins, 733,020 dollars, including Hollywood Oaks); and the half-sisters Admiring (seven wins, 184,581 dollars including Arlington-Washington Lassie Stakes) and Priceless Gem (seven wins, 209,267 dollars, including Futurity Stakes and Frizette Stakes; dam of Allez France).

Hail to Reason also sired three winners of American Classics: Hail to All (eight wins, 494,150 dollars including Belmont Stakes; met with fatal accident in August 1972); Proud Clarion (six wins, 218,730 dollars including Kentucky Derby), and Personality (eight wins, 462,603 dollars, including Preakness Stakes). Other good winners in America by Hail to Reason were Regal Gleam (eight wins, 246,793 dollars including Frizette Stakes, Selima Stakes); Mr. Leader (ten wins, 219,803 dollars); Bold Reason (seven wins, 304,082 dollars including American Derby; third in Kentucky Derby and Belmont Stakes), and Halo (out of Cosmah; nine wins, 259,553 dollars).

From few runners in Europe, Hail to Reason was responsible for Hippodamia (Criterium des Pouliches; second in French 1,000 Guineas) and Hail the Pirates (eight wins in England and Ireland; two wins in America after export at five years). Hail to Reason was put down at the age of 18 in 1976.

Nothirdchance, the dam of Hail to Reason, was proof of the theory that hard racing need not be detrimental to the breeding performance of a mare. She had 93 races between the ages of two and seven, and won 11 of them for a total of 112,660 dollars. She bred three winners apart from Hail to Reason, including Treachery (11 wins, 182,071 dollars). When carrying her third foal, Nothirdchance was the subject of an inquiry to her owner Hirsch Jacobs, who was willing to sell her for 30,000 dollars. The buyer did not take up the offer, and thus missed the chance of acquiring for a relatively small sum the foal she was carrying, Hail to Reason!

Nothirdchance was by Blue Larkspur's son Blue Swords, a stakes winner who totalled five middle-distance victories for 58,065 dollars and at stud was probably better known for his broodmares than as a sire of winners.

Galla Colors, the dam of Nothirdchance, was an unraced mare who bred four other winners but none in the class of Hail to Reason's dam. Galla Colors was by Sir Gallahad III out of Rouge et Noir, who gained two wins at two and three years, finished third in the CCA Oaks but managed only one winning produce, Caillou Rouge, winner of 12 races.

BRAMALEA (1959)

Roberto's dam Bramalea was a very useful racemare who won eight races, including the CCA Oaks over a mile and a quarter, the Jasmine Stakes and the Gazelle Handicap. She was also second in the Delaware Oaks and won a total of 192,396 dollars. She bred three foals before Roberto: Logan Elm (by Swaps), an undistinguished performer who was placed at three and four; Village Street (by Chateaugay), who was unraced; and Glorious Spring (a filly by Hail to Reason and so sister to Roberto), who won three races at two and three years for 30,395 dollars.

After Roberto, Bramalea bred Balkan Knight (by Graustark), who made one appearance when trained in Ireland and won a six-furlong maiden race as a three-year-old; and Cambrian (by Ribot), who was second in two maiden races as a two-year-old, his only season to race in Ireland.

NASHUA

NASHUA (1952)

Bramalea's sire, Nashua, was one of the best horses to have raced in America post-war. He was rated only 1lb behind the top horse in the two-year-old rankings after winning six of his eight races for 192,865 dollars including the Belmont Futurity and the Hopeful Stakes. At three years he won ten of his 12 races for a further 752,550 dollars, including the Preakness Stakes in record time and the Belmont Stakes (by eight lengths). He was beaten into second place in the Kentucky Derby by Swaps, whom he went on to beat by eight lengths in a specially-arranged match.

At three years Nashua also won the Florida Derby and Arlington Classic, and at four he won six of his ten races for 343,150 dollars, including the Widener Handicap and Jockey Club Gold Cup. In all he won 22 of his 30 races for 1,288,565 dollars.

A very sound horse, Nashua sired tough stock. His best son in America was Diplomat Way, winner of 14 races up to nine furlongs for 493,760 dollars. Diplomat Way sired two useful 1972 two-year-olds who ran in England, Fiery Diplomat and Fabled Diplomat, who between them won six races that year.

Nashua's best daughter was Shuvee, a top-class, tough racemare who ran 13 times as a two-year-old when rated the joint top filly. Shuvee stayed in training until she was five years old, in all winning 16 races for 890,445 dollars as the top money-spinning mare in world racing. Her wins included the CCA Oaks (to emulate her dam Levee) and the Jockey Club Gold Cup. Shuvee's sister Nalee was a high-class winner of three stakes for 141,631 dollars and bred six winners from six foals before Meneval, who won the Irish St. Leger.

Nashua sired another CCA Oaks winner in Marshua, and his other winners in Britain included Guillaume Tell (three wins from four races, £7,970, including Gordon Stakes) and Kesar Queen (won three races, including Coronation Stakes; third in 1,000 Guineas).

Nashua's dam Segula, who was by Johnstown, won nine races at three and four years for 35,015 dollars, and was also third in the CCA Oaks. She bred six foals, four winners including one over hurdles, and in financial terms the best was Sabette (seven wins at two and three, 80,755 dollars; second in CCA Oaks).

Segula was one of nine winners bred by the Sardanapale mare Sekhmet. British breeders will be familiar with Sekhmet's daughter Booklet (by Sir Gallahad III), who won two minor races at two and three years in England. At stud Booklet's six winners included Near Way (six wins, £4,837, including Zetland Cup); Lombardo (six wins, £3,635); and Diary (useful winner of two races over five furlongs at two years, only season to race).

RARELEA (1949)

Rarelea, the dam of Roberto's dam Bramalea, won three minor races at two years. She bred seven winners, the next best after Bramalea being Rhodora II (foaled when her dam was 17; won two small races and finished third in Irish Oaks of 1969) and High and Dry (ran 148 times over seven seasons and won 19 races).

Rarelea was a sister to Delta Queen, one of six foals, all winners, bred by the

Blue Larkspur mare Bleebok, who was also responsible for River Gate (24 wins from two years to nine, 88,302 dollars), and Blue Whirl (seven wins; dam of 14-race winner Dip and Whirl). Delta Queen won four races for 16,045 dollars and was second in the Monmouth Oaks before becoming the dam of eight winners, including Advocator (nine wins, 325,761 dollars; second in Kentucky Derby and third in Belmont Stakes). Delta Queen was also the grandam of Delta Judge (eight wins, 159,762 dollars, including Sapling Stakes).

Bleebok was out of the unraced mare Forteresse, who bred four winners including Broker's Tip, who took the 1933 Kentucky Derby.

Roberto was inbred to both Nearco and Blue Larkspur at the fourth remove.

A footnote to the 1972 Derby is that Riva Ridge, like Roberto a grandson of Turn-To, won the Kentucky Derby that year. Riva Ridge is by First Landing.

CHAPTER 29

1973 – 1976
The Mid-1970s

1973

1 A. M. Budgett's ch c Morston (Ragusa – Windmill Girl).
2 Capt. M. Lemos's b c Cavo Doro (Sir Ivor – Limuru).
3 R. B. Moller's b c Freefoot (Relko – Close Up).
   25 ran. Time: 2 min 35.92 sec

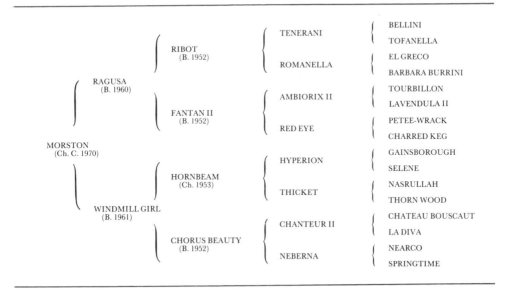

				BELLINI
			TENERANI	
		RIBOT		TOFANELLA
		(B. 1952)		EL GRECO
			ROMANELLA	
	RAGUSA			BARBARA BURRINI
	(B. 1960)			TOURBILLON
			AMBIORIX II	
		FANTAN II		LAVENDULA II
		(B. 1952)		PETEE-WRACK
			RED EYE	
MORSTON				CHARRED KEG
(Ch. C. 1970)				GAINSBOROUGH
			HYPERION	
		HORNBEAM		SELENE
		(Ch. 1953)		NASRULLAH
			THICKET	
	WINDMILL GIRL			THORN WOOD
	(B. 1961)			CHATEAU BOUSCAUT
			CHANTEUR II	
		CHORUS BEAUTY		LA DIVA
		(B. 1952)		NEARCO
			NEBERNA	
				SPRINGTIME

A BIG COLT—bigger than his half-brother, the neat Blakeney—Morston brought immortality to his owner-breeder-trainer Arthur Budgett and his dam Windmill Girl with their second Derby success within four years. Unlike the two previous colts out of Windmill Girl, Blakeney and Alderney, Morston was never intended to pass through the sales ring. He was foaled in France when Windmill Girl was

439

visiting Exbury, and was backward as a two-year-old, when he did nothing to help his condition by refusing to exert himself on the gallops. The consequence was that he did not race at two years and became the first since Bois Roussel in 1938 to win the Derby on a one-race preparation, his single outing resulting in success worth £690 in a Lingfield Park maiden event over a mile and a quarter in May, when he won by three lengths at odds of 14-1.

Circumstance meant that he tackled the Derby without his jockey Hide ever having seen him before they met in the parade ring. In the race Morston was held up in the middle of the field until the leading bunch dropped back in the straight, and having drawn away with Cavo Doro from the distance, he held his half-length advantage to the post, depriving the runner-up's jockey Piggott the distinction of winning the Derby on a horse he bred himself.

None of the Derby field went on to show top Classic form, and four of them never ran again, including the 2,000 Guineas winner Mon Fils and Morston himself. The St. Leger was Morston's objective but in mid-August he sprained a near-fore tendon and was retired to stand at stud in England at a syndication value of £1 million.

Victory for Morston at Epsom meant that only two people had owned, bred and trained two Derby winners. The other was the Scot William I'Anson, who won with Blink Bonny (1857) and her son Blair Athol (1864). Arthur Budgett thus became the first Englishman to accomplish the feat. His mare, Windmill Girl, became the 11th to breed two Derby winners, but the first for 71 years. The previous ones were:

Six sets of brothers

MARE	SIRE	DERBY WINNERS & YEAR
Flyer	Justice	Rhadamanthus 1790 Daedalus 1794
Horatia	Sir Peter	Archduke 1799 Paris 1806
Arethusa	Sir Peter	Ditto 1803 Pan 1808
Penelope	Waxy	Whalebone 1810 Whisker 1815
Canopus Mare	Whalebone	Lapdog 1826 Spaniel 1831
Perdita II	St. Simon	Persimmon 1896 Diamond Jubilee 1900

Four sets of half-brothers

MARE		DERBY WINNERS & YEAR
Mr. Tattersalls' Highflyer mare	Volunteer	Spread Eagle 1795
	Trumpeter	Didelot 1796
Arctic Lass	Tramp	St. Giles 1832
	Mulatto	Bloomsbury 1839
Emma	Catton	Mundig 1835
	Touchstone	Cotherstone 1843
Morganette	Kendal	Galtee More 1897
	St. Florian	Ard Patrick 1902

RAGUSA (1960)

Ragusa's finest moment, in siring a Derby winner, came too late for his own sake; he died after an operation a month before the Derby. He had done well at his owner's stud in Ireland, with such winners as Ballymore (Irish 2,000 Guineas on his first outing; promising stallion who took his own sire's place in Ireland); Ragstone (seven wins, £28,590, including Ascot Gold Cup; died young); Caliban (three wins, £15,826, including Coronation Cup; exported to Denmark after comparatively short spell at stud in England); and Homeric (three races including Prix Kergorlay; showed best form in defeat, when second in St. Leger and Coronation Cup, and third in Arc de Triomphe).

Ragusa was also responsible for Classic-placed horses in Agricultore (third in Irish 2,000 Guineas); Duke of Ragusa (third in St. Leger); and Flair Path (third in Irish 2,000 Guineas and French Derby), as well as Lombardo and Ragapan, who failed by one place to emulate their sire's success in the Irish Sweeps Derby.

Ragusa had been a top-class racehorse, making his public auction price of 3,800 guineas look most generous with seven wins from his 12 races over three seasons. He won a newcomers' race over seven furlongs on his only outing at two years, and the following year, after finishing third to Relko in the Derby, won four straight races—the Irish Sweeps Derby, King George VI and Queen Elizabeth Stakes, Great Voltigeur Stakes and St. Leger. At four years he was a good winner of the Eclipse Stakes but ran badly on his last outing, in the Arc de Triomphe.

His seven wins earned £146,650, and for a sound stayer he had a good turn of foot, as he showed by casting aside Baldric II in the Eclipse Stakes over a distance on the short side for him.

RIBOT (1952)

Morston's victory introduced to the Derby story for the first time the name of Ribot, one of the greatest racers and stallions of the 20th Century. It was remarkable that 17 years after Nearco had been foaled, Federico Tesio produced another champion in Ribot, whose career further emphasized the quirk of fate that allowed Italy a place in the limelight that could not be earned by size alone, since her annual production of horses is a mere tenth of that in Britain and Ireland.

Ribot was foaled at the English National Stud. He was a product of a covering by Tenerani during his last season in Italy, and was foaled when his dam followed Tenerani, who had been dispatched to England, for a further service that produced Raffaelina, a filly who died young and without issue.

Tesio did not live to see Ribot run. He died two months before Ribot had his first outing, in Milan in July 1954. Ribot won that race, and his next 15, spread over three seasons, from five furlongs to 15 furlongs, and on going which ranged from hard to heavy. He won in three countries, though his only appearance outside Italy in 13 races before the King George VI and Queen Elizabeth Stakes was in the 1955 Arc de Triomphe, which he won easily by three lengths. He prepared for his trip to Ascot by taking on the best Italian three-year-olds over 15 furlongs at weight for age in the Gran Premio di Milano, which he won by eight lengths. He won the Ascot race, and then prepared for a second successful crack at the Arc by taking on the Italian 2,000 Guineas winner over nine furlongs, and beat him by eight lengths!

An offer of £575,000 for Ribot to go to America was turned down after his second Arc success and he stood his first season at stud in 1957 in Newmarket. He then returned to Italy, only to be tempted away to Kentucky in 1960, when leased for five years to John W. Galbreath for 1,350,000 dollars. Satisfactory arrangements for his transport back to Italy could not be made at the end of the period of lease and Ribot stayed in America until his death because of a twisted intestine on April 28, 1972.

He was an outstanding success at stud, where his winners came in most shapes and sizes but rarely could be accused of showing unwillingness in a finish, though some of his produce were not so inclined to get themselves into the firing line in the first place. Few of his stock were precociously brilliant, but rarely did a Ribot crop go by without at least one champion emerging, either in America or Europe, or even further afield. He was leading sire in Britain twice, from a limited number of opportunities—1963, thanks largely to Ragusa who gave Ribot the distinction of becoming only the second foreign-based stallion to head the list, after Flageolet in 1879; and 1968, when his six individual winners included Ribero and Ribofilio. He is best known for:

YEAR FOALED	PRODUCE AND RECORD
1958 (English crop)	Molvedo—top of Italian Free Handicap after winning Gran Criterium; went on to win Grand Prix de Deauville and Arc de Triomphe.
1959 (First crop in Italy)	Romulus—Sussex Stakes, Queen Elizabeth II Stakes, Prix du Moulin; second in 2,000 Guineas.
1960	Ragusa—see above.
1961	Prince Royal II—Arc de Triomphe, Gran Premio di Milano. Alice Frey—Italian Oaks.
1962 (First crop in America)	Tom Rolfe—16 wins, 671,297 dollars, including Preakness Stakes, American Derby, Arlington Classic; second in Belmont Stakes, third in Kentucky Derby. Dapper Dan—five wins, 112,102 dollars; second in Kentucky Derby and Preakness Stakes. Long Look—Oaks.
1963	Graustark—won seven out of eight races, 75,904 dollars, at six and seven furlongs.
1964	Ribocco—five wins, £140,531 and 209,214 francs, including Irish Sweeps Derby, St. Leger, Observer Gold Cup; second in Derby, third in King George VI and Queen Elizabeth Stakes and Arc de Triomphe.
1965	Ribero—brother to Ribocco; three wins, £92,343, including Irish Sweeps Derby, St. Leger.
1966	Arts and Letters—11 wins, 632,404 dollars, including Belmont Stakes; second in Kentucky Derby. Ribofilio—five races, £19,264, including Champagne Stakes, Dewhurst Stakes; second in St. Leger and Irish Sweeps Derby; beaten favourite in four Classics.

YEAR FOALED	PRODUCE AND RECORD
1969	Boucher—six races, £49,584, including St. Leger; made Ribot first since St. Simon to sire four St. Leger winners. Regal Exception—two races, including Irish Oaks.
1970	Filiberto—ran only three times at two years and won Prix Morny before finishing lame in Grand Criterium.

Ribot was also responsible for the British-raced sister and brother Arkadina (second in Irish Oaks and Irish 1,000 Guineas, and third in Epsom Oaks) and Blood Royal (unbeaten in four races including Jockey Club Cup and Queen's Vase).

It has been a surprise, since Ribot was considered not to have been suited by America's faster-maturing conditions, that his sons have not done better at stud in Europe. Ragusa's record has been set out, and cannot be crabbed.

Molvedo went back to Italy and got Red Arrow (Italian Derby) and Gallio (Italian St. Leger) to top the sires' list there in 1976. He also sired Orange Triumph, whose son Orange Bay won the Italian Derby before being sent to England, where he won the Hardwicke Stakes.

Ribero went to France and sired Ribecourt (Prix Kergorlay and Gran Premio d'Italia), while Romulus was only moderately successful in England, with Popkins (six wins including Princess Elizabeth Stakes) and Petty Officer (ten wins, £31,032) his only outstanding runners in Britain before he was sent to Japan at the end of 1969.

Andrea Mategna remained in his native Italy to sire Antelio (Italian 2,000 Guineas) and Garvin (Gran Premio di Milano), and was sent to England in 1974 but stayed only a short time before being shipped off to West Germany.

Prince Royal II had a most interesting stud career. His best-known produce were Royal Conductor, winner of the Norwegian 2,000 Guineas, and Unconscious, winner of the Californian Derby.

In America two sons of Ribot stand out at stud, Graustark and Tom Rolfe, while His Majesty (five wins from 22 starts) made a promising start by siring the Jersey Derby winner Cormorant.

Graustark, a brother to His Majesty, was syndicated for what was at the time a record, 2.4 million dollars. He sired Jim French (Santa Anita Derby; second in Kentucky Derby and Belmont Stakes); Avatar (Belmont Stakes); Key to the Mint (champion three-year-old); Ruritania (second in Belmont Stakes); Prove Out (Jockey Club Gold Cup), and Groshawk (champion two-year-old). Graustark's best offspring to run in Europe was the French Derby winner Caracolero.

Tom Rolfe sired Run the Gantlet (559,079 dollars; top grass horse of 1971; won Washington International and Man O'War Stakes; stud in Ireland); Hoist the Flag (unbeaten 1970 champion two-year-old; broke leg early following year); Droll Role (545,497 dollars, including Washington International); and London Company (478,910 dollars). Tom Rolfe's best produce to run in Europe was Manitoulin, the Blandford Stakes winner.

Another American-raced son of Ribot who caused a brief flurry of excitement in Europe was the curiously-named Yrrah Jr., who was out of the Italian Oaks winner Ola and won two small races. He stood in Ireland for two seasons at 98 guineas, and sired Hurry Harriet (Champion Stakes) and three other individual winners before being sent back to America.

Sons of Ribot have done well abroad as stallions. Latin Lover (three wins, £4,210, including Manchester Cup) was sold for 6,000 guineas to go to stud in Australia, and there sired Rain Lover, the first horse for 100 years to win the Melbourne Cup two years in succession.

Con Brio (two wins, £4,941, including Brighton Derby Trial) stood for three seasons in England, where his best produce was the very fast filly Cawston's Pride (unbeaten in eight races at two years including the Molecomb and Cornwallis Stakes; dam of the Coventry Stakes winner Cawston's Clown). Con Brio was then exported to Argentina, where he became a stallion of Classic status.

Marot (Gran Premio Citta di Napoli in native Italy) was exported to Japan and became a class-winning sire, as did Ribotlight (two wins, £723) in New Zealand.

It will be a considerable surprise if the influence of Ribot does not endure; so many of his sons showed top-class form and have been given the best opportunities at stud that the chances are some will have lasting influence.

The significance of mares by Ribot has already been set down by such as Queen Sucree, whose son Cannonade put right the Ribot record in the Kentucky Derby, following the misfortune of Dapper Dan and Arts and Letters (both beaten a neck), Tom Rolfe (third) and Graustark (broke down when odds-on in early betting). Other successful broodmares by Ribot are Formentera (dam of the 1,000 Guineas and Champion Stakes winner Flying Water); Ripeck (dam of the Coronation Cup and Yorkshire Cup winner Buoy, and the Jersey Stakes winner Fluke); Diagonale (dam of the Grand Prix de Deauville winner and French Derby third Djakao); Ofa (dam of the South African 2,000 Guineas and Derby winner Politician); and Irradiate (dam of the Man O'War Stakes winner Majestic Light).

ROMANELLA (1949)

Ribot's dam, Romanella, was a fast and precocious racehorse, who won five of her seven races at two years, including one of Italy's most important juvenile races, the Criterium Nazionale. But at three years she developed a ringbone—and a temper—and did not race again. She bred ten winners in a stud career stretching over 20 years. The most notable apart from Ribot was his five-years younger sister Rossellina (three wins at two and three years including the Italian 1,000 Guineas) and Raeburn (winner of Italian 2,000 Guineas and Coppa d'Oro). Rossellina became the dam of Ruysdael, who beat Raeburn when they were first and second in the Italian Derby. Raeburn stood three seasons at stud in Ireland before being exported to Japan.

Romanella also bred Rabirio, a winner in Italy at four years over six furlongs after he had missed a season because of injury. Rabirio went to stud in Ireland at a fee of 148 guineas, and was modestly successful as a National Hunt stallion.

Romanella was by El Greco, who won 17 middle-distance races in Italy and was sent to France in 1949 but died a year later. She was out of Barbara Burrini, one of two foals sent to the Newmarket December Sales in 1937 by Basil Jarvis, who trained her sire Papyrus. She was sold to Federico Tesio for 350 guineas, while the other, a colt by Lemnarchus, fetched 70 guineas.

Barbara Burrini won six races in Italy at three and four years, and bred five winners, none of the others being in the same class as Romanella. Barbara Burrini was out of Bucolic, a half-sister to Cyclonic (King Edward VII Stakes), Typhonic (Park Hill Stakes) and Panic (dam of eight winners of 18 races including the King Edward VII Stakes winner Solfo and the Nassau Stakes winner Solfatara).

FANTAN II (1952)

Turning to the distaff side of Ragusa's pedigree, his dam Fantan II was by Ambiorix II, whose own dam Lavendula appeared in the previous year's account as the grandam of Turn-To. Ambiorix II, a half-brother to Turn-To's dam Source Sucree, won the Prix Lupin and was second in the French Derby before becoming a leading sire in America.

Fantan II was a winner in America at two and three years of races worth 7,175 dollars. Ragusa was her third foal and second winner, following a colt by Alycidon called To Fortune, who won six races in America at three and four years for 28,365 dollars. Fantan II had been imported to Europe as a four-year-old by her owner, Mr. H. Guggenheim.

After Ragusa, Fantan II bred four more winners, including Ela Marita, who was bought even more cheaply for Ragusa's owners than Ragusa himself. Ela Marita cost 500 guineas within a couple of months of Ragusa's winning his only outing as a two-year-old, though well before he had shown what he could really do. Ela Marita, unraced as a two-year-old, won two of her four races at three years for £5,695, the Fred Darling Stakes and Musidora Stakes. At stud, her first foal was Augustus (by Ribot), who won seven races in France for the equivalent of £14,582 including the Grand Prix de Clairefontaine twice; and her second was Mariel, winner of three races for £4,802, beaten a neck in the Irish 1,000 Guineas, and fourth in both the English and Irish Oaks. Then followed three more winners for Ela Marita—Marie Curie, a filly who won the first of her two races at two years, her only season to race; Carnera, a colt who won twice and was second in the Nijinsky Stakes and third in the Gallinule Stakes; and Stella Mar, a filly who won her final race as a two-year-old.

The nature of the bargain struck when Ela Marita was bought became even more evident when Mariel went to stud, for though she died in 1975 after producing only three foals, all fillies, her first produce was Sarah Siddons, who won a newcomers' race on her sole outing at two years and developed into the best three-year-old filly in England and Ireland. Sarah Siddons won the Irish 1,000 Guineas and Musidora Stakes, finished second in the Irish Oaks and Prix Vermeille, and totalled £42,280 from her three victories.

Fantan II was one of only two foals produced by Red Eye, the other managing to be placed as a five-year-old. Red Eye herself won six races at two and three years for 16,640 dollars including the Gazelle Stakes and Ladies Handicap. She was by Petee-Wrack, whose sire Wrack boasted Chelándry as his grandam, out of Charred Keg, who bred 12 winners from 13 foals. The most successful of Charred

Keg's offspring was Lift, who won nine races including the Washington Derby. Charred Keg also produced Sister Bry (three wins; dam of Sunny Boy S, whose 25 wins earned 74,590 dollars) and Corn Likker (two-year-old winner; dam of Hangover, whose 22 wins earned 89,940 dollars).

1974

1   Mrs. N. Phillips' ch c Snow Knight (Firestreak – Snow Blossom).
2   Col. F. Hue-Williams' b c Imperial Prince (Sir Ivor – Bleu Azur).
3   C. St. George's ch c Giacometti (Faberge II – Naujwan).
        18 ran. Time: 2 min 35.04 sec

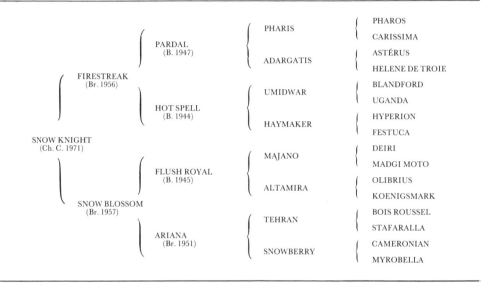

Snow Knight was the first Derby winner since Santa Claus to go through the sales ring as a yearling; and was the first ever to continue his racing career in the top class in America. In between, he was beaten in more races than he won.

Bought for 5,200 guineas as a yearling by his trainer's wife, Mrs. "Mac" Nelson on behalf of a Canadian owner, he won two of his five races at two years, including the Donnington Castle Stakes by five lengths, but ran his best race in defeat, when beaten a short head into second place behind Giacometti in the Champagne Stakes.

After being second in the Sandown Park Classic Trial and third in the Lingfield Trial, both times behind Bustino, Snow Knight started 50-1 for the Derby. This time Bustino was only fourth, beaten just over three lengths by Snow Knight, who encountered firmer ground than he had run on before and proved himself well suited by it, especially in a strongly-contested race where stamina could be brought into play. He went on five furlongs out, led round Tattenham Corner (the only one apart from Nimbus since the war to do so) and won by two lengths from Imperial Prince, with Giacometti a length away third.

That it was one of the poorer Derbys was soon made evident, when Snow Knight was beaten in his next five races that season. While those behind him improved—Giacometti to win the Champion Stakes and Bustino to win the St. Leger—Snow Knight did not, and he finished sixth in the King George VI and Queen Elizabeth Stakes, and third in the Benson and Hedges Gold Cup, his last race in England.

In the summer of 1974 it was announced by Mr. E. P. Taylor, the Canadian breeder of Nijinsky, that Snow Knight had been syndicated for more than £520,000 and he himself had taken up four shares. It was said the colt—a big, rangy, attractive sort—would remain in training in England with the Canadian International Championship as his autumn objective, but at the end of August he was scratched from the St. Leger, packed off to Canada, and Mr. Taylor was announced as the major shareholder in the Snow Knight syndicate.

Snow Knight, whose winning earnings in England amounted to £91,261 from three races, all but £2,031 coming in the Derby, ran three times in Canada towards the end of 1974 but cut little ice, once wearing blinkers and giving rise to the possibility that a suspect temperament, which had been evident in England but did not affect his performances, might be getting the better of him. He stayed in training as a four-year-old and re-established his reputation, reaching even greater heights than in England. The best turf horse in North America in 1975, he was campaigned in the United States, where his wins included the Man O'War Stakes. He also won the Canadian International Championship.

He retired to stud in 1976 and stands in Maryland at a fee of 10,000 dollars per live foal. Whether he proves the ideal stallion for American purposes remains to be seen.

FIRESTREAK (1956)

Snow Knight's starting price of 50-1 was not the only surprise of the 1974 Derby, since his sire Firestreak was not previously regarded as a stallion likely to get a winner of the race.

Bought for 9,500 guineas—4,300 guineas more than his Derby-winning son fetched 15 years later—as a yearling, Firestreak won nine races for £11,459 at two, three and four years. His best season was his last, when he won all three races—the City and Suburban under top weight of 9st 5lb, the Craven Stakes and the Rous Memorial Stakes. His last two seasons were under the training of Peter Nelson, trainer of Snow Knight, and his latest wins suggested that though he stayed a mile and a quarter, he was best at a mile.

When Snow Knight came along, Firestreak had had only 11 individual winners at a mile and a half or more. His best winners were Hotfoot (nine wins, £33,062, from six furlongs to ten, including Peter Hastings Stakes, Players-Wills Stakes, Rous Memorial Stakes; second in Irish 2,000 Guineas, third in Champion Stakes); Workboy (six wins, £13,387, at five and six furlongs; second in King George Stakes; third in Palace House Stakes) and Maystreak (eight wins, £13,495, at five furlongs to a mile).

If Firestreak, who was by Psidium's sire Pardal, was a miler, so was his dam,

Hot Spell, who ran only at three years and was cleverly placed to win five of her 11 races for £1,625. They were at seven furlongs to a mile and a quarter, but the best performances were at a mile.

Bred to staying stallions, Hot Spell had produced two winners in Britain before Firestreak, both successful over middle distances. They were the filly Dust Storm (by Precipitation), who won the Princess Royal Stakes at Ascot and finished third in the Ribblesdale Stakes, and the even better top-of-the-ground gelding Manati (by Nimbus), whose six wins, worth £5,816, included the Gordon Stakes and Old Newton Cup. Dust Storm died of grass sickness before she could go to stud.

Hot Spell bred one winner after Firestreak, Pinzari, who as a four-year-old managed to win a £256 apprentice maiden race before collecting two hurdle races in Ireland, where he retired to stud at a fee of £30.

Hot Spell was by Udaipur's brother Umidwar, who finished second in the Eclipse Stakes and won the Jockey Club Stakes and Champion Stakes on soft ground. Umidwar's best winners at stud were Umberto (eight wins, £8,327); Ujiji (Newmarket Gold Cup; third in Derby, Champion Stakes and Coronation Cup); and Norseman (sire of King George VI and Queen Elizabeth Stakes winner Montaval). Umidwar was by Blandford out of the French Oaks and French St. Leger winner Uganda, who also bred Ut Majeur (Cesarewitch and Newmarket St. Leger), Una (dam of Palestine) and Udaipur (Oaks, Newmarket Oaks and dam of Derby runner-up Ummidad).

Hot Spell was one of only two winners bred by her dam, Haymaker. The other was Rough Caps, who won three of her eight races at two years, all at five furlongs, for £939. Rough Caps did not run again, and at stud bred three minor winners.

Haymaker herself failed to win, as did her dam, the Hyperion mare Festuca, the dam of two minor winners. Festuca was out of Picardel, whose seven winning offspring included Herbalist (Victoria Cup); Pricket (successful at least once in nine consecutive seasons between 1930 and 1938), and Colorow (July Stakes). Picardel was also the dam of Devonshire House, fourth dam of Martial, Skymaster and El Gallo.

SNOW BLOSSOM (1957)

Neither of Snow Knight's first two dams saw a racecourse, and Snow Blossom, his dam, saw no other stallion but Constable in her first six years at stud. Constable, who stood alongside Snow Blossom's sire Flush Royal at the stud where her owner boarded his mares, provided her with three Flat-race winners, including Snow Vista (three wins at a mile) and Good Apple (two wins, £5,089, of which £4,613 came when he was awarded a valuable ten-furlong handicap on a disqualification).

While carrying her sixth foal, Snow Blossom was sold at the Newmarket dispersal of her late owner's breeding interests, comprising five mares and their accompanying foals at foot. Snow Blossom was sold to Mr. J. McAllister for 1,100 guineas. The other lots included Snow Blossom's dam, the 15-year-old Ariana (for 740 guineas) and the 11-year-old mare Royal Cham, who fetched 160 guineas in foal to Fury Royal. Royal Cham's resulting produce, named Tartar Prince, won seven races in 1970–71 for £11,135 including the Great Metropolitan and Northumberland Plate. Even Royal Cham's colt foal, sold at Newmarket for 75 guineas and named Burydell, won two races on the Flat for £1,196 and one over hurdles.

For Mr. McAllister, Snow Blossom foaled Young Dedham, who was placed on the Flat and won over hurdles after being sold for 1,400 guineas as a yearling, followed by two fillies who each won one race at two miles—Blossom Forth (by Celtic Ash, sold for 1,500 guineas as a foal) and Lenton Rose (by Aggressor, sold for 1,700 guineas as a foal). Mr. McAllister then sent Snow Blossom to Firestreak; the result was Snow Knight, but much happened between conception and achievement.

Safely in foal to Firestreak, Snow Blossom went up to the Newmarket December Sales and was sold for 940 guineas to Mr. J. A. C. Lilley, who gets the credit for breeding Snow Knight but who sent Snow Blossom, in foal to Mandamus, to the Newmarket December Sales a year later, in 1971, where she was sold for 1,000 guineas. She had little time for more service, and none at all to bask in the Derby glory of her son, because she broke a leg in the autumn of 1973 and was put down.

FLUSH ROYAL (1945)

Snow Blossom's sire Flush Royal was best known in Britain for his exploits in handicaps, and he ended his racing career at the age of seven with 18 wins, including the Cesarewitch, for £11,342. But this tough horse, who ran 51 times, and showed boundless stamina, had earlier proved himself in the highest class in France, where he was third in the Derby and second in the Grand Prix de Paris, before being bought for £35,000 to go to England. He also won over hurdles.

His most notable produce at stud were under National Hunt Rules, the best-known being Duke of York, Quelle Chance, Spanish Steps and Royal Relief. He sired a Danish Derby winner in Royal Scarlett, but it is safe to say he failed to produce a horse of his own class on the Flat, the best being Snow Blossom's brother Aria Royal. Snow Knight's dam was far and away his best daughter at stud.

Majano, Flush Royal's sire, made two appearances in England, on either side of 2,000 Guineas day at Newmarket in 1940. He won a ten-furlong stakes race for £380 and two days later was third of five to Hypnotist in better class. Majano was by the French St. Leger winner Deiri, who bred Deux Pour Cent (Grand Prix de Paris), Primera's dam Pirette (French Oaks) and Bey (French Derby). Flush Royal's dam, Altamira (by Olibrius), won over hurdles.

Aria Royal was one of three winners bred by Snow Blossom's dam, the Tehran mare Ariana, who was unraced. Aria Royal won six races, for £5,774, up to 15 furlongs, before he was sold as a four-year-old for 2,600 guineas and renamed Rupert, after which he showed no form and was gelded. Ariana, who also bred two sprint winners to the inevitable mating with Constable, was out of the Queen Mary Stakes winner Snowberry, dam of Chamossaire.

1975

1   Dr. C. Vittadini's ch c Grundy (Great Nephew – Word From Lundy).
2   N. B. Hunt's ch f Nobiliary (Vaguely Noble – Goofed).
3   R. Tikkoo's b c Hunza Dancer (Hawaii – Oonagh).
     18 ran. Time: 2 min 35.35 sec

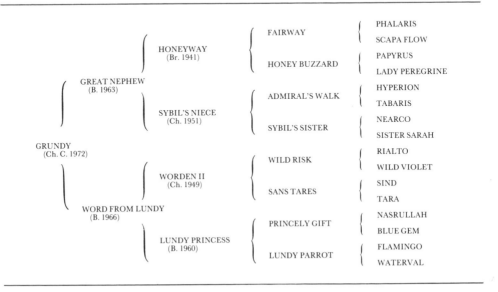

Grundy was the second Derby winner in succession to have been sold in public as a yearling, fetching 11,000 guineas to the bid of agent Keith Freeman on behalf of the Italian owner Dr. Carlo Vittadini. He soon repaid the confidence that not every bloodstock buyer would have in a flashy chesnut and went through his two-year-old career unbeaten, being the best juvenile trained in England by virtue of four wins at six or seven furlongs, including the Champagne Stakes by half a length from Whip It Quick, and the Dewhurst Stakes by six lengths from Steel Heart. The Jockey Club Handicapper placed Grundy top of the Free Handicap on 9st 7lb, 1lb in front of the Observer Gold Cup winner Green Dancer, though theoretically the third-placed filly Cry of Truth, who had 9st 5lb, could be regarded as the best two-year-old, taking into account the usual 3lb sex allowance.

Grundy had a most eventful start to his three-year-old career, when a stable-mate aimed a kick in his face that resulted in an injury which held up his early preparation but could have been much more serious. He was second in his first two races, including to Bolkonski in the 2,000 Guineas, but won his next four races in a row—the Irish 2,000 Guineas; the Derby by three lengths from the filly Nobiliary in a fast-run race; the Irish Sweeps Derby by two lengths; and the King George VI and Queen Elizabeth Stakes in record time by half a length from Bustino in an unforgettable event that some classed The Race of the Century.

There was a degree of licence in describing the Ascot contest such, as if any one race can unequivocally and objectively be given that title, but it was certainly memorable. It was, for whatever reason, also the last time that either Grundy or Bustino won. Grundy bowed out with a miserable fourth in the Benson and Hedges Gold Cup, being beaten two furlongs out and coming home well behind Dahlia, who had been third at Ascot. Grundy retired having won eight of his 11 races for £326,421.

Soon after the Derby, the Horserace Betting Levy Board acquired a three-quarter share in Grundy for £750,000. He was to stand at the National Stud at Newmarket, and was bought on condition that he would retire at the end of his three-year-old career after a maximum of four more races, all of which had to be in Britain. It was the first time the Levy Board had stepped in since it had established the Stallion Advisory Committee under the chairmanship of Lord Porchester in October 1974, with the aim of intervening with finance to the point where top-class horses trained in England would not be sold to go to stud abroad.

This is a laudable policy, always remembering that someone has to take the decision about which horses should be retained and how much should reasonably be paid, and that there is no guarantee the purchase will be successful at stud, whoever makes the decision and however much is paid. What might be questioned, especially in the case of Grundy, is the setting of conditions, since in this instance it meant Grundy would not be kept in training as a four-year-old, and would not run in what has become Europe's most competitive race, the Arc de Triomphe. It could be argued that the Levy Board were aiding and abetting a policy in which finance and its many-sided considerations were all-important in determining decisions to do with breeding and bloodstock. Not everyone connected with the business would say this was correct or desirable.

Be that as it may, Grundy stood his first season in 1976. Of slightly flashy appearance, he was a bright chestnut with flaxen mane and tail, and carried a curiously-shaped white blaze down one side of his face. Nevertheless he was an attractive colt, and his racecourse performances—the last one apart—gave the lie to the view that such horses are soft.

GREAT NEPHEW (1963)

Great Nephew, who sired Grundy from his fourth crop, was a good-class racehorse who had an unusual career. After failing to start on terms in his first two races from the barrier gate, and then winning one of his seven races at two years, he fell on his first outing as a three-year-old and gained a single win that season, in a minor race at Deauville. He did have claims to be regarded among the best milers in Europe by virtue of being second in the 2,000 Guineas and Prix du Moulin, but between these two placings he lost form and was sent to be trained in France. There he improved enormously and remained as a four-year-old to win three of his seven races, including the Prix du Moulin. He was also second in the Eclipse Stakes, which provided £6,670 of the considerable sum of £54,649 Great Nephew picked up in 1967.

He improved physically during his career and his English owner turned down an offer from America to save him for stud in England, where he started at a valuation of £100,000. Until Grundy came along, he looked no better than an average stallion, his best produce being Red Berry (four wins at two and second in the Cheveley Park Stakes) and Alpine Nephew (two wins at two). From his first five crops that raced up to 1976, Grundy stood way, way above the rest.

Sybil's Niece, the dam of Great Nephew, has been discussed in relation to St. Paddy. She was by the St. James's Palace Stakes winner Admiral's Walk. Great Nephew was the fifth winning produce out of Sybil's Niece. There were to be two more, only one of which won at further than a mile. Great Nephew apart, the best get of Sybil's Niece were four fillies who won as two-year-olds—Martial Maiden, Sybil's Comb, Crepello's Daughter and Another Daughter—though none trained on properly, and none made a mark in this country at stud.

WORD FROM LUNDY (1966)

Grundy's dam, Word From Lundy, needed a distance of ground when she was racing. She showed that as a two-year-old, when her three races were at seven furlongs, one of which she won. At three years she won two races at around a mile and a half, and stayed two miles well on the only occasion she tackled the trip.

Grundy was Word From Lundy's second foal. The first, Whirlow Green (by Crocket), was sold for 8,000 guineas and won nine races but could gain no reflected stud glory from his younger half-brother as he was gelded before Grundy's finest hour.

Twice in the four years after Grundy was foaled Word From Lundy failed to produce. She was barren to Tower Walk in 1973 and slipped her foal on a return to Great Nephew in 1976. In between she bred a filly, Tudor Whisper (by Henry the Seventh) in 1974, and a colt, Piece of Lundy (by Pieces of Eight), in 1975 who was sold for 60,000 guineas as a yearling and went to France.

Lundy Princess, the Princely Gift mare who produced Word From Lundy, is unusual in being much younger than the other members of Grundy's second generation. There were 19 years between her and Great Nephew's sire Honeyway. Lundy Princess won two races at three years over a mile, her best trip, for a total of £790. Word From Lundy was her second foal and first winner; and she bred one more winner, Baby Princess, whose five wins before she was sold to race in Jamaica amounted to £2,838 and were gained at six furlongs to a mile.

The fact that Lundy Princess was sent to Worden II, for a mating that produced Word From Lundy, was deliberate policy, since the female family had tended towards speed, rarely shining beyond a mile. The intended injection of stamina through Worden II was achieved.

LUNDY PARROT (1938)

Lundy Parrot, the dam of Lundy Princess, must rank with the best bargains of all time. She never ran, and as a three-year-old was sold for 35 guineas at the 1941 Newmarket December Sales. The Holland-Martin family which bought her can never have regretted the outlay, since Lundy Parrot herself bred eight winners, of which Lundy Princess was her last and Velvet Mist (five wins in South Africa) numerically the most successful.

Lundy Parrot's first winner, Scorned, was very useful at around a mile as a three-year-old, had two years off the course and returned to win an amateurs' race over a mile and a half. Two of Lundy Parrot's other winners were over middle distances—Wee McGregor (by Jock Scot), who won at a mile and a half and two miles; and Honey Parrot (by Honeyway, the sire of Great Nephew and therefore very closely related to Grundy), who won at a mile and a quarter. But Lundy Parrot's best offspring was the six-furlong sprinter Parakeet.

Parakeet won four races for £1,088, including the Challenge Stakes, and became the dam of six winners on the Flat and the 11-race hurdles winner Go-Gailey. Parakeet's first winning produce, Lorrikeet, won a Liverpool maiden race worth £207, was best at up to ten furlongs, and bred four winners including Tower Walk (seven wins, £40,480, at five and seven furlongs, including the Norfolk Stakes, Prix de l'Abbaye and Nunthorpe Stakes; sire). Parakeet was also the dam of Guinea Sparrow (four wins up to six furlongs; second in Cherry Hinton and Lowther Stakes).

Lundy Parrot was by the 2,000 Guineas winner Flamingo, who won six races for £20,925 and finished second in the Derby. Flamingo's dam, Lady Peregrine, also bred a top-class horse in Horus, who won four races for £4,657 including the King Edward VII Stakes and finished third in the St. Leger. Lady Peregrine's daughter Honey Buzzard became the dam of nine winners, including the smart miler Welsh Honey, the National Breeders Produce Stakes winner Tudor Honey, and Honeyway. Lady Peregrine therefore appears in the fourth and fifth generations of Grundy's pedigree—as grandam of his maternal grandsire Honeyway, and as dam of his third dam's sire Flamingo.

WATERVAL (1923)

Waterval, the dam of Lundy Parrot, was out of the mare Lilaline, who won as a two-year-old and recovered form over longer distances when trained in the North as a three-year-old. Lilaline visited the sprinter Friar Marcus six times in nine years immediately after the First World War. Her only colt died, but the five fillies—Morals of Marcus, Madwaska, Marjolaine, Little Mark and Waterval—became a most prolific source of winners.

Morals of Marcus won nine races, including seven over five furlongs as a two-year-old, and was second to Mumtaz Mahal in the Queen Mary Stakes. Morals of Marcus bred three winners, easily the best being Marcus Superbus (six wins, £2,520, including Derby Cup over a mile and threequarters). Royal Step, another winner out of Morals of Marcus, bred Royal Quadrille (four races including Arundel Castle Private Sweepstakes).

Madwaska won one race worth £137, and became the dam of the Queen Anne Stakes winner Madagascar. Marjolaine could not match even that modest racing record, failing to win at all, but she bred three minor winners including Thornbush, dam of Careless Lad, who ran second to Honeyway in the Victoria Cup and won five races. Marjolaine was the grandam of Castleton (King Edward VII Stakes and Blue Riband Trial Stakes).

Little Mark was also a non-winner, though she was highly tried. She made a slow start at stud and when sent to the December Sales in 1933, at the age of 11, had not bred a winner. She was bought, covered by Black Watch, from an East Yorkshire stud by Sir Laurence Philipps for 75 guineas. Her colt foal went elsewhere for 20 guineas and named Black Speck he won 13½ races for £8,441 between 1935 and 1941, including the Coventry Stakes and Newbury Summer Cup. Black Speck's year-younger sister Mark Time won six races and bred the National Hunt sire King's Approach; and there was another winner from the Black Watch–Little Mark union in Blot (two small races).

Little Mark also bred Pecked (by Flamingo), whose three winning offspring, all fillies, included Blue Mark, dam of seven winners here and abroad including Lalibela, winner of the Cheveley Park Stakes. Lalibela carried the colours of Mr. J. P Philipps, son of Sir Laurence Philipps and owner-breeder of Grundy's sire Great Nephew. The Philipps' family also came into the story of Waterval, who herself came third in the list of matings between Friar Marcus and Lilaline.

Waterval was placed as a two-year-old and won three sprints as a three-year-old for £2,062, including the Granville Stakes at Royal Ascot, and was second in the Nunthorpe Stakes. She bred two winners, including the useful five-furlong winner Water-Way, who when mated with Flamingo bred the top-class sprinter Waterbird, winner of seven races for £4,048, including the King's Stand Stakes. Water-Way was sold privately in 1936 when Waterbird was a two-year-old to Sir Laurence Philipps for £2,000, and sent back to Flamingo, she bred Glossy Ibis, who won one race and bred six winners including Rose Linnet (Queen Mary Stakes), The Scarab (good handicapper) and Dorothy Belle (dam of the very speedy Gold Bell). Rose Linnet, who failed to train on at three years, bred three winners in England, the best of whom, the Rous Memorial Stakes winner Red Cardinal, can be said to have been one of the few horses who frightened himself to death. Red Cardinal was sold to America after winning three of his four races as a two-year-old but he took fright at the prospect of entering the aircraft, became very ill after injuring himself and died. Another daughter of Rose Linnet, Scarlet Plume, who like her dam failed to make normal progress between two and three years, bred the very fast colt Sica Dan.

In the meantime, Waterval's daughter Lundy Parrot, bred from a mating with the ever-popular, and ever-influential Flamingo, went her separate way—only to join again the families of Holland-Martin and Philipps in years to come.

1976

1　N. B. Hunt's b c Empery (Vaguely Noble – Pamplona II).
2　Lady Beaverbrook's b c Relkino (Relko – Pugnacity).
3　A. Oldrey's b c Oats (Northfields – Arctic Lace).
　　23 ran. Time: 2 min 35.69 sec

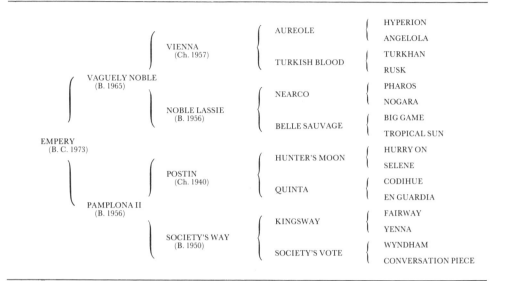

As an illustration of how the international boundaries of modern racing have been opened up, Empery's Derby success could hardly be bettered. Trained in France by an Egyptian who set himself up on the proceeds of betting, and owned by an American whose family was fabulously rich, even by Texan standards, Empery was by a British stallion out of a mare born in Peru. At Epsom he was ridden by one of the great international jockeys of the day, Lester Piggott.

Yet for all that, the notable aspect of Empery's career was his comparative lack of success, since he won only twice, once in a newcomers' race and once in the premier English Classic.

He won his first race as a two-year-old over a mile, and then was sixth in the Grand Criterium and third in the Prix Thomas Bryon, resulting in his being placed 10lb behind the top horse in the French Free Handicap. At three years he had five races, and in the Derby was always close behind the leaders before taking over at the distance and beating Relkino by three lengths. Before that, he had shown himself to be a stayer, being fourth in the French 2,000 Guineas, fifth in the Prix Daphnis and third in the Prix Lupin over eight, nine and ten furlongs respectively.

After Epsom, Empery went for the Irish Sweeps Derby, and was beaten an easy two and a half lengths by Malacate, who had been third to Youth (owned like Empery by Nelson Bunker Hunt) in the French Derby. Empery did not run again after his trip to Ireland. A slight injury kept him out of the Benson and Hedges

Gold Cup in the autumn, and loss of condition after being quarantined meant he missed the Man O'War Stakes in America. Meanwhile, Youth added to his honours by winning the Canadian International Championship by four lengths and the Washington International by ten lengths, and finishing third in the Arc de Triomphe.

Both Empery and Youth were syndicated by the end of June as three-year-olds, for six million dollars each to stand at the same stud in Kentucky. Those who took up the 20 shares in each released by Mr. Hunt took a gamble, which in the case of Youth seemed to come off, but with Empery was hardly made to look good value by subsequent events.

VAGUELY NOBLE (1965)

On the subject of gambles, Empery's sire Vaguely Noble was an example of one that came off. Second in his first two races as a two-year-old, he ended his first season by winning the Sandwich Stakes at Ascot by 12 lengths and the Observer Gold Cup by seven lengths (though it looked further). He then went up to the Newmarket December Sales, for one of the most publicized auctions for many a year. He was sold largely to pay death duties incurred by the estate of his breeder, the late Major L. B. Holliday, and as he held no Classic engagements in 1968, his sale for 136,000 guineas represented a major gamble by his new owner, the Californian plastic surgeon Dr. Franklyn. His sale price was more than three times the previous highest for a horse in training sold at public auction in Britain, the 37,500 guineas that Flying Fox made in 1900.

The fact that Vaguely Noble had shown his best form on soft going merely added to the nature of the gamble, but it came off spectacularly. Third in the Grand Prix de Saint-Cloud, he won his four other races as a three-year-old while trained in France, including the Arc de Triomphe. He earned more than £130,000 on the racecourse, but of more concern to connections, it was possible to retire him to stud in America with a valuation of £2 million.

Certain aspects of Vaguely Noble's career were not so pleasing, though it must be admitted they reflect the passage of time and the appearance of new factors in a new age. The death of Major L. B. Holliday and the subsequent pruning of his bloodstock empire was a further blow to the dwindling ranks of important owner-breeders; and the disparity in prize money between England and France at that time meant it was almost certain Vaguely Noble would not be seen in England again. So it proved, and he did not even stay long in Ireland, where it was originally intended he would be trained, since Nelson Bunker Hunt, underbidder at Newmarket, later took a half share in him and he was sent to be trained in France.

France was always likely to provide him with the easy surface on which he appeared to thrive, but racing there also proved he was not a one-paced stayer, as seemed likely at two years. In the Arc he beat the best middle-distance horses in Europe, winning by three lengths from the Derby winner Sir Ivor, who was the same distance in front of Luthier (soon to emerge as a fine stallion) and Roseliere (later the dam of the Champion Stakes winner Rose Bowl). A most attractive horse, with size and strength but no coarseness, Vaguely Noble was one of the best

Arc winners since the war; probably only Ribot and Sea-Bird II were better. Like those two great horses, Vaguely Noble was a late-maturing type whose stock would probably be better suited by European conditions. But again like those two, the mighty dollar sent Vaguely Noble off to a stud career in America.

By the time of the Arc, American breeder John Gaines had taken an option to buy a quarter-share in Vaguely Noble for 1.25 million dollars. As a result he retired to stud at the end of his three-year-old career in 1968, when Mr. Hunt became the major shareholder. It was particularly fortunate that Mr. Hunt did take such an interest in him, because he had extensive interests in Europe and was able to afford his stock the best opportunity to show their worth. The best of Vaguely Noble's stock which raced exclusively in America was Royal and Regal (six wins, including Florida Derby before import to stud in Ireland in 1974), who came from his first crop. And only one of Vaguely Noble's biggest successes in Europe before 1976 did not carry the Hunt colours. That was Duke of Marmalade, who came from his sire's second crop, was trained in England, France and Italy in his career and won the Grand Prix de Deauville and Premio Roma.

Thanks to the one horse that won for him in England in 1972, Noble Decree, Vaguely Noble was that year's leading first-season sire. Noble Decree had three wins including the Observer Gold Cup, but he injured his back in the Derby and retired without adding to his gains, having earlier been beaten only a head in the 2,000 Guineas. Noble Decree retired to stand alongside his sire in America.

Also from the first crop and carrying the Hunt colours was Dahlia, who won top-class races in five countries—the Prix Saint-Alary in France; Irish Guinness Oaks; Canadian International Championship; the Washington International in America; and the King George VI and Queen Elizabeth Stakes and Benson and Hedges Gold Cup (twice) in England.

Ace of Aces (six wins including Sussex Stakes) was also from Vaguely Noble's first crop, and from the second came Mississippian (half-brother to Youth) who won the Grand Criterium, was second in the Observer Gold Cup, French 2,000 Guineas and Prix Lupin, and fourth in the French Derby and Irish Sweeps Derby. From Vaguely Noble's third crop came the filly Nobiliary (Prix Saint-Alary and Washington International; second in Derby and Prix Vermeille), and along with Empery in the fourth was Exceller (Grand Prix de Paris and French St. Leger).

There must be some concern for the future of Vaguely Noble as a possible sire of sires in that all his eggs seemed to have been placed in the one basket owned by Mr. Hunt. Only time will tell, of course, but it might be as well to remember that Vaguely Noble's own sire, Vienna, was a "one-horse wonder".

VIENNA (1957)

Vienna was not in the highest class as a racehorse, though he did finish third in the St. Leger behind St. Paddy, who was never out of a canter. Vienna did prove the virtue of soundness, since he raced until the age of five, when he was retired to stud in Ireland. As a stallion he can be termed a failure, since apart from Vaguely Noble he sired nothing of particular note. Sold to go to France after the 1968 breeding season, he was exported to Japan in 1971.

Vienna was by Aureole out of the Turkhan mare Turkish Blood, who won twice over five furlongs as a two-year-old but stayed well enough to finish third in the Irish Oaks the following season, though she did not add to her successes. She bred five winners in Britain, all colts and of which Vienna was easily the most successful. The next best was Blood Test, the exception among her offspring in that he was a sprinter. Blood Test (by Fair Trial) won three races for £2,669, including the Cork and Orrery Stakes, form which he had not revealed before nor was to show afterwards.

Turkish Blood's dam Rusk won one race for £151 and bred two other minor winners, both at up to a mile and a half. Rusk was out of Baby Polly, who managed one placing in her career, finishing second in the valuable International Stakes over five furlongs at Kempton Park as a two-year-old, her only season to race. A daughter of Pretty Polly, Baby Polly bred six winners, including Colorado Kid (ten wins including Royal Hunt Cup, Jubilee Stakes and Doncaster Cup) and Precious Polly, dam of five winning fillies. Precious Polly's winners included Sea Parrot, winner of the Nassau Stakes and Yorkshire Oaks, and dam of six winners including Shearwater (Craven Stakes) and Green Opal (Sandleford Priory Stakes), as well as Big Romance, who bred the July Stakes and Richmond Stakes winner Romantic.

NOBLE LASSIE (1956)

Noble Lassie, the dam of Vaguely Noble, was an exceptionally attractive filly who won two of her 14 races, including the Lancashire Oaks. She usually made the running in her races and was well suited by firm ground. She bred a winner either side of Vaguely Noble, her third foal. They were the colt Attractive, who won once over 13 furlongs, and the filly Regal Lady who won at ten and 13 furlongs.

Noble Lassie was one of five winners bred by the Big Game mare Belle Sauvage, who won once for £207 over a mile and a half as a three-year-old. Belle Sauvage's other important winner was Pandour (four wins, £3,302, including Dee Stakes).

Belle Sauvage was out of the Oaks third Tropical Sun, whose own dam won that Epsom race as well as the Goodwood Cup and Prix du Cadran. Tropical Sun bred three other winners, but it was through her unraced daughter Open Court that she is also well known. Open Court bred six winners in Britain, the most significant being the brother and sister Miletus (three wins including Kempton 2,000 Guineas Trial) and Maeander (four wins; third in Cheveley Park Stakes). Open Court was also grandam of the very useful sprinting filly Laxmi.

PAMPLONA II (1956)

The distaff side of Empery's pedigree contains some less familiar names, beginning with his dam Pamplona II, who in her native Peru became the first filly to win the local Triple Crown. She proved her status beyond her own boundaries by finishing third to the colts Escorial and Farwell in Buenos Aires. In all she won 14 races from 19 starts, and was placed in three others, up to 15 furlongs.

She was secured by Mr. Hunt and sent to America, where Empery was her seventh foal. There had been five previous winners, including two stakes winners in America, of whom Sports Event foaled to Nijinsky a colt called Sportsky who won five smallish races in England up to a mile before being exported to South Australia as a stallion. Pamplona II's previous winners also included Pampered Miss, winner of four races from 4½ furlongs to a mile including the French 1,000 Guineas and third in the French Oaks.

Postin, the sire of Pamplona II, was considerably older than the others in the second generation of Empery, being ten years older than Pamplona II's dam and 17 years older than Vaguely Noble's sire. Postin was an outstanding racehorse in Peru, and leading Peruvian sire from 1954 and 1959, with four winners of the Peruvian Derby. Postin was by Hyperion's half-brother Hunter's Moon.

Society's Way, the dam of Pamplona II, was responsible for introducing the family to South America. She won five races for £2,598, including the Melrose Handicap and two more at York as a four-year-old when trained in the north of England. She was submitted to the 1954 Newmarket December Sales as a four-year-old, failed to make her reserve but was sold privately for 4,400 guineas to the bid of the BBA, who sent her to Peru.

Society's Way was by Kingsway out of the unraced Wyndham mare Society's Vote, who bred one other winner, the sprint selling-plater Doctor Bother, who took from 1953 to 1960 to amass his five wins for a total of £1,112. Society's Vote was a daughter of the Irish Oaks winner Conversation Piece, whose grandam Bill and Coo appears as the seventh dam of Busted (through Booktalk, a half-sister to Conversation Piece), Secretariat and Sir Gaylord, and eighth dam of Nonoalco and Alleged.

CHAPTER 30

1977 – 1981
Postscript

1977

1 R. Sangster's ch c The Minstrel (Northern Dancer – Fleur).
2 Lord Leverhulme's b c Hot Grove (Hotfoot – Orange Grove).
3 H.H. Aga Khan's ch c Blushing Groom (Red God – Runaway Bride).
    22 ran. Time: 2 min 36.44 sec

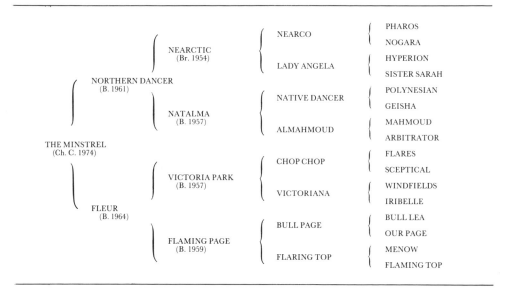

THE POWERFUL COMBINATION of Northern Dancer and a half-sister to Nijinsky (who was by Northern Dancer) produced The Minstrel, and the powerful influence of Mr. Robert Sangster and his trainer Vincent O'Brien brought him to Europe after he had passed into the hands of the BBA (Ireland) Ltd. for 200,000 dollars (then the equivalent of £85,470) at the 1975 Keeneland July Sales. He more than repaid their faith, winning seven of his nine races for a total of £333,197. Then came the pay-off; he was syndicated to stand at stud in Maryland at a valuation of nine million dollars! The Minstrel's Canadian breeder, Mr. E. P. Taylor, purchased 18 of the 36 shares offered in him.

It took The Minstrel a little time as a three-year-old to earn his high valuation. He had been undefeated as a two-year-old in three races, ending his first season by winning the Dewhurst Stakes in good style by four lengths from Saros, a performance that the Jockey Club Handicapper judged to be worth 8st 13lb in the Free Handicap, 8lb behind the top horse J. O. Tobin. The Minstrel reappeared to win the Ascot 2,000 Guineas Trial, but was beaten in the first Classics themselves, being two lengths behind Nebbiolo when third at Newmarket, and a short head behind Pampapaul when second at The Curragh.

Thereafter, The Minstrel ended his career as he had begun it, unbeaten in three races, and it was the opportunity to race at a mile and a half which helped. He won the Derby with a combination of stamina and guts, getting up under the strong handling of Lester Piggott to beat Hot Grove by a neck; he won the Irish Sweeps Derby by a length and a half from Lucky Sovereign, despite drifting off a true line in the straight; and he battled like a lion to hold off Orange Bay by a short head in the King George VI and Queen Elizabeth Diamond Stakes.

The second flashy chestnut to win the Derby inside three years, The Minstrel was in training for the Arc de Triomphe, but import restrictions in America, following outbreaks of contagious equine metritis, meant he was shipped out of Europe at short notice, and he did not race again. His first crop went to the sales as yearlings in 1980, and the following year, before he had had time to set down his merit, he had his first million-dollar produce, a colt who trod his sire's path to the Keeneland July Sales and was bought by English interests.

It is interesting—and at times frightening—for breeders to contemplate how inflation and international influences have pushed up prices in America, where the choicest bloodstock markets are, for the moment, to be found. The Minstrel himself helps to provide a link.

When Mr. Sangster bought The Minstrel in 1975, he paid 200,000 dollars; when he bought a full brother to The Minstrel at the same yearling sales in 1980, he paid 1,250,000 dollars (then the equivalent of £563,063). In 1980, also at Keeneland in July, the world record price for a yearling sold at public auction was set at 1.7 million dollars (£765,765) for a colt by Lyphard bought on behalf of European-based shipping millionaire Mr. Stavros Niarchos. A year later that record was left far behind in the same sales ring when a colt by Northern Dancer made 3.5 million dollars (then approximately £1,850,000) to the bid of Mr. Sangster. The latest world record holder was a brother to the European-raced colt Storm Bird, who had cost Mr. Sangster a million dollars as a yearling in 1979 but just before the Keeneland Sales changed hands for 30 million dollars.

It is hard to comprehend such huge sums, or to relate them to traditional values, unless one happens to be among the few fortunate enough, or brave enough, to be dealing in them. The rest must take heart from the presence of men who retain faith in British racing, even if the finance which sustains that faith is not necessarily generated from inside Britain. As an example, one further set of statistics from the

1981 Keeneland July Sales is worth recording; namely that of the nine yearlings who made a million dollars or more, eight were bought by European clients.

Mr. E. P. Taylor not only bred The Minstrel, Nijinsky and Northern Dancer, but also Fleur, a half-sister to Nijinsky who bred four foals before The Minstrel. Best of them were her first foal Flaming Ace (by Maribeau), who won ten races for 48,970 dollars from three to five years, and The Minstrel's brother Far North, a useful horse who probably never fully recovered from an illness at three years but won three races in France for the equivalent of £19,879. Fleur had to be put down in March 1981 because of an incurable foot disease. She showed modest race-course ability, winning three races for 9,235 dollars in Canada at two and three years, and was once placed in a stakes race. Her sire, Victoria Park, was Canada's Horse of the Year in 1960, when he won the Queen's Plate, a race which also fell to Fleur's dam, Flaming Page.

1978

1   Lord Halifax's b c Shirley Heights (Mill Reef – Hardiemma).
2   R. Sangster's b c Hawaiian Sound (Hawaii – Sound of Success).
3   Mrs D. Jardine's ch c Remainder Man (Connaught – Honerone).
    25 ran. Time: 2 min 35.30 sec

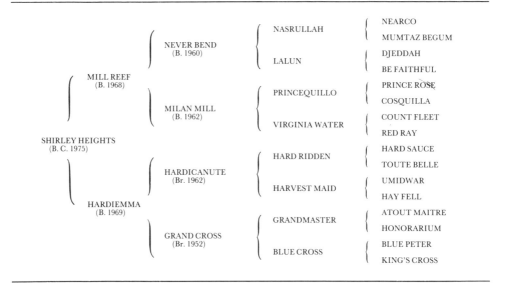

It did not take Mill Reef long to prove the merit of striving to retain him for stud purposes following his terrible accident on the gallops as a four-year-old. From his first full crop he produced the Derby winner Shirley Heights and the French Derby winner Acamas. Mill Reef could hardly have made a more spectacular start to his stud career, joining Never Say Die as the only post-war Derby winners to sire a similar winner.

Shirley Heights ran more times as a two-year-old than most recent Derby winners, winning two of his six races and being placed in three others. The only time he was out of the money was when he finished fifth on his debut. He followed by running his best races when stamina counted most, being placed in the Seaton Delaval Stakes and Solario Stakes over seven furlongs and ending his first season with a win in the Royal Lodge Stakes over a mile. He was placed 6lb behind the top horse, Try My Best, in the Free Handicap on 9st 1lb.

Backward when beaten ten lengths into second place behind Whitstead in the Sandown Classic Trial, Shirley Heights won his next four races. He beat the subsequent King George VI and Queen Elizabeth Diamond Stakes winner Ile de Bourbon in a small race at Newmarket, and decisively claimed the Mecca-Dante Stakes before going to Epsom, where he was in front nowhere but in the last strides, after Hawaiian Sound and his American jockey Shoemaker had attempted to defy convention by leading throughout. There was another close call for Shirley Heights when he started favourite for the Irish Sweeps Derby, and again he caught Hawaiian Sound close home, after drifting off a straight line in the straight, though Exdirectory stayed on to split the pair.

Plans were to train Shirley Heights for the St. Leger, a race his Yorkshire-based owner was particularly anxious to win, but a few days before his preparatory race in the Great Voltigeur, he injured a tendon and was retired to stud to stand his first season in 1979. He was syndicated at a total valuation of £1.6 million, and became only the fourth Derby winner in ten years to stay in England, the other six having left for America. The valuation was almost certainly less than if Shirley Heights had been put on the open market, but 1978 still provided an opportunity to test his value.

Lord Halifax and his son Lord Irwin bought the Hardicanute mare Hardiemma at the 1972 Newmarket December Sales in order to send her to Mill Reef. She had shown only average form on the racecourse when trained in the North, winning a seven-furlong maiden event and an 11-furlong handicap, and as a back-end three-year-old fetched 9,000 guineas. Four years later she returned to the same sale-ring and, covered by Mill Reef, she made 15,000 guineas to the Ballyrogan Stud in Ireland. At that time she had bred three foals for her owners but only one had run, an unsuccessful two-year-old by Upper Case called Hit the Roof. In due course she foaled to the Mill Reef service, and her new owners sent the produce, a filly, to the 1978 September yearling sales in Ireland. By this time, Hardiemma's first foal, Hit the Roof, had won two sellers on the Flat and was embarking on a modest hurdles career, and her third foal, a filly called Bempton (by Blakeney), was unraced. But her second foal, Shirley Heights, had won £205,211 with six wins from 11 outings—and the yearling made 250,000 guineas. Hardiemma followed by foaling Regal Heiress, a filly by English Prince who managed to win a mile and a half maiden race as a three-year-old, and Inviting, a colt by Be My Guest who made 70,000 Irish punts when sold as a yearling in 1980.

Prior to Shirley Heights there had been a noticeable lack of Classic quality among the female side of his pedigree. The nearest equivalent emanated from his third dam Blue Cross (a two-year-old winner, dam of three winners and half-sister to the 2,000 Guineas second King's Bench) and his second dam Grand Cross (winner of a £361 plate, dam of 11 winners, and half-sister to the grandam of the Irish St. Leger runner-up General Ironside).

Largely as a result of Shirley Heights' success, Mill Reef was leading sire in 1978, when his winners also included Idle Waters (Park Hill Stakes), English Harbour (third in Gordon Stakes) and Main Reef (Chesham Stakes, July Stakes). Acamas, also winner of the Prix Lupin, beat Shirley Heights by three days to become Mill Reef's first Classic-winning son.

1979

1 Sir Michael Sobell's b c Troy (Petingo – La Milo).
2 Mme. J. P. Binet's ch c Dickens Hill (Mount Hagen – London Life).
3 Mme. A-M d'Estainville's b c Northern Baby (Northern Dancer – Two Rings).

　　　　23 ran. Time: 2 min 36.59 sec

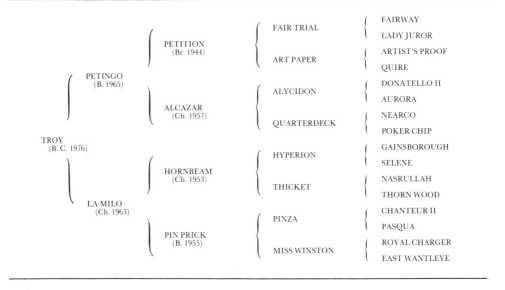

TROY (B. C. 1976)

PETINGO (B. 1965)
- PETITION (Br. 1944)
  - FAIR TRIAL
    - FAIRWAY
    - LADY JUROR
  - ART PAPER
    - ARTIST'S PROOF
    - QUIRE
- ALCAZAR (Ch. 1957)
  - ALYCIDON
    - DONATELLO II
    - AURORA
  - QUARTERDECK
    - NEARCO
    - POKER CHIP

LA MILO (Ch. 1963)
- HORNBEAM (Ch. 1953)
  - HYPERION
    - GAINSBOROUGH
    - SELENE
  - THICKET
    - NASRULLAH
    - THORN WOOD
- PIN PRICK (B. 1955)
  - PINZA
    - CHANTEUR II
    - PASQUA
  - MISS WINSTON
    - ROYAL CHARGER
    - EAST WANTLEYE

Troy gave indication as a two-year-old that he might develop into a formidable contender for the 200th Derby. He won two of his four races—a maiden race at Newmarket and the Lanson Champagne Stakes at Goodwood—and was second in the other two, gaining his high rating of 9st by running Ela-Mana-Mou to threequarters of a length in the Royal Lodge Stakes, the race that revealed Shirley Heights' merit the year before. However, it was not possible to foretell that Troy would put at least seven lengths between himself and the rest of the Derby field, passing eight horses in the last two furlongs after he had had trouble holding his place in the early stages.

The Derby was the third of six straight wins Troy gained as a three-year-old. The first two had been in the Sandown Classic Trial and Goodwood's Predominate Stakes; the last three were in the Irish Sweeps Derby, King George VI and

Queen Elizabeth Diamond Stakes, and the Benson and Hedges Gold Cup. Only the Arc de Triomphe was missing from a glamorous clutch of races, and Troy was beaten there by three lengths and a length into third place behind Three Troikas and Le Marmot. The Troy who ran in the Arc was not the Troy who ran in the Derby, and it is perhaps best to judge him on his form in Britain, which entitles him to be regarded in the highest class, not far behind the best Derby winners of the post-war period and well in front of most.

One criterion on which it would be dangerous to judge Troy is finance. Aided by a record Derby prize of £153,980, Troy comfortably raced to the head of the European list for earnings by a colt. His winning tally of £415,735 from eight victories was bettered only by the fillies Dahlia (£497,741) and Allez France (£493,100), and put him ahead of the colts Youth (£366,624) and Alleged (£327,315). The figures admirably illustrate how the real value of money has fallen rapidly in recent years.

In a similar way, Troy's eventual syndication for stud was a revelation. Before he ran in the Benson and Hedges Gold Cup, attempts to retain him in England were successfully concluded at a total valuation of £7.2 million, shares being paid for at the rate of £90,000 as a down-payment and two annual payments of £45,000. It made him the most valuable horse ever retired to stud in Britain, and allowing for an exchange rate of 2.2 dollars to the pound, put him close to the record syndication drawn up in the 1978 American Triple Crown winner Affirmed, who was valued at 16 million dollars compared with Troy's 15.84 million dollars. The previous American Triple Crown winner, Seattle Slew, was syndicated for 12 million dollars; while that year's Derby winner, The Minstrel, was valued at nine million dollars. Ten years before Troy, the National Stud had bought a half-share in the Derby winner Blakeney for £250,000!

There was some criticism of the decision to retire Troy as a three-year-old; certain observers suggested it was faint-hearted of his owners not to race him as a four-year-old, when he would have the opportunity of being tested against a younger generation. In fact, the retirement of Troy was an indication of how the considerations of commerce have overtaken the sporting element in the treatment of the choicest stud prospects. Troy could not enhance his reputation as a four-year-old, and stood to lose some of it should he be beaten in the top middle-distance races. His owners were balancing that fact, and the possibility of winning a maximum of around £300,000 from the major races of his third season, against the reputation he had already gained, and the guaranteed million pounds he would earn in stud fees from his first season as a stallion. There seems little prospect of this situation changing in the near future, and breeders must accustom themselves to judging the best three-year-olds on that basis.

A strong, good-looking colt, and an impressive mover, Troy was the third Classic winner for his sire Petingo, who himself finished second to Sir Ivor in the 2,000 Guineas. Petingo had been the outstanding two-year-old of 1967, when he won the Gimcrack Stakes and Middle Park Stakes, having been bought as a yearling for 7,800 guineas. He developed into a top-class miler, with wins in the St. James's Palace Stakes and Sussex Stakes, and retired to stud in 1969 with six wins to his name for a total of £30,993. Despite his own apparent stamina limitations, his two Classic winners before Troy were English Prince (Irish Sweeps Derby) and Fair Salinia (Epsom Oaks, Irish Oaks and Yorkshire Oaks). Petingo's sire, Petition, won the Victoria Cup and Eclipse Stakes, stayed ten furlongs and bred Petite Etoile, while his dam Alcazar was by Alycidon and was a half-sister to the Irish Oaks winner Ambergris.

Petingo also sired two of the leading first-season sires of 1978, Pitcairn (who topped the table) and Pitskelly. That year saw his last crop racing as two-year-olds, and included the William Hill Futurity winner Sandy Creek. Pitcairn owed his prominent position mainly to the Royal Lodge Stakes success of Ela-Mana-Mou, who two years later was bought by the owners of Troy. Pitcairn was exported to Japan in 1978.

Not only was Troy a member of Petingo's last crop—he died in February 1976 after only seven complete seasons at stud—he was also the last foal of his dam La Milo, who bred seven foals, all winners and each to a different stallion. Laristan (by Tamerlane) and Cleon (by Major Portion) won small races in England before Admetus (by Reform) came along to prove that geldings are worth a chance in the biggest races, his 12 wins amassing £119,033 from such races as the Washington International, the Grand Prix d'Evry and Royal Ascot's Prince of Wales Stakes. La Milo's fourth and fifth foals, Capuccio (by Zeddaan) and Silk Rein (by Shantung), each won one small race, then came Tully (by Tudor Melody), whose three wins were worth £20,112 and included the White Rose Stakes. Troy completed the list, as La Milo was put down, because of constant leg trouble, within months of his being born.

La Milo herself took some time to reach her peak, needing both time and distance but ending her nine-race career by winning on her last four starts as a three-year-old for £2,945. La Milo, who was by Hornbeam, mirrored the career of her own dam, the Pinza mare Pin Prick, who was unraced at two but as a three-year-old won four staying races. It was Pin Prick who introduced this family to the stud of Troy's breeder Sir Michael Sobell, when he bought the broodmares, horses in training, yearlings, foals and Ballymacoll Stud property of the late Miss Dorothy Paget in 1960. Pin Prick, who was then five and had had one foal, went on to produce five winners for her new owners, including two useful two-year-olds in Kesh and Tamerlina.

Pin Prick, whom Miss Paget bought as a yearling for 2,100 guineas, was one of three minor winners out of the Royal Charger mare Miss Winston, winner of four races at five furlongs and second in the Queen Anne Stakes and third in the Cork and Orrery Stakes. Miss Winston's dam, East Wantleye (by Foxlaw), failed to win and as a four-year-old was sold for 240 guineas. East Wantleye's most significant produce was Miss Winston's brother Royal Hamlet, who won seven times at five to seven furlongs in four seasons' racing. East Wantleye was out of Tetrill, whose half-sister Deva (by Gainsborough) was bought for stud by Miss Paget and bred Nivea, dam of Rowston Manor (Lingfield Derby Trial), French Beige (Doncaster Cup) and French Cream (Irish Oaks).

Troy's first foals were born in 1981.

1980

1   Mme. A. Plesch's b c Henbit (Hawaii – Chateaucreek).
2   W. Barnett's ch c Master Willie (High Line – Fair Winter).
3   R. Fennell's ro c Rankin (Owen Dudley – Cup Cake).
    24 ran. Time: 2 min 34.77 sec

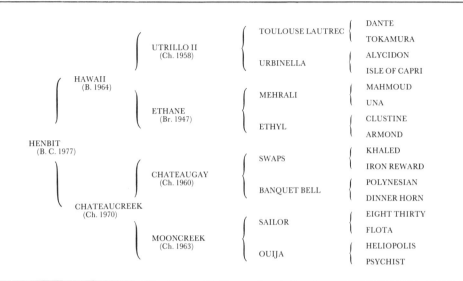

			DANTE
		TOULOUSE LAUTREC	TOKAMURA
	UTRILLO II		ALYCIDON
	(Ch. 1958)	URBINELLA	ISLE OF CAPRI
HAWAII			MAHMOUD
(B. 1964)		MEHRALI	UNA
	ETHANE		CLUSTINE
	(Br. 1947)	ETHYL	ARMOND
HENBIT			KHALED
(B. C. 1977)		SWAPS	IRON REWARD
	CHATEAUGAY		POLYNESIAN
	(Ch. 1960)	BANQUET BELL	DINNER HORN
CHATEAUCREEK			EIGHT THIRTY
(Ch. 1970)		SAILOR	FLOTA
	MOONCREEK		HELIOPOLIS
	(Ch. 1963)	OUIJA	PSYCHIST

For the second year in succession the Derby winner was trained by Dick Hern and ridden by Willie Carson, the first such instance since trainer George Manning and jockey John Wells won with Beadsman and Musjid, owned by Sir John Hawley, in 1858 and 1859. It was also the seventh time in 12 years that the Derby had been won by an American-bred horse, but Henbit could not be described as typical of the rest. He was a staying type of horse, more European than American in make-up, if it is possible to label in such a generalized way.

Henbit was also unusual in being bought as a yearling for 24,000 dollars (then £12,500), an absurdly small sum compared with those discussed in recent pages. He came from the same Kentucky Yearling Sales that produced the French 2,000 Guineas winner In Fijar for 36,000 dollars (£18,750) and the Kentucky Derby winner Genuine Risk for 32,000 dollars (£16,666), so even in America it remains possible to buy big-race winners without paying "telephone numbers" for the pleasure.

Fourth over six furlongs first time out in the Chesham Stakes, Henbit showed his liking for a distance of ground when winning a 25-runner Newmarket maiden race over a mile on his second appearance, and was then a respectable fourth to Monteverdi in the seven-furlong Dewhurst Stakes, a performance that enabled the handicapper to allot him 8st 12lb in the Free Handicap, 9lb behind Monteverdi's top place.

As a three-year-old, Henbit won all his three races. He emulated his stablemate

Troy by winning the Sandown Park Classic Trial; he took the Chester Vase by four lengths; and in the Derby—which he contested only after original plans to send him for the French equivalent had been amended—he emerged from a pack of about nine horses who held a chance two furlongs out and fought off Master Willie by threequarters of a length, with Rankin a length and a half away third.

As Henbit went into the lead under pressure over a furlong out at Epsom, he faltered for a moment and hung right, before his jockey straightened him for the final run to the line. It soon became apparent that Henbit was lame, and a five-inch crack in his off-fore cannon bone was later diagnosed. A light plaster was put on the injury and he was confined to his box for three months. He was back in light work by October, with the intention of keeping him in training as a four-year-old, but although this bold, sporting plan was carried out, it did not earn the rewards that those who applauded the triumph of sportsmanship over commercialism would have wished. He was last of six in the Jockey Club Stakes, almost 35 lengths behind the winner Master Willie, and was fifth of seven in a smallish race in Ireland, almost ten lengths behind the three-year-old winner. Henbit had not recovered his form, though there was nothing wrong with his physical condition, and he was retired forthwith, his four wins from eight outings having yielded £188,758.

Henbit's pedigree provides an interesting combination of English, Irish, Italian, South African and American factors. The lack of names familiar to American buyers might explain why he went so cheaply when put up for public auction, for his sire, Hawaii, has hardly caught the imagination from his Kentucky base, being available at 10,000 dollars for a nomination, and neither his dam, Chateaucreek, nor grandam, Mooncreek, had much to recommend them.

Hawaii made one reputation in South Africa and another in America. Few horses bred in South Africa, whose thoroughbred population rivals that of Britain in numbers at least, have made an international name but Hawaii was given the chance on account of being owned by Mr. Charles Engelhard, for whom he raced throughout his career. He won 15 of his 18 races in his native country, where he had been bought as a yearling; he was champion two-year-old and at three won the South African Guineas. After two outings as a four-year-old, he was sent to America.

Hawaii's ten races in America brought six wins for 279,280 dollars, and he was able to put behind the idea that he was nothing more than a miler with two wins over a mile and a half. His best wins were in the United Nations Handicap and Man O'War Stakes, and he was second to English-trained Karabas in the Washington International as a five-year-old. His emergence as a middle-distance horse was not surprising since there was staying influence in abundance in the pedigree of his sire, the massive Italian horse Utrillo II, whose best son he was.

Utrillo II, Tesio-bred, won twice in his native country but coughing kept him out of the Italian Derby, for which he would probably have started favourite. He was then sent to England where he won two small races as a three-year-old but none out of eight as a four-year-old. Utrillo II was by Dante's son Toulouse Lautrec out of Alycidon's daughter Urbinella.

Hawaii's dam Ethane was by Mahmoud's son Mehrali, who raced in England

and won six times, only once beyond five furlongs, before being exported to stud in South Africa in 1946. Ethane was a winner of no great ability but she bred 12 winners from 13 foals. Hawaii was easily the best, and Henbit's Derby success earned him a place in the race's history, since with his seventh crop he had produced a winner to follow the third place of Hunza Dancer in 1975 and the second place of Hawaiian Sound in 1978. These three top-class colts comprised exactly half the total number of winners in Britain sired by Hawaii in that time.

Henbit's dam Chateaucreek was a sprinter who won six small races for 24,203 dollars, none over further than six furlongs. She had bred one foal before Henbit, a colt by Mr. Leader called Lead Reek, who was placed on the Flat in France and won over jumps there. Her third foal, Airgator (a colt by Dewan), made 22,000 dollars as a yearling, but her fourth produce, a filly by Stop the Music, reaped the reward of Henbit's success and fetched 260,000 dollars as a yearling.

Chateaucreek was by the 1963 Kentucky Derby and Belmont Stakes winner Chateaugay out of the unraced Sailor mare Mooncreek, who bred five winners before being sold for 18,000 dollars in 1976 and sent to Venezuela. Mooncreek's dam Ouija introduces the first piece of class into the female line, since she was a stakes winner whose six successes included the Diana Handicap, and also finished second in the Monmouth Oaks. Ouija also bred Ouija Board (National Stallion Stakes and two other races, and second ten times from 19 starts) and Sailor's Hunch, who bred Limit To Reason (Champagne Stakes and Pimlico-Laurel Futurity Stakes) and Box the Compass (Sorority Stakes and National Stallion Stakes). Ouija's grandam Handcuff was the champion three-year-old filly of 1938, when her wins included the Alabama Stakes and Acorn Stakes and she was second in the CCA Oaks. The family was introduced to America in 1917 through the import as a yearling of Handcuff's third dam Kiss Again, a half-sister to the Oaks winner Straitlace. It took until 1980 and Henbit to bring in another English Classic winner.

1981

1 H.H. Aga Khan's b c Shergar (Great Nephew – Sharmeen).
2 P. Mellon's b c Glint of Gold (Mill Reef – Crown Treasure).
3 K. Dodson's b c Scintillating Air (Sparkler – Chantal).
   18 ran. Time: 2 min 44.21 sec

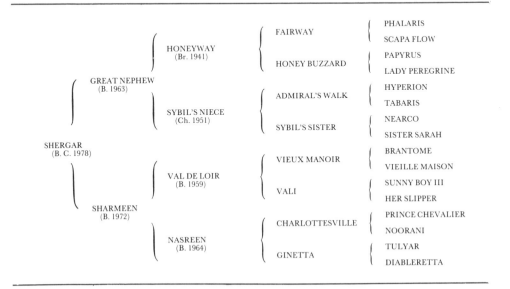

For the first time since 1952 the colours of the Aga Khan were carried to success in the Derby, though those worn by Shergar were slightly different from those successful five times for the grandfather of Shergar's owner. The familiar chocolate and green stripes were replaced by a green jacket with red epaulets when the Aga Khan revived the family tradition of having horses trained in England. Those are academic considerations; the important point was that a horse owned and bred by the Aga Khan had again won the Derby, and it emanated from the superb breeding that had done so well for so many years. By coincidence, the Aga Khan came within an ace of completing the English–French Derby double, since his Akarad was second in Paris, but in contrast to Shergar, Akarad was a product of breeding cultivated by M. Marcel Boussac and taken over on his bankruptcy in the late 1970s.

Shergar's winning margin of ten lengths—following victories by ten lengths in the Guardian Newspaper Classic Trial at Sandown and by 12 lengths in the Chester Vase—was the largest since Manna won by eight lengths in 1925, though Troy won by seven lengths only two years previously. Shergar recorded the slowest time since Airborne in 1946, the going being officially recorded as "good to soft", an unusual occurrence for this race.

As a two-year-old Shergar had been rated at 8st 9lb, 12lb behind Storm Bird, the top horse in the newly-extended European Free Handicap. Shergar had had only two races, both over a mile, winning a 23-runner race at Newbury on his debut and being beaten two and a half lengths into second place behind Beldale Flutter in the William Hill Futurity.

He was never entered for the 2,000 Guineas, and after achieving his objective in the Derby, he went on to win the Irish Sweeps Derby (by a margin of four lengths over Cut Above that owed much to the restraint of his jockey Piggott) and the King George VI and Queen Elizabeth Diamond Stakes (by four lengths from Madam Gay). Shergar had exercised comprehensive midsummer superiority over his middle-distance contemporaries, and the only horse of his own age who took him on at Ascot was the filly Madam Gay, winner of the French Oaks. With Master Willie seeming to run below form and Pelerin failing to fight back after a brush with the rails at halfway, Shergar might have had less to do than was at first thought, but he won with authority, fought clear when sent on over a furlong out, and could claim to have beaten his elders. He was hailed a champion, a horse in a million, a colt to compare with anything that had won the Derby since the War; he might have been overestimated in the heat of excitement.

In the week before his first race against horses other than those of his own age, the syndication of Shergar to stand at his owner's stud in Ireland from 1982 was finalized. The Aga Khan retained six shares and offered 34 for sale at £250,000 each, placing on Shergar a valuation of £10 million. It was the highest ever asked for a Derby winner but was below what had been offered from America. The number of European bids received was twice what could be accommodated. It is anybody's guess what the reaction would have been if syndication had been delayed until later in the year, when Shergar was well beaten into fourth place behind Cut Above, Glint of Gold and Bustomi in the St. Leger. Until then, Shergar's six wins from seven races had been worth £392,245. He did not race again.

Shergar was the second Derby winner sired by Great Nephew, following Grundy. Indeed, Grundy marked a significant turning point in the fortunes of Great Nephew, whose efforts reached another peak in 1979 with the Italian 2,000 Guineas success of Good Times, and were followed in 1980 with the victories of Nikoli (Irish 2,000 Guineas) and Mrs Penny (French Oaks and Prix Vermeille).

Though Shergar traces to the great Mumtaz Mahal (his seventh dam), the immediate female side of his pedigree is an influence for stamina. His dam Sharmeen is by the 1962 French Derby winner Val de Loir out of a mare by another French Derby winner, Charlottesville. Sharmeen was undistinguished on the racecourse, winning one race, a maiden event over an extended ten furlongs, as a three-year-old. Her first foal was a Kalamoun filly called Shaiyneen, who won over seven and nine furlongs in France as a three-year-old; her second foal was Shergar.

Nasreen, the dam of Sharmeen, won over 12½ furlongs but had limited ability. At stud she bred six winners, the best being Nassiri (Prix Greffulhe and second in Prix Hocquart and Grand Prix de Deauville). Nasreen was a sister to Grand Roi, winner of four races in France before he was retired to stud in England with modest results.

Ginetta, dam of Nasreen, is the closest influence for speed on the distaff side of Shergar's pedigree, since her six victories included the French 1,000 Guineas and the Prix du Moulin. Ginetta, who was also second in the five-furlong Prix de l'Abbaye, died in 1965 at the age of nine after producing only four foals. Three of Ginetta's produce were winners; the other, the Aureole filly Jasseem, did not race. Ginetta was easily the best European-raced produce of her dam Diableretta, who was exported to America as a ten-year-old. Diableretta was a precociously speedy two-year-old who won seven races in a row including the Queen Mary Stakes and July Stakes, but though she won one of her two races as a three-year-old, she did not truly hold her form and was retired to stud. Diableretta was closely related to Nasrullah since Nearco sired both Nasrullah and Dante, sire of Diableretta, and Mumtaz Begum was the dam of both Nasrullah and Diableretta.

DERBY UPDATE

PINZA

Pinza died late in 1977 at the age of 27. His last Flat-race winner in Britain and Ireland was Pin Cry, who at the age of six won a two-mile maiden event at Roscommon in September 1975.

NEVER SAY DIE

Never Say Die had his last success on the Flat with Man of Vision, who was awarded a mile race at Kempton Park on a disqualification in April 1979. Man of Vision went on to be fourth in the Prix Lupin and second to Ela-Mana-Mou in the King Edward VII Stakes before being sold for £250,000 to stand at stud in Japan.

RELKO

Relko produced his best English-trained produce in the shape of Relkino, second in the 1976 Derby who came into his own the following year by winning the Benson and Hedges Gold Cup. Olwyn's surprise victory in the Irish Guinness Oaks in 1977 helped put Relko into fourth place in the British list of leading sires. Relko's later successes were Relfo (1978 Ribblesdale Stakes and second in Prix Vermeille); Pragmatic (1979 Yorkshire Cup, 1980 Sagaro Stakes), and Idelka (1981 Gran Premio Citta di Napoli).

Relko's three-parts brother Reliance II was put down in August 1979 suffering from the effects of chronic sinus infection. At his death his ten crops had realized 350 wins for over £900,000. He was the maternal grandsire of the tip-top miler Kris, and his sons Tug of War (Northumberland Plate twice, Goodwood Cup) and Rymer (Brigadier Gerard Stakes) entered stud in 1979.

SEA-BIRD II

An outstanding Derby winner deserved to have an outstanding racer as his last product in Britain, and Sea-Bird II got it in Sea Pigeon, a remarkable horse who ran seventh in Morston's Derby, was then bought privately by Mr. Pat Muldoon for not much more than £10,000, and proceeded to get better and better, running first over hurdles and then returning with renewed vigour to Flat racing.

By the end of the 1980 Flat season Sea Pigeon had won £96,985 in first-prize money to beat Boldboy's record for a gelding, with 16 wins (one as a two-year-old before he was gelded), including the Vaux Gold Tankard three times, the Ebor

Handicap and the Chester Cup. And in 1981 he won his second Champion Hurdle in a row to become the highest money-earner in National Hunt history, with £130,395 from 21 wins.

CHARLOTTOWN

Charlottown was put down in August 1979 after an accident in Australia, where he had three seasons at stud. Two months later, his sire Charlottesville's half-brother Sheshoon was put down as a result of deteriorating health.

ROYAL PALACE

The disappointing stud career of Royal Palace was reflected in a drop in his stud fee from 5,000 guineas in 1972 to only 1,000 guineas in 1976. The slide was temporarily halted in 1977, when he suddenly jumped to second place in the leading sires' list, thanks to the exploits of Dunfermline and Royal Hive. Royal Palace trailed Northern Dancer by the respectful distance of £349,739 to £163,223, but got there largely because of the three wins worth £105,876 by the Queen's filly Dunfermline, who in the year of her owner's Silver Jubilee rose to the occasion by winning the Oaks and St. Leger, the first to do so since Meld in 1955. Royal Hive won the Park Hill Stakes.

This upsurge in Royal Palace's fortunes was not sustained, and in June 1980 his owners made a present of him to the National Stud. He was dispatched to Northumberland, and his stud fee was further reduced to 500 guineas.

Royal Palace's sire Ballymoss died after a heart attack in July 1979, at the age of 25, and his dam Crystal Palace was put down in November 1980, at the age of 24, after she had cut herself badly in a paddock accident. Crystal Palace had already been retired from stud duties. She bred one more winner to add to those already discussed, the fairly useful miler handicapper Owen Jones (by Tudor Melody), who chipped a sesamoid bone as a two-year-old and had his career held up, and jarred the same bone as a four-year-old and had his prospects so seriously hampered that he was sold for only 2,100 guineas at the end of his racing days.

Crystal Palace, who died the dam of eight winners, came back into the Classic picture through the produce of her daughter Glass Slipper, who bred Light Cavalry (by Brigadier Gerard), winner of five races for £118,692, including the St. Leger, King Edward VII Stakes and Princess of Wales's Stakes before being sold to America in mid-1981; and Fairy Footsteps (by Mill Reef), winner of three races for £69,212, including the 1000 Guineas, but retired before the Oaks.

SIR IVOR

The view that Sir Ivor's fillies were more outstanding than his colts was strengthened by four horses that ran in the colours of Mr. Robert Sangster—Lady Capulet (winner of Irish 1,000 Guineas on her racecourse debut but retired to stud after only two more runs); Turkish Treasure (out of a half-sister to Lady Capulet; won three races at two years, including Cherry Hinton Stakes, and sent to stud immediately); Godetia (Irish 1,000 Guineas and Oaks; sold for one million dollars, after costing 60,000 dollars as a yearling); and Calandre (won Pretty Polly Stakes at The Curragh).

Balanced against these in the years 1977 to 1981 were two colts of decent quality—Super Asset (Horris Hill Stakes) and Gielgud (Champagne Stakes; exported to Australia at three years).

BLAKENEY

Set Free, the first mare covered by Blakeney and dam of the dual Oaks winner Juliette Marny as a result, continued to play an important part in his story. The mating was repeated in 1974; a colt was born in 1975, and in 1978 Julio Mariner won the St. Leger. Several mares have bred two Classic winners, but only two had previously bred a brother and sister successful in such races this century. They were Scapa Flow (dam of the St. Leger winner Fairway and the 1,000 Guineas winner Fair Isle) and Ranai (dam of the Derby winner Watling Street and the 2,000 Guineas-winning filly Garden Path). Set Free was soon to put herself even further forward into the breeding spotlight, when in 1979 her daughter Scintillate (by Sparkler) won the Oaks. Set Free thus became the seventh mare to breed three British Classic winners and the first since Araucaria, who was born in 1862, 102 years before Set Free. The other mares on this illustrious list are Penelope (born 1798), Pope Joan (1809), Fillagree (1815), mare by Rubens out of Tippety-Witchet (1819) and Cobweb (1821).

The Blakeney–Set Free mating was used again three years in a row after Scintillate was born, resulting in Saviour (fourth in the St. Leger, third in the Great Voltigeur Stakes and winner of a small maiden race before being sent to the Far East); Adamson (sold for 175,000 guineas as a yearling and an expensive purchase on the evidence of his first and second seasons) and a colt who went into training as a two-year-old in 1981.

In the meantime, Blakeney had done enough to show he was not necessarily a "one-mare" stallion. Far from it, he established himself as the type to get winners and plenty of them. In 1979, for instance, he had 26 individual winners of 35 races, the best being Sexton Blake, who in three seasons won three races for £24,252, including the Champagne Stakes and Seaton Delaval Stakes at two, the Gordon Stakes at three and the Westbury Stakes at four. Another 1979 winner sired by Blakeney was Tyrnavos, who in 1980 won the Irish Sweeps Derby and Craven Stakes for £151,867 to help his sire into third place in the leading money-earners' list. Tyrnavos, Blakeney's third Classic winner, was also from a prepotent mare, Stilvi, whose first five foals were: Tachypous (two wins, £35,462, including Middle Park Stakes; second in 2,000 Guineas); Taxiarchos (three wins, £12,490; third in Craven Stakes); Tromos (two wins, £36,894, including Dewhurst Stakes, leading two-year-old of 1978; second in Craven Stakes); Tyrnavos; and Tolmi (three wins, £26,409, including Coronation Stakes; second in 1,000 Guineas).

NIJINSKY

The bright start that Nijinsky made at stud continued in 1977 and flourished even brighter the following year. In 1977 he was represented by Bright Finish (Yorkshire Cup), Cherry Hinton (the season's leading staying two-year-old filly), Lucky Sovereign (whose sole win in a 15-race career was in the Mecca-Dante Stakes, and who was near top-class when in the mood but unbelievably bad when not), and Valinsky (Geoffrey Freer Stakes and second in the Grand Prix de Paris).

In 1978 it was the turn of Ile de Bourbon, whose career total was five wins for £169,694. He was well beaten in the only Classic he contested, the St. Leger, but at three years he won the King George VI and Queen Elizabeth Diamond Stakes, King Edward VII Stakes and Geoffrey Freer Stakes, and at four won the Coronation Cup by seven lengths before being syndicated for £4 million. The one sadness was that he marked the end of the Engelhard racing empire. He was the product of two of the best Engelhard runners (Nijinsky and Roseliere), but when he was sold to an English-based syndicate, the Engelhard colours were to be seen no more.

Nijinsky was credited with his third Classic winner in 1979, when Niniski, third in the English St. Leger, went on to win the Irish and French equivalents, the latter in the first year the race was opened to four-year-olds and upwards. At four years Niniski won the John Porter Stakes and Ormonde Stakes. Also in 1979, Shining Finish gained five wins, the last in the St. Simon Stakes before joining his brother Bright Finish at stud in Australia, and the filly Princess Lida won the Prix Morny, before running third in the 1980 French 1,000 Guineas.

In 1980 Night Alert, who had won the previous year's Houghton Stakes, won the Prix Jean Prat and the Gladness Stakes before returning to his native America, and in 1981 Nijinsky was represented by Leap Lively, the Lingfield Oaks Trial winner, and Kings Lake, who got the Irish 2,000 Guineas on a disqualification but needed no such intervention to win the Sussex Stakes and Joe McGrath Memorial Stakes.

The popularity of Nijinsky goes on undimmed, though he remains in America and his biggest conquests continue to come in Europe. His stud fee for 1980 was 100,000 dollars, and the 14 yearlings by him which changed hands at the major Keeneland Sales in Kentucky in July 1981 averaged 349,286 dollars.

MILL REEF

To update the record of Million, Mill Reef's record-breaking yearling sold in 1975 for 202,000 guineas, he raced for three seasons and won two races—over a mile and a half at Newmarket for £2,004, and over two miles at Haydock Park for £2,372—but was out of his depth against top-class horses, wore blinkers three times, and as a five-year-old was exported to Wagga Wagga in Australia. For the sake of his sire, there were others who put the record straight, notably Shirley Heights, who helped put Mill Reef into top spot in the sires' list in 1978.

Mill Reef's best winners in 1979 were Milford (three wins including Lingfield Derby Trial and Princess of Wales's Stakes) and Main Reef (who repeated his feat of three wins at two years with victories that included the Cumberland Lodge Stakes and St. Simon Stakes). By his new standards, 1980 was a modest year, but in 1981 he was represented by two more Classic winners—Fairy Footsteps (1,000 Guineas) and Glint of Gold (Italian Derby, Grand Prix de Paris and Preis von Europa). Glint of Gold also finished second in the Derby and St. Leger. As an illustration of Mill Reef's popularity, the nine yearlings by him sold at the Newmarket Premier Sales in 1981 averaged 228,000 guineas, and included another record-breaker, a colt out of Arkadina (by Ribot), who fetched 640,000 guineas.

ROBERTO

Roberto made a marvellous start to his stud career, being the leading first-season sire in Britain in 1977, when his four individual winners of nine races included Sookera (three wins including Cheveley Park Stakes) and Octavo (three wins from four races, including Athford Castle Stakes, and third-best in Irish Free Handicap). In America Roberto's winners included Pirateer (To Market Stakes) and Darby Creek Road (Saratoga Special Stakes).

The next two years were unremarkable, but in 1980 came the exploits of the two-year-olds Robellino (four wins, £43,326, including Royal Lodge Stakes) and Critique (short-head second in Grand Criterium), as well as the champion Canadian three-year-old Driving Home, whose victory in the Queen's Plate gave Roberto his first Classic success as a sire. There was also Fool's Prayer, who won four races in England before being exported in 1980 to America, where his first six wins earned 193,076 dollars.

MORSTON

The earliest indications are that Morston will be an influence for stamina. His first crop included Whitstead, winner of the Great Voltigeur Stakes and Sandown Classic Trial, and third in the Grand Prix de Paris. His second crop included More Light, winner of the Gordon Stakes and Jockey Club Stakes before being exported to Argentina in 1980.

SNOW KNIGHT

Snow Knight had his first two-year-olds running in 1980, and they provided one winner in Britain, Pellegrini, who took the Fenwolf Stakes. That same year, in August, Snow Knight's sire Firestreak died at the age of 24 after being in semi-retirement. Firestreak's male line in Britain depends on Hotfoot, who himself has three sons at stud—Free State, Hot Grove and Tachypous.

GRUNDY

Though Grundy had his first runners in 1979, with four winners, it was the following year, when his first crop were three-years-old, that brought recognition of his quality and high expectations for his future. Bireme brought him Classic success at the first time of asking, winning the Oaks among her three successes from only four outings for £88,301. She was injured when getting loose on the roads at home soon after the Oaks and did not run again, but she had laid the foundation for interest in Grundy at the 1980 yearling sales, where only nine stallions with ten or more entries at public auction made a greater average than Grundy's 29,525 guineas from 16 lots. The highest price, 100,000 guineas, fell way short of the 264,000 guineas that his son Mushoor fetched as the highest-priced yearling sold in Britain in 1978, but it was enough to keep Grundy above all but Artaius and Be My Guest among the unproven stallions. The leading stallion averages for 1980 were: Mill Reef 73,900 guineas; Northfields 59,323 guineas; Habitat 53,659 guineas; Be My Guest 39,626 guineas; Nonoalco 35,887 guineas; Busted 34,612 guineas; Blakeney 34,064 guineas; High Top 32,360 guineas; Artaius 31,791 guineas; Grundy 29,525 guineas.

In 1981 Grundy was represented by the three-year-old Kirtling, winner of the Dee Stakes and Gran Premio d'Italia, and the two-year-old Zilos, winner of the seven-furlong Seaton Delaval Stakes.

Grundy's dam Word From Lundy died of a twisted gut in April 1980, two days after foaling a brother to Grundy and Centurius (her 1978 foal who held the European yearling record for less than 72 hours in 1979, when sold for 270,000 guineas). Her orphan foal was her eighth. Those not already discussed were the Habitat colt Marisco, who made 60,000 guineas as a yearling and was sent to Italy after two unsuccessful outings in England as a three-year-old; Centurius, who showed plenty of merit in winning the Blue Riband Trial at Epsom and finishing second though disqualified in the King Edward VII Stakes, but was not an easy ride: and the Blakeney colt Chronicle, who made 122,000 guineas as a yearling and won on his two-year-old debut in Ireland.

Little Mark, a sister to Grundy's fourth dam Waterval, made sure of a small share of the limelight in 1981, as the fourth dam of Aldaniti, the Grand National winner.

EMPERY
The first yearlings by Empery came up for sale in Britain in 1979, and the five that changed hands realized an average of 33,880 guineas, with the most expensive fetching 70,000 guineas. The following year the three for sale made a less impressive average of 11,500 guineas. His son Paradis Terrestre was narrowly beaten into second place in the William Hill Futurity in 1981, after making a winning debut in the Hyperion Stakes at Ascot.

Empery's sire Vaguely Noble joined Nijinsky in 1980 as being rated at a stud fee of 100,000 dollars, being raised from 80,000 dollars the previous year. That year he gained Classic success with the Irish St. Leger victory of Gonzales, who at 750,000 dollars was the fourth most expensive yearling ever sold at that time, and subsequently highest-priced horse to win a Classic. Both achievements were loosely-held at a time when prices go ever higher and higher. Gonzales is a brother to Mississippian, who made little impact in his early years at stud in America.

THE MINSTREL
Four yearlings by The Minstrel sold in Britain in 1980, the first year they appeared on the market, made an impressive total of 692,000 guineas, including a half-brother to Robellino which fetched 350,000 guineas, the second highest price of the year. The Minstrel had his first runners in 1981 and they included Peterhof, winner of the Curragh Stakes and Flying Childers Stakes in the same colours carried by his sire. The Minstrel's two-year-old brother, Pilgrim, won on his debut in Ireland late in 1981.

CHAPTER 31

Some Conclusions Concerning Derby Winners 1900 – 1976

TIMES MAY CHANGE and fashions may alter but the attraction of the Derby seems unshakeable. There can be little doubt that the dream of all bloodstock breeders in Britain and several throughout the rest of the world includes winning the Derby at Epsom. It might no longer be true that the Derby winner is the most prized thoroughbred in the world, but it remains a fact that few races bestow a bigger increase in valuation on a horse as soon as he has passed the winning post than the English Derby.

Whether one agrees with this situation is immaterial; and the commercial breeder ignores the fact at his peril. The breeder who races his own stock can swim against the tide if he so wishes, but he too would be advised not to overlook the more general implications of Derby-winning pedigrees.

If we review the Derby winners of this century, the first impression we get is that they are bred in a heterogeneous manner. Some are bred this way and some that way, but is there anything in common in their ancestral make-up? Bearing in mind that during the period under review there have been an average 500 stallions at stud each season in Britain and Ireland alone, it is remarkable that the names of only about a couple of dozen of these are continually appearing in the pedigrees of Derby winners. The influence of St. Simon, Cyllene, Galopin, Phalaris, Vatout, Chaucer, Blandford, Pharos, Gainsborough, Hurry On, Polymelus, Hyperion, Fairway, Prince Rose, Nearco, etc., is manifest time and time again. Sometimes they appear in the tail male lines, at others in one of the quarterings of the grandparents or great-grandparents, etc., but only in the cases of Spearmint and his son Spion Kop is one or other absent.

We must conclude therefore that a small band of sires have shown a capacity for transmitting racing merit to their stock vastly superior to the other sires of their time. If we turn to the winners of the other Classics, or races of comparable class, the same names in varying incidence will be encountered in their ancestral trees. It is true that other names also occur and re-occur, but not with the same frequency.

It is often misleading to classify stallions by groups based on the amount of stakes or even class of races won during their Turf careers. These results are frequently dependent on the strength or weakness of the opposition, and in either case may bear no reflection to the winner's merit. As the real value of money has fallen in post-war years, the amount of prize money has risen dramatically, so the Derby won by Lavandin in 1956 was worth £17,982 to his owner, while the same race won by Shergar in 1981 was worth fractionally short of £150,000. Shergar would almost certainly be regarded as a better horse than Lavandin, but not to the extent that the figures alone might suggest. Meanwhile, the Derby second of one year may be a better racer than the Derby winner of another year. For instance, there are grounds for regarding Meadow Court, Gyr and Linden Tree the equal of, if not better than half a dozen Derby winners in the last 20 years, after they had run second to Sea-Bird II, Nijinsky and Mill Reef; but while the winners went on to stud fame, the runners-up have sunk almost into obscurity.

Nevertheless, in order to attempt to investigate how the Derby winners of this century have been bred, it is necessary to make some kind of classification, using as a base the winner's sire. The following attempts should provide food for thought; whether they provide the basis for anything further is left to the reader.

I have divided the Derby winners' sires into three classes, depending on races which they won or in which they were placed. Originally Class One was confined to stallions who themselves had won an English or French Classic, the Ascot Gold Cup or the Grand Prix de Paris; Class Two was for stallions placed in one or more of those races; and Class Three was for other stallions. The Ascot Gold Cup and Grand Prix de Paris were included with the Classics because some horses gained distinction in these races but did not run for the Classics, although they were undoubtedly the best of their age. Cyllene and St. Simon are two examples of this.

Under these conditions, the statistics for Derby winners up to 1955 showed that of 56 winners, 36 were sired by Class One stallions. It was only fair to add Spearmint to this classification as his sire Carbine raced in the Antipodes and proved himself fully up to the Class One standard. This brought the successes achieved by Class One up to 37, or almost double the number recorded by those of Class Two and Class Three combined, which added up to 19. Of the remainder, eight were sired by Class Two stallions, but it was right to increase their number by the addition of Orby, whose sire Orme could hardly have escaped being placed, and would probably have won a Classic, had he run. This left ten stallions in Class Three, as follows:

YEAR	WINNER	SIRE	SIRE'S RECORD
1901	Volodyovski	Florizel II	Goodwood Cup winner; high-class handicapper. Brother to two Classic winners
1902	Ard Patrick	St. Florian	Very moderate racer
1908	Signorinetta	Chaleureux	Cesarewitch winner
1911	Sunstar	Sundridge	Best sprinter of his time
1913	Aboyeur	Desmond	Far from top-class racer
1925	Manna	Phalaris	Best sprinter of his time
1929	Trigo		
1930	Blenheim	Blandford	Not run in the Classics, but probably up to Classic form.
1934	Windsor Lad		
1935	Bahram		

Of these ten, it was fair to say that seven were by very good racehorses, and three—Ard Patrick, Signorinetta and Aboyeur—by indifferent ones.

Since 1955 a change of emphasis can be detected and has been commented upon at various stages in this narrative. The trend away from staying races, while to be deplored, must be accepted, and the greater emphasis placed on middle-distances must be recognized. The Ascot Gold Cup and the Grand Prix de Paris no longer carry such prestige as they did, at least not among those most closely concerned with attempting to breed top-class horses, and the King George VI and Queen Elizabeth Stakes and Prix d l'Arc de Triomphe have largely taken their place in stature. Account must be taken of these changes when assessing a sire's racing performance, and consequently these two events have been added to the list of qualifying races used previously. Furthermore, the incidence of American breeding on the European scene since the mid-1960s must also be recognized, and to the English and French Classics must be added the American equivalents.

Two stallions which would otherwise have been in Class Three move into Class One as a result of these revised conditions—Northern Dancer (winner of the Kentucky Derby and Preakness Stakes); and Vaguely Noble (who was not entered for the Classics but won the Arc de Triomphe). Aureole moves into Class One from Class Two by virtue of his victory in the King George VI and Queen Elizabeth Stakes, and Never Bend goes into Class Two for having been placed in the Kentucky Derby and Preakness Stakes. Sir Gaylord, ante-post favourite for the Kentucky Derby before he broke down, would probably qualify for at least Class Two, but he has been retained in Class Three on the grounds of strict interpretation.

The new set of figures, for the 77 Derby winners between 1900 and 1976, shows that 47 were by Class One stallions, 13 by Class Two and 17 by Class Three. The figures have levelled out slightly since 1956 but the balance remains heavily weighted towards Class One, going from 37–19 in their favour in 1956 to 47–30 in 1976. The breakdown of stallions between 1956 and 1976 is as follows:

CLASS ONE		CLASS TWO	CLASS THREE
Verso II	Aureole	Donatello II	Hard Sauce
Never Say Die	Charlottesville	Tanerko	Persian Gulf
Chamossaire	Northern Dancer	Dan Cupid	Pardal
Ballymoss	Ragusa	Never Bend	Sir Gaylord
Hethersett	Vaguely Noble		Hail To Reason
			Firestreak
			Great Nephew

The question now arises as to whether these results show that Class One stallions are themselves better sires of Derby winners than those of Classes Two and Three, or whether their success is attributable to being mated with superior mares. To sort out the answer, we must consider the various classes of sires as teams and not as individuals.

Are there annually more potential dams of Derby winners dotted about among the mates of all Class One stallions at stud than among all the Class Two and Three stallions? There are incomparably more Class Two and Three horses at stud than Class One, and I venture the opinion that between them, although more scattered, they get as many good mares as Class One sires. Apart from the difficulty of defining which mares are, and which mares are not potential dams of Derby winners, I have no statistics to work on—and they would probably be available only after considerable study, not to mention assistance from the most elaborate computer—so I am expressing purely a personal opinion in a broad way.

Allowing for the shortcomings of the investigation, it is hard for me to believe that the Class One team gets over one and a half times the number of high-grade mares as the combined Class Two and Three teams, yet Class One stallions have sired over one and a half times the number of Derby winners this century.

Even though some of the Class One stallions may have appeared to have been fortunate winners of one of the races named above, I can only conclude that they are, in fact, markedly better sires of Derby winners than those in Class Two and Class Three. In my notes about stallions, I have stressed the importance of their possession of racing merit, and I consider that while the Derby results must not be taken in isolation, they are but one of several proofs of this.

I now come to the question of which parents have had the greatest influence in the build-up of these Derby winners. A straightforward addition of the number of times the various names appear in these pedigrees would lead to a false conclusion. A grandparent or great-grandparent has not the same influence as a parent. I have therefore worked out the matter on a points system, counting eight points for a parent, four for a grandparent, two for a great-grandparent, and one for a great-great-grandparent. I have ignored strains beyond the fourth generation, and when a name occurs twice or more in the same pedigree, I have awarded the full entitlement of points according to its position. I have then added up the total number of points earned by each of the horses named in the pedigrees, and have arranged them in order, to form a table of incidence. To give a comparison with earlier research, the table reproduced below also includes positions which obtained in 1956.

STALLION AND YEAR FOALED	1900–76 PTS PLACE		1900–55 PLACE PTS		CLASS—ON RACING MERIT	DEGREE OF INBREEDING
St. Simon (1881)	72	I	I	72	I	4th remove, to Voltaire
Cyllene (1895)	64	2	2	64	I	4th remove, to Stockwell
Polymelus (1902)	52	3	3	52	2	Outbred
Nearco (1935)	48	4	11	28	I	4th remove, to St. Simon
Galopin (1872)	47	5	4	47	I	3rd remove, to Voltaire
Blandford (1919)	46	6	5	41	3	4th remove, to Isonomy
Phalaris (1913)	45	7	6	35	3	3rd and 4th remove, to Springfield

STALLION AND YEAR FOALED	1900–76 PTS	PLACE	1900–55 PLACE	PTS	CLASS—ON RACING MERIT	DEGREE OF INBREEDING
Pharos (1920)	41	8	10	30	2	3rd and 4th remove, to St. Simon
Gainsborough (1915)	40	9	8	32	1	3rd and 4th remove, to Galopin
Hurry On (1913)	37	10	8	32	1	4th remove, to Hermit
Bayardo (1906)	35	11	7	33	1	2nd and 4th remove, to Galopin; 4th remove, to Sterling
Swynford (1907)	30	12	11	28	1	Outbred
Hyperion (1930)	27	13	—	8	1	3rd and 4th remove, to St. Simon
Vatout (1926)	26	14	16	22	1	3rd and 4th remove, to Gallinule
Chaucer (1900)	25	15	13	24	3	Outbred
Spearmint (1903)	24	16	14	23	1	4th remove, to Stockwell
Fairway (1925)	24	16	19	20	1	3rd and 4th remove, to Stockwell
Marcovil (1903)	23	18	17	21	3	3rd remove, to Hermit
Solario (1922)	23	18	—	18	1	4th remove, to Hampton
Hampton (1872)	23	18	14	23	3	Outbred
Blenheim (1927)	21	21	—	12	1	4th remove, to Isinglass
Carbine (1885)	21	21	17	21	1	3rd and 4th remove, to Black Bess
Nasrullah (1940)	20	23	—	8	2	Outbred
Sundridge (1898)	20	23	19	20	3	4th remove, to both Stockwell and Newminster
Vatellor (1933)	20	23	19	20	2	Outbred

Total number of points:	1900–55		1900–76	
Class One stallions	460	(63%)	539	(63%)
Class Two ,, ...	133	(16%)	110	(15%)
Class Three ,, ...	182	(21%)	164	(22%)

The total number of points attributed to each team of stallions, while showing a remarkable consistency in breakdown between groups from the original research in 1956 to the present day, again draws attention to the superior results achieved by stallions with Class One racing ability.

A noteworthy item in the list is the very prominent position held by Cyllene, who spent only nine seasons at stud in England before export to Argentina, whereas St. Simon held court at home for 22 years. The total number of Cyllene's English-sired stock was comparatively small, but their influence was great.

That St. Simon, Cyllene and Polymelus should retain their positions in the table is interesting. That Nearco should force his way into the top four is remarkable, since the competition he has faced has been much stronger—in terms of numbers—a fact that can be gauged from the dates of foaling of the sires listed above. The next youngest stallion to Nearco (born 1935) in the top dozen is Pharos, 15 years Nearco's senior. Out of the leading 12, those nearest in age to Nearco are Hyperion (born 1930), his own son Nasrullah (1940) and Vatellor (1933). Whether Nearco can emulate St. Simon in this field remains to be seen.

If calculations are confined to the 41 Derby winners from 1936 to 1976, a useful picture emerges showing the most important progenitors of recent years, confirming the rise of Nearco and the promise of his son Nasrullah.

STALLION AND YEAR FOALED	PTS	CLASS, ON RACING MERIT
Nearco (1935)	48	1
Pharos (1930)	33	2
Phalaris (1913)	31	3
Gainsborough (1915)	27	1
Hyperion (1930)	27	1
Vatout (1926)	26	1
Fairway (1925)	24	1
Solario (1922)	23	1
Blenheim (1927)	21	1
Vatellor (1933)	20	2
Nasrullah (1940)	20	2

There are exactly 300 points at stake in this calculation, and Class One stallions account for 196 (or 65⅓ per cent), Class Two 73 (24⅓ per cent) and Class Three 31 (10⅓ per cent). Again Class One stallions are shown securing better results than the combined efforts of Class Two and Class Three, in almost exactly the same ratio as was established in the overall list for 1900 to 1976, that is something over 6:4.

It should be reiterated that both Fairway and Pharos were by Phalaris; that Pharos sired Nearco, who in turn sired Nasrullah; that Vatout got Vatellor; and that Gainsborough sired Solario and Hyperion. It will therefore be clear that the most successful blood for the production of recent Derby winners is interrelated to a remarkable degree.

Both Fairway and Pharos were out of the Chaucer mare Scapa Flow, who thus takes high position in a table compiled by extending the above points system to broodmares alone. The table is as follows:

MARE AND YEAR FOALED	POINTS	DAM OF . . .
Arcadia (1887)	31	Cyllene
St. Angela (1865)	31	St. Simon
Scapa Flow (1914)	27	Pharos, Fairway
Blanche (1912)	25	Blandford
Rosedrop (1907)	25	Gainsborough
Nogara (1928)	22	Nearco
Selene (1919)	22	Hyperion
Canterbury Pilgrim (1893)	21	Swynford, Chaucer
Maid Marian (1886)	20	Polymelus
Plucky Liege (1912)	19	Bois Roussel
Bromus (1905)	18	Phalaris
Tout Suite (1904)	18	Hurry On
Malva (1919)	17	Blenheim
Galicia (1898)	17	Bayardo, Lemberg
Windmill Girl (1961)	16	Blakeney, Morston

In Chapter 2 I compared the percentage of inbred Derby winners of this century with an estimate of the normal proportion of thoroughbreds bred annually to corresponding degrees of inbreeding. I also included a similar calculation in respect of the 25 most influential sires in the production of these Derby winners. The names and particulars of inbreeding of the latter will be found in the three tables detailed above. I will now list the inbred and outbred Derby winners from 1900 to 1976.

OUTBREDS (Name and year foaled)

Diamond Jubilee (1897)	Bahram (1932)	St. Paddy (1957)
Volodyovski (1898)	Mahmoud (1933)	Relko (1960)
Ard Patrick (1899)	Mid-day Sun (1934)	Santa Claus (1961)
Cicero (1902)	Pont l'Eveque (1937)	Sea-Bird II (1962)
Signorinetta (1905)	Ocean Swell (1941)	Charlottown (1963)
Sunstar (1908)	Dante (1942)	Royal Palace (1964)
Aboyeur (1910)	Airborne (1943)	Sir Ivor (1965)
Durbar II (1911)	Pearl Diver (1944)	Nijinsky (1967)
Humorist (1918)	Arctic Prince (1948)	Mill Reef (1968)
Captain Cuttle (1919)	Never Say Die (1951)	Morston (1970)
Coronach (1923)	Lavandin (1953)	Snow Knight (1971)
Call Boy (1924)	Crepello (1954)	Grundy (1972)
Trigo (1925)	Hard Ridden (1955)	Empery (1973)
Windsor Lad (1931)	Parthia (1956)	

Total: 41 out of 77 (53.2%)
Estimated percentage of outbreds born yearly is 56.8%

INBRED AT THE FOURTH REMOVE

NAME AND YEAR FOALED	COMMON ANCESTOR AT FOURTH REMOVE, AND INCIDENCE	NAME AND YEAR FOALED	COMMON ANCESTOR AT FOURTH REMOVE, AND INCIDENCE
Rock Sand (1900)	Stockwell (3)	Blue Peter (1936)	St. Simon (2)
St. Amant (1901)	King Tom (2)	Straight Deal (1940)	St. Frusquin (2)
Spearmint (1903)	Stockwell (2)	Nimbus (1946)	Bromus (2)

INBRED AT THE FOURTH REMOVE (*contd.*)

NAME AND YEAR FOALED	COMMON ANCESTOR AT FOURTH REMOVE, AND INCIDENCE	NAME AND YEAR FOALED	COMMON ANCESTOR AT FOURTH REMOVE, AND INCIDENCE
Orby (1904)	Doncaster (2)	Galcador (1947)	Bayardo (2)
Minoru (1906)	Sterling (2)	Tulyar (1949)	Phalaris (2)
Fifinella (1913)	Isonomy (2)	Pinza (1950)	Blandford (2)
Grand Parade (1916)	Galopin (3) St. Angela (2)	Psidium (1958)	Phalaris (2)
		Larkspur (1959)	Vatout (2)
Spion Kop (1917)	Skirmisher Mare (2)	Blakeney (1966)	Nearco (2)
Blenheim (1927)	Isinglass (2)	Roberto (1969)	Nearco (2) and Blue Larkspur (2)
Cameronian (1928)	St. Simon (2)		

Total: 20 out of 77 (26.0%)
Estimated percentage of thoroughbreds born yearly inbred at the fourth remove is 25.6%

INBRED AT THE THIRD AND FOURTH REMOVE

NAME AND YEAR FOALED	COMMON ANCESTOR	INCIDENCE OF COMMON ANCESTOR AT 3RD REMOVE	4TH REMOVE
Tagalie (1909)	Isonomy	1	1
Gainsborough (1915)	Galopin	1	1
Papyrus (1920)	St. Simon	1	1
Manna (1922)	St. Simon	1	2
Felstead (1925)	Carbine	1	1
April the Fifth (1929)	Cyllene	1	1
Hyperion (1930)	St. Simon	1	1
Bois Roussel (1935)	St. Simon	1	1
Owen Tudor (1938)	Chaucer	1	1
Watling Street (1939)	St. Simon	1	1
My Love (1945)	Teddy	1	1
Phil Drake (1952)	Teddy	1	1

Total: 12 out of 77 (15.6%)
Estimated percentage of thoroughbreds born yearly inbred at the third and fourth removes is 14.8%

INBRED AT THE THIRD REMOVE (those horses having seven, instead of eight great-grandparents)

NAME AND YEAR FOALED	COMMON ANCESTOR AT THIRD REMOVE, AND INCIDENCE	NAME AND YEAR FOALED	COMMON ANCESTOR AT THIRD REMOVE, AND INCIDENCE
Lemberg (1907)	Isonomy (2)	Gay Crusader (1914)	Galopin (2)
Pommern (1912)	Hampton (2) and inbred at third and fourth remove to Distant Shore	Sansovino (1921)	Pilgrimage (2)

Total: 4 out of 77 (5.2%)
Estimated percentage of thoroughbreds born yearly inbred at the third remove is 2.6%

In 1956 the overall figures showed that 57.1 per cent of the Derby winners of the century were inbred at the fourth remove or closer; now the figure obtained from the tables above is 46.8 per cent, reflecting how the international boundaries of thoroughbred breeding have been widened in recent years. This can be seen from the incidence of Derby winners added to the table of outbreds since 1956, with outbreds outweighing the rest 15 to four. None of the majority was inbred closer than the fourth remove, leaving the 1955 Derby winner Phil Drake as the last inbred at the third and fourth removes, and the 1924 winner Sansovino the last inbred at the third remove.

The incidence of Derby winners inbred at the fourth remove or closer has come much nearer to the estimate of 43.2 per cent as the normal incidence of matings annually effected to produce similar degrees of inbreeding. But complete reliance cannot be placed on the estimate which produced this last-named figure. This is based on the dissection of the pedigrees of 1,000 thoroughbreds taken at random. If the family trees of another cross-section of the breed were examined, a different ratio might result. Perhaps someone, with the help of a computer, might one day consider running through the pedigrees of all thoroughbreds born in a particular year with a view to finding a more reliable figure. Until then, we must take the estimates given here for what they are worth and bear in mind the reservations.

Since the dams of these Derby winners are considered no less important than the sires, it is worth looking at the tail female achievements and relations to see if they have anything in common. To restrict the research to the most meaningful details, I have tabled those dams who can claim an association with the Classics only as far back as their grandams. In other words, I have examined the first three dams of each Derby winner under the heading of achievement and offspring, but only in relation to the Classics.

YEAR	WINNER	DAM AND HER CLASSIC CONNECTION
1900	Diamond Jubilee	Perdita II: dam of Persimmon (Derby)
1902	Ard Patrick	Morganette: dam of Galtee More (Triple Crown); out of half-sister to Marie Stuart (Oaks and St. Leger)
1903	Rock Sand	Roquebrune: half-sister to Seabreeze (Oaks and St. Leger)
1904	St. Amant	Lady Loverule: grandam won Oaks
1905	Cicero	Gas: half-sister to Ladas (Derby) and Chelandry (1,000 Gns)
1907	Orby	Rhoda B: dam of Rhodora (1,000 Gns)
1908	Signorinetta	Signorina: second in Oaks; dam of Signorino (placed Derby and 2,000 Gns)
1910	Lemberg	Galicia: dam of Bayardo (Leger)
1911	Sunstar	Doris: dam of Princess Dorrie (1,000 Gns and Oaks) and Bright (placed 1,000 Gns and Oaks)
1912	Tagalie	Tagale: sister to dam of Mary Legend (French Oaks)
1916	Fifinella	Silver Fowl: dam of Classic-placed Silver Tag, Soubriquet and Silvern
1917	Gay Crusader	Gay Laura: out of Galeottia (Oaks)
1918	Gainsborough	Rosedrop: won Oaks
1919	Grand Parade	Grand Geraldine: out of sister to dam of Minoru (Derby)
1920	Spion Kop	Hammerkop: sister to dam of Electra (1,000 Gns)
1921	Humorist	Jest: won 1,000 Gns and Oaks; half-sister to Black Jester (St. Leger)
1924	Sansovino	Gondolette: dam of Ferry (1,000 Gns) and Derby-placed Let Fly and Great Sport
1925	Manna	Waffles: dam of Sandwich (St. Leger); grandam third 1,000 Gns
1926	Coronach	Wet Kiss: sister to Soldennis (Irish 2,000 Gns)
1929	Trigo	Athasi: out of Athgreany (Irish Oaks)
1930	Blenheim	Malva: dam of King Salmon (second 2,000 Gns and Derby)

YEAR	WINNER	DAM AND HER CLASSIC CONNECTION
1931	Cameronian	Una Cameron: out of Cherimoya (Oaks)
1933	Hyperion	Selene: dam of Sickle (third 2,000 Gns); three-parts sister to Tranquil (1,000 Gns and St. Leger)
1934	Windsor Lad	Resplendent: won Irish 1,000 Gns and Oaks, and second English Oaks; half-sister to Irish Classic winners Sol Speranzo and Soldumeno, and to Ferrybridge (third 1,000 Gns)
1935	Bahram	Friar's Daughter: dam of Dastur (second 2,000 Gns, Derby and St. Leger); out of half-sister to Plucky Liege (see 1938, below)
1936	Mahmoud	Mah Mahal: half-sister to 2,000 Gns-placed Mirza II and Badruddin; out of Mumtaz Mahal (second 1,000 Gns)
1937	Mid-day Sun	Bridge of Allan: sister to Knockando (second 2,000 Gns)
1938	Bois Roussel	Plucky Liege: dam of Admiral Drake (Grand Prix de Paris) and Sir Gallahad III (French 2,000 Gns); half-sister to grandam of Bahram (Triple Crown)
1940	Pont l'Eveque	Ponteba: half-sister to Pervencheres (second French 1,000 Gns)
1941	Owen Tudor	Mary Tudor II: won French 1,000 Gns; dam of Solar Princess (third Oaks)
1942	Watling Street	Ranai: dam of Garden Path (2,000 Gns)
1944	Ocean Swell	Jiffy: dam of Iona (placed 1,000 Gns and Oaks)
1945	Dante	Rosy Legend: dam of Sayajirao (St. Leger); out of half-sister to Sans Souci II (Grand Prix de Paris)
1947	Pearl Diver	Pearl Cap: won French 1,000 Gns and Oaks; half-sister to Bipearl (French 1,000 Gns) and Pearlweed (French Derby)
1950	Galcador	Pharyva: dam of Galgala (French 1,000 Gns)
1951	Arctic Prince	Arctic Sun: half-sister to Solar Slipper (third St. Leger); out of Solar Flower (third 1,000 Gns and Oaks)
1952	Tulyar	Neocracy: dam of Saint Crespin III (third Derby); out of sister to Trigo (Derby)
1953	Pinza	Pasqua: half-sister to Pasch (2,000 Gns) and Chateau Larose (second St. Leger); grandam second Oaks
1954	Never Say Die	Singing Grass: out of half-sister to Galatea II (1,000 Gns and Oaks)
1955	Phil Drake	Philippa: out of sister to dam of Le Petit Prince (French Derby)
1956	Lavandin	Lavande: dam of Lavarede (second Grand Prix de Paris)
1957	Crepello	Crepuscule: dam of Honeylight (1,000 Gns)
1961	Psidium	Dinarella: fourth Italian Oaks; dam of Thymus (French 2,000 Gns); out of Dagherotipia (Italian 1,000 Gns)
1962	Larkspur	Skylarking: dam of Rising Wings (fourth Irish Oaks) and Ballymarais (fourth Irish Sweeps Derby and English St. Leger)
1963	Relko	Relance III: dam of Match III (French St. Leger, second French Derby) and Reliance II (French Derby, Grand Prix de Paris and French St. Leger)
1966	Charlottown	Meld: won 1,000 Guineas, Oaks and St. Leger
1967	Royal Palace	Crystal Palace: dam of Prince Consort (third St. Leger); half-sister to dam of Photo Flash (second 1,000 Gns)
1968	St Ivor	Attica: grandam second Kentucky Oaks
1969	Blakeney	Windmill Girl: second Oaks, third Irish Oaks; dam of Morston (Derby)
1970	Nijinsky	Flaming Page: won Queen's Plate and Canadian Oaks, second Kentucky Oaks
1971	Mill Reef	Milan Mill: half-sister to Berkeley Springs (second 1,000 Gns and Oaks)
1972	Roberto	Bramalea: won Coaching Club of America Oaks; half-sister to Rhodora II (third Irish Oaks)
1973	Morston	Windmill Girl: see 1969, above

YEAR	WINNER	DAM AND HER CLASSIC CONNECTION
1976	Empery	Pamplona II: won Peruvian Triple Crown
1977	The Minstrel	Fleur: half-sister to Nijinsky (Triple Crown); out of Flaming Page (Queen's Plate and Canadian Oaks)

In tabling the above, reference to grandams of the Derby winners has been restricted to their own racecourse achievements, not to their immediate relationship to Classic winners or Classic-placed horses. Had the qualifications been opened to include relatives of the grandams, several more Derby winners would have been included. For instance, Aboyeur's grandam was half-sister to the 2,000 Guineas winner Fitz Roland; Pommern's grandam was half-sister to the dam of unbeaten Ormonde; Blue Peter's grandam was half-sister to the Oaks winner Musa, who in turn produced another Oaks winner in Mirska; Sea-Bird II's grandam was half-sister to the French 1,000 Guineas winner Camaree; and Snow Knight's grandam was half-sister to the St. Leger winner Chamossaire.

As it is, the qualifications still produce the interesting statistics that from the 82 Derby winners this century, 55 (or 67 per cent) have an immediate Classic connection through their dam or grandam. On examination in this way, it is fair to say that the great majority of dams of these Derby winners come from tail female lines who have produced in close relationship other high-grade horses up to Classic-winning or near-Classic-winning calibre. The far more numerous stirps who throw up good handicapper after good handicapper rarely achieve Derby-winning honours.

Though all Classic-winning lines must start from somewhere, it is usual for an upstart stirp to trace back in tail female to a mare who produced other stock of Classic ability, even though it may have a poor record for some recent generations. This point illustrates the fact that the genes which produce superior racing merit are sometimes carried forward latently in parental germ plasms for generation after generation until they suddenly reassert themselves—or, of course, the results may be due to the almost complete dominance of a sire's influence over a mare's.

The next consideration involves the ages, at the time of mating, of the sires and dams of this century's Derby winners.

AGE	NUMBER OF DERBY WINNERS	
	SIRES	DAMS
4	2	5
5	6	10
6	9	9
7	12	11
8	5	11
9	7	7
10	14	10
11	6	3
12	5	3
13	5	3
14	4	2
15	2	2
16	0	3
17	2	1

Some Conclusions Concerning Derby Winners 1900–1976   489

AGE	NUMBER OF DERBY WINNERS	
	SIRES	DAMS
18	1	1
19	1	0
20	1	0
21	0	0
22	0	1

It will be seen that the best results have come from parents in the age groups six to ten years, which is only to be expected as at that period they are in the prime of life. Furthermore, a stallion or broodmare who has not produced anything of value in the first few years of his or her stud life is not infrequently mated with cheaper and inferior partners later, and so can hardly be expected then to throw a Derby winner. There is a possible case for arguing, on the strength of figures for the ages of sires and dams of Derby winners, that the modern fashion is away from older stallions, since in the 21-year period 1939 to 1959 inclusive, nine winners were by stallions under ten years of age at mating, while in the 22 years since, 14 were under the age of ten. It is dangerous to be dogmatic on such evidence alone, but the suggestion is worth pondering.

The mean age of the sires of these 82 winners is nine, and the mean age of the dams eight—one year lower in each case than that which obtained in 1956. I would hazard that if the mean ages of the parents of every thoroughbred foal born were investigated, the result would be about the same. Thus there is nothing magical about the ages of nine for a sire and eight for a mare being likely to produce a Derby winner. Only once since 1939 has the exact mating of a nine-year-old sire with an eight-year-old mare had this result, that coming with Morston's victory in 1973.

The oldest mare who produced a Derby winner was Plucky Liège, who was 23 years old when Bois Roussel was foaled. As a 19-year-old, she threw the Grand Prix de Paris winner Admiral Drake, and at the age of 20 Admiral Drake sired the Derby winner Phil Drake. Curiously, Admiral Drake was the oldest stallion to get a Derby winner in this century. Earlier there were a few of greater age, the all-time record being held by Muley, who was 26 when he sired the 1840 winner The Little Wonder. It is highly unlikely that either this or Admiral Drake's achievement will be emulated.

It is impossible to define within narrow limits how Derby winners or other high-class racehorses are bred, but from the above details it is possible to suggest that the following points are relative:

1   Practically all Derby winners have been sired by stallions who were themselves of Classic-class racing ability. Those stallions designated Class One and Class Two, on their racing worth as having won or been placed in the English, French or American Classics, the King George VI and Queen Elizabeth Stakes and the Prix de l'Arc de Triomphe, clearly come within this classification. Horses who have won one or more of these races show the best results as a team.

2   The great majority of the dams of these Derby winners come from strains

which have produced in tail female other high-class performers, often up to Classic-class.

3   Broadly speaking, the average Derby winner is the product both on his sire's and dam's side of a limited band of parental stock with a Classic-winning, or near-Classic-winning background.

4   Certain sires of the highest class in most other respects—such as Buchan, Son-in-Law, Tetratema, Big Game, Fair Trial and Court Martial—and who were all champion sires at one time or another, have played very little part in the production of Derby winners. This cannot be entirely ascribed to distance limitations, as Son-in-Law and Buchan were above suspicion in this respect. Moreover, Phalaris, Vatout, Pharos, Sundridge and more recently Nasrullah, all of limited stamina, show up well.

5   Young parents can show better results than old ones, but sires up to 20 years of age and dams up to 22 (at the time of mating) have produced Derby winners since 1900.

6   The percentage of Derby winners inbred at the fourth generation or closer has drawn considerably nearer to the figure that could be expected from an estimate of the percentage of thoroughbred foals produced to similar degrees of inbreeding, perhaps illustrating how the breeding of potentially high-class thoroughbreds has become more international in its use of the raw materials. But the estimate may be faulty, and the point requires a very wide investigation, to embrace all the winners of all the English, French and American Classics, the King George VI and Queen Elizabeth Stakes and the Prix de l'Arc de Triomphe, and then to compare these with the inbred and outcross ratio of tens of thousands of other horses.

As has already been intimated, perhaps one day someone will have the time, energy and necessary finance to undertake a study of the great mass of thoroughbred pedigrees. Until then, we must rely on other evidence, of which the interpretation of past results provides just one element. In this respect, two features which have been developed in the last 30 years are worth considering, since each can be used to measure racing merit, one of the key factors which has been stressed throughout this book. They involve the greater use of handicap systems to grade horses and compare one generation with another, and the introduction of Pattern races to denote the most important events for individual age-groups and distances.

Ever since handicaps were introduced, horses have been graded so as to provide more competitive sport, and individuals have done their best to work out ways of finding flaws in the official assessments. Those ratings which have best stood the test of time are published by Timeform, the company formed by Mr. Phil Bull which since the beginning of the 1947 season has kept its own universal handicap. Each horse with form worthy of an assessment is given a rating of merit expressed in the number of pounds which Timeform reckon the horse would be entitled to receive in an average Free Handicap. A horse regarded as worth 9st 7lb in an average Free Handicap is rated at its equivalent in pounds, 133; one regarded as worth 6st (84lb) is rated at 84, and so on. At the end of each season the general level of the handicap is adjusted so that the mean of all the ratings is kept at the

STONE = 12#

same standard level from year to year, as a result of which it is possible to compare directly horses of one year with those of another, whether they have competed against each other or not.

This method also enables two-year-olds to be compared with their elders, and sprinters with Cup horses. I am grateful for Timeform's permission to quote their figures for Derby winners, though before looking at those, it is interesting to note that from the top horses rated by Timeform since 1947—Sea-Bird II, Brigadier Gerard, Tudor Minstrel, Abernant, Ribot, Windy City, Mill Reef, Vaguely Noble, Pappa Fourway, Alleged, Alycidon, Exbury, Nijinsky and Star of India— four (Alycidon, Ribot, Vaguely Noble and Mill Reef) went on to become champion sire in Britain at least once, and both Sea-Bird II and Nijinsky covered themselves with considerable glory.

The assessment of Derby winners, which is readily available in the Timeform publications, has also been studied closely by Major David Swannell, the senior Jockey Club Handicapper, who since 1959 has worked out a figure in relationship to a norm of 100, which corresponds to an average year, with each digit representing a difference of 1lb in weight.

If breeding is an inexact science, so too is handicapping, and the expressions of both Timeform and Major Swannell are opinions—better than many of the next man's for their diligence, independence and expertise, but opinions nevertheless. This explains why there are discrepancies between their respective ratings for the Derby winners since 1959. They use the same broad principles, but in certain instances come up with different answers.

YEAR	WINNER	SWANNELL RATING	TIMEFORM	ORDER OF MERIT SWANNELL		ORDER OF MERIT TIMEFORM	
1959	Parthia	98	132	Sea-Bird II	110	Sea-Bird II	145
1960	St. Paddy	100	133	Mill Reef	106	Mill Reef	141
1961	Psidium	97	130	Nijinsky	104	Nijinsky	138
1962	Larkspur	94	128	Relko	103	Grundy	137
1963	Relko	103	136	Royal Palace	101	Troy	137
1964	Santa Claus	100	133	Sir Ivor	101	Relko	136
1965	Sea-Bird II	110	145	Grundy	101	Sir Ivor	135
1966	Charlottown	95	127	Santa Claus	100	The Minstrel	135
1967	Royal Palace	101	131	St. Paddy	100	Santa Claus	133
1968	Sir Ivor	101	135	Parthia	98	St. Paddy	133
1969	Blakeney	96	123	Roberto	98	Parthia	132
1970	Nijinsky	104	138	Psidium	97	Roberto	131
1971	Mill Reef	106	141	Snow Knight	97	Royal Palace	131
1972	Roberto	98	131	Blakeney	96	Henbit	130
1973	Morston	96	125	Morston	96	Psidium	130
1974	Snow Knight	97	125	Troy	96	Shirley Heights	130
1975	Grundy	101	137	Charlottown	95	Empery	128
1976	Empery	90	128	The Minstrel	95	Larkspur	128
1977	The Minstrel	95	135	Larkspur	94	Charlottown	127
1978	Shirley Heights	92	130	Shirley Heights	92	Morston	125
1979	Troy	96	137	Empery	90	Snow Knight	125
1980	Henbit	88	130	Henbit	88	Blakeney	123

The differences of opinion are obvious from study of the order of merit into which each party places the Derby winners. Perhaps we should be more surprised at the degree of uniformity which can also be gauged. The exact nature of the differences become more evident when the mean rating for each set is worked out, being the median of all the ratings. The mean Swannell rating is 97-98; the mean Timeform rating is 132.

| | | ABOVE OR BELOW MEAN | |
YEAR	WINNER	SWANNELL	TIMEFORM
1959	Parthia	—	—
1960	St. Paddy	+2	+1
1961	Psidium	—	−2
1962	Larkspur	−3	−4
1963	Relko	+5	+4
1964	Santa Claus	+2	+1
1965	Sea-Bird II	+12	+13
1966	Charlottown	−2	−1
1967	Royal Palace	+3	−1
1968	Sir Ivor	+3	+3
1969	Blakeney	−1	−9
1970	Nijinsky	+6	+6
1971	Mill Reef	+8	+9
1972	Roberto	—	−1
1973	Morston	−1	−7
1974	Snow Knight	—	−7
1975	Grundy	+3	+5
1976	Empery	−7	−4
1977	The Minstrel	−2	+3
1978	Shirley Heights	−5	−2
1979	Troy	−1	+5
1980	Henbit	−9	−2

On seven occasions in the 22 examples, there is a discrepancy of 3lb or more. In the case of Royal Palace it is 4lb; Blakeney 8lb; Morston 6lb, and Snow Knight 7lb. These are all rated higher by Major Swannell than by Timeform. In the other cases—The Minstrel 5lb; Troy 6lb, and Henbit 7lb—they are all rated higher by Timeform. In chronological order, the four cases rated higher by Major Swannell came first, and the three rated lower by him have arisen within the last four years under scrutiny. Perhaps this signifies a change of method by either Major Swannell or Timeform.

As a guide to the relative merit of Derby winners these figures are most useful, since they put more meaning into the accolade than might otherwise be available. And the extension of the work of both Timeform and the official handicappers is correspondingly valuable to anyone who wishes to assess racing merit. Timeform's ratings are renowned and accepted the world over; and the official handicapper's pronouncements have assumed greater importance since the idea of the norm system was brought into use for horses of three years and upwards which ran in Britain in 1975.

For the first time the official Free Handicaps—other than for two-year-olds—were related to previous years, with the norm of 100 adopted as the fulcrum on which the system balanced. Thanks to the co-operation of Turf authorities in Britain, Ireland and France, an international classification for three-year-olds and upwards was brought out for the first time in 1977; it was extended to two-year-olds in 1978, and was expanded in coverage in 1980 to allow for the provision of a European Free Handicap to be run in England in 1981. The idea of foreign three-year-olds running in the English Free Handicap failed to take off in its first year, but the fact that an assessment was down on paper was more important.

It will take some time to build up a comprehensive assessment of respective generations, and observers must be careful to note when the norm is changed, in the light of experience. But the international classification is a handy device in the breeder's armoury.

Another useful measuring-stick was introduced in the wake of the Duke of Norfolk's Committee on the Pattern of Racing, which reported in 1965 and led to the birth of Pattern races in 1968. These races, agreed by the Flat Race Planning Committee under the chairmanship of Lord Porchester, formed what was officially described as a pattern "designed to provide a complete series of tests for the best horses of all ages and over all the accepted distances".

In the early years, there were 147 races in the Pattern scheme; since then they have been trimmed and for several years were restricted to 100, until changes were made for 1981 and the package involved 101 races.

The events are divided into three groups, and Pattern-race prize money is itself graded so as to give the greatest rewards to the best horses, those expected to contest Group One races, whether two-year-olds or older, sprinters or stayers. In general the system works well, in diverting horses and prize money into those events in which quality is most important.

The statistical evidence gleaned from results of Pattern races is useful to breeders in pinpointing the best horses, those which in theory should make the best stallions and the best mares. Once the Pattern system went international, with Britain, France, Germany, Ireland and Italy agreeing a European framework of top races, and races in North and South America, Australia, Japan, New Zealand, Scandinavia, South Africa and Spain were designated similar status, its value was greatly enhanced. There have been arguments about the exact status accorded to certain races, but on the whole the system has been seen to be valuable, with a horse that has won a Group race, whether it be Group One, Two or Three, increasing his stature over one who has not won a Group race.

Expanding on the ratings quoted earlier for Derby winners since 1959, it is interesting to note that only three are by stallions which did not win a Group One Pattern race. They are Psidium, whose sire Pardal won the Group Two Princess of Wales's Stakes; Sea-Bird II, whose sire Dan Cupid was beaten a short head in the Group One French Derby; and Snow Knight, who won the Group Three Craven Stakes. The rest, even those bred outside Europe, showed themselves capable of winning at least one of the best races in the country where they raced.

Analysis of Pattern-race results can take many forms, ranging from investigation of individual merit to trends which may be observed among stallions, mares or even whole breeding areas. Perhaps this is another area in which the computer might be used to telling effect. In the meantime, Pattern races must be regarded as another pointer to excellence, and whatever changes may have taken place this century, that still remains the thoroughbred breeder's goal.

Index

495